CISM COURSES AND LECTURES

The series presents lecture notes, monographs, edited works and proceedings in the field of Mechanics, Engineering, Computer Science and Applied Mathematics.
Purpose of the series is to make known in the international scientific and technical community results obtained in some of the activities organized by CISM, the International Centre for Mechanical Sciences.

INTERNATIONAL CENTRE FOR MECHANICAL SCIENCES

COURSES AND LECTURES - No. 436

MODERN PROBLEMS OF STRUCTURAL STABILITY

EDITED BY

ALEXANDER P. SEYRANIAN
MOSCOW STATE LOMONOSOV UNIVERSITY, RUSSIA

ISAAC ELISHAKOFF
FLORIDA ATLANTIC UNIVERSITY, USA

Springer-Verlag Wien GmbH

This volume contains 111 illustrations

Originally published by Springer-Verlag Wien New York in 2002

SPIN 10903315

In order to make this volume available as economically and as rapidly as possible the authors' typescripts have been reproduced in their original forms. This method unfortunately has its typographical limitations but it is hoped that they in no way distract the reader.

ISBN 978-3-211-83697-2 ISBN 978-3-7091-2560-1 (eBook)
DOI 10.1007/978-3-7091-2560-1

PREFACE

Stability of structures is one of the most important and interesting fields in mechanics. This book is dedicated to fundamental concepts, problems and methods of structural stability along with qualitative understanding of instability phenomena. It includes modern theoretical and experimental static and dynamic stability studies.

The book presents some recent achievements and knowledge on bifurcation theory, novel closed-form solutions, sensitivity analysis of stability characteristics, general aspects of non-conservative stability problems, study of pre- and post-critical behavior, analysis of singularities of the stability boundaries, stability analysis of multi-parameter linear periodic systems, and optimization of structures under stability constraints. Both systems with finite degrees of freedom and continuous models are considered. It is intended to combine within this book mathematical foundation and interesting classical and modern mechanical problems.

A number of mechanical problems on interesting physical phenomena are discussed. Among these problems we consider planar and spatial buckling of elastic rods, buckling and post-buckling behavior of anisotropic shells, dynamic stability of pipes conveying fluid, elastic columns losing stability under action of periodic and non-conservative forces (e.g. rocket thrust), bending-torsional flutter of a wing braced by a strut, optimization problems for conservative and non-conservative systems etc. The methods presented in the book are constructive and easy to implement in computer programs. Recent exciting experiments on dynamic stability of non-conservative systems are described and shown by many photographs.

The course on structural stability was given in Udine, 16-20 July 2001, by seven scientists active in this field, namely Professors J. Arbocz (Delft University of Technology, The Netherlands), T. Atanackovic (University of Novi Sad, Yugoslavia), G. Domokos (Technical University of Budapest, Hungary), I. Elishakoff (Florida Atlantic University, USA), P. Pedersen (Technical University of Denmark, Denmark), A.P. Seyranian (Moscow State Lomonosov University, Russia), and Y. Sugiyama (Osaka Prefecture University, Japan). The course was coordinated by Alexander P. Seyranian and Isaac Elishakoff.

The book is divided into seven parts.

Part I, by T. Atanackovic, is devoted to planar and spatial buckling of elastic rods with generalized constitutive equations. Post-buckling behavior of structures is studied, and types of bifurcations are analyzed. Optimization problems for elastic rods are formulated and solved numerically. The influence of shear deformation and axis extension on the stability boundary are examined.

Part II, by G. Domokos, deals with classical Euler's stability problem of a beam. A new type of bifurcation at the second critical load is uncovered, and it is shown that the only globally stable solution beyond this load is the bar in tension. Investigation of the discretized problem with several rigid links is given.

Part III, by I. Elishakoff, discusses non-classical problems in the theory of elastic stability. The exact methods dealing with buckling loads are developed. Novel closed form solutions for inhomogeneous columns with distributed load are found and analyzed, and the effect of boundary conditions is investigated, over 250 years after Euler's classical works.

In Part IV J. Arbocz, J.M.A.M. Hol, and J. de Vries consider buckling and post-buckling behavior of anisotropic cylindrical shells under combined loading. The effects of general imperfections and elastic boundary conditions are investigated, and reliability of buckling load predictions is estimated. The description of the computer code "DISDECO" accumulating theoretical, numerical and practical knowledge on buckling of shell structures is given.

Part V, by A.P. Seyranian, is devoted to bifurcations of eigenvalues of nonsymmetric operators depending on parameters with applications to different mechanical problems. Singularities of boundaries of stability domains for fluid-structure interaction problem are analyzed, and the effect of transference of instability between eigenvalue branches is explained. Instability domains for Hill's equation with damping in the three-dimensional parameter space is described.

Part VI, by P. Pedersen, deals with sensitivity analysis and optimization methods for dynamic stability problems. Sensitivity analysis of critical loads for non-conservative stability problems is developed. Stability analysis for periodic systems dependent on multiple parameters is described with applications to mechanical systems. Optimization problems with several eigenvalue constraints are formulated and solved numerically.

Part VII, by Y. Sugiyama, is devoted to theoretical and experimental studies of non-conservative stability problems. Flutter of cantilevered columns subjected to a non-conservative force produced by a rocket thrust is analyzed both numerically and experimentally. The dynamic stability of pipes conveying fluid is considered and influence of problem parameters on stability and flutter regions is studied. The results for columns optimized for the maximum flutter load are presented.

The book is addressed to graduate students, researchers and engineers working in aerospace, naval, civil and mechanical engineering areas. No special background is needed except basic knowledge in mathematics and mechanics.

Finally, the efforts of the CISM Rector, Prof. S. Kaliszky, CISM Secretary General G. Bianchi, the Editor of the Series, Prof. C. Tasso, and all the extremely dedicated CISM Staff in Udine are gratefully acknowledged.

Alexander P. Seyranian and Isaac Elishakoff

CONTENTS

Part I. Stability Bounds and Optimal Shape of Elastic Rods

Teodor M. Atanackovic

Faculty of Technical Sciences, University of Novi Sad, Yugoslavia
e-mail: atanackovic@uns.ns.ac.yu

Abstract. In these Lecture notes we treat two problems of determining stability boundary for elastic rods (columns) and two optimization problems in which we determine the shape of the lightest rod, stable against buckling. In the problems of determining stability boundary we shall use generalized rod equations (corresponding to planar and spatial deformations) that take into account both shear deformation and axis extension and examine the influence of these effects on the stability boundary.

1 Stability Bounds of Elastic Rods with Generalized Constitutive Equations

In this section we shall present the stability analysis for a rotating rod and for a rod loaded by a force and a couple (the Greenhill problem). Both problems may be considered classic in the stability theory of elastic rods. We shall treat generalized versions of those problems. The generalization consists in using generalized constitutive equations that take into account both axis compressibility and the influence of shear stresses. We shall analyze how the generalization influences the critical values and the type of bifurcation pattern.

1.1 Rotating Elastic Rod

Consider an elastic rod, fixed at the one end and free at the other end. Suppose that the rod is rotating with constant angular velocity ω about its axis. At certain velocity the rod could lose stability and deform so that its axis is bent under the action of centrifugal forces. In the simplest case of a prismatic rod, the deformed axis of the rod is a plane curve. However during the *motion* between two relative equilibrium configuration (one corresponding to the initial state in which rod axis is straight and one in which rod axis is bent) the axis of a rod is, in general *not* a plane curve. This is because of the action of Coriolis inertial forces. If the bending stiffness of the rod about the principal axis of the cross−section that is in the rotating plane is much larger than the bending stiffness about the principal axis orthogonal to the rotating plane, the motion of the rod between the relative equilibria will be such that during this motion the axis of the rod is, approximately, a plane curve. It will be exactly a plane curve if the rod is constrained to move between to smooth rotating planes. The problem of determining the critical rotation speed and the post−critical behavior of the rod described has been subject of many investigations. We mention works of Stodola (1906) Odeh and Tadjbakhsh (1965), Bazely and Zwahlen (1968), Parter (1970), Atanackovic (1987,1991), and Clément and Descloux (1991). In all of these works the classical Bernoulli−Euler theory of rods was used. We shall analyze the rotating rod problem with generalized constitutive equations that take into account the influence of shear stresses and axis extensibility. We shall take the linear constitutive equations in the specially selected strain measures. The strain measures that we use are the same as those used by Antman and Nachman (1980). In this respect the constitutive equations that we use can be viewed as a first approximation of the more complicated non–linear theories. The use of linear constitutive equations can lead to solutions that, from a physical point of view, are singular. Namely linear constitutive equations do not satisfy coercivity conditions (the second basic restriction – strict monotonicity is satisfied).

We shall use variational methods to determine the bifurcation points of the equations describing the relative equilibrium configurations of the rotating rod. The variational principle for relative equilibrium equations will be formulated for the case when the rod is not initially straight (the rod axis is a known plane curve when the rod is not rotating) and when the rod is loaded by a distributed forces along its axis. We shall also give the estimate of the maximal deflection of the rod in the post–critical state. The estimate is given for the special case when the rod is initially straight, there is no distributed load along rod axis and when the constitutive equations reduce to the Bernoulli–Euler model. The estimate is based on an inequality for convex functions. It can be used to obtain the unknown constant in a first integral of the equilibrium equations.

Model. Consider an elastic rod, whose axis is a plane non–intersecting curve C lying in a plane Π. Suppose that one end of the rod, say B, is fixed while the other end C is free (see Figure 1).

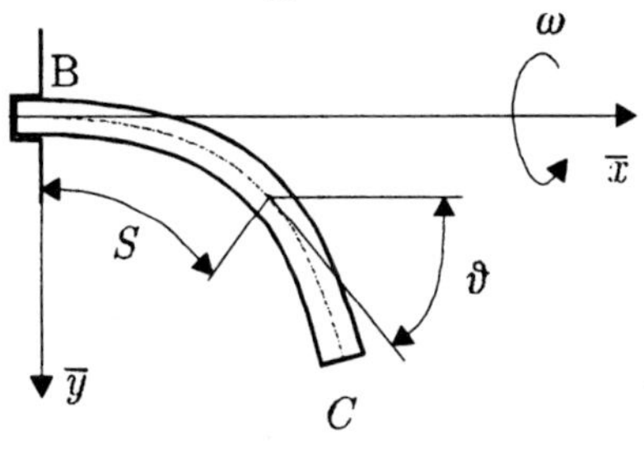

Figure 1. Coordinate system for the rotating rod.

Let $\bar{x}$ be the axis of the rotation of the plane Π. We assume that $\bar{x}$ passes through the point B. Suppose further that Π is rotating about $\bar{x}$ with constant angular velocity ω. Let $\bar{y}$ be an axis, perpendicular to $\bar{x}$, laying in Π and passing through B. We assume that the points on the curve C, in the coordinate system $\bar{x}$– B –$\bar{y}$, can be represented as

$$X = \widehat{X}\,(S); \qquad Y = \widehat{Y}\,(S), \tag{1}$$

where S is the arc–length of C and $\widehat{X}\,(S), \widehat{Y}\,(S)$ are smooth functions. We assume that $S \in [0, L]$, where L is the length of the rod in the natural state. Since C is non–intersecting it has a well defined tangent at each point. Let $\mathbf{e}_1$ and $\mathbf{e}_2$ be the unit vectors along the $\bar{x}$ and $\bar{y}$ axis respectively. Then, if $\mathbf{T}(S)$ is the unit tangent vector at the point (X, Y) the relations

$$\cos\Theta(S) = \mathbf{T}\cdot\mathbf{e}_1; \qquad \sin\Theta(S) = \mathbf{T}\cdot\mathbf{e}_2, \tag{2}$$

determine a unique $\Theta(S)$. We assume that the rod is so fixed that $\Theta(0) = 0$. With $\Theta(S)$ so determined the curvature of C is $K_0 = d\Theta/dS$. The natural configuration of the rod is defined by two vector functions $\mathbf{R}(S) = X\mathbf{e}_1 + Y\mathbf{e}_2$ and $\mathbf{B}(S)$ characterizing the points on the rod axis and orientation of the cross–section, respectively. $\mathbf{B}(S)$ is assumed to be of unit length and to belong to Π. We assume that the natural configuration is stress free so that $\mathbf{B}(S)$ is orthogonal to $\mathbf{T}(S)$. In this case

$$\mathbf{B}(S) = -\sin\Theta(S)\mathbf{e}_1 + \cos\Theta(S)\mathbf{e}_2. \tag{3}$$

When Π rotates about the $\bar{x}$ axis the rod will assume the relative (with respect to Π) equilibrium configuration. Let c be the curve corresponding to the points on the rod axis in the deformed state. We assume that c is a plane curve lying in Π (thus we exclude the possible lateral buckling). If the rod is prismatic, as we assume it is, an equilibrium configuration with c laying in Π will be always possible. The equilibrium and geometrical equations, in the system $\bar{x}$– B –$\bar{y}$ are

$$\begin{aligned} &H' = 0; \quad V' = -\rho\omega^2 \,\widehat{y} - q; \\ &M' = V(1+\varepsilon)\cos\vartheta - H(1+\varepsilon)\sin\vartheta; \\ &\widehat{x}\,' = (1+\varepsilon)\cos\vartheta; \quad \widehat{y}\,' = (1+\varepsilon)\sin\vartheta. \end{aligned} \tag{4}$$

where the differentiation with respect to S is denoted by prime, $\mathbf{F} = H\mathbf{e}_1 + V\mathbf{e}_2$ is the contact force representing the influence of the part of the rod $[0, S]$ on the part $(S, L]$, M is resultant couple, $\widehat{x}$ and $\widehat{y}$ are the coordinates in the deformed state of a point on the rod axis that was at the point (X, Y) in the natural state, ϑ is the angle between the tangent to c at $(\widehat{x}, \widehat{y})$ and the $\bar{x}$ axis, ρ is the mass of the rod per unit length of the rod axis in the natural state, ε is the axial strain, i.e., if $\mathbf{r} = \widehat{x}\,\mathbf{e}_1 + \widehat{y}\,\mathbf{e}_2$ is the position vector of an arbitrary point on the rod axis in the deformed state, then $1 + \varepsilon = |d\mathbf{r}/dS|$ and q is the intensity of the distributed forces along the rod axis. To (4) we adjoin the following boundary conditions

$$H(L) = 0; \quad V(L) = 0; \quad \widehat{x}\,(L) = 0; \quad \widehat{y}\,(L) = 0; \quad M(L) = 0. \tag{5}$$

The deformed configuration is specified when two vector functions, i.e., position vector of an arbitrary point on the rod axis $\mathbf{r}(S)$ and the director $\mathbf{b}(S)$ are specified. Let φ be the rotation angle of the cross–section. Then, $\mathbf{b}(S)$ becomes

$$\mathbf{b}(S) = -\sin(\Theta + \varphi)\mathbf{e}_1 + \cos(\Theta + \varphi)\mathbf{e}_2. \tag{6}$$

Let $\mathbf{e}_3 = \mathbf{e}_1 \times \mathbf{e}_2$. We define $\mathbf{a} = \mathbf{b} \times \mathbf{e}_3$, so that

$$\mathbf{a}(S) = \cos(\Theta + \varphi)\mathbf{e}_1 + \sin(\Theta + \varphi)\mathbf{e}_2. \tag{7}$$

The strains are taken to be the derivatives of the rotation angle φ' and two functions $\xi(S)$ and $\eta(S)$ defined by

$$\mathbf{t} = \mathbf{r}' = [1 + \xi(S)]\mathbf{a} + \eta(S)\mathbf{b}. \tag{8}$$

Introducing the shear angle γ as the angle between $\mathbf{a}(S)$ and $\mathbf{t}$, we obtain the following relation

$$\gamma = (\vartheta - \Theta) - \varphi. \tag{9}$$

With (9) the strains $\xi(S)$ and $\eta(S)$ become Libai (1992)

$$\xi = (1+\varepsilon)\cos\gamma; \qquad \eta = (1+\varepsilon)\sin\gamma. \tag{10}$$

Let the contact (also called internal) force $\mathbf{F}$ and the resultant couple $\mathbf{M}$ in an arbitrary cross–section of the rod be represented as $\mathbf{F} = Q\mathbf{b}+N\mathbf{a}$ and $\mathbf{M} = M\mathbf{e}_3$. Then we assume that (see Libai (1992))

$$N = EA[(1+\varepsilon)\cos\gamma - 1]; \quad Q = GA(1+\varepsilon)\sin\gamma; \quad M = -EI\varphi', \tag{11}$$

where EA, GA and EI are extensional, shear and bending rigidities, respectively. They are all taken as positive constants. From the definitions of Q and N we have

$$Q = -H\sin(\Theta+\varphi) + V\cos(\Theta+\varphi); \quad N = H\cos(\Theta+\varphi) + V\sin(\Theta+\varphi). \tag{12}$$

Thus H and V can be expressed in terms of ξ and η. Before doing this we note that from $(4)_1$, $(5)_1$ we get $H(S) = 0$. Then, introducing a new variable $\psi = \Theta + \varphi$ in (4), we obtain

$$\begin{aligned} V'' &= -\rho\omega^2[\sin\psi + V(\frac{\sin^2\psi}{EA} + \frac{\cos^2\psi}{GA})] - q; \\ \psi'' &= -\frac{V}{EI}[1 + V(\frac{1}{EA} - \frac{1}{GA})\sin\psi]\cos\psi + K_0'. \end{aligned} \tag{13}$$

Finally we introduce the following non–dimensional quantities

$$\begin{aligned} W &= \frac{VL^2}{EI}; \quad \lambda^2 = \frac{\omega^2\rho L^4}{EI}; \quad t = \frac{S}{L}; \quad \beta = \frac{EI}{EAL^2}; \quad y = \frac{\hat{y}}{L}; \\ \delta &= \frac{EI}{GAL^2}; \quad f_1 = \frac{qL^4}{EI}; \quad f_2 = K_0'L^2, \end{aligned} \tag{14}$$

and a new dependent variable $u = -W/\lambda$. Then, (13) transforms to

$$\begin{aligned} \ddot{u} &= \lambda[\sin\psi - \lambda u(\beta\sin^2\psi + \delta\cos^2\psi)] - \frac{f_1}{\lambda}; \\ \ddot{\psi} &= \lambda u[1 - \lambda u(\beta - \delta)\sin\psi]\cos\psi + f_2. \end{aligned} \tag{15}$$

The boundary conditions corresponding to (15) read

$$\dot{u}(0) = 0; \quad u(1) = 0; \quad \psi(0) = 0; \quad \dot{\psi}(1) = 0, \tag{16}$$

where $(\cdot)^{\cdot} = d(\cdot)/dt$. In writing $(16)_3$ we assumed that the rod is welded at $t = 0$. We call the rod described by (15),(16) the *imperfect rod* and initially straight rod that is not loaded by external load the *perfect rod*. In the case of perfect rod system (15) becomes

$$\begin{aligned} \ddot{u} &= \lambda[\sin\psi - \lambda u(\beta\sin^2\psi + \delta\cos^2\psi)]; \\ \ddot{\psi} &= \lambda u[1 - \lambda u(\beta - \delta)\sin\psi]\cos\psi. \end{aligned} \tag{17}$$

The functions $u = \psi = 0$ are solutions to (17),(16) for all values of λ. We call this solution the *trivial* solution for the perfect rod. Note that (17) possesses the following first integral of energy type

$$K = \frac{\dot{u}^2}{2} + \frac{\dot{\psi}^2}{2} - \lambda[u \sin\psi - \frac{\lambda u^2}{2}(\beta \sin^2\psi + \delta \cos^2\psi)] = const. \tag{18}$$

From the boundary conditions it follows that $K = \dot{u}^2(1)/2$. Using this value and (see definition of u and $(4)_1$)

$$u = -\lambda \int_t^1 y(p)dp, \tag{19}$$

we get

$$\frac{\dot{u}^2}{2} + \frac{\dot{\psi}^2}{2} - \lambda[u \sin\psi - \frac{\lambda u^2}{2}(\beta \sin^2\psi + \delta \cos^2\psi)] = \frac{\lambda^2}{2} y^2(1). \tag{20}$$

For the perfect rod without influence of shear and extensibility ($\delta = \beta = 0$) the system (15) reduces to

$$\ddot{u} = \lambda \sin\psi; \quad \ddot{\psi} = \lambda u \cos\psi, \tag{21}$$

a problem that was treated in Odeh and Tadjbakhsh (1965), Bazely and Zwahlen (1968), Atanackovic (1987), and Clément and Descloux (1991), for example. In this case the first integral (20) becomes

$$\frac{\dot{u}^2}{2} + \frac{\dot{\psi}^2}{2} - \lambda u \sin\psi = \frac{\lambda^2}{2} y^2(1). \tag{22}$$

Later, we shall estimate the constant on the right hand side of (22).

Variational formulation. Let $H^1((0,1), R^2)$ be the space of square integrable vector functions $w = (u, \psi)$ mapping interval $(0, 1)$ into $\mathbb{R}^2$ satisfying boundary conditions $(16)_{2,3}$ and having first derivative, in the sense of distributions, that is square integrable

$$H^1 = \{w : \int_0^1 w w^T dt < \infty; \quad \int_0^1 \dot{w} \dot{w}^T dt < \infty; \quad u(1) = \psi(0) = 0\}, \tag{23}$$

where w^T denotes the transpose of w. The norm on H^1 is

$$\|w\|_1 = \left[\int_0^1 (u^2 + \psi^2 + \dot{u}^2 + \dot{\psi}^2) dt \right]^{1/2}. \tag{24}$$

The space H^1 is a separable reflexive Banach space. The Sobolev imbedding theorem implies that $H^1 \subset C((0,1), \mathbb{R}^2)$, where $C((0,1), \mathbb{R}^2)$ is the space of continuous functions. Therefore, $u(1)$ and $\psi(0)$ in (23) make sense. H^1 is also a real Hilbert space with the inner product $\langle w_1, w_2 \rangle$ given by

$$\langle w_1, w_2 \rangle = \int_0^1 (\dot{u}_1 \dot{u}_2 + \dot{\psi}_1 \dot{\psi}_2 + u_1 u_2 + \psi_1 \psi_2) dt \tag{25}$$

Consider the functional $\Delta : H^1 \to \mathbb{R}$ defined by

$$\Delta = \int_0^1 \{\frac{\dot{u}^2}{2} + \frac{\dot{\psi}^2}{2} + \lambda[u\sin\psi - \frac{\lambda u^2}{2}(\beta\sin^2\psi + \delta\cos^2\psi)] - \frac{f_1 u}{\lambda} + f_2\psi\}dt \tag{26}$$

and $\Delta_0 : H^1 \to \mathbb{R}$

$$\Delta_0 = \int_0^1 \{\frac{\dot{u}^2}{2} + \frac{\dot{\psi}^2}{2} + \lambda[u\sin\psi - \frac{\lambda u^2}{2}(\beta\sin^2\psi + \delta\cos^2\psi)]\}dt. \tag{27}$$

The functional Δ and Δ_0 represent the total potential energy of outer forces (loads) and inner forces for rotating rod and perfect rotating rod, respectively. We treat first the case of a perfect rod described by (17),(16) by the use of Δ_0. Thus we have the following result:

If $\lambda^2\overline{\beta} < 2$, where $\overline{\beta} = \max\{\beta, \delta\}$, then there exists an element $w_0 = (u_0, \psi_0) \in H^1$ such that $\Delta_0(u_0, \psi_0) \leq \Delta_0(u, \psi)$ for all $w = (u, \psi) \in H^1$. Moreover $w_0 = (u_0, \psi_0) \in C^\infty((0,1), \mathbb{R}^2)$ and satisfies (17),(16).

The prove the result we note that under stated conditions the functional (27) is sequentially weakly lower semicontinuous and coercive. Let $\|u\|_0 = \left[\int_0^1 u^2 dt\right]^{1/2}$. Then, Δ_0 can be estimated as

$$\Delta_0 \geq \frac{1}{2}\left\|\dot{\psi}\right\|_0^2 + \frac{1}{2}\|\dot{u}\|_0^2 - \lambda\frac{1}{2}\|u\|_0 - \frac{\lambda^2}{2}\overline{\beta}\|u\|_0^2. \tag{28}$$

However, from the boundary conditions $u(1) = 0$ and $\psi(0) = 0$, it follows that

$$\|\psi\|_0^2 \leq \frac{1}{2}\left\|\dot{\psi}\right\|_0^2; \quad \|u\|_0^2 \leq \frac{1}{2}\|\dot{u}\|_0^2. \tag{29}$$

Let $\varepsilon_1 > 0$ be such that $\lambda^2\overline{\beta} + \varepsilon_1 < 2$. Then from (28),(29) we get

$$\begin{aligned}\Delta_0 &\geq \frac{1}{2}\left\|\dot{\psi}\right\|_0^2 + \frac{1}{4}\varepsilon_1\|\dot{u}\|_0^2 + \frac{1}{2}\left(1 - \frac{\varepsilon_1}{2}\right)\|\dot{u}\|_0^2 - \lambda\|u\|_0 - \frac{\lambda^2}{2}\overline{\beta}\|u\|_0^2 \\ &\geq \frac{1}{2}\left\|\dot{\psi}\right\|_0^2 + \frac{1}{4}\varepsilon_1\|\dot{u}\|_0^2 + \frac{1}{4}(2 - \varepsilon_1 - \lambda^2\overline{\beta})\|\dot{u}\|_0^2 - \lambda\|u\|_0.\end{aligned} \tag{30}$$

Transforming (30) we obtain

$$\begin{aligned}\Delta_0 &\geq \frac{1}{2}\left\|\dot{\psi}\right\|_0^2 + \frac{1}{8}\varepsilon_1\|\dot{u}\|_0^2 + \frac{\varepsilon_1}{4}\|u\|_0^2 - \lambda\|u\|_0 \\ &= \frac{1}{2}\left\|\dot{\psi}\right\|_0^2 + \frac{1}{8}\varepsilon_1\|\dot{u}\|_0^2 + \left[\frac{\sqrt{\varepsilon_1}}{2}\|u\|_0 - \frac{\lambda}{\sqrt{\varepsilon_1}}\right]^2 - \frac{\lambda^2}{\varepsilon_1} \\ &\geq \frac{1}{4}\left\|\dot{\psi}\right\|_0^2 + \frac{1}{4}2\|\psi\|_0^2 + \frac{1}{16}\varepsilon_1\|\dot{u}\|_0^2 + \frac{1}{16}\varepsilon_1 2\|u\|_0^2 - \frac{\lambda^2}{\varepsilon_1} \\ &\geq \frac{\varepsilon_1}{16}\left[\left\|\dot{\psi}\right\|_0^2 + \|\psi\|_0^2 + \|\dot{u}\|_0^2\|u\|_0^2\right] - \frac{\lambda^2}{\varepsilon_1} = \frac{\varepsilon_1}{16}\|w\|_1^2\frac{\lambda^2}{\varepsilon_1}.\end{aligned} \tag{31}$$

From (31) we conclude that Δ_0 is coercive over H^1. Further, let $\Delta_0 = \int_0^1 \mathbf{L}(w, \dot{w})\, dt$. Then, from (27) we have

$$\mathbf{L}(w, \dot{w}) = \frac{\dot{u}^2}{2} + \frac{\dot{\psi}^2}{2} + \lambda[u \sin\psi - \frac{\lambda u^2}{2}(\beta \sin^2\psi + \delta\cos^2\psi)]. \tag{32}$$

Note that $\mathbf{L}(w, \cdot)$ is a convex function. It follows (see Chow and Hale, 1982 and Dacorogna, 1989)) that Δ_0 is a weakly lower semicontinuous functional. Since Δ_0 is coercive over H^1 and H^1 is a reflexive Banach space, Δ_0 attains a minimum on H^1. Let $w_0 = (u_0, \psi_0) \in H^1$ be a minimizer of Δ_0. Then the first Fréchet derivative of Δ_0 at the point $w_0 = (u_0, \psi_0)$ is zero, i.e.,

$$\langle D\Delta_0(w_0, \lambda), w\rangle = 0. \tag{33}$$

for all $w \in H^1$. Thus, $w_0 = (u_0, \psi_0)$ is a weak solution to (17),(16). Since the system (17) is analytic, it follows that $w_0 = (u_0, \psi_0)$ is the classical solution.

Remark 1. From (16),(17) we conclude that when (u_0, ψ_0) is a solution then $(-u_0, -\psi_0)$ is a solution too. Also since $\Delta_0(u_0, \psi_0) \leq \Delta_0(u, \psi)$ for all $(u, \psi) \in H^1$ it follows that (u_0, ψ_0) and $(-u_0, -\psi_0)$ are global minimizers.

Remark 2. For the inextensible unshearable rod $(\delta = \beta = 0)$ the condition $\lambda^2\bar{\beta} < 2$ is trivially satisfied for all λ so that (27) attains minimum on the solution of (21)

Bifurcation points for a rod without imperfections. We examine now bifurcation points for the perfect rod. From the results of the previous section we conclude that the equilibrium configuration of the rod is described by (17),(16) and that on this solution the functional (27) attains the minimum. We determine now the bifurcation points for (17),(16). The main result is stated as:

There exists a unique sequence of real numbers $\{\lambda_n\}_{n=1}^{\infty}, \lambda_n > 0$ and λ_n increasing as n increases, such that: each point $(0, \lambda_n) \in H^1 \times R$ is a bifurcation point for the system (17),(16).

To prove the statement we note that system (17),(16) can be written as

$$G(u, \psi, \lambda) = 0 \tag{34}$$

with the non–linear operator $G : H^1 \to L_2$. Note that $H^1 \subset L_2$ as a consequence of Sobolev imbedding Theorem. Since (17),(16) is equivalent to

$$\langle D\Delta_0(w_0, \lambda), w_1\rangle = 0, \tag{35}$$

we obtain

$$D\Delta_0(w) = G(w, \lambda) \tag{36}$$

with $G : H^1 \subset L_2 \to L_2$. The bifurcation theory for the nonlinear equations of the type (36) was studied in .. To apply the results of Rabinowitz (1977) we write

$$D\Delta_0(w) = G(w, \lambda) = B(\lambda)w + \hat{\mathbf{H}}(w, \lambda) \tag{37}$$

where $\overset{\wedge}{\mathrm{H}}(w,\lambda) = o(\|w\|_1)$ at $w = 0$, and $B(\lambda)$ is a linear bounded operator between H^1 and L_2. The condition $B(\lambda)w = 0$ is equivalent to

$$\ddot{u} = \lambda(\psi - \lambda u\delta); \qquad \ddot{\psi} = \lambda u. \tag{38}$$

Eigenvalues of (38),(16) are solutions of the following equations

$$\lambda^2 - \frac{\lambda^3\delta}{2}\sin\left[\frac{\lambda^2\delta + \sqrt{\lambda^4\delta^2 + 4\lambda^2}}{2}\right]^{1/2}\sinh\left[\frac{-\lambda^2\delta + \sqrt{\lambda^4\delta^2 + 4\lambda^2}}{2}\right]^{1/2}$$

$$+(\lambda^2 + \frac{\lambda^4\delta}{2}\cos\left[\frac{\lambda^2\delta + \sqrt{\lambda^4\delta^2 + 4\lambda^2}}{2}\right]^{1/2}\cosh\left[\frac{-\lambda^2\delta + \sqrt{\lambda^4\delta^2 + 4\lambda^2}}{2}\right]^{1/2} = 0. \tag{39}$$

Equation (39) has countable number of positive solutions. To see this, note that (39) can be written as $F_1(\lambda) = F_2(\lambda)$ where

$$F_1(\lambda) = \lambda^2;$$

$$F_2(\lambda) = \frac{\lambda^3\delta}{2}\sin\left[\frac{\lambda^2\delta + \sqrt{\lambda^4\delta + 4\lambda^2}}{2}\right]^{1/2}\sinh\left[\frac{-\lambda^2\delta + \sqrt{\lambda^4\delta + 4\lambda^2}}{2}\right]^{1/2}$$

$$+(\lambda^2 + \frac{\lambda^4\delta}{2}\cos\left[\frac{\lambda^2\delta + \sqrt{\lambda^4\delta + 4\lambda^2}}{2}\right]^{1/2}\cosh\left[\frac{-\lambda^2\delta + \sqrt{\lambda^4\delta + 4\lambda^2}}{2}\right]^{1/2}. \tag{40}$$

Since $F_1(\lambda)$ is increasing and $F_2(\lambda)$ is oscillating with increasing amplitude, that is larger that the value of $F_1(\lambda)$, it follows that (39) has a countable increasing sequence of eigenvalues. We take the sequence $\{\lambda_n\}_{n=1}^{\infty}$ as the sequence in the theorem The eigenfunctions u_n and ψ_n corresponding to λ_n are

$$u_n = C_n\{\cos s_{1n}t + \frac{s_{1n}^3 - \lambda_n^2\delta}{s_{2n}^3 + \lambda_n^2\delta}\cosh s_{2n}t + [\frac{s_{2n}}{s_{1n}\sinh s_{2n} - s_{2n}\sin s_{1n}}$$

$$\times\frac{(s_{2n}^3 + \lambda_n^2\delta)\cos s_{1n} + (s_{1n}^3 - \lambda_n^2\delta)\cosh s_2}{s_{2n}^3 + \lambda_n^2\delta}][\sin s_{1n}t - \frac{s_{1n}}{s_{2n}}\sinh s_{2n}t]\};$$

$$\psi_n = \frac{\ddot{u}_n}{\lambda_n} + \lambda_n\delta u_n. \tag{41}$$

In (41) we used the following notation

$$s_{1n} = \left[\frac{\lambda_n^2\delta + \sqrt{\lambda_n^4\delta + 4\lambda_n^2}}{2}\right]^{1/2}; \qquad s_{2n} = \left[\frac{-\lambda_n^2\delta + \sqrt{\lambda_n^4\delta + 4\lambda_n^2}}{2}\right]^{1/2}. \tag{42}$$

Also in (41) C_n denotes an arbitrary constant. Since λ_n are isolated and simple it follows (see Rabinowitz, 1977, Theorem 0.2) that each point $(0, \lambda_n)$ is a bifurcation point for (17),(16). This completes the proof.

Remark 3. The critical value of the rotation speed follows from the smallest eigenvalue λ_1. Note that the critical value does not depend on the extensional rigidity (β). Numerical examination of (39) shows that with increasing δ the critical parameter λ_1 decreases (see 13 (13) where different rod model was used but the characteristic equation was the same as (39). For the unshearable rod $(\delta = 0)$ the equation (39) reduces to the results presented in 39 (39).

We show now that for $\lambda > \lambda_1$ the trivial solution of equilibrium equations is *not* a minimizer of Δ_0, that is, the functional Δ attains a minimum on the nontrivial (bifurcated) solution of equilibrium equations.

Consider the second Fréchet derivative of Δ_0 at the trivial solution $w_0 = (u_0, \psi_0) = (0,0)$ and for $\lambda = \lambda_1 + \bar{\lambda}$, with $\bar{\lambda} \ll 1$, in the direction $w_1 = (u_1, \psi_1)$ where u_1 and ψ_1 are eigenfunctions of linearized equilibrium equations (38) corresponding to λ_1 (i.e., u_1 and ψ_1 are given by (41) for $n = 1$). From (27) we obtain

$$< D^2 \Delta_0 (w_0 = 0, \lambda = \lambda_1 + \bar{\lambda}), w_1, w_1 >$$
$$= \int_0^1 \{\dot{u}_1^2 + \dot{\psi}_1^2 + (\lambda_1 + \bar{\lambda})[2u_1\psi_1 - \delta(\lambda_1 + \bar{\lambda})^2 u_1^2]\} dt. \tag{43}$$

However, from (73) follows

$$\int_0^1 \{\dot{u}_1^2 + \dot{\psi}_1^2 + \lambda_1[2u_1\psi_1 - \delta\lambda_1^2 u_1^2]\} dt = 0. \tag{44}$$

By using (44) in (43) we have

$$< D^2 \Delta_0 (w_0 = 0, \lambda = \lambda_1 + \bar{\lambda}), w_1, w_1 >$$
$$= \bar{\lambda} \int_0^1 \{2u_1\psi_1 - 2\delta\lambda_1 u_1^2 - \delta\,\bar{\lambda}\,u_1^2 - \delta(\lambda_1 + \bar{\lambda})^2 u_1^2\} dt. \tag{45}$$

In the first mode (i.e., for $\lambda = \lambda_1$) equations (41) imply

$$u(t) \leq 0; \qquad \psi(t) \geq 0. \tag{46}$$

The estimates (46) can be obtained by numerical examination of (41) or from a result of Parter (see Parter, 1970, p.444). Using (46) in (45) we conclude that $< D^2 \Delta_0 (w_0 = 0, \lambda = \lambda_1 + \bar{\lambda}), w_1, w_1 >$ is less than zero. Since the necessary condition for $w_0 = 0$ to be the minimizing element is violated, and since Δ_0 attains minimum on H^1 it follows that the minimum is attained on nontrivial (bifurcated) solution.

Remark 4. According to the energy criteria of stability, in a stable configuration the (total) potential energy has a weak local minimum. From the fact that for $\lambda > \lambda_1$ (i.e., $\bar{\lambda} > 0$) the trivial solution is not a minimizer we conclude that for $\lambda > \lambda_1 = \lambda_{cr}$ the trivial configuration $w_0 = (u_0 = 0, \psi_0 = 0)$ is unstable. In Table 1 we give λ_{cr} determined as the smallest root of (39) for a few values of δ.

Table 1. Critical values of λ.

δ	λ_{cr}^2
0.0	12.362
0.01	11.812
0.05	10.005
0.1	8.377
0.5	3.584
1.0	2.079
2.0	1.129

Estimates of maximal deflection for a rod without imperfections. In this section we analyze the perfect inextensible rod, i.e.,

$$\ddot{u} = \lambda \sin\psi; \qquad \ddot{\psi} = \lambda u \cos\psi, \tag{47}$$

subject to

$$\dot{u}\,(0) = 0; \quad u(1) = 0; \quad \psi(0) = 0; \quad \dot{\psi}\,(1) = 0. \tag{48}$$

For the problem (47),(48) we derive an estimate of the maximal deflection $y(1)$ in the first buckling mode. The estimate of $y(1)$ will also give a bound on the constant in the first integral (22). We first consider linearization of (47)

$$\ddot{u} = \lambda\psi; \qquad \ddot{\psi} = \lambda u. \tag{49}$$

Let λ_{cr} be the smallest eigenvalue to (49),(48). Our main result is the following estimate:

For $\lambda > \lambda_{cr}$ the maximal deflection $y(1)$ satisfies the inequality

$$y^2(1) \leq \frac{\lambda^2 - \lambda_{cr}^2}{\frac{1}{8}(\frac{\pi}{2})^4 + \frac{\lambda^2}{3}}. \tag{50}$$

To prove (50) we write the system (47),(48) as

$$\dot{u} = \lambda y; \qquad \dot{m} = -\lambda u \cos\psi;$$

$$\dot{y} = \sin\psi; \qquad \dot{\psi} = -m, \tag{51}$$

with

$$y(0) = 0; \qquad \psi(0) = 0; \qquad u(1) = 0; \qquad m(1) = 0. \tag{52}$$

From (51),(52) we have

$$u = -\int_t^1 y(p)dp; \qquad m = -\int_t^1 \lambda^2 [\int_\xi^1 y(p)dp] \sqrt{1 - \dot{y}^2\,(\xi)} d\xi;$$

$$\dot{\psi} = \frac{\ddot{y}}{\sqrt{1 - \dot{y}^2}}. \tag{53}$$

Combining (51),(53) we get

$$\frac{\ddot{y}(t)}{\sqrt{1-\dot{y}^2(t)}} = \lambda^2 \int_t^1 [\int_\xi^1 y(p)dp]\sqrt{1-\dot{y}^2(\xi)}d\xi. \tag{54}$$

Multiplying (54) by $\ddot{y}$ and integrating (partial integration and the use of boundary conditions), we obtain

$$\int_0^1 \frac{\ddot{y}^2\, dt}{\sqrt{1-\dot{y}^2}} = \lambda^2 \int_0^1 [\ddot{y}(p)\sqrt{1-\dot{y}^2(p)} \int_p^1 y(k)dk]dp. \tag{55}$$

Using the inequalities (see Mitrinovic, 1970)

$$\frac{1}{\sqrt{1-\dot{y}^2}} \geq 1+\frac{1}{2}\dot{y}^2; \qquad \sqrt{1-\dot{y}^2} \leq 1-\frac{1}{2}\dot{y}^2, \tag{56}$$

in (54) we find

$$\int_0^1 \ddot{y}^2\, dt + \frac{1}{2}\int_0^1 \dot{y}^2 \ddot{y}^2\, dt \leq \lambda^2 \int_0^1 y^2 dt - \frac{\lambda^2}{2}\int_0^1 \dot{y}^3(t)[\int_t^1 y(k)dk]dt, \tag{57}$$

or

$$\frac{\int_0^1 \ddot{y}^2\, dt}{\int_0^1 y^2 dt} + \frac{1}{2}\frac{\int_0^1 \dot{y}^2\ddot{y}^2\, dt}{\int_0^1 y^2 dt} \leq \lambda^2 - \frac{\lambda^2}{2}\frac{\int_0^1 \dot{y}^3(t)[\int_t^1 y(k)dk]dt}{\int_0^1 y^2 dt}. \tag{58}$$

The various terms in (58) can be estimated as follows:

i) Let

$$\Phi = \frac{\dot{y}^2}{2}. \tag{59}$$

Then, from (52) follows

$$\Phi(0) = 0; \qquad \dot{\Phi}(0) = 0; \qquad \dot{\Phi}(1) = 0. \tag{60}$$

Using Troesch's inequality (see Troesch, 1967 and Mitrinovic, 1970) we obtain

$$\begin{aligned} \int_0^1 \dot{\Phi}^2\, dt &= \int_0^1 \dot{y}^2\ddot{y}^2\, dt \geq (\frac{\pi}{2})^2 \int_0^1 \Phi^2 dt \\ &= (\frac{\pi}{2})^2 \int_0^1 (\frac{\dot{y}^2}{2})^2 dt \geq \frac{1}{4}(\frac{\pi}{2})^2 \left\|y\right\|_{L_2}^4, \end{aligned} \tag{61}$$

where $\|\cdot\|_{L_2}$ denotes the L_2 norm and in the last step we used Cauchy inequality. Also from the Troesch inequality, applied to y, we have

$$\int_0^1 \dot{y}^2\, dt \geq (\frac{\pi}{2})^2 \int_0^1 y^2 dt. \tag{62}$$

Observing (61) and (62) the second term on the left hand side of (58) can be estimated as

$$\frac{\int_0^1 \dot{y}^2 \ddot{y}^2 \, dt}{\int_0^1 y^2 dt} \geq \frac{1}{4}\left(\frac{\pi}{2}\right)^4 \left\|\dot{y}\right\|_{L_2}^2 . \tag{63}$$

ii) To estimate the second term on the right hand side of (58) note that in the first mode $\dot{y} \geq 0$ and $\ddot{y} \geq 0$ (see (46) and (51)). Differentiating (54) we obtain

$$\begin{aligned} \dddot{y} &= -\lambda^2 [1 - \dot{y}^2(t)] \int_t^1 y(p)dp \\ &\quad -\lambda^2 \frac{\dot{y}\ddot{y}}{\sqrt{1-\dot{y}^2(t)}} \int_t^1 \left[\int_\xi^1 y(p)dp\right] \sqrt{1-\dot{y}^2(\xi)}\, d\xi, \end{aligned} \tag{64}$$

so that $\dddot{y} \leq 0$. Thus $\dot{y}$ is a concave function. Let

$$f_3 = \dot{y} \int_t^1 y(p)dp. \tag{65}$$

Differentiation shows that f_3 is also a concave function. The following inequality for concave functions was proved in Karlin and Ziegler (1975)

$$\frac{\left[\frac{n}{2}\right]! \left[\frac{n+1}{2}\right]!}{(n+1)!} 2^n \prod_{i=1}^{n} \left(\int_0^1 f_i(t)dt \right) \leq \int_0^1 \prod_{i=1}^{n} f_i(t)dt. \tag{66}$$

In writing (66) we corrected the constant on the left hand side, according to the result of Brenner and Alzer (1991). Using (66) with $f_1 = \dot{y}$; $f_2 = \dot{y}$ and f_3 given by (65), we get

$$\int_0^1 \dot{y}^3(t) \left[\int_t^1 y(k)dk\right] dt \geq \frac{2}{3} y^2(1) \int_0^1 y^2(t)dt. \tag{67}$$

With (67) the second term on the right hand side of (58) can be estimated as

$$\frac{\int_0^1 \dot{y}^3(t) \left[\int_t^1 y(k)dk\right] dt}{\int_0^1 y^2 dt} \geq \frac{2}{3} y^2(1). \tag{68}$$

iii) Finally, from the variational characterization of eigenvalues, the first term on the left hand side of (58) can be estimated as

$$\frac{\int_0^1 \ddot{y}^2 \, dt}{\int_0^1 y^2 dt} \geq \lambda_1^2 = \lambda_{cr}^2, \tag{69}$$

where λ_1 is the smallest eigenvalue of (49),(48).

Using the estimates (63),(68) and (69) in (58) we obtain

$$\lambda_{cr}^2 + \frac{1}{8}\left(\frac{\pi}{2}\right)^4 \left\|\dot{y}\right\|_{L_2}^2 \leq \lambda^2 - \frac{1}{3}\lambda^2 y^2(1). \tag{70}$$

The boundary condition $y(0) = 0$ implies

$$y^2(1) \leq \left\| \dot{y} \right\|^2_{L_2}. \tag{71}$$

Combining (70) and (71) we obtain (50).

Remark 5. With the estimate (57) the first integral (1.1.1-22) becomes

$$\frac{\lambda^2}{2} \frac{\lambda^2 - \lambda^2_{cr}}{\frac{1}{8}(\frac{\pi}{2})^4 + \frac{\lambda^2}{3}} \geq \frac{\dot{u}^2}{2} + \frac{\dot{\psi}^2}{2} - \lambda u \sin \psi. \tag{72}$$

Also, with $\lambda = \lambda_{cr} + \bar{\lambda}$ and $\bar{\lambda} \ll 1$ the expression (57) leads to the conclusion that maximal deflection is proportional to the square root of $\bar{\lambda}$. This also follows from the Liapunov–Schmidt method applied to the same problem (see 12 (12)).

The system (47),(48) can be treated variationally by using variational principle developed in Atanackovic (2001). Following Atanackovic (2001) we consider the functional

$$\mathbb{I}_0 = \int_0^1 [\frac{d\vartheta}{dt}\frac{d\xi}{dt} + \frac{du}{dt}\frac{d\eta}{dt} + \lambda \eta \sin \vartheta + \lambda u \xi \cos \vartheta] dt. \tag{73}$$

The stationarity condition for the functional (73) reads $D\mathbb{I}_0 (\vartheta, \xi) = 0$. It is equivalent to

$$\begin{aligned} &\frac{d^2 u}{dt^2} - \lambda \sin \vartheta = 0; &&\frac{d^2 \vartheta}{dt^2} - \lambda u \cos \vartheta = 0, \\ &\frac{d^2 \xi}{dt^2} - \lambda [\eta \cos \vartheta - u \xi \sin \vartheta] = 0; &&\frac{d^2 \eta}{dt^2} - \lambda \xi \cos \vartheta = 0, \end{aligned} \tag{74}$$

where $\mathbf{w} = (\vartheta, u)$ and $\mathbf{v} = (\xi, \eta)$ are elements of the space $\mathbf{X}$ defined as

$$\mathbf{X} = \{\mathbf{w} = (\vartheta, u) : (0,1) \to \mathbb{R}^2;\ \mathbf{w} \in C^1((0,1); \mathbb{R}^2);$$

$$\vartheta(0) = 0; \frac{d\vartheta(1)}{dS} = 0; \frac{du(0)}{dS} = 0; u(1) = 0\}. \tag{75}$$

Note that the stationarity condition for (73) reproduces equations of relative equilibrium (47) *and* equations of small deformation (ξ, η) superposed on the finite one (u, ϑ). The functional (73) can be used for bifurcation analysis, as is shown in Atanackovic (2001). Note also that the system (74) possesses a first integral of Jacobi type given by

$$\frac{d\vartheta}{dt}\frac{d\xi}{dt} + \frac{du}{dt}\frac{d\eta}{dt} - \lambda \eta \sin \vartheta - \lambda u \xi \cos \vartheta = const. \tag{76}$$

1.2 Stability Bounds for a Rod Subjected to a Couple and a Force at Its Ends

The problem of determining the stability boundary of a twisted and axially compressed rod is indeed an old one. Suppose that the rod is loaded by a compressive force of intensity P and a torsion couple of intensity M_t. For the case of a rod described by Bernoulli-Euler rod theory the critical value of the load parameters (P_{cr}, M_{tcr}) was determined by Greenhill in 1883 (see Timoshenko and Gere, 1961). Later the problem was treated by many authors. We mention the works of Biezano and Grammel (1953), Beck (1955), Kovari (1969), Antman and Kenney (1981), Béda et al. (1992), Atanackovic and Glavardanov (1996). Besides the value of critical load parameters (P_{cr}, M_{tcr}) another important result concerns the *type* of bifurcation at the critical point. It is known that in the case of Benoulli-Euler model the rod loaded with compressive force only exhibits supercritical bifurcation while the rod loaded by torque only exhibits subcritical bifurcation. In our previous work Atanackovic and Glavardanov (1996) we determined, analytically, the value of load parameters (P_{cr}^*, M_{tcr}^*) separating sub from the supercritcial bifurcation. Also in Atanackovic and Glavardanov (1996) the values (P_{cr}, M_{tcr}) are determined for a rod described by Harnigx's type of constitutive equation.

Our intention here is to reconsider the problem of stability and bifurcation of a rod, described by Haringx's type of constitutive equation and loaded by a force and a couple. We shall reduce the system of equilibrium, geometrical and constitutive equations (altogether 18 equations) to a *single* nonlinear second order equation. This is a significant reduction, since in Béda et al. (1992), Atanackovic and Glavardanov (1996) the bifurcation analysis was performed on two second order nonlinear differential equations. We shall determine the critical load parameters as well as type of bifurcation. Also the reduced system, when linearized, has a null space of dimension one. Thus, we obtain single bifurcation equation for the reduced system. In earlier approach (see Béda et al., 1992, Atanackovic and Glavardanov, 1996) the null space of dimension two led to two bifurcation equations, one of which was eliminated by symmetry considerations. The analysis presented here is taken from Atanackovic and Glavardanov (2001). Another characteristic of our approach is that the second order equation that we deal with is Euler-Lagrange equation of an energy type functional and possesses a first integral. We shall use this first integral to express the solution in terms of elliptic integrals. Thus, our results represent a generalization of the results of Iljohin (1979 p. 38) obtained for Bernoulli-Euler rod. On the other hand, our results may be viewed as a specialization of the results of Antman and Kenney Antman and Kenny (1981) to a specific constitutive equation. Due to this specialization we were able to obtain more information about the solution. Namely, we obtained a first integral and representation of the solution in terms of elliptic integrals and we obtained the type of bifurcation for *all* values of load and material parameters.

Equilibrium equations. Consider an elastic rod O_1O_2 shown in Figure 2. The rod is naturally straight and loaded by a concentrated compressive force and a couple at its end O_2. The compressive force $\mathbf{P} = -P\mathbf{e}_{10}$ is of constant intensity $P = const.$ and oriented along the $\bar{x}_{10}$ axis of a fixed rectangular Cartesian coordinate system $\bar{x}_{10}, \bar{x}_{20}, \bar{x}_{30}$ with the unit vectors $\mathbf{e}_{10}, \mathbf{e}_{20}, \mathbf{e}_{30}$, respectively. The couple is given as $\mathbf{T} = M_t\mathbf{e}_{10}$ with $M_t = const.$

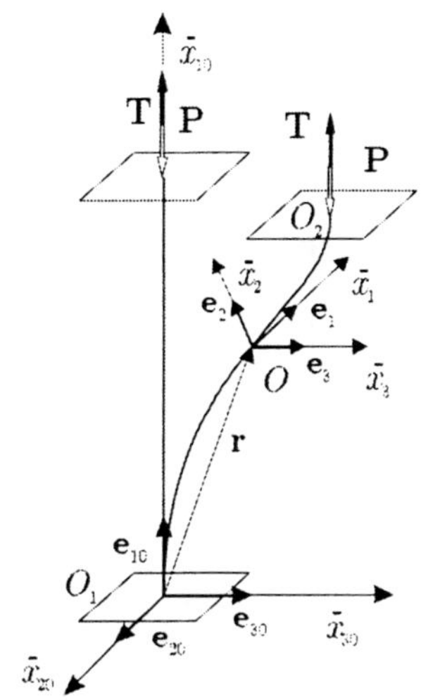

Figure 2. Coordinate system and load configuration.

The end O_1 of the rod is fixed to a unmovable rigid plate, laying in $\bar{x}_{20} - \bar{x}_{30}$ coordinate plane so that the cross-section of the rod that is in contact with the rigid plate does not have any rotation or translation (welded end). For interpretation of this boundary condition as well as for the review of other possibilities of specifying boundary condition for shearable rods, see Antman and Christoforis (1997). At the end O_2 the rod is welded to a movable rigid plate that can move freely but remains parallel to coordinate plane $\bar{x}_{20} - \bar{x}_{30}$.

Let S be the arc−length of the rod axis in the undeformed state. $S \in [0, L]$ where L is the length of the rod.

We describe the configuration of the rod by one vector function $\mathbf{r}(S)$ specifying the position of a point on the rod axis and by orientation of the Cartesian coordinate system with axes $\bar{x}_1, \bar{x}_2, \bar{x}_3$ oriented along the normal to the cross-section and along the principal directions of the rod cross-section at an arbitrary point O of the rod axis, respectively. Thus we have

$$\mathbf{r}(S) = x_{10}\mathbf{e}_{10} + x_{20}\mathbf{e}_{20} + x_{30}\mathbf{e}_{30}. \tag{77}$$

Let $\mathbf{e}_1, \mathbf{e}_2, \mathbf{e}_3$ be the unit vectors along the $\bar{x}_1, \bar{x}_2, \bar{x}_3$ respectively. The orientation of the system $\mathbf{e}_1, \mathbf{e}_2, \mathbf{e}_3$ with respect to the unit vectors parallel with $\mathbf{e}_{10}, \mathbf{e}_{20}, \mathbf{e}_{30}$ and passing through point O is given by three Euler type angles. We take ship angles (see Lurie, 1961 and Atanackovic, 1997) that bring $\mathbf{e}_{10}, \mathbf{e}_{20}, \mathbf{e}_{30}$ to $\mathbf{e}_1, \mathbf{e}_2, \mathbf{e}_3$ by the sequence of three rotations. The first is rotation of amount θ_1 about $\bar{x}_{10}$ axis. The next rotation is about ξ axis for an amount θ_3 and the last rotation is of amount θ_2 about the axis $\bar{x}_2$. All rotations are performed counterclockwise.The vector $\boldsymbol{\omega}$ (the angular velocity vector) is

$$\boldsymbol{\omega} = \theta_1^{'}\mathbf{e}_{10} + \theta_3^{'}\boldsymbol{\mu} + \theta_2^{'}\mathbf{e}_2, \tag{78}$$

where $(\cdot)^{'} = d/dS(\cdot)$ and $\boldsymbol{\mu}$ is the unit vector along ξ axis. From (78) we obtain the components of $\boldsymbol{\omega}$ in the local coordinate system $\mathbf{e}_1, \mathbf{e}_2, \mathbf{e}_3$. Thus, $\boldsymbol{\omega} = \omega_1\mathbf{e}_1 + \omega_2\mathbf{e}_2 + \omega_3\mathbf{e}_3$ with

$$\begin{aligned} &\omega_1 = \theta_1^{'}\cos\theta_2\cos\theta_3 - \theta_3^{'}\sin\theta_2; \quad \omega_2 = \theta_2^{'} - \theta_1^{'}\sin\theta_3; \\ &\omega_3 = \theta_1^{'}\cos\theta_3\sin\theta_2 + \theta_3^{'}\cos\theta_2. \end{aligned} \tag{79}$$

Let $\mathbf{\Gamma}$ be a vector defined by

$$\mathbf{\Gamma} = \mathbf{r}' - \mathbf{e}_1. \tag{80}$$

In the local coordinate system $\mathbf{e}_1, \mathbf{e}_2, \mathbf{e}_3$ the vector $\mathbf{\Gamma}$ may be expressed as

$$\mathbf{\Gamma} = \Gamma_1 \mathbf{e}_1 + \Gamma_2 \mathbf{e}_2 + \Gamma_3 \mathbf{e}_3. \tag{81}$$

Then the six quantities $(\omega_1, \omega_2, \omega_3)$ and $(\Gamma_1, \Gamma_2, \Gamma_2)$ are *strains* in the rod theory that we use (see Antman, 1995 and Atanackovic, 1997).

The equilibrium equations for the rod can be written as

$$\mathbf{F}' = 0; \quad \mathbf{M}' = -\mathbf{r}' \times \mathbf{F}, \tag{82}$$

where $\mathbf{F} = F_1 \mathbf{e}_1 + F_2 \mathbf{e}_2 + F_3 \mathbf{e}_3$ is the contact force, $\mathbf{M} = M_1 \mathbf{e}_1 + M_2 \mathbf{e}_2 + M_3 \mathbf{e}_3$ is the contact couple and we assumed that there are no distributed forces or couples. Following Eliseyev Eliseyev (1988) we assume the constitutive equations for the contact force and contact couple in the form

$$\mathbf{F} = B_{11}\Gamma_1 \mathbf{e}_1 + B_{22}\Gamma_2 \mathbf{e}_2 + B_{33}\Gamma_3 \mathbf{e}_3; \quad \mathbf{M} = A_{11}\omega_1 \mathbf{e}_1 + A_{22}\omega_2 \mathbf{e}_2 + A_{33}\omega_3 \mathbf{e}_3, \tag{83}$$

where $B_{11}, B_{22}, ..., A_{33}$ are constants. Note that (80) my be written as

$$\mathbf{r}' = \mathbf{\Gamma} + \mathbf{e}_1. \tag{84}$$

By using (83) in (79), (82) and (84)and by solving (79) for $\theta_1', \theta_2', \theta_3'$ we obtain

$$\begin{aligned}
& F_1' + F_3 \frac{M_2}{A_{22}} - F_2 \frac{M_3}{A_{33}} = 0; \quad F_2' + F_1 \frac{M_3}{A_{33}} - F_3 \frac{M_1}{A_{11}} = 0; \\
& F_3' + F_2 \frac{M_1}{A_{11}} - F_1 \frac{M_2}{A_{22}} = 0; \\
& M_1' + \left(\frac{1}{A_{22}} - \frac{1}{A_{33}} \right) M_2 M_3 = F_2 F_3 \left(\frac{1}{B_{33}} - \frac{1}{B_{22}} \right); \\
& M_2' + \left(\frac{1}{A_{33}} - \frac{1}{A_{11}} \right) M_1 M_3 = F_3 \left[1 + \left(\frac{1}{B_{11}} - \frac{1}{B_{33}} \right) F_1 \right]; \\
& M_3' + \left(\frac{1}{A_{11}} - \frac{1}{A_{22}} \right) M_1 M_2 = -F_2 \left[1 + \left(\frac{1}{B_{11}} - \frac{1}{B_{22}} \right) F_1 \right]; \\
& \theta_1' = \frac{c_2}{c_3} \frac{M_1}{A_{11}} + \frac{s_2}{c_3} \frac{M_3}{A_{33}}; \\
& \theta_2' = \frac{c_2 s_3}{c_3} \frac{M_1}{A_{11}} + \frac{M_2}{A_{22}} + \frac{s_2 s_3}{c_3} \frac{M_3}{A_{33}}; \\
& \theta_3' = -s_2 \frac{M_1}{A_{11}} + c_2 \frac{M_3}{A_{33}}; \\
& x_{10}' = c_2 c_3 \left(1 + \frac{F_1}{B_{11}} \right) - s_3 \frac{F_2}{B_{22}} + c_3 s_2 \frac{F_3}{B_{33}};
\end{aligned}$$

$$x_{20}^{'} = (c_1c_2s_3 + s_1s_2)\left(1 + \frac{F_1}{B_{11}}\right) + c_1c_3\frac{F_2}{B_{22}}$$
$$+ (c_1s_2s_3 - c_2s_1)\frac{F_3}{B_{33}};$$
$$x_{30}^{'} = (c_2s_1s_3 - c_1s_2)\left(1 + \frac{F_1}{B_{11}}\right) + s_1c_3\frac{F_2}{B_{22}}$$
$$+ (s_1s_2s_3 + c_1c_2)\frac{F_3}{B_{33}}, \qquad (85)$$

where $c_1 = \cos\theta_1, s_1 = \sin\theta_1, ..., s_3 = \sin\theta_3$. Note also that in writing (85) we assumed that $c_3 \neq 0$. The boundary conditions corresponding to the rod shown in Fig. 2 are

$$\begin{aligned}
&F_1(L) = -P; \quad F_2(L) = 0; \quad F_3(L) = 0;\\
&M_1(L) = M_t; \quad \theta_2(L) = 0; \quad \theta_3(L) = 0;\\
&\theta_1(0) = 0; \quad \theta_2(0) = 0; \quad \theta_3(0) = 0;\\
&x_{10}(0) = 0; \quad x_{20}(0) = 0; \quad x_{30}(0) = 0.
\end{aligned} \qquad (86)$$

In the analysis that follows we assume that the rod has axial symmetry so that $A_{22} = A_{33}; B_{22} = B_{33}$. It is easy to see that the system (85), (86) has the following first integrals

$$F_1M_1 + F_2M_2 + F_3M_3 = -PM_t; \quad F_1^2 + F_2^2 + F_3^2 = P^2. \qquad (87)$$

The first expression in (87) is the scalar product of vectors **F** and **M**, while the second one expresses the fact that the intensity of contact force in an arbitrary cross-section is constant. Introducing the dimensionless quantities

$$\begin{aligned}
&\bar{F}_1 = \frac{F_1L^2}{A_{22}}; \quad \bar{F}_2 = \frac{F_2L^2}{A_{22}}; \quad \bar{F}_3 = \frac{F_3L^2}{A_{22}}; \quad \lambda = \frac{PL^2}{A_{22}};\\
&\bar{M}_1 = \frac{M_1L}{A_{22}}; \quad \bar{M}_2 = \frac{M_2L}{A_{22}}; \quad \bar{M}_3 = \frac{M_3L}{A_{22}}; \quad m = \frac{M_tL}{A_{22}};\\
&\bar{x}_{10} = \frac{x_{10}}{L}; \quad \bar{x}_{20} = \frac{x_{20}}{L}; \quad \bar{x}_{30} = \frac{x_{30}}{L}; \quad t = \frac{S}{L};\\
&\alpha = \frac{A_{22}}{B_{11}L^2}; \quad \beta = \frac{A_{22}}{B_{22}L^2}; \quad \gamma = \alpha - \beta,
\end{aligned} \qquad (88)$$

we obtain from (85),(86)

$$\dot{\bar{F}}_1 + \bar{F}_3\bar{M}_2 - \bar{F}_2\bar{M}_3 = 0; \; \dot{\bar{F}}_2 + \bar{F}_1\bar{M}_3 - \frac{A_{22}}{A_{11}}\bar{F}_3\bar{M}_1 = 0;$$
$$\dot{\bar{F}}_3 + \frac{A_{22}}{A_{11}}\bar{F}_2\bar{M}_1 - \bar{F}_1\bar{M}_2 = 0; \quad \dot{\bar{M}}_1 = 0;$$
$$\dot{\bar{M}}_2 + \left(1 - \frac{A_{22}}{A_{11}}\right)\bar{M}_1\bar{M}_3 = \bar{F}_3\left[1 + \gamma\bar{F}_1\right];$$
$$\dot{\bar{M}}_3 + \left(\frac{A_{22}}{A_{11}} - 1\right)\bar{M}_1\bar{M}_2 = -\bar{F}_2\left[1 + \gamma\bar{F}_1\right];$$

$$\dot{\theta}_1 = \frac{c_2}{c_3}\frac{A_{22}}{A_{11}}\bar{M}_1 + \frac{s_2}{c_3}\bar{M}_3;$$
$$\dot{\theta}_2 = \frac{c_2 s_3}{c_3}\frac{A_{22}}{A_{11}}\bar{M}_1 + \bar{M}_2 + \frac{s_2 s_3}{c_3}\bar{M}_3;$$
$$\dot{\theta}_3 = -s_2\frac{A_{22}}{A_{11}}\bar{M}_1 + c_2\bar{M}_3;$$

$$\dot{\bar{x}}_{10} = c_2 c_3\left(1+\alpha\bar{F}_1\right) - \beta s_3\bar{F}_2 + \beta c_3 s_2\bar{F}_3;$$
$$\dot{\bar{x}}_{20} = \left(c_1 c_2 s_3 + s_1 s_2\right)\left(1+\alpha\bar{F}_1\right) + \beta c_1 c_3\bar{F}_2 + \beta\left(c_1 s_2 s_3 - c_2 s_1\right)\bar{F}_3;$$
$$\dot{\bar{x}}_{30} = \left(c_2 s_1 s_3 - c_1 s_2\right)\left(1+\alpha\bar{F}_1\right) + \beta s_1 c_3\bar{F}_2 + \beta\left(s_1 s_2 s_3 + c_1 c_2\right)\bar{F}_3, \tag{89}$$

and

$$\bar{F}_1(1) = -\lambda;\quad \bar{F}_2(1) = 0;\quad \bar{F}_3(1) = 0;$$
$$\bar{M}_1(1) = m;\quad \theta_2(1) = 0;\quad \theta_3(1) = 0;$$
$$\theta_1(0) = 0;\quad \theta_2(0) = 0;\quad \theta_3(0) = 0;$$
$$\bar{x}_{10}(0) = 0;\quad \bar{x}_{20}(0) = 0;\quad \bar{x}_{30}(0) = 0. \tag{90}$$

Let $\bar{F}_1, \bar{F}_2, \bar{F}_3, ..., \bar{x}_{30}$ be a solution to $(89), (90)$. Then arbitrary rotation about x_{10} axis leads to another solution, i.e., $\bar{F}_1, \bar{F}_2\cos\delta + \bar{F}_3\sin\delta, -\bar{F}_2\sin\delta + \bar{F}_3\cos\delta, ..., -\bar{x}_{20}\sin\delta + \bar{x}_{30}\cos\delta$ is a solution to $(89), (90)$ for arbitrary δ. Note that $(89)_4$ together with $(90)_4$ lead to

$$\bar{M}_1 = m. \tag{91}$$

For further analysis we shall need the first six equations of the system (89), that after the use of (91) become

$$\dot{\bar{F}}_1 + \bar{F}_3\bar{M}_2 - \bar{F}_2\bar{M}_3 = 0;$$
$$\dot{\bar{F}}_2 + \bar{F}_1\bar{M}_3 - \frac{A_{22}}{A_{11}}m\bar{F}_3 = 0;$$
$$\dot{\bar{F}}_3 + \frac{A_{22}}{A_{11}}m\bar{F}_2 - \bar{F}_1\bar{M}_2 = 0;$$
$$\dot{\bar{M}}_2 + \left(1 - \frac{A_{22}}{A_{11}}\right)m\bar{M}_3 = \bar{F}_3\left[1+\gamma\bar{F}_1\right];$$
$$\dot{\bar{M}}_3 + \left(\frac{A_{22}}{A_{11}} - 1\right)m\bar{M}_2 = -\bar{F}_2\left[1+\gamma\bar{F}_1\right]; \tag{92}$$

Also by the use of (88) and (91) the first integrals (87) read

$$\bar{F}_1 m + \bar{F}_2\bar{M}_2 + \bar{F}_3\bar{M}_3 = -\lambda m;$$
$$\bar{F}_1^2 + \bar{F}_2^2 + \bar{F}_3^2 = \lambda^2. \tag{93}$$

We simplify system (92). Thus, we introduce the variables

$$\begin{aligned} X_2 &= \bar{M}_2^2 + \bar{M}_3^2; \\ X_3 &= \bar{F}_2\bar{M}_2 + \bar{F}_3\bar{M}_3; \\ X_4 &= \bar{F}_2\bar{M}_3 - \bar{F}_3\bar{M}_2. \end{aligned} \tag{94}$$

By differentiating $(93)_1$ and by using (92) we get

$$\dot{\bar{F}}_1 = X_4. \tag{95}$$

Further by differentiating (94) and by using (92) we obtain

$$\begin{aligned} \dot{X}_2 &= -2X_4\left[1+\gamma\bar{F}_1\right]; \\ \dot{X}_3 &= -mX_4; \\ \dot{X}_4 &= -X_2\bar{F}_1 + mX_3 + \left(\left(\bar{F}_1\right)^2 - \lambda^2\right)\left(1+\gamma\bar{F}_1\right) \end{aligned} \tag{96}$$

It can be seen that the system (95), (96) has the following first integrals

$$\begin{aligned} X_3^2 + X_4^2 &= \left(\lambda^2 - \left(\bar{F}_1\right)^2\right)X_2; \\ X_3 &= -m\left(\bar{F}_1 + \lambda\right); \\ X_2 + 2\bar{F}_1 + \gamma\left(\bar{F}_1\right)^2 &= C = const. \end{aligned} \tag{97}$$

From system $(95)-(97)$ we derive the following second order equation

$$\ddot{\bar{F}}_1\left(\lambda^2 - \left(\bar{F}_1\right)^2\right) + \left(\dot{\bar{F}}_1\right)^2\bar{F}_1 + \lambda m^2\left(\lambda+\bar{F}_1\right)^2 + \left(\lambda^2 - \left(\bar{F}_1\right)^2\right)^2\left(1+\gamma\bar{F}_1\right) = 0. \tag{98}$$

The boundary conditions corresponding to (98) follow from $(90)_1$ and the condition of global equilibrium of forces in x_{10} direction. Thus, we have

$$\bar{F}_1(0) = -\lambda; \quad \bar{F}_1(1) = -\lambda. \tag{99}$$

Following Antman and Kenney (1981) we introduce the angle θ (the angle of nutation) between $\mathbf{e}_1$ and $\mathbf{e}_{10}$. Then,

$$\bar{F}_1 = -\lambda\cos\theta. \tag{100}$$

With (100) equation (98) becomes

$$\lambda^2\sin^3\theta\left[\ddot{\theta} + m^2\frac{\sin\theta}{\left(1+\cos\theta\right)^2} + \lambda\sin\theta\left(1-\gamma\lambda\cos\theta\right)\right] = 0. \tag{101}$$

We are going to use (101) for the *local* bifurcation analysis. Thus we are interested in solution $\theta(t)$ of (101) that are $C^2(0,1)$ functions, small in sup norm. Therefore we assume $|\theta| < \pi$. Then from (101) we obtain

$$\ddot{\theta} + m^2\frac{\sin\theta}{\left(1+\cos\theta\right)^2} + \lambda\sin\theta\left(1-\gamma\lambda\cos\theta\right) = 0. \tag{102}$$

The boundary conditions corresponding to (102) follow from (99) and (100) and read

$$\theta(0) = 0; \quad \theta(1) = 0. \tag{103}$$

The system (102), (103) will be the basis for our bifurcation analysis. Equations (102), (103) may be obtained from the variational principle that we formulate next. Consider the problem of determining the minimum of the functional

$$I = \int_0^1 \left(\frac{\dot{\Theta}^2}{2} - \frac{m^2}{(1+\cos\Theta)} + \lambda \left[\cos\Theta + \gamma \frac{\lambda}{2} \sin^2\Theta \right] \right) dt. \tag{104}$$

where $\Theta \in Y$ with

$$Y = \left\{ \Theta : \Theta \in C^2(0,1), \quad \Theta(0) = \Theta(1) = 0 \right\}. \tag{105}$$

Then on the solution of (102) the necessary condition for the minimum of I is satisfied, i.e.,

$$\delta I(\theta, w) = 0, \tag{106}$$

where $w = \Theta - \theta$. From (95), (97) we can obtain an important relation that we shall use later. Namely, by substituting in $(97)_1$ the variables X_3, X_2 and X_4 from $(97)_2$, $(97)_3$ and (95) respectively, we obtain

$$\left(\dot{\bar{F}}_1\right)^2 = (\lambda + \bar{F}_1)\left\{(\lambda - \bar{F}_1)\left(C - 2\bar{F}_1 - \gamma(\bar{F}_1)^2\right) - m^2(\lambda + \bar{F}_1)\right\}. \tag{107}$$

The boundary conditions corresponding to (107) are given by (99). We shall be concerned with the solutions of (107) that are symmetric with respect to the middle point $t = 1/2$. In this case we have $\dot{\bar{F}}_1(1/2) = 0$. Note also that (107) is a generalization of the result presented in 30 (30) where the classical Bernoulli-Euler rod is treated. Our first integral reduces to the one presented in Iljohin (1979) when we set $\gamma = 0$ in (107).

Bifurcation analysis. The bifurcation analysis will be based on the *second* order nonlinear boundary value problem (102),(103). First, we write (102),(103) in compact form. Thus, we define a nonlinear operator $\mathcal{F}(\Theta, m^2)$ mapping $Y \times \mathbb{R}_+ \to Z$, where Y is given by (105) and $Z = C(0,1)$ is the space of continuous functions. Note that we shall treat λ and γ as fixed and given in advance while m^2 will be considered as a bifurcation parameter. With

$$\mathcal{F}(\Theta, m^2) = \ddot{\Theta} + m^2 \frac{\sin\Theta}{(1+\cos\Theta)^2} + \lambda \sin\Theta(1 - \gamma\lambda\cos\Theta), \tag{108}$$

system (102),(103) becomes

$$\mathcal{F}(\theta, m^2) = 0. \tag{109}$$

Note that $\theta_0 = 0$ satisfies (109) for *all* values of $m^2 > 0$. The first Fréchet derivative of $\mathcal{F}(\Theta, m^2)$ calculated at $\theta_0 = 0$ reads

$$D^{(1)}\mathcal{F}(\theta_0 = 0, m^2)\, w = B(m^2)\, w = \ddot{w} + w\left[\frac{m^2}{4} + \lambda(1 - \gamma\lambda)\right] \tag{110}$$

The lowest eigenvalue m_1 and corresponding eigenfunction w_1 of the equation $B\left(m^2\right)w=0$ are

$$\frac{m_1^2}{4}+\lambda\left(1-\gamma\lambda\right)=\pi^2;\qquad w_1\left(t\right)=D\sin\pi t,\tag{111}$$

where D is an arbitrary constant. Without loss of generality we take $D=1.$. Next we show that $\left(0\times m_1^2\right)\in Y\times\mathbb{R}_+$ is a bifurcation point of $\mathcal{F}\left(\theta,m^2\right)=0$. Note that $B\left(m^2\right)$ is self-adjoint linear operator $B\left(m^2\right)=B^*\left(m^2\right)$. Therefore the eigenvector of the adjoint operator $B^*\left(m^2\right)$ corresponding to m_1^2 is

$$q_1\left(t\right)=Q\sin\pi t,\tag{112}$$

where Q is an arbitrary constant that we shall, again, take as $Q=1$. From (108) we conclude that

$$\mathcal{F}\left(\theta,m^2\right)=-\mathcal{F}\left(-\theta,m^2\right).\tag{113}$$

For $m_1^2-\Delta m_1^2\leq m^2\leq m_1^2+\Delta m_1^2$ with $\left|\Delta m_1^2\right|$ small, we assume the solution to (109) in the form

$$\theta=a\sin\pi t+u^*\left(a,\Delta m_1^2,t\right)\tag{114}$$

The equation (113) implies that $u^*\left(a,\Delta m_1^2,t\right)$ is at least of the order $O\left(a^3\right)$ (see Golubitsky and Scheaffer, 1985, p.300). The parameter a (the amplitude in the terminology of Troger and Steindl, 1991) is determined from the bifurcation equation

$$f\left(a,\Delta m_1^2\right)=\int_0^1\left[\mathcal{F}\left(a\sin\pi t+u^*\left(a,\Delta m_1^2,t\right),m_1^2+\Delta m_1^2\right)\right]q_1\left(t\right)dt.\tag{115}$$

The equation (115) can be written as

$$f\left(a,\Delta m_1^2\right)=c_1a\Delta m_1^2+c_3a^3+c_5a^5+c_7a^7+h.o.t.,\tag{116}$$

where $h.o.t.$ denotes terms of the order $O\left(a^9,\left(\Delta m_1^2\right)a^3\right)$ and

$$c_1=\frac{1}{4}\int_0^1 w_1^2\left(t\right)dt=\frac{1}{8};\quad c_3=-\frac{k}{6}\int_0^1 w_1^4\left(t\right)dt=-\frac{k}{16};$$

$$c_5=\int_0^1\left[-\frac{k}{12}w_1^3\left(t\right)U_{aaa}\left(t\right)+\frac{l}{240}w_1^6\left(t\right)\right]dt;$$

$$c_7=\int_0^1\left\{-k\left[\frac{1}{72}w_1^2\left(t\right)U_{aaa}^2\left(t\right)+\frac{1}{240}w_1^3\left(t\right)U_{aaaaa}\left(t\right)\right]\right.$$
$$\left.+\frac{l}{288}w_1^5\left(t\right)U_{aaa}\left(t\right)+\frac{r}{5040}w_1^8\left(t\right)\right\}dt.\tag{117}$$

In (117) we used k,l and r to denote the following constants

$$k=3\lambda-6\gamma\lambda^2-2\pi^2;\quad l=-15\lambda-15\gamma\lambda^2+17\pi^2;\quad r=-63\lambda+126\gamma\lambda^2+62\pi^2,\tag{118}$$

and the functions $U_{aaa}(t)$ and $U_{aaaaa}(t)$ are solutions of the linear system of differential equations (see Golubitsky and Scheaffer, 1985, p.33)

$$\begin{aligned} &B\left(m_1^2\right)U_{aaa}+ED^{(3)}\mathcal{F}\left(w_1,w_1,w_1\right)=0; \\ &B\left(m_1^2\right)U_{aaaaa}+10ED^{(3)}\mathcal{F}\left(w_1,w_1,U_{aaa}\right)+ED^{(5)}\mathcal{F}\left(w_1,w_1,w_1,w_1,w_1\right)=0. \end{aligned} \tag{119}$$

In (119) we used $D^{(3)}\mathcal{F}$ and $D^{(5)}\mathcal{F}$ to denote the third and fifth Fréchet derivative of $\mathcal{F}\left(\Theta,m^2\right)$ with respect to Θ, calculated at the point $\left(\theta_0=0,m_1^2\right)$. Those derivatives are calculated in the directions $\left(w_1,w_1,w_1\right)$, $\left(w_1,w_1,U_{aaa}\right)$ and $\left(w_1,w_1,w_1,w_1,w_1\right)$ respectively. Also in (119) we used E to denote the projection operator mapping Z onto the range of $B\left(m_1^2\right)$. With $\mathcal{F}$ and B given by (108) and (110) we obtain

$$\begin{aligned} \ddot{U}_{aaa}+\pi^2U_{aaa}&=-\frac{k}{4}\sin 3\pi t; \\ \ddot{U}_{aaaaa}+\pi^2U_{aaaaa}&=\frac{5}{32}\left\{\frac{k^2}{\pi^2}+l\right\}\sin 3\pi t-\frac{1}{64}\left\{\frac{5k^2}{\pi^2}+2l\right\}\sin 5\pi t. \end{aligned} \tag{120}$$

Note that (see Golubitsky and Scheaffer, 1985, p.32)

$$\begin{aligned} U_{aaa}&=\left[\partial^3u^*\left(a,\Delta m_1^2,t\right)/\partial a^3\right]_{a=\Delta m_1^2=0}; \\ U_{aaaaa}&=\left[\partial^5u^*\left(a,\Delta m_1^2,t\right)/\partial a^5\right]_{a=\Delta m_1^2=0}. \end{aligned} \tag{121}$$

Since u^* does not belong to the null space of $B\left(m_1^2\right)$ the solutions to (120) read

$$\begin{aligned} U_{aaa}&=\frac{k}{32\pi^2}\sin 3\pi t; \\ U_{aaaaa}&=-\frac{5}{256\pi^2}\left\{\frac{k^2}{\pi^2}+l\right\}\sin 3\pi t+\frac{1}{1536\pi^2}\left\{\frac{5k^2}{\pi^2}+2l\right\}\sin 5\pi t. \end{aligned} \tag{122}$$

By using (122) in $(117)_{3,4}$ we obtain

$$c_5=\frac{k^2+4l\pi^2}{3072\pi^2};\quad c_7=\frac{4r\pi^4-k^3-2kl\pi^2}{73728\pi^4}. \tag{123}$$

For the qualitative study of solutions, equation (116) may be simplified by neglecting $h.o.t.$ Note that $c_1>0$ for all values of parameters. Then in the analysis we distinguish the following special cases:

1. Suppose that $k\neq 0$. Then from $(117)_2$ we have $c_3\neq 0$ and (116) is contact equivalent (see Keyfitz, 1986) to
$$a\Delta m_1^2-\operatorname{sgn}(k)\,a^3=0. \tag{124}$$
From $(118)_1$ we conclude that k as a function of λ has two zeros, that separate sub from super-critical bifurcation (see Atanackovic and Glavardanov, 1996).

2. Suppose that, $k = 0$ and $l \neq 0$. Then $c_3 = 0$ and $c_5 \neq 0$ so that (116) is contact equivalent to
$$a\Delta m_1^2 + \text{sgn}(l)\, a^5 = 0. \tag{125}$$
Again the bifurcation could be both sub and super critical. There exist a single point λ^* where l as a function of λ vanishes (remember that $k = 0$)
$$\lambda^* = \frac{44}{45}\pi^2. \tag{126}$$
This point separates sub from the super critical bifurcation. For $\lambda < \lambda^*$ we have sub critical bifurcation while for $\lambda > \lambda^*$ the bifurcation is super critical.
3. Suppose that $k = l = 0$. Thus, $c_3 = c_5 = 0$ and
$$c_7 = \frac{5}{4608}\pi^2. \tag{127}$$
Therefore (116) is contact equivalent to
$$a\Delta m_1^2 + a^7 = 0, \tag{128}$$
i.e., the bifurcation is sub-critical.

Next we examine the stability of the bifurcating branches. We shall use the functional (104). Consider the bifurcation diagram shown in Figure 3. The second variation of I calculated on the trivial solution $\theta_0 = 0$ reads
$$\delta^2 I\,(\Theta = 0, w, w) = \int_0^1 \left[\dot{w}^2 - m^2 w^2 - \lambda\,(1 - \gamma\lambda)\, w^2\right] dt. \tag{129}$$

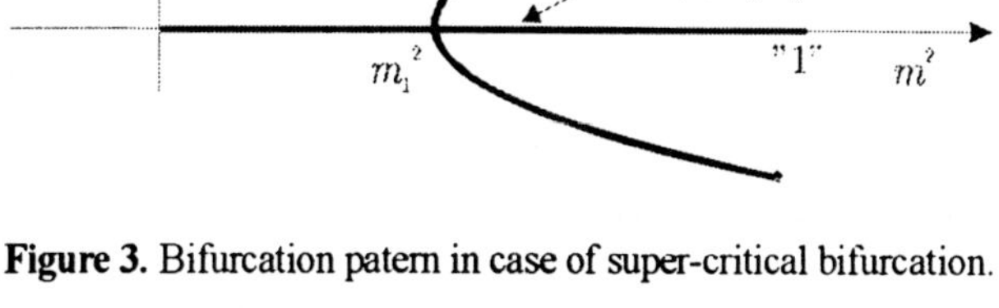

Figure 3. Bifurcation patern in case of super-critical bifurcation.

Note that the variation w satisfies $w\,(0) = w\,(1) = 0$. Suppose that $m^2 = m_1^2 + \Delta m_1^2$. Then, we have
$$\int_0^1 \left[\dot{w}_1^2 - m_1^2 w_1^2 - \lambda\,(1 - \gamma\lambda)\, w_1^2\right] dt = 0 \tag{130}$$

We calculate the second variation (129) in the direction w_1 and for $m^2 = m_1^2 + \Delta m_1^2$. The result is

$$\delta^2 I\left(\theta_0 = 0, w_1, w_1\right) = -\Delta m_1^2 \int_0^1 w_1^2\left(t\right) dt < 0, \tag{131}$$

where we used (130). The inequality (131) shows that on the branch "1" in Fig. 3 the functional I is *not* in minimum since it does not satisfy the necessary condition for the minimum $\delta^2 I\left(\Theta = 0, w, w\right) > 0$ for *all* w.

Note that the integrand in functional (104) does not depend on t. Therefore, there is Jacobi type first integral of the Euler-Lagrange equations (102) that reads (see Vujanovic and Jones, 1989)

$$\frac{\dot{\theta}^2}{2} + \frac{m^2}{\left(1+\cos\theta\right)} - \lambda\left[\cos\theta + \gamma\frac{\lambda}{2}\sin^2\theta\right] = const. \tag{132}$$

The integral (132) can be obtained from (107) by using the relation (100).

Solution of equation (31). With $G = \bar{F}_1$, $\bar{F}_1\left(1/2\right) = G_1$ and by using $\dot{\bar{F}}_1\left(1/2\right) = 0$, to determine C, we obtain from (107)

$$C = \gamma G_1^2 + 2G_1 + \frac{m^2\left(\lambda + G_1\right)}{\lambda - G_1}. \tag{133}$$

Since $-\lambda$ and G_1 are zeros of the right hand side of (107) it follows that (107) may be written as

$$\dot{G}^2 = \left(\lambda + G\right)\left(G - G_1\right)\left\{\gamma G^2 + \left(2 + \gamma G_1 - \gamma\lambda\right)G - \lambda\left(2 + \gamma G_1 + \frac{2m^2}{\lambda - G_1}\right)\right\}. \tag{134}$$

From (134) we obtain

$$\dot{G}^2 = \gamma\left(\lambda + G\right)\left(G - G_1\right)\left(G - b\right)\left(G - c\right), \tag{135}$$

where

$$b = \frac{-\left(2 + \gamma G_1 - \gamma\lambda\right) - \sqrt{\left(2 + \gamma G_1 + \gamma\lambda\right)^2 + 8\gamma\lambda\frac{m^2}{\lambda - G_1}}}{2\gamma};$$

$$c = \frac{-\left(2 + \gamma G_1 - \gamma\lambda\right) + \sqrt{\left(2 + \gamma G_1 + \gamma\lambda\right)^2 + 8\gamma\lambda\frac{m^2}{\lambda - G_1}}}{2\gamma}. \tag{136}$$

Note that $t = 0$ corresponds to $G = -\lambda$ and $t = 1/2$ corresponds to $G = G_1$. Therefore if there is a solution to (135) we must have $\gamma\left(\lambda + G\right)\left(G - G_1\right)\left(G - b\right)\left(G - c\right) > 0$ when $G \in \left(-\lambda, G_1\right)$. In what follows we assume that parameters in (135),(136) are chosen so that this condition holds.

By separating variables in (135) we obtain

$$\frac{1}{2}=\int_{-\lambda}^{G_1}\frac{dG}{\sqrt{\gamma\left(G+\lambda\right)\left(G-G_1\right)\left(G-b\right)\left(G-c\right)}}. \tag{137}$$

The integral in (137) can be expressed in terms of complete elliptic integral of the first kind, so that

$$\frac{1}{2}=\frac{2}{\sqrt{\gamma\left(\lambda+c\right)\left(G_1-b\right)}}F\left(\frac{\pi}{2},q\right), \tag{138}$$

where $F\left(\phi,q\right)=\int_0^{\phi}\frac{dx}{\sqrt{1-q^2\sin^2 x}}$ and

$$q=\sqrt{\frac{\left(\lambda+G_1\right)\left(c-b\right)}{\left(\lambda+c\right)\left(G_1-b\right)}}. \tag{139}$$

Note that when $\theta\rightarrow 0$ (the rod axis becomes straight) from (24) we have $G_1\rightarrow-\lambda$ and from (139) $q\rightarrow 0$. Using this in (136) we find

$$b=\frac{-1}{\gamma}\left[1-\gamma\lambda+\sqrt{1+\gamma m^2}\right];\quad c=\frac{1}{\gamma}\left[-\left(1-\gamma\lambda\right)+\sqrt{1+\gamma m^2}\right]. \tag{140}$$

With (140) equation (138) leads to

$$\frac{m^2}{4}+\lambda\left(1-\gamma\lambda\right)=\pi^2, \tag{141}$$

in agreement with (111). Thus the solution (138) confirms the bifurcation analysis presented in the previous Section.

Equation (138) can be used to calculate G_1 if m,λ and γ are prescribed. We shall compare so determined G_1 with the value obtained from numerical integration of the system (89),(90).

Numerical results. We present now numerical solution of the system (89),(90) for specific values of parameters $\lambda, m, \alpha, A_{22}/A_{11}$ and β. Note that, in engineering terms, A_{22}/A_{11} represents the ratio between bending (A_{22}) and torsional (A_{11}) stiffness. As stated earlier (see comment after equation (90)) the system (89),(90) has rotational symmetry so that, without loss of generality, we shall choose a solution that satisfies $\bar{M}_3\left(0\right)=0$. Once the solution to (89),(90) is obtained we compare $\bar{F}_1\left(1/2\right)$ with G_1 determined from (138) with given λ, m and $\gamma=\alpha-\beta$.

In Figure 4*a* and 4*b* we show the projections of the rod axis on the $\overline{x}_{10}-\overline{x}_{20}$ and $\overline{x}_{10}-\overline{x}_{30}$ coordinate planes, respectively. The shape is determined for $\lambda=9, A_{22}/A_{11}=1.3$ and $m=2$ and for four different choices of α and β. The curve 1 corresponds to Bernoulli-Euler rod, i.e., $\left(\alpha=0,\beta=0\right)$, the curve 2 corresponds to $\left(\alpha=0.00111,\beta=0\right)$, the curve 3 corresponds to $\left(\alpha=0,\beta=0.00288\right)$ and finally the curve 4 corresponds to $\left(\alpha=0.00111,\beta=0.00288\right)$.

Table 2. Some numerical results.

α	β	$\bar{F}_1\,(1/2)$	$\bar{M}_2\,(0)$	C
0	0	−8.32746	1.22488	−16.49966
0.00111	0	−8.78083	0.69520	−17.42669
0	0.00288	−7.32887	1.95715	−14.40354
0.00111	0.00288	−7.68626	1.72666	−15.16263

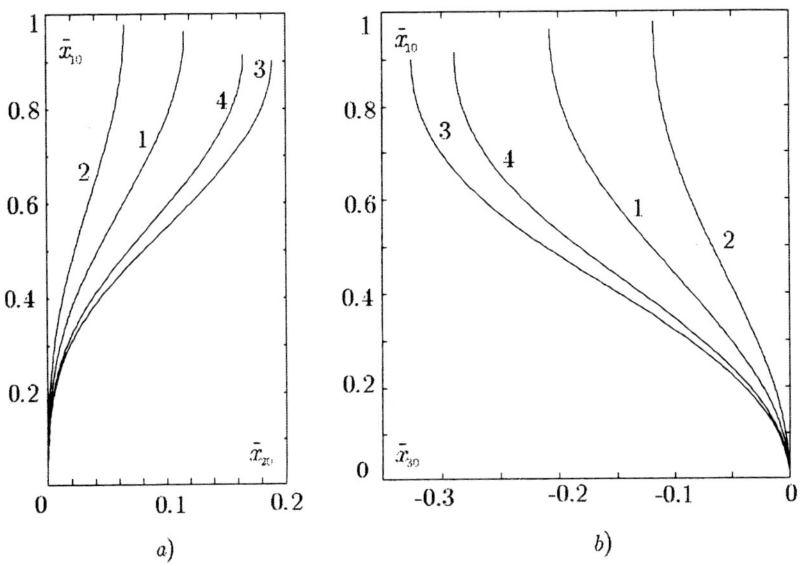

Figure 4. Post-critical shape of the rod.

In Table 2 we present the values of $\bar{F}_1\,(1/2)$, $\bar{M}_2\,(0)$ and the constant C in the first integral $(97)_3$ for different values of α and β. Note that, because of global equilibrium, we have $F_2\,(0) = F_3\,(0) = 0$ so that the only unknown dependent variable at $t = 0$ is $\bar{M}_2\,(0)$. In the process of numerical integration of the system (89),(90) we calculated the first integrals (93) at each step of integration. They were constant up to the terms of the order 10^{-8}.

2 Optimal Shape of Elastic Columns

In this Section we shall treat two concrete problems of determining optimal (against buckling) shape of elastic columns. The problem of determining the shape of the column of greatest efficiency is indeed an old one. For simply supported column loaded by concentrated forces at its ends it was formulated by Lagrange (see Lagrange, 1867). For historical account see Cox (1992). The solution of the problem was obtained by many authors, see Clausen (1851), Blasius (1914) and Ratzersdorfer (1936). We mention also the work of Keller (1960) that became classic. Tadjbakhsh and Keller (1962) treated other boundary conditions by the method developed in Keller

(1960). It was, for the first time, pointed out by Olhoff and Rasmussen (1977) that the change of boundary conditions can lead to multiple eigenvalues of the equation relevant for determining the optimal cross-sectional area function. In this case one is faced with a so called multiple eigenvalue optimization problem, treated by Seyranian (1984), Seyranian et al. (1994), Seyranian (1995) and the analysis becomes more delicate. In the examples that we treat here all eigenvalues are simple.

2.1 Optimal Shape of a Pflüger Column

As a first example we study the problem of determining the shape of a Pflüger column of greatest efficiency Atanackovic and Simic (1999). A Pflüger column is a simply supported column loaded by uniformly distributed follower type of load (see Pflüger, 1975). The uniformly distributed follower load is a non-conservative load. It is interesting, however, that despite the non-conservative character of the load the stability analysis for Pflüger column can be based on static (Euler) method. In the first part of this paper we shall formulate the nonlinear system of equations describing the equilibrium configuration of a rod. On the basis of these equations we shall derive the linear boundary value problem that determines the stability boundary. By analyzing the multiplicity of the lowest eigenvalue of this equation we shall confirm that it represents the bifurcation point of the nonlinear system. Then we shall formulate an optimization problem. The necessary conditions for the minimum of volume will be reduced to a single nonlinear differential equation. The rest of the paper is devoted to analysis of this equation, both numerical and variational. As results of this analysis we shall determine the optimal distribution of the material along the column axis and show that the optimal column has 19% smaller volume then the corresponding column with constant cross section.

Formulation of the problem. Consider a column shown in Figure 5. The column is simply supported at both ends with the movable end C. The axis of the column is initially straight and the column is loaded by uniformly distributed follower type of load of constant intensity q_0. We shall assume that the column axis has length L and that it is inextensible.

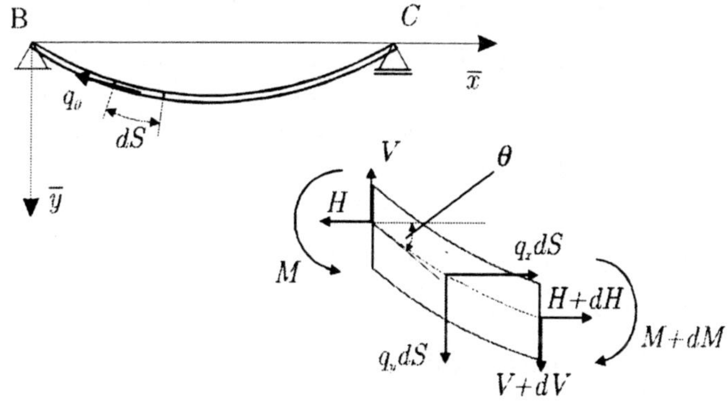

Figure 5. Coordinate system and load configuration for Pflüger column.

Let $\bar{x}-\mathrm{B}-\bar{y}$ be a Cartesian coordinate system with the origin at the point B and with the $\bar{x}$ axis oriented along the column axis in the undeformed state. The equilibrium conditions yield

(see Atanackovic, 1997)

$$\frac{dH}{dS} = -q_x; \qquad \frac{dV}{dS} = -q_y; \qquad \frac{dM}{dS} = -V\cos\theta + H\sin\theta, \tag{1}$$

where H and V are components of the contact force along the $\bar{x}$ and $\bar{y}$ axis, respectively, M is the bending moment and θ is the angle between the tangent to the column axis and $\bar{x}$ axis. Also in (1) q_x and q_y are components of the distributed forces along the $\bar{x}$ and $\bar{y}$ axes respectively. Since the distributed force is tangent to the column axis we have

$$q_x = -q_0\cos\theta; \qquad q_y = -q_0\sin\theta. \tag{2}$$

To the system (1) we adjoin the following geometrical

$$\frac{dx}{dS} = \cos\theta; \quad \frac{dy}{dS} = \sin\theta, \tag{3}$$

and constitutive relation

$$\frac{d\theta}{dS} = \frac{M}{EI}. \tag{4}$$

In (3) and (4) we use x and y to denote coordinates of an arbitrary point of the column axis and EI to denote the bending rigidity. The boundary conditions corresponding to the column shown in Fig. 6 are

$$x(0) = 0; \quad y(0) = 0; \quad M(0) = 0; \quad y(L) = 0; \quad M(L) = 0; \quad H(L) = 0. \tag{5}$$

The system (1)–(5) possesses a trivial solution in which column axis remains straight, i.e.,

$$\begin{aligned} &H^0(S) = -q_0(L-S); \quad V^0(S) = 0; \quad M^0(S) = 0; \\ &x^0(S) = S; \quad y^0(S) = 0; \quad \theta^0(S) = 0. \end{aligned} \tag{6}$$

In order to formulate the minimum volume problem for the column we take the cross-sectional area $A(S)$ and the second moment of inertia $I(S)$ of the cross-section in the form

$$A(S) = A_0 a(S); \qquad I(S) = I_0 a^2(S), \tag{7}$$

where A_0 and I_0 are constants (having dimensions of area and second moment of inertia, respectively) and $a(S)$ is cross-sectional area function. For the case of a column with circular cross section we have the connection between A_0 and I_0 given by $I_0 = (1/4\pi)A_0^2$. Let $\Delta H, ..., \Delta\theta$ be the perturbations of $H, ..., \theta$ defined by

$$\begin{aligned} &H = H^0 + \Delta H; \qquad V = V^0 + \Delta V; \qquad M = M^0 + \Delta M; \\ &x = x^0 + \Delta x; \qquad y = y^0 + \Delta y; \qquad \theta = \theta^0 + \Delta\theta. \end{aligned} \tag{8}$$

Then, by introducing the following dimensionless quantities

$$\begin{aligned} &h = \frac{\Delta H L^2}{EI_0}; \qquad v = \frac{\Delta V L^2}{EI_0}; \qquad m = \frac{\Delta M L}{EI_0}; \\ &\xi = \frac{\Delta x}{L}; \qquad \eta = \frac{\Delta y}{L}; \qquad t = \frac{S}{L}; \qquad \lambda = \frac{q_0 L^3}{EI_0}, \end{aligned} \tag{9}$$

and by substituting (7) in (1)–(5) we arrive at the following nonlinear system of equations describing nontrivial configuration of the column

$$\begin{aligned}
&\dot{h} = -\lambda(1-\cos\theta); \quad \dot{v} = \lambda \sin\theta;\\
&\dot{m} = -v\cos\theta + [-\lambda(1-t)+h]\sin\theta;\\
&\dot{\xi} = 1-\cos\theta; \quad \dot{\eta} = \sin\theta;\\
&\dot{\theta} = \frac{m}{a^2},
\end{aligned} \tag{10}$$

where $(\dot{\cdot}) = d(\cdot)/dt$. The boundary conditions corresponding to (10) are

$$\xi(0)=0; \quad \eta(0)=0; \quad m(0)=0; \quad \eta(1)=0; \quad m(1)=0; \quad h(1)=0. \tag{11}$$

Note that the system (10)–(11) has the solution $h(t) = 0, ..., \theta(t) = 0$ for all values of λ. Next we linearize (10) to obtain

$$\begin{aligned}
&\dot{h} = 0; \quad \dot{v} = \lambda\theta; \quad \dot{m} = -v - \lambda(1-t)\theta; \quad \dot{\xi} = 0;\\
&\dot{\eta} = \theta; \quad \dot{\theta} = \frac{m}{a^2}.
\end{aligned} \tag{12}$$

By using boundary conditions (11) in (12) we conclude that $h(t) = \xi(t) = 0$ and the rest of equations (12) can be reduced to

$$\ddot{m} + \frac{\lambda}{a^2}(1-t)m = 0, \tag{13}$$

subject to

$$m(0) = m(1) = 0. \tag{14}$$

The system (13)–(14) constitutes a linear spectral problem. For the case when $0 < a(t) < \infty$, $t \in (0,1)$ the eigenvalues $0 < \lambda_1 < \lambda_2 < ..\lambda_i < \infty$, are simple and the only accumulation point is at infinity (see Churchill, 1969, p.73). In accordance with this fact and Krasnoselskii's theorem (see Rabier, 1985) we have the following result:

The bifurcation points of the system (10),(11) are of the form $(0, \lambda_n)$ *where* λ_n *are eigenvalues of the system (13)–(14).*

The column of greatest efficiency is the one that has minimal volume for fixed lowest eigenvalue λ_1 of (13),(14). Thus, for the column of the greatest efficiency we have to minimize

$$J = \int_0^1 a(t)dt, \tag{15}$$

subject to

$$\ddot{m} + \frac{\lambda_1}{a^2}(1-t)m = 0, \tag{16}$$

with boundary conditions (14). In (16) we consider λ_1 to be known. J in (15) has meaning of the dimensionless volume of the column.

Optimization problem and its solution. To determine the optimal shape of the column, i.e., $a(t)$ such that J is minimal, we shall use the Pontryagin's maximum principle (Sage and White, 1977 and Alekseev et al., 1979). Let us rewrite optimization problem (15)–(16) as:

Find continuous function $u(t) = a(t)$, *usually termed control, such that* $0 < u(t) < \infty$, $t \in (0,1)$, *for which the optimality criterion*

$$J = \int_0^1 u(t)dt, \tag{17}$$

attains minimum value. The governing differential equations are

$$\dot{x}_1 = x_2; \qquad \dot{x}_2 = -\frac{\lambda_1}{u^2}(1-t)x_1, \tag{18}$$

subject to

$$x_1(0) = x_1(1) = 0. \tag{19}$$

The optimization problem just stated can be modified in many ways. For example, we can allow the piecewise continuous controls satisfying inequality type of constraints. The Pontryagin's maximum principle can also be applied to these types of problems (see Atanackovic, 1997, p. 217), but we shall not be concerned with them[1].

For system (17)-(18) the Hamiltonian function $\mathcal{H}$ is

$$\mathcal{H} = u + p_1 x_2 - p_2 \frac{\lambda_1}{u^2}(1-t)x_1, \tag{20}$$

where costate variables p_1 and p_2 have to satisfy the following system of differential equations

$$\dot{p}_1 = -\frac{\partial \mathcal{H}}{\partial x_1} = \frac{\lambda_1}{u^2}(1-t)p_2; \quad \dot{p}_2 = -\frac{\partial \mathcal{H}}{\partial x_2} = -p_1, \tag{21}$$

subject to

$$p_2(0) = p_2(1) = 0. \tag{22}$$

The optimality condition $\min\limits_u \mathcal{H}(t, x_1, x_2, p_1, p_2, u)$ leads to

$$\frac{\partial \mathcal{H}}{\partial u} = 1 + \frac{2\lambda_1}{u^3}(1-t)x_1 p_2 = 0. \tag{23}$$

By solving (23) for u we obtain

$$u = -\left[2\lambda_1(1-t)x_1 p_2\right]^{1/3}. \tag{24}$$

Note that the boundary value problems (18)-(19) and (21)-(22) are not coupled. Moreover, their solutions are related by the following equations

$$p_1(t) = x_2(t); \qquad p_2(t) = -x_1(t). \tag{25}$$

[1] Without constraints the problem can be treated by classical variational calculus methods as given in Prager and Taylor (1968) for example.

Therefore the control variable $u(t)$ given by (24) becomes

$$u = \left[2\lambda_1(1-t)x_1^2\right]^{1/3}. \tag{26}$$

By using (25) in (20) we have $(\partial^2\mathcal{H}/\partial u^2) > 0$ so that the necessary condition for minimum of $\mathcal{H}$ is satisfied. By substituting (26) into (18) we arrive at the single differential equation

$$\ddot{x} + \bar{\lambda}(1-t)^{1/3}x^{-1/3} = 0, \tag{27}$$

subject to

$$x(0) = x(1) = 0. \tag{28}$$

In (27) and (28) we used $x(t) = x_1(t)$, $\bar{\lambda} = (\lambda_1/4)^{1/3}$. Note that the end points $t = 0$ and $t = 1$ are singular points of the problem (27)–(28). To examine the local behavior of the solution near end points we first transform the independent variable t to $\zeta = 1 - t$. Then (27) becomes

$$\ddot{x} + \bar{\lambda}\zeta^{1/3}x^{-1/3} = 0, \tag{29}$$

subject to

$$x(0) = 0; \qquad x(1) = 0. \tag{30}$$

In (29) we used again $\dot{(\cdot)} = d(\cdot)/d\zeta$. Suppose that the solution $x(\zeta)$ of (29)–(30) in the vicinity of $\zeta = 0$ behaves as

$$x(\zeta) \sim C\zeta^{\alpha}\left[1 + a_1\zeta + a_2\zeta^2 + a_3\zeta^3 + a_4\zeta^4 + \ldots\right]; \quad \zeta \to 0, \tag{31}$$

where C, α and $a_i, i = 1, 2, \ldots$ are constants. By substituting (31) in (29) we conclude that

$$\begin{aligned} &\alpha = 1; \quad a_1 = -\frac{1}{2}\bar{\lambda}C^{-4/3}; \quad a_2 = -\frac{1}{36}\bar{\lambda}^2C^{-8/3}; \\ &a_3 = -\frac{7}{1296}\bar{\lambda}^3C^{-4}; \quad a_4 = -\frac{23}{15552}\bar{\lambda}^4C^{-16/3}. \end{aligned} \tag{32}$$

In (32) the constant C remains undetermined. It is related to the first derivative of x at $\zeta = 0$. Namely $C = \dot{x}(0)$. Note that from (29) and (32) it follows that

$$\lim_{\zeta\to 0}\ddot{x}(\zeta) = -\frac{\bar{\lambda}}{C^{1/3}}. \tag{33}$$

To solve (29) numerically we shall write the first order system of differential equations

$$\dot{x}_1 = x_2; \quad \dot{x}_2 = -\bar{\lambda}\zeta^{1/3}x_1^{-1/3}. \tag{34}$$

Then, we choose $x_2(0) = C$ and solve (34) as an initial value problem. The boundary condition $x_1(1) = 0$ will be satisfied by the shooting method. In the first integration step the right hand side of $(34)_2$ will be equated with (33).

Variational principles for (27)—(28). In this section we shall formulate complementary variational principles (see Arthurs, 1980) for the boundary value problem (27)–(28). Let $\mathbf{X}$ be the following space

$$\mathbf{X} = \{X(t) : X(t) \in C^2(0,1); X(0) = X(1) = 0\}, \tag{35}$$

where $C^2(0,1)$ is the space of continuous functions mapping $(0,1)$ into set of real numbers $\mathbb{R}$, having continuous first and second derivatives. Note that the assumption (see section 3) that $u(t)$ is continuous, together with (18), implies that $\ddot{x}\ (t)$ in (27) is a continuous function. Thus, the solution of (27)-(28) belongs to $\mathbf{X}$. Next we consider the variational problem

$$I(X) = \int_0^1 \left[\frac{1}{2}\dot{X}^2 - \frac{3}{2}\bar{\lambda}(1-t)^{1/3}X^{2/3}\right] dt \to extr, \tag{36}$$

for $X(t) \in \mathbf{X}$. It is easy to see that the minimizing element of (36), if it exists, satisfies (27)-(28). Therefore

$$\delta I(x,w) = 0, \tag{37}$$

where the "variation" w is $w = X - x$. The variational principle (37) is called primary variational principle. Calculating the second variation of (36) on $X = x$ we obtain

$$\delta^2 I(x,w) = \int_0^1 \left[\dot{w}^2 + \frac{1}{3}\bar{\lambda}(1-t)^{1/3}x^{-4/3}w^2\right] dt. \tag{38}$$

From (38) it follows that

$$\delta^2 I(x,w) \geq \left\|\dot{w}\right\|^2_{L_2}, \tag{39}$$

where $\left\|\dot{w}\right\|_{L_2} = (\int_0^1 \dot{w}^2\ dt)^{1/2}$. By combining (37) and (39) we conclude that the following statement holds:

On the solution $x(t)$ of the boundary value problem (27)—(28) the functional (36) attains minimum

$$\min_{X \in \mathbf{X}} I(X) = I(x). \tag{40}$$

Now we proceed to construct the dual variational principle. Let us denote by $F(t, X, \dot{X})$ the Lagrangian function of the variational principle (37) that is

$$F(t, X, \dot{X}) = \frac{1}{2}\dot{X}^2 - \frac{3}{2}\bar{\lambda}(1-t)^{1/3}X^{2/3}. \tag{41}$$

We get the canonical form of (27)—(28) by introducing the variable P as

$$P = \frac{\partial F}{\partial \dot{X}} = \dot{X}, \tag{42}$$

and function $K(t, X, P)$, also called Hamiltonian, connected with F by Friedrichs transformation

$$K(t, X, P) = P\dot{X} - F = \frac{1}{2}P^2 + \frac{3}{2}\bar{\lambda}(1-t)^{1/3}X^{2/3}. \tag{43}$$

The system (27)–(28) then becomes

$$\dot{X} = \frac{\partial K}{\partial P} = P; \qquad \dot{P} = -\frac{\partial K}{\partial X} = -\bar{\lambda}(1-t)^{1/3}X^{-1/3}. \tag{44}$$

In order to obtain the dual variational principle we have first to solve (43) for F. Then, in so obtained expression we substitute X in terms of $\dot{P}$ from $(44)_2$ and use the resulting expression in (36). After partial integration and application of the boundary conditions, we obtain

$$G(P) = I(P, X(P)) = -\int_0^1 \left[\frac{1}{2}P^2 + \frac{1}{2}\bar{\lambda}^3 \frac{1-t}{\dot{P}^2}\right] dt. \tag{45}$$

Let $x(t)$ and $p(t)$ be the solution of canonical system (44). Then it is easy to show that the following dual variational principle holds

$$\delta G(p, n) = 0, \tag{46}$$

where the variation n is $n = P - p$ and P is an admissible function, i.e., $P \in \mathbf{Y} = \{P : P \in C^1(0,1); \int_0^1 [(1-t)/\dot{P}^2] dt < \infty\}$. Note that elements of $\mathbf{Y}$ need not satisfy any boundary condition. However, in practical applications P is usually expressed as $P = \dot{X}$, where $X(t) \in \mathbf{X}$. Also from (45) it follows

$$\delta^2 G(p, n) = -\int_0^1 \left[n^2 + 3\bar{\lambda}^3 \frac{1-t}{\dot{P}^4} \dot{n}^2\right] dt \leq -\|n\|_{L_2}^2. \tag{47}$$

By combining (46) and (47) we have the following result:

On $p(t)$ the functional (45) attains maximum

$$\max_{P \in \mathbf{Y}} G(P) = G(p). \tag{48}$$

From (40),(49) the chain of inequalities can be derived

$$G(P) \leq G(p) = I(x) \leq I(X). \tag{49}$$

We shall use (36) and (45) to obtain and estimate error of an approximate solution of (27)–(28). This is achieved by using the inequality

$$\|w\|_{L_\infty} = \sup_{t \in (0,1)} |w(t)| \leq \frac{\|\dot{w}\|_{L_2}}{\sqrt{2}}, \tag{50}$$

valid for any $w(t)$ satisfying $w(0) = w(1) = 0$. From (49) and (50) we obtain

$$\|w\|_{L_\infty} \leq [I(X) - G(P)]^{1/2}. \tag{51}$$

Numerical results. In this section we present results of numerical integration of the system (17)–(18). The value of λ_1 was taken as the lowest eigenvalue of the column with constant cross-section having unit volume. Precisely, we take $a(t) = 1$, $A_0 = \pi R_0^2$, $I_0 = \pi R_0^4/4$ in (7), where R_0 is the radius of the cross–section that we assume to be circular. Then, from (15) we conclude that $J = 1$. With these values λ_1 becomes $\lambda_1 = 18.956266$ (see Atanackovic, 1997).

The solution to (17)–(18) may be obtained numerically. When the volume of the optimal column is calculated from (17) we obtain

$$J_{opt} = \int_0^1 u(t)dt = 0.81051. \tag{52}$$

It is interesting to compare the critical force of the column with constant cross-section having the same volume as the optimal column. The radius of the cross-section of this column is $r_0 = 0.90028R_0$. The corresponding dimensionless critical load is $\lambda_0 = 12.452807$.

We use now complementary variational principles, described in the previous section, together with Ritz method (see Vujanovic and Jones, 1989) to determine an analytical approximate solution of the equation (27)-(28) for $\lambda_1 = 18.956266$, already used in numerical treatment. We assume the solution of the problem in the following form

$$X(C, C_1, C_2, t) = m_3(t) = Ct(1-t)(1 + C_1 t + C_2 t^2), \tag{53}$$

where C, C_1 and C_2 are constants to be determined. Since $x(t)$ in (27) represents the dimensionless moment (see (13)) we used this in writing (53). The index attached to m indicates the number of free constants that are to be determined. Note that $X(t)$ given by (53) is an admissible function since it satisfies the boundary conditions for all values of constants C, C_1 and C_2. By substituting (53) into (2.1.3-36) and minimizing with respect to C, C_1 and C_2 we obtain $C = 1.3036, C_1 = -0.4563, C_2 = 0.2293$ so that (53) and the corresponding value of the functional I become

$$m_3(t) = 1.3036t(1-t)(1 - 0.4563t + 0.2293t^2); \qquad I(m_3) = -0.4052. \tag{54}$$

For the functional G given by (45) we take an admissible function $P(t)$ in the form

$$P(D, D_1, D_2, t) = \dot{X}(D, D_1, D_2, t). \tag{55}$$

Substituting (55) in (45) and maximizing with respect to D, D_1 and D_2 we obtain

$$\begin{aligned} P(D, D_1, D_2, t) &= D(1 + 2D_1 t + 3D_2 t^2) - Dt\left(2 + 3D_1 t + 4D_2 t^2\right); \\ D &= 1.3128; \quad D_1 = -0.4646; \quad D_2 = 0.2353; \\ G(P) &= -0.4088. \end{aligned} \tag{56}$$

From (55), (56) and (51) we obtain the estimate of the error of the solution (54) as

$$\|w\|_{L_\infty} \leq [I(X) - G(P)]^{1/2} = 0.06. \tag{57}$$

2.2 Optimal Shape of a Compressed Pflüger Column

In the previous Section a problem of determining optimal shape of the Pflüger column was formulated. Our aim in this Section is to generalize the optimization problem of the previous Section by introducing compressive force at the end of the rod. Thus, we shall formulate, and solve, the problem of determining optimal shape of a *compressed* Pflüger column. We shall show that the problem of determining the optimal cross-section, again, reduces to a boundary value problem for a single second order ordinary differential equation. For this equation we shall formulate extremum, complementary variationl principles and we prove the existence of a weak solution. The variational principles are also used to derive á priori lower and upper bounds of the L_∞ norm of solution.

Formulation of the problem. Consider an elastic column of length L, initially straight and simply supported at both ends. One end, say B is fixed while the other, that we denote by C, is movable. The column is loaded by concentrated force of intensity F, acting at the point C and having the action line parallel to the column axis in the undeformed state, and by uniformly distributed follower type of load of constant intensity q_0 per unit length of the column axis. We assume that the column axis is inextensible and use the classical Bernoulli–Euler bending theory to describe the behavior of the column.

By using the same method as in the previous problem, we conclude that the linearized equations of equilibrium lead to the following boundary value problem

$$\ddot{m} + \frac{m}{a^2}\left[\lambda_1(1-t) + \lambda_2\right] = 0, \tag{58}$$

subject to

$$m(0) = m(1) = 0 \tag{59}$$

In (58) $a(t)$ is the dimensionless cross sectional area and $\lambda_1 = \frac{q_0}{\alpha EL}$; $\lambda_2 = \frac{F}{\alpha EL^2}$ where $I(S) = \alpha A^2(S)$ is the axial moment of inertia of the cross section and $A = aL^2$. If $a(t)$ in (58) is given, then the values of $(\lambda_1, \lambda_2) \in \mathbb{R}^2$ for which (58),(59) have nontrivial solution define a set of curves C_n, $n = 1, 2, ..$ (see Antman, 1995). Let $(\lambda_1^1, \lambda_2^1)$ be a point on the lowest interaction curve C_1 (i.e., a curve corresponding to first buckling mode). Then a straight line connecting the point $(0,0)$ with the point $(\lambda_1^1, \lambda_2^1)$ does not intersect any other interaction curve.

Suppose now that (λ_1, λ_2) is given. We define *the lightest compressed Pflüger column* as the column so shaped that any other column of the same length (in our case equal to one) and smaller volume will buckle under given load characterized by (λ_1, λ_2). Thus, the problem of determining the shape of the lightest compressed Pflüger column can be stated as:

Given (λ_1, λ_2) *find* $a(t) \in U$ *such that*

$$\ddot{m} + \frac{m}{a^2}\left[\lambda_1(1-t) + \lambda_2\right] = 0, \tag{60}$$

subject to

$$m(0) = m(1) = 0, \tag{61}$$

has (λ_1, λ_2) *as a point on the lowest interaction curve* C_1 *and that at the same time, the integral*

$$V_0 = \int_0^1 a(t)dt, \tag{62}$$

is minimal.

The set **U** consists of admissible cross–sectional area functions. In the works of Clausen (1851), Blasius (1914), Ratzersdorfer (1936), Keller (1960) and Tadjbakhsh and Keller (1962) it was assumed that **U** is the set of continuous functions on the interval $[0, 1]$. Following this, we also assume that $\mathbf{U} = C[0, 1]$.

Optimization problem and its solution. To determine $a(t)$ such that V_0 is minimal we shall use the Pontryagin maximum principle. Let us rewrite optimization problem (60)–(62) as:

Find the control $a(t)$ such that $0 < a(t) < \infty$, $t \in (0, 1)$, for which the optimality criterion

$$V_0 = \int_0^1 a(t)dt, \tag{63}$$

attains minimum value. The governing differential equations are

$$\dot{x}_1 = x_2; \quad \dot{x}_2 = -\frac{x_1}{a^2}\left[\lambda_1(1-t) + \lambda_2\right], \tag{64}$$

subject to

$$x_1(0) = x_1(1) = 0. \tag{65}$$

For the system (63)–(64) the Hamiltonian function $\mathcal{H}$ can be easily constructed as

$$\mathcal{H} = a + p_1 x_2 - p_2 \frac{1}{a^2}\left[\lambda_1(1-t) + \lambda_2\right] x_1, \tag{66}$$

where the costate variables p_1 and p_2 have to satisfy

$$\dot{p}_1 = -\frac{\partial \mathcal{H}}{\partial x_1} = \frac{1}{a^2}\left[\lambda_1(1-t) + \lambda_2\right] p_2; \quad \dot{p}_2 = -\frac{\partial \mathcal{H}}{\partial x_2} = -p_1, \tag{67}$$

subject to

$$p_2(0) = p_2(1) = 0. \tag{68}$$

The optimality condition $\min_u \mathcal{H}(t, x_1, x_2, p_1, p_2, a)$ leads to

$$\frac{\partial \mathcal{H}}{\partial a} = 1 + \frac{2}{a^3}\left[\lambda_1(1-t) + \lambda_2\right] x_1 p_2 = 0. \tag{69}$$

By solving (69) for a we obtain

$$a = -\left\{2\left[\lambda_1(1-t) + \lambda_2\right] x_1 p_2\right\}^{1/3}. \tag{70}$$

Comparison of the boundary value problems (64)–(65) and (67)–(68) leads to the conclusion that their solutions are connected as

$$p_1(t) = x_2(t); \qquad p_2(t) = -x_1(t). \tag{71}$$

By using (71) the control variable $a(t)$ given by (70) becomes

$$a = \left\{2\left[\lambda_1(1-t) + \lambda_2\right] x_1^2\right\}^{1/3}. \tag{72}$$

Note that with (71) substituted in (66) we have $(\partial^2\mathcal{H}/\partial a^2) > 0$ so that the necessary condition for minimum of $\mathcal{H}$ is satisfied. By combining (72) and (64) we arrive at the single differential equation

$$\ddot{x} + \left[\bar{\lambda}_1(1-t) + \bar{\lambda}_2\right]^{1/3} x^{-1/3} = 0, \tag{73}$$

subject to

$$x(0) = x(1) = 0. \tag{74}$$

In (73) and (74) we used $x(t) = x_1(t)$, $\bar{\lambda}_1 = (\lambda_1/4), \bar{\lambda}_2 = (\lambda_2/4)$. The case treated by Keller (1960) corresponds to $\bar{\lambda}_1 = 0; \bar{\lambda}_2 \neq 0$. Note also that the end points $t = 0$ and $t = 1$ are singular points of the problem (73),(74).

The existence of solution of the system (73),(74). In this Section we prove the existence of a weak solution of the boundary value problem (73),(74). The weak solution of (73),(74) is a function $u \in H_0^{1,2}$ where

$$H_0^{1,2} = \left\{z: \quad z(0) = z(1) = 0;\ \|z\|_{L_2} < \infty;\ \|\dot{z}\|_{L_2} < \infty\right\}, \tag{75}$$

that satisfies

$$\int_0^1 \left\{\dot{u}\dot{\varphi}_0 - \left[\bar{\lambda}_1(1-t) + \bar{\lambda}_2\right]^{1/3} u^{-1/3}\varphi_0\right\} dt = 0, \tag{76}$$

for all smooth (infinitely differentiable) functions vanishing at the end points of the interval $[0,1]$ i.e., $\varphi_0 \in C_0^\infty(0,1)$. In (75) we used $\|z\|_{L_2} = \left(\int_0^1 u^2(t)dt\right)^{1/2}$. Note that the Sobolev imbedding theorem implies $H_0^{1,2} \subset C([0,1])$ so that the boundary conditions $u(0) = u(1) = 0$ make sense. We denote the norm on $H_0^{1,2}$ as $\|u\|_{H_0^{1,2}} = \left[\int_0^1 \left(\dot{u}^2 + u^2\right) dt\right]^{1/2}$. Consider the functional

$$I(u) = \int_0^1 \left\{\frac{1}{2}\dot{u}^2 - \frac{3}{2}\left[\bar{\lambda}_1(1-t) + \bar{\lambda}_2\right]^{1/3} u^{2/3}\right\} dt, \tag{77}$$

and the problem of finding its minimum over the set $M \subset H_0^{1,2}$. The set M is defined as

$$M = \left\{u : u \in H_0^{1,2}; \quad \underline{u} \leq u \leq \bar{u} \quad \text{almost evrywhere}\right\}, \tag{78}$$

where $\underline{u}$ and $\bar{u}$ denote weak sub and sup solutions of (73),(74), i.e., $\underline{u} \in H_0^{1,2}$; $\bar{u} \in H_0^{1,2}$ and for all $\varphi_0 \in C_0^\infty(0,1)$; $\varphi_0 \geq 0$, we have

$$\int_0^1 \left\{\dot{\underline{u}}\,\dot{\varphi}_0 - \left[\bar{\lambda}_1(1-t) + \bar{\lambda}_2\right]^{1/3} \underline{u}^{-1/3}\,\varphi_0\right\} dt \leq 0;$$

$$\int_0^1 \left\{\left(\dot{\bar{u}}\right)\dot{\varphi}_0 - \left[\bar{\lambda}_1(1-t) + \bar{\lambda}_2\right]^{1/3} \left(\bar{u}\right)^{-1/3}\varphi_0\right\} dt \geq 0, \tag{79}$$

with $\underline{u}\,(0) \leq 0$; $\underline{u}(1) \leq 0$; $\bar{u}\,(0) \geq 0$; $\bar{u}\,(1) \geq 0$. Note that M is reflexive and that there exists an element $\check{u} \in M$ such that $I(\check{u}) < \infty$. We construct next weak sub and sup solutions. Consider the function

$$\underline{u}= \varepsilon t(1-t), \tag{80}$$

where $\varepsilon > 0$. By substituting (80) into $(79)_1$ we obtain

$$\int_0^1 \left\{ \varepsilon(1-2t)\dot{\varphi}_0 - \left[\frac{\bar{\lambda}_1(1-t)+\bar{\lambda}_2}{\varepsilon t(1-t)}\right]^{1/3} \varphi_0 \right\} dt =$$

$$-2\varepsilon \int_0^1 \left[(t\varphi_0)^{\cdot} - \varphi_0\right] dt - \int_0^1 \left[\frac{\bar{\lambda}_1(1-t)+\bar{\lambda}_2}{\varepsilon t(1-t)}\right]^{1/3} \varphi_0 dt =$$

$$-\int_0^1 \left\{ \left[\frac{\bar{\lambda}_1(1-t)+\bar{\lambda}_2}{\varepsilon t(1-t)}\right]^{1/3} - 2\varepsilon \right\} \varphi_0 dt. \tag{81}$$

Thus, if $\varepsilon \leq \left(\bar{\lambda}_1 + \bar{\lambda}_2\right)^{1/4} / (2)^{3/4}$ the inequality $(79)_1$ holds and (80) is a sub-solution. To construct the sup-solution we consider

$$\ddot{y} + \left[\bar{\lambda}_1 + \bar{\lambda}_2\right]^{1/3} y^{-1/3} = 0, \tag{82}$$

subject to

$$y(0) = y(1) = 0. \tag{83}$$

The solution y of (82),(83) is given in parametric form in Keller (1960). We take this solution as sup-solution $\bar{u}$ so that

$$\bar{u}= u_0 \sin^3\theta; \qquad t = \frac{1}{\pi}\left[\theta - \frac{\sin 2\theta}{2}\right], \tag{84}$$

where $u_0 = \left[4/(3\pi^2)\right]^{3/4} (\bar{\lambda}_1 + \bar{\lambda}_2)^{1/4}$. Since $\bar{u}$ is the the solution of (82),(83) we have

$$\int_0^1 \left\{ \dot{\left(\bar{u}\right)} \dot{\varphi}_0 - \left[\bar{\lambda}_1 + \bar{\lambda}_2\right]^{1/3} \left(\bar{u}\right)^{-1/3} \varphi_0 \right\} dt = 0, \tag{85}$$

for all $\varphi_0 \geq 0$. However, for $\varphi_0 \geq 0$ we obtain

$$\int_0^1 \left[\bar{\lambda}_1 + \bar{\lambda}_2\right]^{1/3} \left(\bar{u}\right)^{-1/3} \varphi_0 dt \geq \int_0^1 \left[\bar{\lambda}_1(1-t) + \bar{\lambda}_2\right]^{1/3} \left(\bar{u}\right)^{-1/3} \varphi_0 dt. \tag{86}$$

Therefore

$$\int_0^1 \left\{ \dot{\left(\bar{u}\right)} \dot{\varphi}_0 - \left[\bar{\lambda}_1(1-t) + \bar{\lambda}_2\right]^{1/3} \left(\bar{u}\right)^{-1/3} \varphi_0 \right\} dt \geq 0. \tag{87}$$

With (80) and (84) the set M is defined. It is closed and convex, hence weakly closed. We now prove the following statement:

The functional (77) attains a relative minimum on M.

To prove the statement recall that set M is closed and convex. Functional I is coercive and sequentially weakly lower semi-continuous on M. To see this note that

$$I(u)=\int_0^1\left\{\frac{1}{2}\,\dot{u}^2-\frac{3}{2}\left[\bar{\lambda}_1(1-t)+\bar{\lambda}_2\right]^{1/3}u^{2/3}\right\}dt\geq$$

$$\frac{1}{2}\left\|\dot{u}\right\|_{L_2}^2-c\geq\frac{1}{4}\left[\left\|\dot{u}\right\|_{L_2}^2+\pi^2\left\|u\right\|_{L_2}^2\right]-c=\frac{1}{4}\left\|u\right\|_{H_0^{1,2}}^2-c, \tag{88}$$

where $c=\frac{3}{2}\int_0^1\left\{\left[\bar{\lambda}_1(1-t)+\bar{\lambda}_2\right]^{1/3}u^{2/3}\right\}dt<\infty$ and we used the Wirtinger's inequality. From (88) it follows that I is coercive on M. We show next that I is sequentially weakly lower semi-continuous. First we write

$$I(u)=I_1-I_2;$$

$$I_1=\int_0^1\left\{\frac{1}{2}\,\dot{u}^2\,dt;\quad I_2=\int_0^1\frac{3}{2}\left[\bar{\lambda}_1(1-t)+\bar{\lambda}_2\right]^{1/3}u^{2/3}\right\}dt. \tag{89}$$

The functional I_1 is sequentially weakly lower semi-continuous (see Dacorogna, 1989, p.75) since the expression under the integral is convex function of $\dot{u}$. Therefore to show that I is sequentially weakly lower semi-continuous it suffices to prove that

$$\lim_{n\to\infty}\int_0^1\frac{3}{2}\left[\bar{\lambda}_1(1-t)+\bar{\lambda}_2\right]^{1/3}u_n^{2/3}dt=\int_0^1\frac{3}{2}\left[\bar{\lambda}_1(1-t)+\bar{\lambda}_2\right]^{1/3}u^{2/3}dt, \tag{90}$$

when the sequence $u_n\in M$, $n=1,2,...,$ $u\in M$ and converges weakly to u in $H_0^{1,2}$i.e., $u_n\overset{w}{\to}u$. Since $\left|\left[\bar{\lambda}_1(1-t)+\bar{\lambda}_2\right]^{1/3}u_n(t)^{2/3}\right|$ is bounded ($u_n(t)$ are continuous) we may use the Lebesque's dominated convergence theorem to conclude that (90) holds. Therefore,

$$\lim_{n\to\infty}I(u_n)\geq I(u),\qquad \text{when}\quad u_n\overset{w}{\to}u. \tag{91}$$

By using the classical arguments (see Struwe, 1996 for example) it follows that I attains its minimum on M. Let $x(t)$ be the minimizing element, i.e., $I(x)=\inf\left\{I(u);\quad u\in M\right\}$.

We state now the main result as:

There exist a weak solution of the problem (73) and (74) and it is equal to $x(t)$.

To prove the statement we proceed as follows. Let $\varphi_0(t)\in H_0^{1,2}$ be arbitrary and let $\varepsilon>0$, be given. Following Struwe (1996) we define $v_\varepsilon=\min\left\{\bar{u},\max\left\{\underline{u},x+\varepsilon\varphi_0\right\}\right\}=x+\varepsilon\varphi_0-\varphi^\varepsilon+\varphi_\varepsilon$, where

$$\varphi^\varepsilon=\max\left\{0,x+\varepsilon\varphi_0-\bar{u}\right\}\geq 0;\quad \varphi_\varepsilon=-\min\left\{0,x+\varepsilon\varphi_0-\underline{u}\right\}\geq 0. \tag{92}$$

Note that $v_\varepsilon\in M$. Since I is Fréchet differentiable we calculate the differential of I at $u=x$ in the direction $v_\varepsilon-\hat{u}$.

$$\langle DI(x),v_\varepsilon-\hat{u}\rangle=\int_0^1\left\{(v_\varepsilon-x)^{\cdot}\,\dot{\varphi}_0-\left[\bar{\lambda}_1+\bar{\lambda}_2\right]^{1/3}(v_\varepsilon-x)^{-1/3}\varphi_0\right\}dt. \tag{93}$$

Since x minimizes I it follows that $\langle DI(x), v_\varepsilon - \hat{u}\rangle \geq 0$. By substituting (92) into (93) we obtain

$$\int_0^1 \left\{\dot{x}\left[\varepsilon\dot{\varphi}_0 - \dot{\varphi}^\varepsilon + \dot{\varphi}_\varepsilon\right] - \left[\bar{\lambda}_1(1-t)+\bar{\lambda}_2\right]^{1/3} x^{-1/3}\left[\varepsilon\varphi_0 - \varphi^\varepsilon + \varphi_\varepsilon\right]\right\} dt \geq 0. \tag{94}$$

From (93) it follows

$$\langle DI(x), v_\varepsilon - \hat{u}\rangle \geq \frac{1}{\varepsilon}\left[\langle DI(x), \varphi^\varepsilon\rangle - \langle DI(x), \varphi_\varepsilon\rangle\right]. \tag{95}$$

By using the fact that $\bar{u}$ is a supersolution, we obtain

$$\langle DI(x), \varphi^\varepsilon\rangle = \left\langle DI(\bar{u}), \varphi^\varepsilon\right\rangle + \left\langle DI(x) - DI(\bar{u}), \varphi^\varepsilon\right\rangle \geq \left\langle DI(x) - DI(\bar{u}), \varphi^\varepsilon\right\rangle =$$

$$\int_{\Omega_\varepsilon} \left\{(x - \bar{u})^{\cdot}(x + \varepsilon\varphi_0 - \bar{u})^{\cdot} - \left[\bar{\lambda}_1(1-t)+\bar{\lambda}_2\right]^{1/3} \times \left[x^{-1/3} - \bar{u}^{-1/3}\right]\left[x + \varepsilon\varphi_0 - \bar{u}\right]\right\} dt \tag{96}$$

$$\geq \varepsilon\int_{\Omega_\varepsilon} \left\{(x - \bar{u})^{\cdot}\dot{\varphi}_0\right\} dt - \varepsilon(\lambda_1+\lambda_2)^{1/3}\int_{\Omega_\varepsilon}\left\{\left|x^{-1/3} - \bar{u}^{-1/3}\right| |\varphi_0|\right\} dt. \tag{97}$$

where $\Omega_\varepsilon = \{t : t \in [0,1] \quad x(t) + \varepsilon\varphi(t) \geq \bar{u}\ (t) \geq x(t)\} \subset [0,1]$. Since $x(t), \varphi_0(t)$ and $\bar{u}\ (t)$ are continuous the measure of Ω_ε tends to zero as $\varepsilon \to 0$. By the absolute continuity of the Lebesque integral, we conclude from (97)

$$\langle DI(x), \varphi^\varepsilon\rangle \geq o(\varepsilon), \tag{98}$$

where $\lim_{\varepsilon\to 0} o(\varepsilon)/\varepsilon = 0$. By using similar arguments it follows

$$\langle DI(x), \varphi_\varepsilon\rangle \leq o(\varepsilon). \tag{99}$$

Combining (98),(99) and (95) we have

$$\langle DI(x), \varphi_0\rangle \geq 0. \tag{100}$$

By reversing the sign of φ_0 we conclude that $\langle DI(x), \varphi_0\rangle = 0$. Finally since $C_0^\infty(0,1)$ is dense in $H_0^{1,2}$ it follows that

$$\int_0^1 \left\{\dot{x}\dot{\varphi}_0 - \left[\bar{\lambda}_1(1-t)+\bar{\lambda}_2\right]^{1/3} x^{-1/3}\varphi_0\right\} dt = 0, \tag{101}$$

for all $\varphi_0 \in H_0^{1,2}$. Thus, x is a weak solution of (73),(74). Also from (101) we have

$$DI(x) = 0. \tag{102}$$

Dual variational principles for (73),(74). In order to obtain the á priori bounds on the solution x and to estimate the accuracy of numerical solution of (73),(74) we shall construct the complementary variational principles. First we show that I given by (77) has global minimum for $u = x$. By calculating the second Fréchet differential of (77) we obtain

$$D^2 I(x,w) = \int_0^1 \left\{ \dot{w}^2 + \frac{1}{3} \left[\bar{\lambda}_1 (1-t) + \bar{\lambda}_2\right]^{1/3} x^{-4/3} w^2 \right\} dt. \tag{103}$$

From (103) it follows that

$$D^2 I(x,w) \geq \left\| \dot{w} \right\|^2_{L_2} \geq \frac{1}{2} \left\| \dot{w} \right\|^2_{H_0^{1,2}}. \tag{104}$$

Therefore $I(x)$ is the global minimum. Next we proceed to construct the dual variational principle. Let $F(t, X, \dot{X})$ denote the Lagrangian function of the variational principle (77) that is

$$F(t, X, \dot{X}) = \frac{1}{2} \dot{X}^2 - \frac{3}{2} \left[\bar{\lambda}_1 (1-t) + \bar{\lambda}_2\right]^{1/3} X^{2/3}. \tag{105}$$

The canonical form of (73),(74) is obtained by introducing the variable P as

$$P = \frac{\partial F}{\partial \dot{X}} = \dot{X}, \tag{106}$$

and the function $K(t, X, P)$, is connected with F by

$$K(t, X, P) = P \dot{X} - F = \frac{1}{2} P^2 + \frac{3}{2} \left[\bar{\lambda}_1 (1-t) + \bar{\lambda}_2\right]^{1/3} X^{2/3}. \tag{107}$$

The system (73),(74) then becomes

$$\dot{X} = \frac{\partial K}{\partial P} = P; \quad \dot{P} = -\frac{\partial K}{\partial X} = - \left[\bar{\lambda}_1 (1-t) + \bar{\lambda}_2\right]^{1/3} X^{-1/3}. \tag{108}$$

In order to obtain the dual variational principle we have to solve (107) for F. Then, in so obtained expression we substitute X in terms of $\dot{P}$ from $(108)_2$ and use the resulting expression in (77). After partial integration and application of boundary conditions, we obtain

$$G(P) = I(P, X(P)) = - \int_0^1 \left\{ \frac{1}{2} P^2 + \frac{1}{2} \frac{\left[\bar{\lambda}_1 (1-t) + \bar{\lambda}_2\right]}{\dot{P}^2} \right\} dt. \tag{109}$$

Let $x(t)$ and $p(t)$ be the solution of canonical system (108). Then it is easy to show that the following dual variational principle holds

$$DG(p, n) = 0, \tag{110}$$

where the variation is $n = P - p$, and P is an admissible function, i.e., $P \in \mathbf{Y}$ where

$$\mathbf{Y} = \{P : P \in C^1(0,1); \int_0^1 \left\{ \left[\bar{\lambda}_1 (1-t) + \bar{\lambda}_2\right] / \dot{P}^2 \right\} dt < \infty\}. \tag{111}$$

Note that elements of $\mathbf{Y}$ need not satisfy any boundary condition. However, in applications P is usually taken as $P = \dot{X}$, where $X(t) \in \mathbf{X}$. Also from (109) it follows

$$D^2G(p,n) = -\int_0^1 \left[n^2 + 3\frac{[\bar{\lambda}_1(1-t)+\bar{\lambda}_2]}{\dot{P}^4}\dot{n}^2 \right] dt \leq -\|n\|^2_{L_2}. \tag{112}$$

By combining (110) and (112) we have the following result:

On the solution $p(t)$ of (108) the functional (109) attains maximum.

From the results proved so far, we have the chain of inequalities

$$G(P) \leq G(p) = I(x) \leq I(X). \tag{113}$$

We can use (77),(109) to obtain and estimate error of an approximate solution of (73),(74) by using the same procedure as in previous example.

Á priori estimates of $\|x(t)\|_{L_\infty}$. In this Section we shall derive estimates of the form $\alpha(\bar{\lambda}_1,\bar{\lambda}_2) \leq \|x(t)\|_{L_\infty} \leq \beta(\bar{\lambda}_1,\bar{\lambda}_2)$ for the solution $x(t)$ of the boundary value problem (73),(74). The form of functions $\alpha(\bar{\lambda}_1,\bar{\lambda}_2)$ and $\beta(\bar{\lambda}_1,\bar{\lambda}_2)$ will be explicitly given. We start with the condition $DI(x) = 0$ (see (102)) that may be written

$$\int_0^1 \left\{ \dot{x}\dot{\varphi}_0 - [\bar{\lambda}_1(1-t)+\bar{\lambda}_2]^{1/3} x^{-1/3}\varphi_0 \right\} dt = 0. \tag{114}$$

By using $x = \varphi_0$ we obtain

$$\int_0^1 \dot{x}^2 dt = \int_0^1 [\bar{\lambda}_1(1-t)+\bar{\lambda}_2]^{1/3} x^{2/3} dt. \tag{115}$$

With (115) used in (77) the inequalities (113) become

$$G(P) \leq I(x) = -\int_0^1 \dot{x}^2 dt = -\int_0^1 [\bar{\lambda}_1(1-t)+\bar{\lambda}_2]^{1/3} x^{2/3} dt = G(p) \leq I(X). \tag{116}$$

From (116) we have $\int_0^1 \dot{x}^2 dt \geq -G(P)$ for any $P \in \mathbf{Y}$. By using the Wirtinger inequality we find

$$\|x(t)\|_{L_\infty} = \sup_{t\in(0,1)} |x(t)| \leq \frac{[-G(P)]^{1/2}}{2} = \frac{\left[\int_0^1 \left\{ \frac{1}{2}P^2 + \frac{1}{2}\frac{[\bar{\lambda}_1(1-t)+\bar{\lambda}_2]}{\dot{P}^2} \right\} dt \right]^{1/2}}{2}. \tag{117}$$

A simple function

$$P = k(1-t)^{1/2} - c, \tag{118}$$

where k and c are constants, when substituted in (109) and minimized with respect to k and c leads to $c = 2k/3; k = (12\lambda_1 + 36\lambda_2)^{1/4}$ and $G(P) = -(12\lambda_1+36\lambda_2)^{1/2}/18$. Therefore (117) becomes

$$\|x(t)\|_{L_\infty} \leq (12\bar{\lambda}_1 + 36\bar{\lambda}_2)^{1/4}\frac{\sqrt{2}}{12}. \tag{119}$$

To derive a lower bound on $\|x(t)\|_{L_\infty}$ we use (116) to get

$$\int_0^1 \left[\bar{\lambda}_1(1-t)+\bar{\lambda}_2\right]^{1/3} x^{2/3} dt \geq -I(X), \tag{120}$$

for any $X \in M$. By using Cauchy inequality in (120) it follows that

$$\|x(t)\|_{L_\infty}^{2/3} \left\{\int_0^1 \left[\bar{\lambda}_1(1-t)+\bar{\lambda}_2\right]^{2/3}\right\}^{1/2} \geq -I(X), \tag{121}$$

or

$$\|x(t)\|_{L_\infty} \geq \frac{[-I(X)]^{3/2}}{\left\{\int_0^1 \left[\bar{\lambda}_1(1-t)+\bar{\lambda}_2\right]^{2/3} dt\right\}^{1/2}}$$
$$= \frac{\left[\frac{1}{2}\int_0^1 \left\{3\left[\bar{\lambda}_1(1-t)+\bar{\lambda}_2\right]^{1/3} X^{2/3} - \dot{X}^2\right\} dt\right]^{3/2}}{\left\{\frac{3}{5}\frac{1}{\bar{\lambda}_1}\left[(\bar{\lambda}_1+\bar{\lambda}_2)^{5/3} - \bar{\lambda}_2^{5/3}\right]\right\}^{3/4}}. \tag{122}$$

By using a trial function $X = ft(1-t)$, where f is a constant, in (77) we obtain

$$-I(X) \geq$$
$$\frac{1}{2}\left\{\frac{3(\bar{\lambda}_1+\bar{\lambda}_2)^{1/3} f^{2/3}\Gamma(\frac{5}{6})}{\sqrt{\pi}\,\Gamma(\frac{4}{3})} - \frac{3\left[(\bar{\lambda}_1+\bar{\lambda}_2)^{1/3} - (\bar{\lambda}_2)^{1/3}\right] f^{2/3}\Gamma(\frac{5}{3})}{3(4)^{1/3}\left(\Gamma(\frac{4}{3})\right)^2} - \frac{f^2\pi^2}{2}\right\}, \tag{123}$$

where $\Gamma(x)$ is the Euler gamma function. The minimum of the right hand side of (123) is attained for

$$f = \left\{\frac{1}{\pi^2}\frac{1}{\Gamma(\frac{4}{3})}\left[\frac{2(\bar{\lambda}_1+\bar{\lambda}_2)^{1/3}}{\sqrt{\pi}}\Gamma(\frac{5}{6}) - \frac{1}{(4)^{1/3}}\left[(\bar{\lambda}_1+\bar{\lambda}_2)^{1/3} - (\bar{\lambda}_2)^{1/3}\right]\frac{\Gamma(\frac{5}{3})}{\Gamma(\frac{4}{3})}\right]\right\}^{3/4}. \tag{124}$$

In performing integration in (77) to obtain (123) we used the inequality

$$\left[\bar{\lambda}_1(1-t)+\bar{\lambda}_2\right]^{1/3} \geq (\bar{\lambda}_1+\bar{\lambda}_2)^{1/3} - t\left\{(\bar{\lambda}_1+\bar{\lambda}_2)^{1/3} - (\bar{\lambda}_2)^{1/3}\right\}, \tag{125}$$

that follows from the fact that $\left[\bar{\lambda}_1(1-t)+\bar{\lambda}_2\right]^{1/3}$ is a concave function. With $I(X)$ given by (123) the inequality (122) becomes

$$\|x(t)\|_{L_\infty} \geq$$
$$\frac{\left[\frac{1}{2}\left\{\frac{3(\bar{\lambda}_1+\bar{\lambda}_2)^{1/3} f^{2/3}\Gamma(\frac{5}{6})}{\sqrt{\pi}\,\Gamma(\frac{4}{3})} - \frac{3\left[(\bar{\lambda}_1+\bar{\lambda}_2)^{1/3} - (\bar{\lambda}_2)^{1/3}\right] f^{2/3}\Gamma(\frac{5}{3})}{3(4)^{1/3}\left(\Gamma(\frac{4}{3})\right)^2} - \frac{f^2\pi^2}{2}\right\}\right]^{3/2}}{\left\{\frac{3}{5}\frac{1}{\bar{\lambda}_1}\left[(\bar{\lambda}_1+\bar{\lambda}_2)^{5/3} - \bar{\lambda}_2^{5/3}\right]\right\}^{3/4}} \tag{126}$$

with f given by (124).

Numerical results. In this section we present results of numerical integration of the system (73),(74). The value of λ_1 was taken as the lowest eignevalue for the Pflüger column with constant cross–section, i.e., $\lambda_1 = 18.956266$ so that $\bar{\lambda}_1 = \lambda_1/4 = 4.7390665$. The value for $\bar{\lambda}_2$ was taken as $\bar{\lambda}_2 = 1$. The cross sectional–area is determined from (72) that for the circular cross–section leads to the following expression for radius R

$$R = \sqrt{\frac{1}{\pi}\left\{2\left[4\bar{\lambda}_1(1-t) + 4\bar{\lambda}_2\right]x^2\right\}^{1/3}}. \tag{127}$$

The estimates (119) and (126) for $\bar{\lambda}_1 = 4.7390665$ and $\bar{\lambda}_2 = 1$ give

$$0.17332 \leq \|x(t)\|_{L_\infty} \leq 0.36585, \tag{128}$$

while the numerical solution yields $\|x(t)\|_{L_\infty} = 0.2997344$.

We compare now the volume of the optimal compressed Pflüger column with the volume of the Pflüger column having constant cross–section and (λ_1, λ_2) on the lowest interaction curve. The linearized equilibrium equation for the Pflüger column with constant cross–section reduces to (see Atanackovic, 1997, p. 184)

$$\ddot{z} + [\hat{\lambda}_1(1-t) + \hat{\lambda}_2]z = 0; \qquad z(0) = z(1) = 0, \tag{129}$$

where

$$\hat{\lambda}_1 = \frac{q_0 L^3}{EI}; \qquad \hat{\lambda}_2 = \frac{FL^2}{EI}. \tag{130}$$

The critical values of $(\hat{\lambda}_1, \hat{\lambda}_2)$ are determined from

$$J_{1/3}(u_1)J_{-1/3}(u_2) - J_{1/3}(u_2)J_{-1/3}(u_1) = 0, \tag{131}$$

with

$$u_1 = \frac{2}{3}\left[\hat{\lambda}_1\left(1 + \frac{\hat{\lambda}_2}{\hat{\lambda}_1}\right)^3\right]^{1/2}; \qquad u_2 = \frac{2}{3}\left[\hat{\lambda}_1\left(\frac{\hat{\lambda}_2}{\hat{\lambda}_1}\right)^3\right]^{1/2}. \tag{132}$$

By using (**??**) and (130) we obtain

$$\frac{\lambda_1}{\hat{\lambda}_1} = \frac{\lambda_2}{\hat{\lambda}_2} = \frac{I}{\alpha L^4}, \tag{133}$$

so that $(\lambda_2/\lambda_1) = (\hat{\lambda}_2/\hat{\lambda}_1)$. Therefore for given (λ_1, λ_2) the corresponding values $(\hat{\lambda}_1, \hat{\lambda}_2)$ are determined as

$$\hat{\lambda}_1 = k; \qquad \hat{\lambda}_2 = k\frac{\lambda_2}{\lambda_1}, \tag{134}$$

where k is the lowest root of the equation

$$J_{1/3}\left(\frac{2}{3}\left[k\left(1 + \frac{\lambda_2}{\lambda_1}\right)^3\right]^{1/2}\right) J_{-1/3}\left(\frac{2}{3}\left[k\left(\frac{\lambda_2}{\lambda_1}\right)^3\right]^{1/2}\right)$$

$$-J_{1/3}\left(\frac{2}{3}\left[k\left(\frac{\lambda_2}{\lambda_1}\right)^3\right]^{1/2}\right)J_{-1/3}\left(\frac{2}{3}\left[k\left(1+\frac{\lambda_2}{\lambda_1}\right)^3\right]^{1/2}\right)=0. \tag{135}$$

With $(\hat{\lambda}_1, \hat{\lambda}_2)$ known equation (133) leads to the volume of the column with constant cross-section given by

$$V_{const.} = L^3\sqrt{\frac{\lambda_1}{\hat{\lambda}_1}}. \tag{136}$$

The volume of the optimal column is

$$V_{optimal} = L^3\int_0^1 a dt = L^3\int_0^1 \left\{2\left[\lambda_1(1-t)+\lambda_2\right]x^2\right\}^{1/3} dt, \tag{137}$$

where we used (72). With (136),(137) we find

$$\frac{V_{optimal}}{V_{const.}} = \frac{\int_0^1 \left\{2\left[\lambda_1(1-t)+\lambda_2\right]x^2\right\}^{1/3} dt}{\sqrt{\frac{\lambda_1}{\hat{\lambda}_1}}}. \tag{138}$$

For the case $\lambda_1 = 18.956; \lambda_2 = 4$ we obtain from (135) $k = 13.597$.Then (134) gives $\hat{\lambda}_1 = 13.597; \hat{\lambda}_2 = 2.8691$. Also from numerical integration $V_{optimal} = L^3\int_0^1 a dt = 0.992894L^3$ so that $(V_{optimal}/V_{const.}) = 0.84091$.

2.3 Optimal Shape of a Rotating Rod

Consider the problem of rotating rod treated in Section 1 when the cross-sectional area of the rod is not constant. Our intention in this Section is to formulate an optimization problem for rotating rod by using the procedure presented in the previous Section. Suppose that the angular velocity with which the rod rotates $\omega = \omega_0 = const.$ and the length of the rod L are given. Let S be the arc–length of the rod axis, so that $S \in [0, L]$. We define *the optimal rotating rod* as the rod so shaped that it will lose stability for $\omega > \omega_0$ while any other rod of the same length (in our case equal to L) and equal or smaller volume will lose stability for $\omega \leq \omega_0$.

Thus, the optimal rotating rod will be the lightest rod of length L that is stable against buckling for $0 \leq \omega \leq \omega_0$. In solving the problem of determining the shape of an optimal rotating rod we shall use the procedure based on Pontryagin's maximum principle. For the relevant equations describing the optimal rotating rod, that we believe are new, we formulate three variational principles. On the basis of these variational principles we shall find a first integral corresponding to Euler–Lagrange equations. This first integral will be used to check the accuracy of numerical integration of the relevant differential equations. In presentation we shall follow the analysis given in Atanackovic (2002).

Model. The equilibrium, geometrical and constitutive equations for the rotating rod, are

$$H' = 0; \qquad V' = -\rho_0\omega^2\bar{y}; \qquad M' = V\cos\theta - H\sin\theta;$$
$$\bar{x}' = \cos\theta; \qquad \bar{y}' = \sin\theta; \quad \theta' = \frac{M}{EI}, \tag{139}$$

where H and V are components of the contact force in an arbitrary cross–section, M is the bending moment, θ is the angle between the axis of rotation and tangent to the rod axis, $\bar{x}$ and $\bar{y}$ are coordinates of an arbitrary point with respect to the rotating Cartesian frame x–B–y. Also in (139) we use ρ_0 to denote line density of the rod (mass per unit length of the rod axis), E is the modulus of elasticity, I is the second moment of the cross–sectional area of the rod and $\dot{(\cdot)} = d(\cdot)/dS$. Note that $\rho_0 = \rho A$ where ρ is the density of the rod (mass per unit volume of the rod). We assume that the cross–section of the rod is circular, so that

$$I = \alpha A^2, \tag{140}$$

where A is the cross–sectional area and $\alpha = (1/4\pi)$. With the notation introduced, the volume of the rod is

$$W = \int_0^L A(S)dS. \tag{141}$$

By using the dimensionless variables and parameters

$$t = \frac{S}{L}; \qquad a = \frac{A}{L^2}; \qquad y = \frac{\bar{y}}{L}; \qquad w = \frac{W}{L^3};$$
$$m = \frac{M}{E\alpha L^3}; \qquad v = \frac{V}{E\alpha L^2}; \qquad \lambda^2 = \frac{\rho\omega^2 L^2}{E\alpha}, \tag{142}$$

and a new dependent variable

$$u = -\frac{w}{\lambda}, \tag{143}$$

we obtain from (139)

$$\dot{u} = \lambda a y; \qquad \dot{m} = -\lambda u \cos\theta; \qquad \dot{y} = \sin\theta; \qquad \dot{\theta} = -\frac{m}{a^2}. \tag{144}$$

The boundary conditions corresponding to the rod shown in Fig. 1 are

$$u(1) = 0; \qquad m(1) = 0; \qquad y(0) = 0; \qquad \theta(0) = 0. \tag{145}$$

The system (144),(145) possesses a trivial solution, in which the axis of the rod remains straight for any value of the dimensionless rotation speed λ, in the form

$$u_0 = 0; \qquad m_0 = 0; \qquad y_0 = 0; \qquad \theta_0 = 0. \tag{146}$$

To examine stability of the equilibrium configuration (146) we use Euler's method. Thus, we assume that

$$u = u_0 + \Delta u; \qquad m = m_0 + \Delta m; ...; \qquad \theta = \theta_0 + \Delta\theta, \tag{147}$$

where $\Delta u, ...\Delta\theta$, denote perturbations of the corresponding variables. Then, by substituting (147) into (144) and linearizing the resulting expressions, we obtain (omitting Δ in front of perturbed variables)

$$\left(\frac{\dot{u}}{a}\right)^{\cdot} = \lambda\theta; \qquad \left(\dot{\theta}a^2\right)^{\cdot} = \lambda u; \qquad \dot{y} = \theta, \tag{148}$$

subject to

$$u(1) = 0; \qquad \dot{u}(0) = 0; \qquad \theta(0) = 0; \qquad \dot{\theta}(1)a^2(1) = 0; \qquad y(0) = 0, \tag{149}$$

For the bifurcation it is enough to consider

$$\left(\frac{\dot{u}}{a}\right)^{\cdot} = \lambda\theta; \qquad \left(\dot{\theta}a^2\right)^{\cdot} = \lambda u, \tag{150}$$

subject to

$$u(1) = 0; \qquad \dot{u}(0) = 0; \qquad \theta(0) = 0; \qquad \lim_{t\to 1} \dot{\theta}(t)a^2(t) = 0. \tag{151}$$

By introducing new variables $u = w_1, \dot{u} = w_2, \theta = w_3, \dot{\theta} = w_4$ and $\mathbf{w} = [w_1, w_2, w_3, w_4]$ system (150),(151) can be written in compact form as

$$\mathbf{F}(\lambda)\mathbf{w} = \frac{d}{dt}\left(\mathbf{A}\frac{d}{dt}\mathbf{w}\right) + \lambda\mathbf{B}\mathbf{w} = 0, \tag{152}$$

where $\mathbf{A}$ and $\mathbf{B}$ are given as

$$\mathbf{A} = \begin{bmatrix} 0 & 1/a & 0 & 0 \\ 0 & 0 & 0 & 0 \\ 0 & 0 & 0 & 0 \\ 0 & 0 & 0 & a^2 \end{bmatrix}; \qquad \mathbf{B} = \begin{bmatrix} 0 & 0 & 1 & 0 \\ 0 & 0 & 0 & 0 \\ 0 & 0 & 0 & 0 \\ 1 & 0 & 0 & 0 \end{bmatrix}. \tag{153}$$

Note that the boundary condition $(151)_4$ that corresponds to $m(1) = 0$ is equivalent to $\dot{\theta}(1) = 0$ *if* $a(1) \neq 0$. However if $a(1) = 0$ (as will be in our analysis) the condition $m(1) = 0$ can be satisfied with $\dot{\theta}(1) \neq 0$. For fixed λ the existence of a nontrivial solution of (150),(151) is a necessary condition for the loss of stability. The dimensionless volume of the rod is given as

$$w = \int_0^1 a(t)dt. \tag{154}$$

We state now the following optimization problem:

Given λ determine $a(t) > 0$ for $t \in (0,1)$ so that λ is the smallest eigenvalue of (152),(153) and that at the same time w given by (154) is minimal.

We call the rod with such $a(t)$ the *optimal rod.* Thus, the optimal rod is so shaped that any other rod with smaller volume w will buckle at the rotation speed that is smaller than λ.

The optimization problem and its solution. Let $x_1, ...x_4$ be a set of dependent variables defined by

$$u = x_1; \qquad (\dot{u}/a) = x_2; \qquad \theta = x_3; \qquad (a^2\dot{\theta}) = x_4. \tag{155}$$

Then, the system (148),(149) becomes

$$\dot{x}_1 = ax_2; \qquad \dot{x}_2 = \lambda x_3; \qquad \dot{x}_3 = \frac{x_4}{a^2}; \qquad \dot{x}_4 = \lambda x_1, \tag{156}$$

subject to

$$x_1(1) = 0; \qquad x_2(0) = 0; \qquad x_3(0) = 0; \qquad x_4(1) = 0. \tag{157}$$

The problem of determining the shape $a(t)$ of the optimal rod may be stated, in terms of the optimal control theory, as:

determine the function $a(t) > 0, t \in (0,1)$ *that minimizes the functional*

$$I = \int_0^1 a(t)dt, \tag{158}$$

when the system is described by (156)(157).

To solve the optimization problem we use, again, Pontryagin's maximum principle. For system (156) the Hamiltonian function $\mathcal{H}$ is

$$\mathcal{H} = a + p_1 a x_2 + p_2 \lambda x_3 + p_3 \frac{x_4}{a^2} + p_4 \lambda x_1, \tag{159}$$

where the costate variables $p_1, \dots p_4$ satisfy

$$\begin{aligned} \dot{p}_1 &= -\frac{\partial \mathcal{H}}{\partial x_1} = -\lambda p_4; & \dot{p}_2 &= -\frac{\partial \mathcal{H}}{\partial x_2} - p_1 a; \\ \dot{p}_3 &= -\frac{\partial \mathcal{H}}{\partial x_3} = -\lambda p_2; & \dot{p}_4 &= -\frac{\partial \mathcal{H}}{\partial x_4} = -\frac{p_3}{a^2}, \end{aligned} \tag{160}$$

subject to

$$p_1(0) = 0; \qquad p_2(1) = 0; \qquad p_3(1) = 0; \qquad p_4(0) = 0. \tag{161}$$

The optimality condition $\min_u \mathcal{H}(t, x_1, x_2, p_1, p_2, a)$ leads to

$$\frac{\partial \mathcal{H}}{\partial a} = 1 + p_1 x_2 - 2p_3 \frac{x_4}{a^3} = 0. \tag{162}$$

By solving (162) for a we obtain

$$a = \left\{ \frac{2p_3 x_4}{1 + p_1 x_2} \right\}^{1/3}. \tag{163}$$

Further procedure simplifies significantly if we make the following observation: solution of the boundary value problem (156),(157) leads to a solution of the boundary value problem (160),(161) if we make the following identification[2]

$$p_1(t) = x_2(t); \qquad p_2(t) = -x_1(t); \qquad p_3 = x_4; \qquad p_4 = -x_3. \tag{164}$$

By using (164) the control variable $a(t)$ given by (163) becomes

$$a = \left\{ \frac{2\,(x_4)^2}{1 + (x_2)^2} \right\}^{1/3}. \tag{165}$$

[2] There is one more possibility to conect the solutions of (156) and (160). This is given by the following identification of dependent variables: $p_1 = -x_2, p_2 = x_1, p_3 = -x_4, p_4 = x_3$. However this identification is not of interest since it does not provide $a(t) \geq 0$ in (163) and $(\partial^2 \mathcal{H}/\partial a^2) > 0$ with $\mathcal{H}$ given by (159).

Note that with (164) substituted in (162) we have $(\partial^2\mathcal{H}/\partial a^2) > 0$ so that the necessary condition for minimum of $\mathcal{H}$ is satisfied. From (165) and the boundary condition $(157)_4$ we conclude that

$$a(1) = 0. \tag{166}$$

Thus: *the optimal rod is tapered so that it has zero cross-sectional area and zero moment of inertia at its free end.*

Also, when the original variables (see (155)) are used in (165) we obtain

$$\frac{a^3\dot{\theta}^2}{a^2 + \dot{u}^2} = \frac{1}{2}. \tag{167}$$

By using the boundary condition (149) in (167) we get

$$a(0)\dot{\theta}^2(0) = \frac{1}{2}. \tag{168}$$

We transform now the conditions of optimality (163). First we write it in the form

$$1 + (x_2)^2 = \frac{2\,(x_4)^2}{a^3} = \frac{2a^4\dot{\theta}^2}{a^3} = 2a\dot{\theta}^2. \tag{169}$$

Next by differentiating (169) and by using (155),(157) we find

$$(a\dot{\theta}^2)^{\cdot} = \lambda\theta\left(\frac{\dot{u}}{a}\right), \tag{170}$$

as the optimality condition. Now we transform the system (148) as follows. First, we integrate $(148)_1$ to obtain

$$\left(\frac{\dot{u}}{a}\right) = \lambda\int_0^t \theta(\xi)d\xi, \tag{171}$$

where we used the fact that $\dot{u}(0) = 0$ and $a(0) \neq 0$ (see $(149)_2$ and (168)). Substituting (171) into (170), integrating and using (168), we get

$$\dot{\theta}a^2 = \lambda^2\int_0^t\left[\theta(\xi)\int_0^\xi \theta(\zeta)d\zeta\right]d\xi + \frac{1}{2}, \tag{172}$$

Since $\dot{y} = \theta$ (see $(148)_3$ the equation (172) that represents the condition of optimality, may be written as

$$a\ddot{y}^2 = \frac{1}{2}\left[1 + \lambda^2 y^2\right]. \tag{173}$$

Finally by differentiating $(148)_2$ and by using (171) we get

$$\left(\dot{\theta}a^2\right)^{\cdot\cdot} = (\ddot{y}a^2)^{\cdot\cdot} = \lambda\dot{u} = \lambda^2 a\int_0^t \theta(\xi)d\xi = \lambda^2 ay. \tag{174}$$

Therefore the optimal shape $a(t)$ of the rotating rod is determined from the solution of the system (173),(174)

$$(\ddot{y}a^2)^{\cdot\cdot} = \lambda^2 ay \qquad a\ddot{y}^2 = \frac{1}{2}\left[1 + \lambda^2 y^2\right], \tag{175}$$

subject to

$$y(0)=0;\qquad \dot y(0)=0;\qquad \lim_{t\to 1}\ddot y(t)a^2(t)=0;\qquad \lim_{t\to 1}\left\{\left[\ddot y(t)a^2(t)\right]^{\cdot}\right\}=0. \tag{176}$$

We analyze next system (175),(176).

Variational principles for the system (175),(176). In this section we shall formulate three different variational principles corresponding to the system (175),(176). Also we shall construct a conservation law (the first integral) corresponding to (175),(176).

The variational principle with two arguments. Let $\mathbf{W}_1$ be the linear function space defined as:

$$\mathbf{W}_1=\left\{\mathbf{w}=(y,a): y\in C^4(0,1); y(0)=\dot y(0)=0; a\in C^2(0,1); a\geq 0, a(1)=0\right\}. \tag{177}$$

Consider the functional

$$I_1(y,a)=\int_0^1 F_1\,dt, \tag{178}$$

with the Lagrangian function

$$F_1=a^2\ddot y^2-\lambda^2 a y^2-1. \tag{179}$$

Suppose further that we want to determine the minimum of I_1 on $\mathbf{W}_1$. We claim that I_1 is stationary on the solution of (175),(176). To prove this note that the condition of stationarity, i.e., vanishing of the first variation δI_1, leads to the following Euler-Lagrange equations

$$\left(\ddot y a^2\right)^{\cdot\cdot}=\lambda^2 a y;\qquad a\ddot y^2-\frac{\lambda^2}{2}y^2-\frac{1}{2}=0, \tag{180}$$

and natural boundary conditions

$$\ddot y(1)a^2(1)=0;\qquad \left\{\left[\ddot y(t)a^2(t)\right]^{\cdot}\right\}_{t=1}=0. \tag{181}$$

The system (180),(181) is equivalent to (175),(176).

The variational principle with one argument. We can write system (175) as a single differential equation of fourth order if we calculate a from $(175)_2$ and then substitute the result in $(175)_1$. Thus we obtain

$$\left[\frac{1}{\ddot y^3}\left(1+\lambda^2y^2\right)\right]^{\cdot\cdot}-2\lambda^2 y\frac{\left(1+\lambda^2y^2\right)}{\ddot y^2}=0, \tag{182}$$

subject to

$$y(0)=0;\qquad \dot y(0)=0,$$

$$\frac{1}{4\ddot y^3(1)}\left[1+\lambda^2y^2(1)\right]^2=0;\qquad \left(\left\{\frac{1}{4\ddot y^3(t)}\left[1+\lambda^2y^2(t)\right]^2\right\}^{\cdot}\right)_{t=1}=0. \tag{183}$$

Consider the space

$$\mathbf{W}_2=\left\{y: y\in C^4(0,1); y(0)=\dot y(0)=0\right\}, \tag{184}$$

and the functional

$$I_2 = \int_0^1 F_2 dt, \tag{185}$$

with Lagrangian function

$$F_2 = \frac{\left(1+\lambda^2 y^2\right)^2}{\ddot{y}^2}. \tag{186}$$

Then the Euler-Lagrange equation corresponding to $\delta I_2 = 0$ is equivalent to (182). Note that the natural boundary conditions for the minimization of I_2 on the set $\mathbf{W}_1$ are identical to (183).

The canonical formalism. The variational principle $\delta I_2 = 0$ can be used to write (175) in canonical form. We define a variable (a "momentum") p as

$$p = \frac{\partial F_2}{\partial \ddot{y}} = -2\frac{(1+\lambda^2 y^2)^2}{\ddot{y}^3}. \tag{187}$$

Then, the Hamiltonian function is

$$H_2 = p\ddot{y} - F_2 = -\frac{3}{2^{2/3}} p^{2/3} \left(1+\lambda^2 y^2\right)^{2/3}. \tag{188}$$

With (188) the canonical equations

$$\ddot{y} = \frac{\partial H_2}{\partial p}; \qquad \ddot{p} = \frac{\partial H_2}{\partial y}, \tag{189}$$

yield

$$\ddot{y} = -\left[2\frac{\left(1+\lambda^2 y^2\right)^2}{p}\right]^{1/3}; \quad \ddot{p} = -2^{4/3}\lambda^2 y \left[2\frac{p^2}{\left(1+\lambda^2 y^2\right)}\right]^{1/3}; \quad t \in (0,1). \tag{190}$$

From (2.3.2-176) and (187) we obtain the boundary conditions corresponding to the system (190) as

$$y(0) = 0; \qquad \dot{y}(0) = 0; \qquad p(1) = 0; \qquad \dot{p}(1) = 0. \tag{191}$$

Consider the space $\mathbf{W}_3$

$$\mathbf{W}_3 = \{\mathbf{w} = (y,p) : y \in C^2(0,1); y(0) = \dot{y}(0) = 0; \;, \quad p \in C^2(0,1); p(1) = 0, \dot{p}(1) = 0\} \tag{192}$$

and the problem of determining the minimum on $\mathbf{W}_3$ of the functional

$$I_3 = \int_0^1 F_3 dt, \tag{193}$$

with,

$$F_3 = \dot{y}\dot{p} - \frac{3}{2^{2/3}} p^{2/3} \left(1+\lambda^2 y^2\right)^{2/3}. \tag{194}$$

It is easy to see that the condition $\delta I_3 = 0$ reproduces the system (190). However, since F_3 does not depend explicitly on t we have Jacobi type of first integral for (190) in the form

$$\dot{y}\dot{p} + \frac{3}{2^{2/3}}p^{2/3}\left(1+\lambda^2 y^2\right)^{2/3} = const. \tag{195}$$

We determine now the constant in (195). By using the boundary conditions (191) it follows that

$$\dot{y}\dot{p} + \frac{3}{2^{2/3}}p^{2/3}\left(1+\lambda^2 y^2\right)^{2/3} = C = \frac{3}{2^{2/3}}\left[p(0)\right]^{2/3}. \tag{196}$$

We shall use (196) to check numerical integration of the system (190). Namely we shall, in each step of numerical integration, calculate the left hand side of (196) and compare so obtained value with the constant on the right hand side.

Finally we note that with (y,p) known, the cross–sectional area $a(t)$ is determined from the equation (173) and (187) so that

$$a(t) = \left[\frac{p^2}{2^5(1+\lambda^2 y^2)}\right]^{1/3}. \tag{197}$$

Numerical results. The system (190) is integrated by using the Runge–Kutta double precision procedure. Note that the point $t = 1$ is a singular point for the system (190) so that the equation $(190)_1$ can not be satisfied at $t = 1$. To avoid this singularity, we proceed as follows: we constructed a sequence of numerical solutions (y_n, p_n), $n = 1, 2, ...$ with $y_n(0) = \dot{y}_n(0) = 0$ and $y_n(t) > 0, \dot{y}_n(t) > 0, p_n(t) < 0, \dot{p}_n(t) > 0$ for $t \in (0,1)$ and $p_n(1) = -\varepsilon_n, \dot{p}(1) = \delta_n$ with the constants $\varepsilon_n > 0, \delta_n > 0$. For each solution (y_n, p_n) the values of variables $p_n(0)$ and $\dot{p}_n(0)$, and the cross–sectional area $a_n = a_n(\varepsilon_n, \delta_n)$ are determined (a_n is determined according to (197)). Then, the optimal cross–sectional area $a(t)$ and corresponding initial values $p(0)$ and $\dot{p}(0)$ are obtained as $\lim a_n$, $\lim p_n(0)$ and $\lim \dot{p}_n(0)$ when $\varepsilon_n \to 0, \delta_n \to 0$, respectively.

We performed the calculations for $\lambda = \sqrt{10}$ and obtained $p(0) = -0.195077$ and $\dot{p}(0) = 0.379104$. The accuracy of integration was controlled by evaluation of the first integral (196) in each step of integration. The left hand side in (196) was constant and equal to $\frac{3}{2^{2/3}}\left[0.195077^2\right]^{1/3} = 0.6356$ up the 10^{-6}. From (197) we obtain $a(0) = (p^2(0)/2^5)^{1/3} = 0.10595$. The dimensionless volume w of the optimal rod is determined by using (154) and (197). For $\lambda = \sqrt[2]{10}$ we obtain

$$w = \int_0^1 a(t)dt. = 0.03712. \tag{198}$$

We show next that with the solution of (190) for single λ we can determine the solution for any λ. Let (y,p) be the solution of the problem (190) and let $a(t)$ be corresponding cross–sectional area determined for the specified value of the dimensionless rotation λ. Let $(\hat{y},\hat{p})$ and $\hat{a}(t)$ be the corresponding functions determined for the dimensionless rotation speed $\Lambda = \beta\lambda$, with the constant β given. By using (190) and (197) it is easy to show that the following relations hold

$$\hat{y} = \frac{1}{\beta}y; \qquad \hat{p} = \beta^3 p; \qquad \hat{a} = \beta^2 a. \tag{199}$$

Thus, with the solution for single λ we have solutions for any Λ.

We compare now the volume of the optimal rod and the rod with constant circular cross–section if both are stable up to the same angular velocity ω. Suppose that both rods are made of the same material, i.e., E, ρ are the same. Also we assume that both rods loose stability at the same angular velocity ω. For the rod with the constant circular cross–section we have Atanackovic (1997)

$$\bar{\lambda} = \frac{\rho A \omega^2 L^4}{EI} = 12.362, \tag{200}$$

where $I = \alpha A^2$ with $A = const$. From (142) we find

$$\lambda^2 = \frac{\rho \omega^2 L^2}{E\alpha} = \sqrt{10}. \tag{201}$$

With (200),(201) we obtain

$$\frac{\lambda^2}{\hat{\lambda}} = \frac{A}{L^2}. \tag{202}$$

The rod with constant cross–section A has the volume $W_{const..} = AL$ while the volume of the optimal rod is $W_{optimal} = wL^3$. By using (201) we obtain

$$W_{const.} = \frac{\lambda^2}{\hat{\lambda}} L^3, \tag{203}$$

so that

$$\frac{W_{optimal}}{W_{const.}} = \frac{\hat{\lambda}}{\lambda^2} w = 0.04589. \tag{204}$$

Therefore, the volume of the optimal rotating rod is just 4.59% of the volume that has the rod with the same length and with constant cross–section! Note also that from (202) the dimensionless cross–sectional area of the rod with constant cross–section $a_{const.} = A/L^2$ becomes 0.809.

3 Conclusions

In these Lecture notes we treated the problem of determining stability boundary and post-critical behavior of elastic rods and three problems of determining optimal (against buckling) shape of elastic rods. Our main results may be stated as:

1. We showed that the axis compressibility and shear stresses influence the critical rotation speed given by the solution of equation (1-39). For the case of the Bernoulli-Euler rod we obtained an estimate of the maximal deflection in the post-critical state, given by (1-50).

2. For the case of a rod loaded by a compressive force and couple we obtained the critical buckling load in the form (1-111) confirming our previous result Atanackovic and Glavardanov (1996). By the suitable transformation of dependent variables, we reduced the initial system of equilibrium equations (eighteen of them) to a single second order nonlinear differential equation (1-102). The linearization of this equation has a null space of dimension one. In previous works (see Béda et al., 1992 and Atanackovic and Glavardanov, 1996) the equilibrium equations were reduced to two second order differential equations that lead to the null space of dimension two.

The bifurcation analysis was escorted on equation (1-102) and we determined the type of bifurcation for *all* values of parameters m, λ, α and β.

The optimization problem for the spatially deformed rod based on the equations derived here will be presented elsewhere Glavardanov and Atanackovic (2001).

3. Next we treated the problem of determining the optimal shape of a Pflüger column. We derived the nonlinear system of equilibrium equations describing equilibrium configuration of a Pflüger column with variable cross-section. We have shown that under the condition that the cross-section does not vanish at any point inside $(0, 1)$, the eigenvalues of the linearized system of equations correspond to the bifurcation points of the full nonlinear system of equilibrium equations. By using Pontryagin's maximum principle we have derived equation (2-27) that determines the cross-sectional area function providing minimum volume (mass) of the column for given load.

4. We have formulated complementary variational principles for equation (2-27). We have also demonstrated that these principles are extremal. They have been used to obtain an approximate solution and to estimate its error.

5. Next we treated the problem of determining the optimal shape of a compressed Pflüger column. We showed that the problem of determining the optimal shape reduces to solving a nonlinear boundary value problem (2-73) and (2-74). We proved the existence of a weak solution of this problem. Also dual complementary variational principles (2-102) and (2-110) for (2-73),(2-74) are formulated. These principles are used to obtain á priori estimates of the solution given by (2-119) and (2-123).

For one set of parameters λ_1 and λ_2 we solved, numerically, the system (2-73),(2-74). In this particular case, the volume of the optimal column is approximately 16% smaller than the volume of the column with constant cross−section.

6. Finally we treated the problem of determining optimal shape of the rotating rod. By using Pontryagin's maximum principle we obtained the relevant equations that determine the optimal shape, as well as variational principles and first integrals, corresponding to those equations.

References

Alekseev, V. M., Tihomirov, V. M. and Fomin, S. V. (1979). *Optimal Control*. Moscow: Nauka, (in Russian).

Antman, S. S., and Kenny, C. S. (1981). Large Buckled States of nonlinearly Elastic Rods under Torsion, Thrust and Gravity. *Archive for Rational Mechanics and Analysis* 76:289−338.

Antman, S. S., and Nachman, A. (1980). Large Buckled States of Rotating Rods. *Nonlinear Analalysis, Theory Applications* 4:303−327.

Antman, S. S., and Christoforis, M. L.(1997). Peculiar Instabilities due to the Clamping of Shearable rods. *International Journal of Non−Linear Mechanics* 32:31−54.

Antman, S. S. (1995). *Nonlinear Problems of Elasticity*. New York: Springer.

Arthurs, A. M. (1980). *Complementary Variational Principles*. Oxford: Clarendon Press.

Atanackovic, T. M., and Glavardanov, V. B.(1996). Twisted axially loaded rod with shear and compressibility. *Acta Mechanica* 119:119−130.

Atanackovic, T. M., and Glavardanov, V. B. (2002) Buckling of a twisted and compressed rod. *International Journal of Solids and Structures* (in press).

Atanackovic, T. M. (2001). On the optimal shape of a rotating rod. *Transactions of ASME Journal of Applied. Mechanics 68: 860-864.*

Atanackovic, T. M. and Simic, S. S. (1999). On the optimal shape of Pflüger column. *European. Journal of Mechanics, A/Solids* 18: 903-913.

Atanackovic, T. M. (1997). *Stability Theory of Elastic rods*. Singapore: World Scientific.

Atanackovic, T. M. (1987). Stability of rotating compressed rod with imperfections. *Mathematical. Procedings of Cambridge Philospical. Society* 101:593–607.

Atanackovic, T. M., Djukic Dj. S., and Jones, S. E. (1991). Effect of shear on the stability and nonlinear behavior of rotating rod. *Archive of. Applied. Mechanics* 61:285–294.

Atanackovic, T. M. (2001). On a Vujanovic–type variational priniciple with application to rod theory. *Quoterly Journal Mechanics and Applied Mathematics* 54:1–11.

Bazely, N., and Zwahlen, B. (1968). Remarks on the Bifurcation of Solutions of a Non-linear Eigenvalue problem. *Archive of Rational Mechanics and Analysis* 28:51–58.

Beck, M., Knickung gerader Stabe durch Druck und conservative Torsion. *Ingenieur Archiv* 23:231–255.

Béda, B. P., Steindl, A., and Troger, H., Postbuckling of a twisted prismatic rod under terminal thrust. *Dynamics and Stability of Systems* 7:219–232.

Biezeno, C. B., and Grammel, R. (1953). *Technische Dynamik*. Berlin: Springer.

Blasius, H. (1914). Träger kleinster Durchbiegung und Stäbe großter Knickfestigkeit bei gegebenem Materialverbrauch. *Zeitsch. Math. Physik* 62:182–197.

Brenner, J. L., and Alzer, H.(1991). Integral inequalities for concave functions with applications to special functions. *Proceedings of Royal Society of Edinburgh* 118A:173–192.

Chow, S-N., and Hale, J. K. (1982). *Methods of Bifurcation Theory*. New York: Springer.

Churchill, R. V. (1969). *Fourier Series and Boundary Value Problems*. New York: McGraw-Hill.

Clausen, T. (1851). Über die Form architektonischer Säulen. *Bull. cl, physico–math. Acad. St. Pétersbourg* 9:369–380.

Clément, Ph., and Descloux, J. (1991). A Variational Approach to a Problem of Rotating Rods. *Archive for Rational Mechanics and Analysis* 114:1–13.

Cox, S. J. (1992). The Shape of the Ideal Column. *The Mathematical Intelligencer,* 14:16–24.

Dacorogna, B., (1989). *Direct Methods in the Calculus of Variations*. Berlin: Springer.

Golubitsky, M., and Schaeffer, D. (1985). *Singularities and Groups in Bifurcation Theory*. vol. I, New York: Springer.

Glavardanov, V. G. and Atanackovic, T. M. (2001). Optimal Shape of a twisted and compressed rod. *European. Journal of Mechanics, A/Solids* 20:795–809.

Eliseyev, V. V. (1988). The non–linear dynamics of elastic rods. *Prikl. Mathem. Mekh.*, 52: 635–641.

Iljohin, A. A. (1979). *Spatial problems of Nonlinear theory of Elastic rods.* Kiev: Naukova Dumka (in Russian).

Karlin, S., and Ziegler, Z. (1975). Some Inequalities for Generalized Concave Functions. *Journal of Approximation Theory* 13:267–293.

Keller, J., The shape of the strongest column. *Archive for Rational Mechanics and Analysis* 5:275–285.

Keyfitz, B. L. (1986). Classification of one–state variable bifurcation problems up to codimension seven. *Dynamics and Stability of Systems* 1:1–41.

Kovari, K., Raumliche Verzweingungsprobleme des dunnen elastischen Stabes mit endlichen Verformungen. *Ing. Archiv* **37**, 393–416 (1969).

Oeuvres ds Lagrange, Serret, J. A. ed. (1867). Paris: Gauthier–Villars.

Libai, A. (1992). Equations for the Nonlinear Planar deformation of Beams. *Transactions of ASME Journal of Applied Mechanics* 59:1028–1030.

Lurie, *A. I. (1961). Analytical Mechanics.* Moscow: F.M. editors (in Russian).

Mitrinovic, D. S. (1970). *Analytic Inequalities. Berlin:* Springer.

Odeh, F., and Tadjbakhsh, I. (1965). A Nonlinear Eigenvalue Problem for Rotating Rods. *Archive of Rational Mechanics and Analysis* 20:81-94.

Olhoff, N., and Rasmussen, S. (1977). On single and bimodal optimum buckling loads of clamped columns. *International Journal of Solids and Structures* 13:605–614.

Parter, S. V. (1970). Nonlinear Eigenvalue Problems for Some Fourth Order Equations: I Maximal solutions, *SIAM Journal of Mathematical Analysis* 1:437–457, and II: Fixed–point methods 1:458–478.

Pflüger, A. (1975). *Stabilitätsprobleme der Elastostatik*. Berlin: Springer.

Prager, W. and Taylor, J. E. (1968). Problems of optimal structural design. *Journal of Applied Mechanics (transactions of ASME)* 35:102-106.

Rabier, P. (1985). *Topics in one-parameter bifurcation problems*. Berlin: Springer.

Rabinowitz, P. D. (1977). Bifurcation Theorem for Potential Operators. *Journal of Functional Analysis* 25:412–424.

Ratzersdorfer, J. (1936). *Die Knickfestigkeit von Stäben und Stabwerken*. Wien: Springer.

Sage, A. P. and White, C. C. (1977). *Optimum System Control*. New Jersey: Prentice-Hall, 2nd edition.

Stodola, A. (1906). *Steam Turbines*. New York: D. Van Nostrand Company.

Struwe, M. (1996). *Variational Methods*. New York: Springer, 2nd edition.

Seiranian, A. (1984). On the Problem of Lagrange. *Inzhenernyi Zh., Mekhanika Tverdoga Tela,* 19:101–111.

Seyranian, A. P., Lund, E. and Olhoff, N. (1994). Multiple eigenvalues in structural optimization problems. *Structural Optimization* 8: 207–227.

Seiranyan, A. P.(1995). New Solutions to Lagrange's Problem. *Physics–Doklady* 40: 251–253.

Tadjbakhsh, I., and Keller, J.(1962). Strongest columns and isoperimetric inequalities for eigenvalues. *Transactions of ASME Journal of Applied. Mechanics* 29:159–164.

Timoshenko, S. P., and Gere, J. M.(1961). *Theory of Elastic Stability*. New York: McGraw-Hill.

Troger, H. and Steindl, A. (1991). *A Nonlinear Stability and Bifurcation Theory*. Wien: Springer.

Troesch, B. A. (1967). Integral inequalities for two functions. *Archive of Rational Mechanics and Analysis* 24:412–424.

Vujanovic, B., and Jones, S. E. (1989). *Variational Methods in Nonconservative Phenomena*. Boston: Academic Press.

Part II. The Odd Stability of the Euler Beam

Gábor Domokos

Department of Strength of Materials,
Budapest University of Technology and Economics,
H-1521 Budapest, Hungary. email: domokos@iit.bme.hu

Abstract

In this part of the lecture notes we investigate a classical chapter of elastic stability from a new perspective. We show that even the first buckling mode in Euler's problem becomes unstable at a second critical load parameter $P_L = 2.183P_{Euler}$ and the only stable solution for $P > P_L$ is the straight bar in tension. We also show that if we discretize the beam into a sequence of n rigid links coupled by linear torsional springs, the described stability behaviour is only reflected correctly if n is an odd number.

1 Introduction

In this part we study the most classical stability problem in structural mechanics: the buckling of the simply supported, slender, planar, elastic beam (Euler's "linea elastica"). Although the problem was solved by Euler (1744), careful analysis still reveals surprising new features. It is generally accepted that the buckling modes have no secondary bifurcations, i.e. the primary sinusoidal modes are the only non-trivial branches of the equilibrium path. An essentially complete proof of this statement was provided by Euler (1744).

However, the proof considers only a restricted version of the problem where vertical reaction forces are neglected. Our aim is to demonstrate that considering all globally possible equilibrium shapes there exists a secondary bifurcation point in each buckling mode. This highly singular bifurcation, already noted by Maddocks (1984), occurs when the hinge and roller supports coincide and the structure becomes statically overdeterminate and kinematically indeterminate. Although the bifurcation corresponds physically to a rigid body rotation and appears to be rather singular, all investigated functions are smooth and thus the results of classical bifurcation theory can be applied to predict stability. We will show that at these points unstable branches bifurcate off the primary ones, and the imperfect structure displays a limit point instability. Our model is far from being an artificial generalization, it is a tool to explain the physical behaviour of a wide class of imperfect problems. Since stable equilibrium has to exist globally, it is an interesting question, to what shape the imperfect structure will snap beyond the limit point: the only remaining solution is the straight bar in tension, and only the extended problem with vertical reaction forces can explain this behaviour.

The discovery of the secondary, unstable bifurcation raises the question of the global connectivity of the bifurcation diagram. We will use a *topologically correct* embedding enabling us to identify the connectivity and show that in the extended problem the straight bar in tension and in compression can be continuously connected via equilibria.

The analogy between the differential equation of the Euler rod and that of the mathematical pendulum was first recognized by Kirchhoff (1859). Since the natural (physical) formulation of the pendulum motion is an initial value problem (IVP), whereas that of the bar shape is a boundary value problem (BVP), there is an obvious discrepancy in this analogy (cf. Domokos and Holmes (1993)). Since only a subset of all possible IVP-s satisfies boundary conditions, moreover the BVP is defined for a finite segment of the independent variable of this subset of IVP-s, one is tempted to believe that the discrepancy just mentioned means that the BVP can display only a subset of the behaviour observable in the IVP. However, the secondary bifurcations studied here are only possible in the BVP and therefore it becomes apparent that the analogy is incomplete both ways in the following manner: there is no *single* IVP to which *all* solutions of a given BVP belong, and vice versa, there is no *single* BVP to which *all* solutions of a given IVP belong.

In general one expects *sufficiently dense* discretizations to display qualitatively the same behaviour as the original structure. We will show that in our problem this is not the case: the discrete system consisting of n rigid links coupled by torsional springs shows a similar stability behaviour if n is odd. This *odd* behaviour can be explained if we investigate the $P \to \infty$ limit of the problem. We introduce the fundamental concepts in section 2, describe the secondary bifurcation in section 3. Section 4 is devoted to stability of the continuous problem: the behaviour of the perfect and imperfect problem will be discussed separately (subsections 4.1 and 4.2). Section 5 discusses the stability properties of the discrete problem. In section 6 we draw conclusions and summarize the results.

2 The fundamental concepts

2.1 Formulation of the equations

The differential equation of the bar is formulated on the basis of the static equilibrium of the differential bar element and the material law. Fig.1. illustrates the bar element in the global $[xy]$ frame.

The x and y components of the internal force are denoted by $H(s)$ and $V(s)$, respectively, $M(s)$ denotes the bending moment, $\alpha(s)$ the slope and s the (inextensional) arc length. The conditions of static equilibrium can be formulated as

$$\begin{aligned} \dot{H} &= 0 \\ \dot{V} &= 0 \\ \dot{M} &= \dot{y}H - \dot{x}V \end{aligned} \tag{1}$$

where $()^{\cdot}$ denotes differentiation with respect to the arc length. Since both the axial and the shear stiffnesses are infinite and the bar is assumed to be relaxed with straight axis, the material law takes the simple form

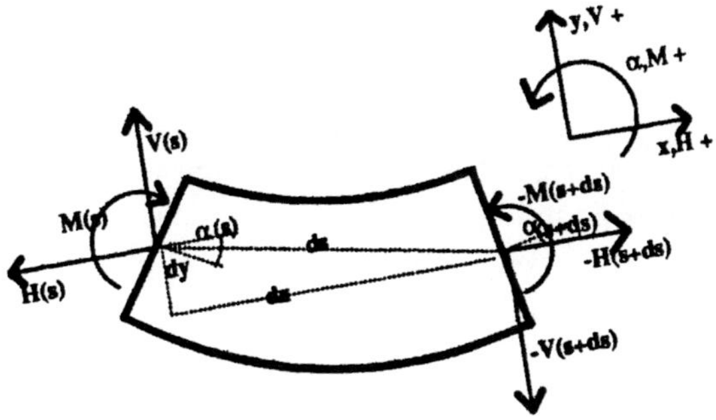

Figure 1: The bar element

$$\begin{aligned} \dot{x} &= \cos(\alpha(s)) \\ \dot{y} &= \sin(\alpha(s)) \\ \dot{\alpha} &= -M(s)/EI \end{aligned} \tag{2}$$

where EI denotes the bending stiffness of the bar. Substituting the material law into the equilibrium conditions yields

$$\ddot{\alpha}EI + H(0)\sin\alpha - V(0)\cos\alpha = 0 \tag{3}$$

The differential equation system (1)-(2) can be discretized using a semi-implicit Euler scheme, which is equivalent to a system of rigid links coupled by linear torsional springs. The equilibrium is described via the recursive discrete scheme

$$\begin{aligned} x_i &= x_{i-1} + l\cos(\alpha_{i-1}) \\ y_i &= y_{i-1} + l\sin(\alpha_{i-1}) \\ \alpha_i &= \alpha_{i-1} - \frac{Hy + Vx_i}{c}, \end{aligned} \tag{4}$$

where c denotes the stiffness of the torsional springs between the rigid elements.

2.2 The continuous IVP and the Kirchhoff analogy

The definition of a (Cauchy-type) IVP for (3) requires the fixing of the scalars $\dot{\alpha}(0), \alpha(0), H(0)$ and $V(0)$. Let P_0 be the vector with components $H(0), V(0)$. By rotating the global frame by $\arctan(V(0)/H(0))$, (3) reduces to

$$\ddot{\alpha}EI + P_0\sin\alpha = 0, \tag{5}$$

which analogous to the pendulum equation

$$\alpha'' mh + mg\sin\alpha = 0 \tag{6}$$

where $()'$ denotes differentation with respect to time, g denotes gravitational acceleration, m the mass of the pendulum and h the length of the rod. Apparently, equations (5) and (6) differ only in the physical interpretation of the variables. This remarkable analogy was first pointed out by Kirchhoff (1859) and has led to several interesting discoveries. Nevertheless, the analogy has to be treated with great caution: the natural formulation for the pendulum is an initial value problem (IVP), while the bar can be only understood physically as a boundary value problem (BVP), as we will discuss in the next section. We remark that (5) can be also written in the first-order form

$$\begin{aligned} \dot{y} &= \sin(\alpha) \\ \dot{\alpha} &= -P_0 y/EI \end{aligned} \tag{7}$$

and the analogous, reduced version of the discrete scheme is

$$\begin{aligned} y_i &= y_{i-1} + l \sin(\alpha_{i-1}) \\ \alpha_i &= \alpha_{i-1} - \frac{P_0 y_i}{c}. \end{aligned} \tag{8}$$

3 The perfect BVP and the corresponding bifurcations

The BVP to be considered is a bar pin-jointed at both ends, one end ($s = 0$) fixed at the origin of the $[xy]$ system, the other end ($s = L$) rolling freely on the x axis. The discretized version consists of n rigid links of length $l = 1/n$ coupled by linear torsional springs with stiffness $c = EI/l = nEI$. Throughout the paper $EI = 1$ will be assumed, so we have $c = n = 1/l$. The compressive load P is applied at $s = L$; at $s = 0$ we have the horizontal and vertical reaction forces $H(0)$ and $V(0)$, respectively. Both structures are illustrated in figure 2.

The boundary conditions for the continuous case can be formulated as

$$y(0) = 0 \qquad\qquad y(L) = 0, \tag{9}$$

and for the discrete case as

$$y_0 = 0 \qquad\qquad y_n = 0. \tag{10}$$

Why is it not possible now to get rid of the parameter $V(0)$, as we did it in the IVP? Because in the case of the IVP the rotation of the global frame was specified for each single direction of the initial force vector P_0. In our case this could be done only if we could guarantee that *for all IVP satisfying the boundary conditions (9) the ratio* $(V(0)/H(0))$ *does not change.* By making this assumption (i.e. V(0)/H(0)=constant=0) we could prove that no secondary bifurcations occur. This 'reduced', classical case is illustrated in figure 3, showing the $[\alpha, y]$ phase space of equation 7, the physical shapes and (half of) the bifurcation diagram. This problem has been completely solved by Euler (1744).

The discrete case is much more complicated. Even the BVP corresponding to the reduced equation 8 produces a vastly complex bifurcation diagram (cf. Domokos and Holmes (1993)). In this case the phase portrait is very complex, here we only illustrate the $n = 4$ case with part of the bifurcation diagram and some physical shapes, see figure 4.

Both of the illustrated bifurcation diagrams are valid only if we assume $V(0) = 0$. However, the assumption $V(0) = 0$ holds for only a subset of IVP-s satisfying the boundary conditions. In particular, if $x(0) = x(L) = 0$ (or, in the discrete case $x_0 = x_n = 0$), then the structure

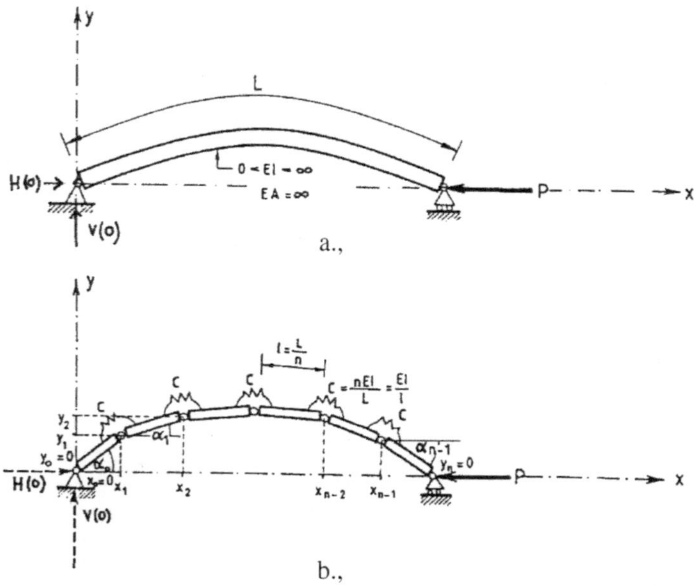

Figure 2: The perfect BVP: a) the continuous, b) the discretized beam.

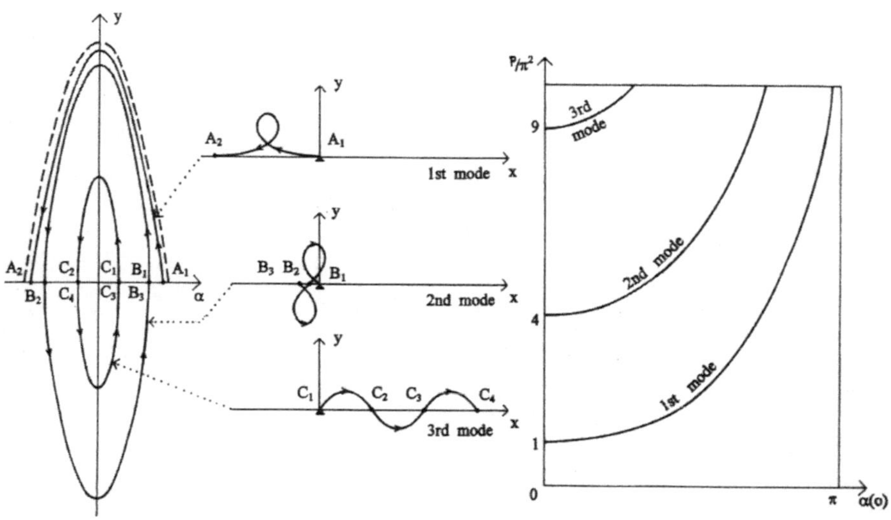

Figure 3: Phase portrait, physical shapes and bifurcation diagram for the reduced ( ), continuous BVP

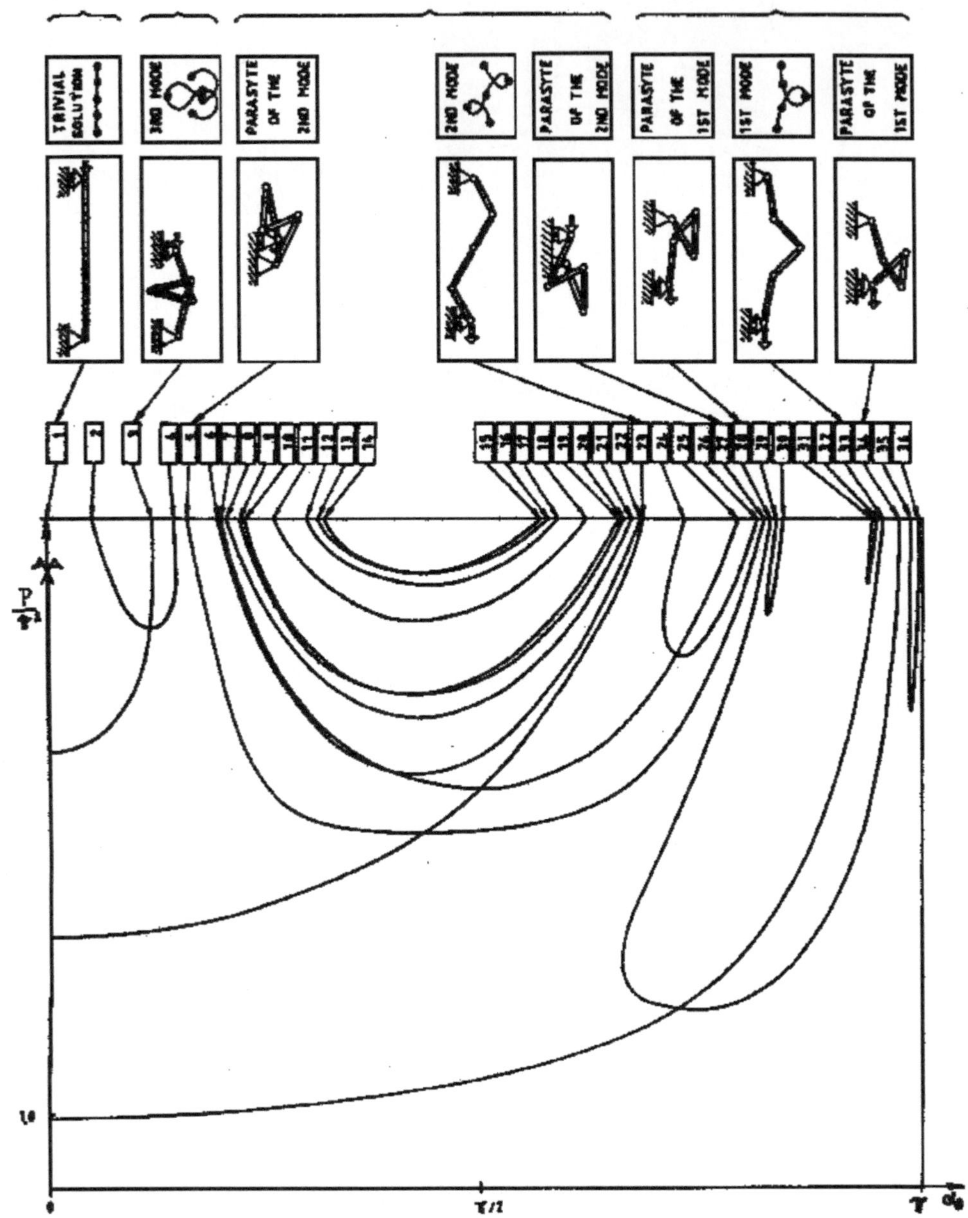

Figure 4: Bifurcation diagram and some physical shapes for the reduced ($V(0) = 0$), discrete BVP with $n = 4$.

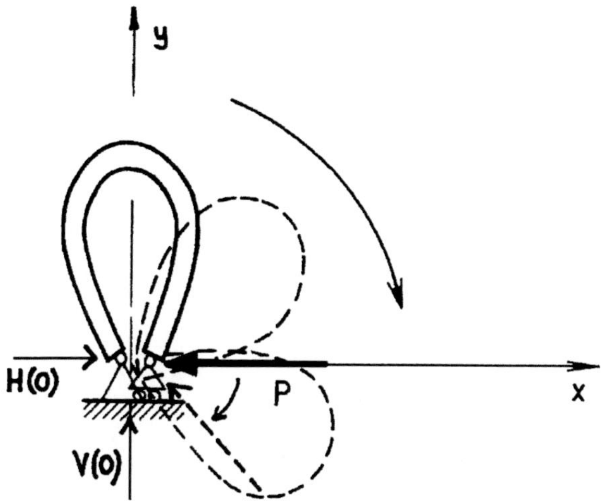

Figure 5: Rigid body rotations at $x(L) = 0$

becomes kinematically indeterminate and can perform rigid body rotations around the origin. This situation is illustrated in figure 5 for the continuous beam; the discrete structure behaves analogously.

During this rotation the magnitude of the force P_0 acting at $s = 0$ remains constant ($P_0 = \sqrt{H(0)^2 + V(0)^2}$) but both $H(0)$ and $V(0)$ are changing continuously. This implies that the magnitude of the load $P = H(0)$ is changing during the rotation, as well. Our next aim is to fit this behaviour into the global picture of the equilibrium path (bifurcation diagram). We are primarily interested in the *global connectivity* of this diagram, i.e. we need a *topologically correct* picture. Since solutions depend uniquely on the initial conditions, this topologically correct global picture may be plotted in the space of the initial conditions and parameters not fixed by the boundary conditions defined at the origin.

In the continuous case the complete set of initial conditions and parameters of (3) is (as already mentioned): $\dot{\alpha}(0), \alpha(0), H(0), V(0)$. Since the boundary conditions (9) fix $\dot{\alpha}(0)$, three of them remain free. The global equilibrium path is plotted in this three-dimensional space in figure 6. Figure 7 helps to identify the physical shape of the bar along the equilibrium path.

As indicated before, the spirals corresponding to the rotating loops move on a cylinder with axis $H(0) = V(0) = 0$ and radius $r = P_L$, where P_L is the load parameter corresponding to $x(L) = 0$ on the primary branch. The value of P_L has been computed numerically as $P_L = 2.183 P_{Euler}$. Figure 6 shows only the first mode and the corresponding spirals. For higher modes the same picture repeats at larger scale. In the discrete case the scenario is similar, all 'primary' branches (bifurcating off the trivial solution) will exhibit similar behaviour.

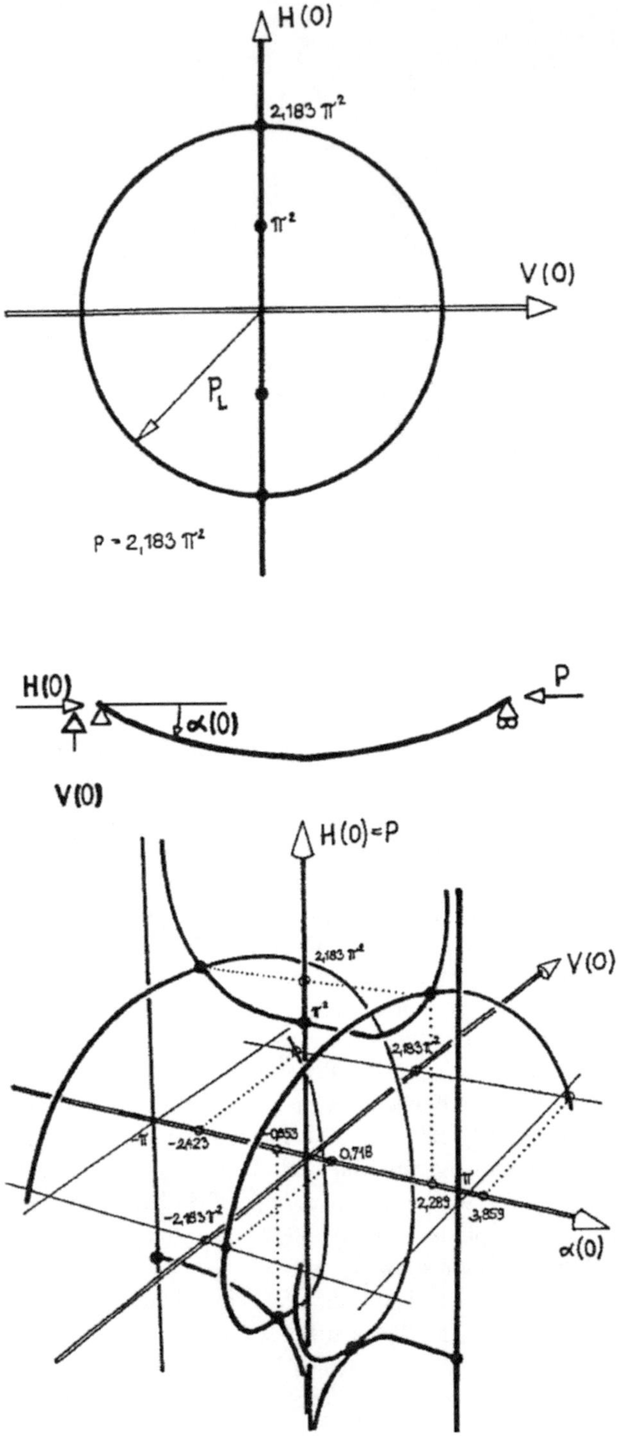

Figure 6: Global behaviour of the continuous, perfect BVP

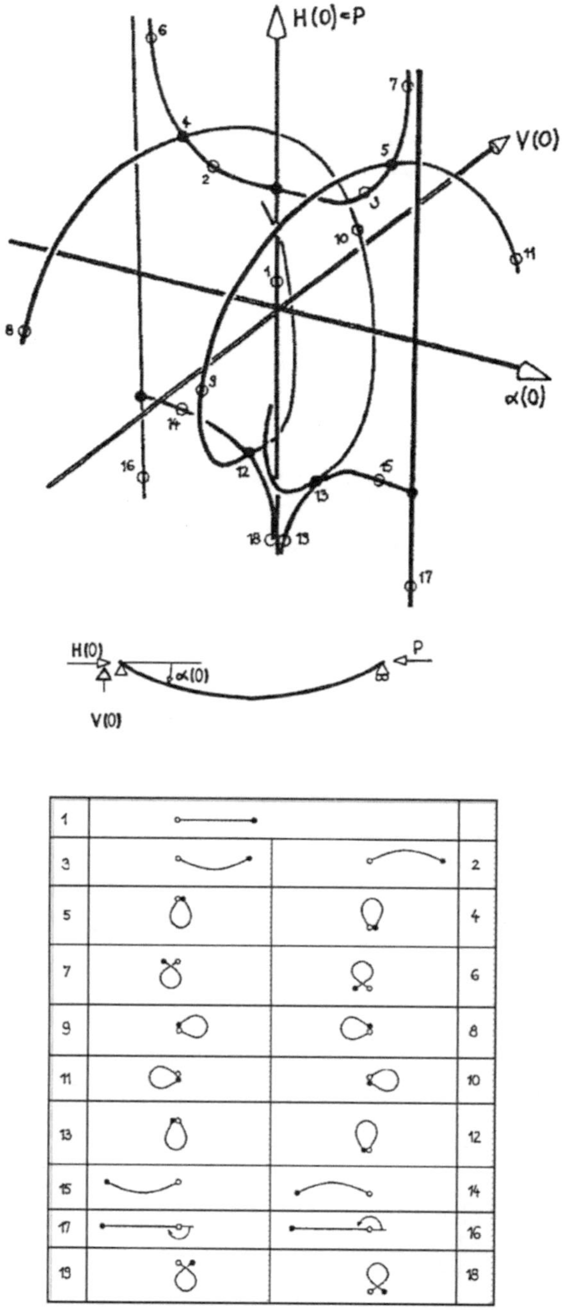

Figure 7: Identification of physical shapes on the bifurcation diagram of the continuous, perfect BVP

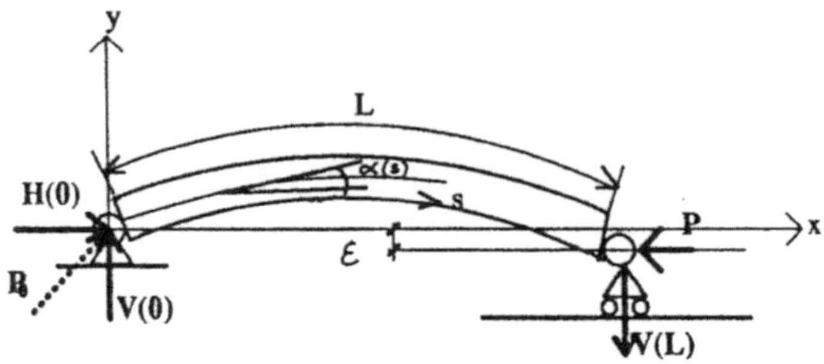

Figure 8: Geometry of the imperfect BVP

4 Stability of the first mode in the continuous BVP

4.1 Stability of the continuous, perfect BVP

As figure 6 shows, the first buckling mode exhibits a secondary bifurcation point at $P = P_L$. Elementary bifurcation theory tells us that at that point the stability of the equilibrium has to change. Since the first mode was stable at its origin at $P = P_{Euler}$, *the first mode branch becomes unstable beyond* $P = P_L$. The remaining question is the stability of the secondary 'spiral' branch. Since the load $P = H(0)$ is decreasing along the secondary spiral branch after the bifurcation point, the engineer's intuition tells that this branch has to be unstable. This intuition proves to be correct, since we have evidence for the instability of the secondary branch: Take an arbitrary point on this branch. By rotating the loop by an infinitesimal angle and keeping the load constant we do not change the total potential energy, since rigid body rotation does not involve elastic deformation. The new position is a non-equilibrium position, infinitesimally close to the equilibrium ('variation'), and on the same energy level. Thus the equilibrium position cannot be an extremal point of the energy, i.e. it cannot be stable.

4.2 Stability of the imperfect, continuous BVP

The interesting stability results related to the perfect structure can be easily verified on the imperfect system. As an imperfection parameter we choose a vertical offset of the rolling support by the amount ϵ, so the boundary conditions (9) are modified as

$$y(0) = 0 \qquad\qquad y(L) = -\epsilon \,. \tag{11}$$

The imperfect structure is illustrated in figure 8.

In the imperfect problem there is no 'trivial' solution: the equilibrium path intersects the $[\alpha(0), V(0)]$ plane at $[\epsilon/L, 0]$, but it is tilted in the $V(0)$ direction. (For any $P \neq 0$ $V(0) \neq 0$.) By moving the rolling bar-end along the x axis the end load P necessary to maintain equilibrium

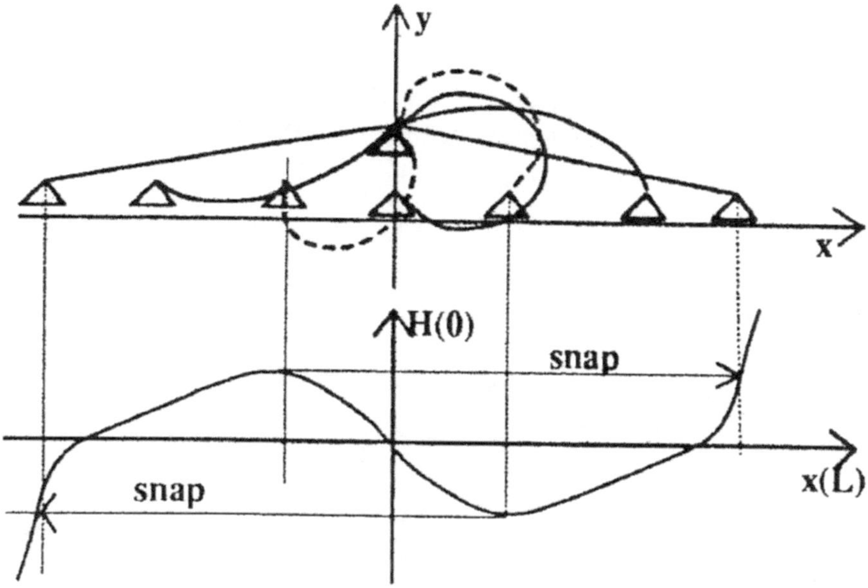

Figure 9: Global behaviour of the imperfect BVP

initially increases, but there has to be a maximum, since at $x(L) = 0$ the load P has to be zero, as well. This situation is illustrated in figure 9.

The existence of the limit point supports our previous conclusion, i.e. the secondary branch is unstable. We remark, that the loading history illustrated in figure 9 can be easily veryfied experimentally. Since unstable equilibria can hardly be realized in an experiment, after reaching the limit point, the structure will snap through to the opposite side of the origin. This means, that the globally stable equilibrium is a straight bar in tension. The equilibrium path of the imperfect system is illustrated in figure 10.

5 Stability of the discrete BVP

As we saw in figure 4, the bifurcation diagram of the discrete problem is very complex. Here we do not attempt to describe the stability of all solutions (cf. Károlyi and Domokos (1999)), we are merely interested in the stability of the first mode primary branch. Even this task poses a considerable difficulty, since we can see in figure 4 that there can be secondary bifurcations off the first mode primary branch *even in the restricted,* $V(0) = 0$ *problem.*Following the ideas of Domokos (1997), we will investigate the stability by looking at the $P/EI \to \infty$ limit of the bifurcation diagram. Physically, this limit can be either realized by increasing P to infinity, which is hard to imagine, or, by decreasing EI to zero while holding P constant. The latter limit

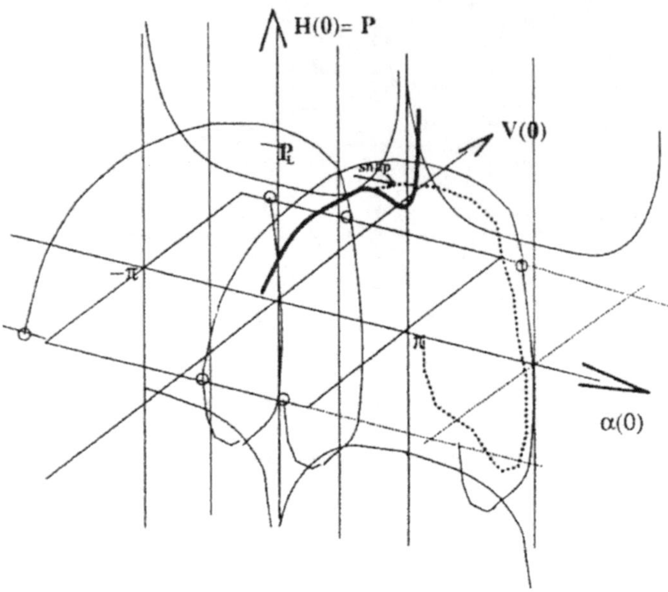

Figure 10: Bifurcation diagram for the imperfect BVP

is familiar, since it corresponds physically to an *inelastic chain coupled by frictionless hinges.* If we apply an axial force to such a chain, the only equilibria are those where all links are lined up with the applied force. Since each link can be either in tension or in compression independently of each other, there are 2^n possible limit configurations: the $n = 4$ case is illustrated in figure 11, where we used the new parameter $u = arctg(P/EI)$ to visualize the limiting behaviour.

The 2^n limiting configurations can be identified by the *binary labels* $B_n = \{b_1, b_2, \dots b_n\}$, which are binary strings with n digits: the kth digit is 0 if the kth link is in compression and it is 1 if the link is in tension. The stability of these 16 limiting configurations is easily understood: only the chain in tension can be stable, so we have 15 unstable and 1 stable configuration, the latter has the binary label $B_4 = \{1111\}$. (To each of these configurations infinitely many branches of the discrete BVP will converge.) The stability can be also understood via the number E of negative eigenvalues of the Hessian matrix associated with the total potential energy; each configuration has $E = n - \sum_{i=1}^{n} b_i$ negative eigenvalues. In case of the stable solution we have $b_i = 1$, thus $\sum_{i=1}^{n} b_i = n$ and $E = 0$. Now we have to identify the limiting configuration corresponding to the first mode primary branch. The global behaviour of the $n = 4$ and $n = 5$ primary first mode branches is shown in figure 12.

As we can observe, in the case of $n = 4$, after passing the secondary bifurcation corresponding to rigid body rotation the branch becomes unstable ($E = 1$). Since at $P = \infty$ we have the binary label $B_4 = \{1111\}$, the limiting configuration is stable, hence there has to be at least one secondary bifurcation after the rigid body rotation.

The $n = 5$ case is radically different, since here the limiting configuration corresponds to the

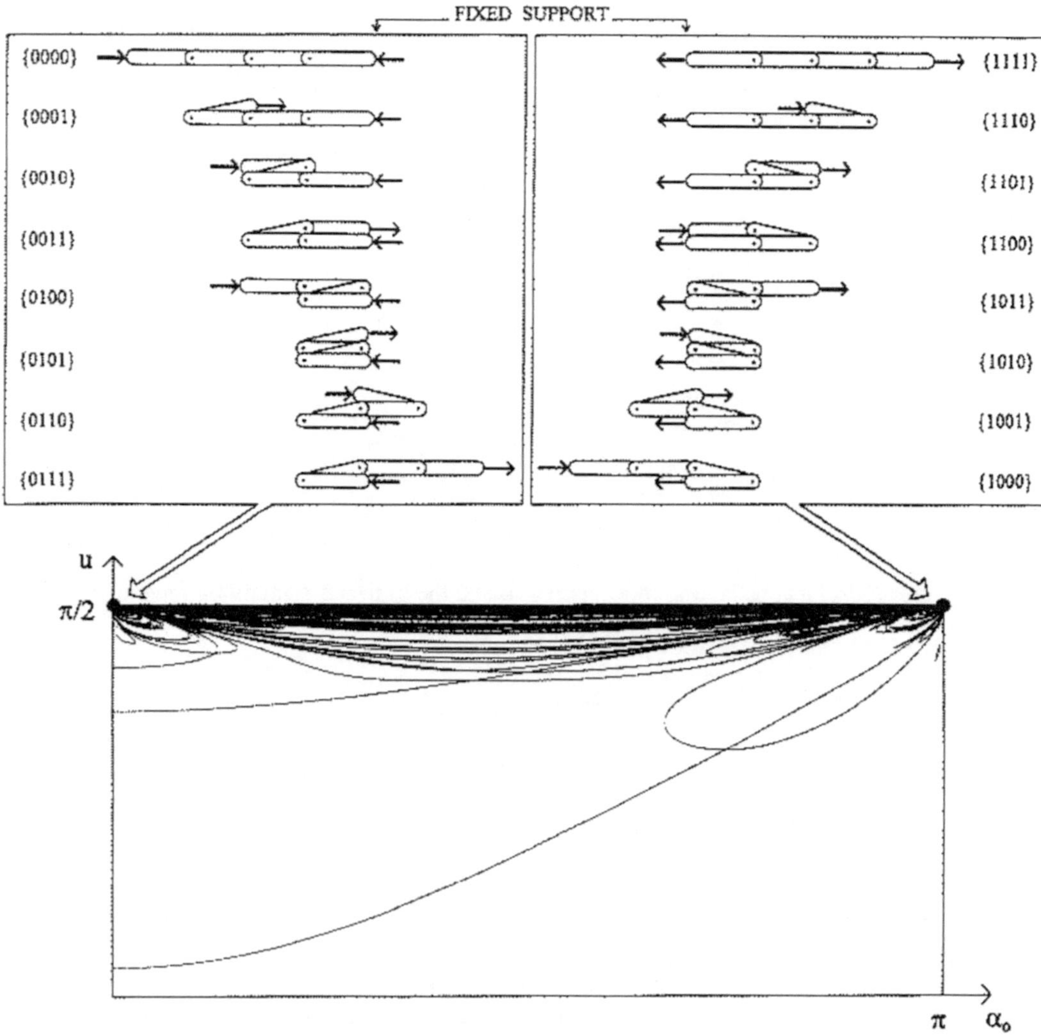

Figure 11: The $P/EI \to \infty$ limit for the $n = 4$ discrete problem: $P/EI = \tan(u)$, $2^4 = 16$ limiting configurations shown on top.

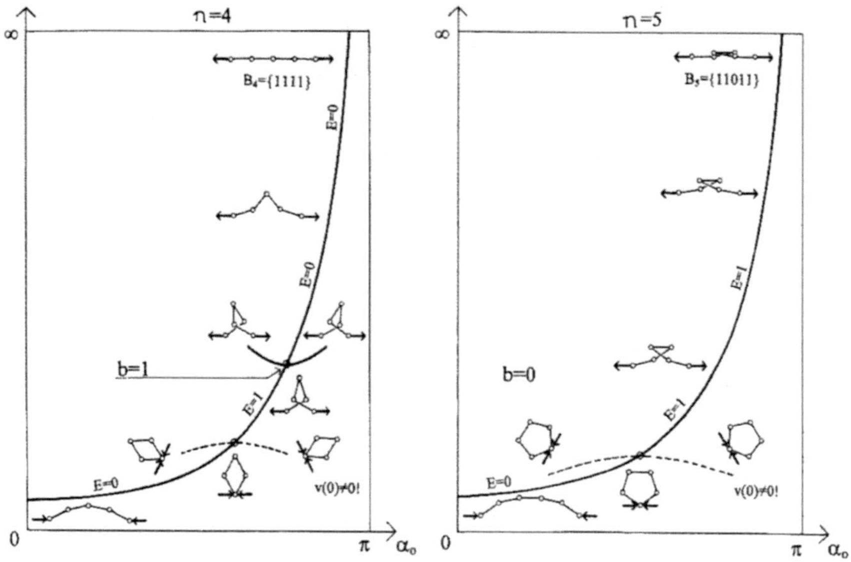

Figure 12: Global behaviour of a) the $n = 4$ and the b) $n = 5$ primary first mode branches.

label $B_5 = \{11011\}$ and thus it is unstable. Hence, no further secondary bifurcation is needed to explain the stability.

As we can see, in case of *even* values of n, the stability of the discrete system at large P is different from the stability of the continuous system, *even for arbitrarily high* n. If n is *odd,* then the primary first mode branch behaves globally exactly as in the continuous model.

6 Summary

In this paper we investigated the classical buckling problem first solved by Euler. We showed that by considering the possibility of vertical reaction forces a new type of bifurcation at $P = P_L$ can be discovered. This bifurcation, corresponding to the rigid body-like rotation of the beam, changes the stability of the first mode branch, which is the only stable solution of Euler's problem. By simple inspection of the imperfect problem we showed that the only globally stable solution is the bar in tension and all solutions beyond $P = P_L$ snap to this configuration. The global connectivity of the bifurcation diagram revealed that the two trivial solutions (straight bar in tension and in compression) are continuously connected via equilibria: this is *not* the case in the classical, restricted Euler problem. We investigated the imperfect structure with slightly offset supports, and the hysteresis in the equilibria of this physical problem can only be explained by looking at the extended problem.

Investigation of the discretized problem with n rigid links revealed that one can expect to model the stability characteristics of the continuous case correctly only for *odd* values of n, no matter how large n is. This paradox behaviour indicates that there are differences between arbitrarily fine discretizations and the continuous structure, so finite element models should be applied with great caution.

Acknowledgements: This work was supported by OTKA grant T 031744, FKFP grant 0177/2001 and the János Bolyai Research Fellowship.

References

Domokos, G., and Holmes, P. (1993). Euler's problem and Euler's method, or the discrete charm of buckling. *J. Nonlin. Sci.* 3:109–151.

Domokos, G. (1997). Static solitary waves as limits of discretization. *Phil. Trans. R.Soc. London* 355:2099–2116.

Euler, L. (1744). Additamentum I de curvis elasticis, methodus inveniendi lineas curvas maximi minimivi proprietate gaudentes. In *Opera Omnia I, vol. 24*. Lausanne: Bousquet. 231–297.

Károlyi, G., and Domokos, G. (1999). Symbolic dynamics of infinite depth. *Physica D* 134:316–336.

Kirchhoff, G. (1859). Üeber das Gleichgewicht und die Bewegung eines unendlich dünnen elastischen Stabes. *J.f.Reine angew. Math (Crelle)* 56:285–313.

Maddocks, J. (1984). Stability of nonlinearly elastic rods. *Arch. Rat. Mech. Anal.* 85:311–354.

Part III. Some Unconventional Elastic Stability Problems

Isaac Elishakoff[1]

Florida Atlantic University,
Boca Raton, FL 33431-0991, USA
e-mail: ielsihak@me.fau.edu

Abstract: In this paper new closed – form solutions are obtained for instability of inhomogeneous columns. The problem is posed as the semi-inverse eigenvalue determination task ; namely the mode shape is formulated as the fourth order polynomial that satisfies all boundary conditions. Then the following question is posed: Find such variation of the flexural rigidity along the axial coordinate, that corresponds, in exact terms, to the postulated mode shape. Whereas usually the polynomial mode shapes are utilized in the approximate methods of Rayleigh, Rayleigh – Ritz or Boobnov – Galerkin, here the closed form solution is derived. Moreover, even for the columns with uniform cross-section, the closed-form solutions are obtained in terms of irrational expressions for the buckling load, here rational expressions are obtained. These semi-inverse problems can be used as benchmark solutions. In addition, when the technology will become available of producing the arbitrarily varying flexural rigidity, the derived solutions can be utilized for design purposes. One will be able to construct columns with pre-selected buckling loads. In the second part the effect of boundary conditions is investigated. It is intriguing that for the inhomogeneous column in question a linear relationship is established between the natural frequency squared and the applied load, exactly as it happens for the simply – supported uniform column. These results show that even the classical problems may present rich opportunities for deriving unconventional solutions.

[1] Acknowledgement: Discussions with Professor Oren Masory of the Florida Atlantic University, Professor Arnold Kerr of the University of Delaware, Mr. Zakoua Guédé of IFMA, France, Professor Teodor M. Atanackovic of the University of Novi Sad, Yugoslavia and Professor Alexander P. Seyranian of Insitute of Mechanics, Moscow, Russia are gratefully acknowledged. This work has been supported by the National Science Foundation by the grant 99 – 10195 (Program Director Dr. K. P. Chong). Opinions, findings and conclusions expressed are those by the writer and do not necessarily reflect the views of the sponsor.

1 Introduction

The exact solutions for buckling load of uniform beams are treated in almost any textbook on mechanics of solids. Exact solutions for non-uniform columns is the subject of several works (see, e.g. the textbook by Timoshenko and Gere, 1961). These solutions are derived in terms of Bessel or Lommel functions, or some other, elementary or transcendental functions. As far as the closed – form solutions are concerned the results are much more restricted. All closed – form solutions obtained between the years 1744 to 1999 are apparently listed, along with some new solutions, in the study by Elishakoff and Rollot (1999). New closed form solutions include four recent papers by this writer (Elishakoff, 2000, 2001a, 2001b, 2002).

This study is devoted to obtaining novel closed – form solutions, which are posed as semi-inverse problems. Namely, the formulation of the problem is as follows : Find the polynomial distribution of the Young modulus *E(x)* of an inhomogeneous column of the uniform cross – section with specified boundary conditions, so that the buckling mode will be a pre – selected polynomial function. It turns out that, this seemingly simple formulation allows to derive new unconventional closed – form solutions. The obtained solutions appear to be of much importance, since once the technologies will be available to construct columns with given variation of modulus of elasticity, any pre – selected buckling load can be achieved for the appropriate design of the structure.

2 Formulation of the Problem

The differential equation that governs the buckling of the column under axial load *P*, reads

$$\frac{d^2}{dx^2}\left[D(x)\frac{d^2w}{dx^2}\right]+P\frac{d^2w}{dx^2}=0 \qquad (1)$$

where $w(x)$ = transverse displacement, $D(x)=E(x)I(x)$ is the stiffness, $E(x)$ = modulus of elasticity, $I(x)$ = moment of inertia, x = axial coordinate. We consider four sets of boundary conditions. For the column that is simply supported at its both ends the simplest polynomial that satisfies the boundary conditions at $x = 0$ and $x=L$, with L = length of the column, reads

$$\psi(\xi)=\xi-2\xi^3+\xi^4, \qquad \xi=x/L \qquad (2)$$

We pose the following question : Is there a column with polynomial variation of $E(\xi)$ that possesses the function in Eq. (2) as its fundamental buckling mode ? Indeed, if the sought solution exists, it corresponds to the fundamental buckling load, since $\psi(\xi)$ in Eq. (2) does not have internal nodes (results can also be verified by the use of numerical techniques such as FDM or FEM). This problem differs from the *direct* buckling problem, which presupposes the knowledge of the stiffness D and requires the determination of the mode $w(x)$ and the buckling load P. Here we are looking for the *cause*, i.e. the distribution of the stiffness, knowing the *effect*, the buckling mode.

We are looking for the stiffness $D(\xi)$ represented as follows :

$$D(\xi) = b_0 + b_1\xi + b_2\xi^2 \tag{3}$$

where b_0, b_1 and b_2 are sought constants. The inverse problems may have no solution, multiple solutions or unique solution. It turns out that in this case under study, unique solution exists for reconstructing the column, that possesses the function in Eq. (2) as its buckling mode. We consider now the different sets of boundary conditions.

3 Column Simply Supported at Its Both Ends

Utilization of the non-dimensional axial coordinate ξ, defined in Eq. (2) reduces the governing Eq. (1) to

$$\frac{d^2}{d\xi^2}\left[D(\xi)\frac{d^2\psi}{d\xi^2}\right] + PL^2\frac{d^2\psi}{d\xi^2} = 0 \tag{4}$$

With the buckling mode postulated in Eq. (2) we have for the term in Eq. (4)

$$PL^2\frac{d^2\psi}{d\xi^2} = PL^2\left(-12\xi + 12\xi^2\right) \tag{5}$$

whereas the first differential expression in Eq. (4) reads

$$\frac{d^2}{d\xi^2}\left[D(\xi)\frac{d^2\psi}{d\xi^2}\right] = -12\left[2(b_1 - b_0) + 6\xi(b_2 - b_1) - 12b_2\xi^2\right] \tag{6}$$

The sum of expression in the right – hand sides in Eqs. (5) and (6) must vanish, due to Eq. (4). Since the above sum must equal to zero identically, for any value of ξ, we get following expressions :

$$2(b_1 - b_0) = 0 \tag{7}$$

$$-72(b_2 - b_1) - 12PL^2 = 0 \tag{8}$$

$$144b_2 + 12PL^2 = 0 \tag{9}$$

Solution of Eq. (9) yields

$$P = -12\frac{b_2}{L^2} \tag{10}$$

In order the load P to remain compressive, it must be positive. We conclude, therefore, that the coefficient b_2 must be negative. The Equations (7) and (8) lead, with Eq. (10) taken into account, to :

$$b_0 = b_1 = -b_2 \tag{11}$$

Thus, the stiffness is defined up to the coefficient b_2 :

$$D(\xi) = \left(-1 - \xi + \xi^2\right)b_2 \tag{12}$$

We have already established that b_2 must take an negative value. Hence Eq. (12) can be rewritten as

$$D(\xi) = \left(1 + \xi - \xi^2\right)|b_2| \tag{13}$$

This function is in agreement with physical realizability condition, namely, with the requirement of positivity of the function $D(\xi)$ in the interval [0 ;1]. We thus have found the function $D(\xi)$ that corresponds to the postulated buckling mode in Eq. (2). As is seen, the solution of the posed problem is an unique one in the class of polynomially varying stiffnesses. Note that Eq. (10) coincides with Eq. (56) in the paper by Elishakoff and Rollot (1999). It pertains to the column that is simply supported at its both ends. We will show that Eq. (10) is valid for two other sets of boundary conditions.

4 Column Clamped at Its Both Ends

The boundary conditions

$$w = 0, \quad \frac{dw}{d\xi} = 0, \quad at \quad \xi = 0, \quad \xi = 1 \tag{14}$$

are satisfied for the following polynomial function :

$$\psi(\xi) = \xi^2 - 2\xi^3 + \xi^4 \tag{15}$$

We are interested in establishing if this polynomial function may serve as a buckling shape of any inhomogeneous column. The expression for $PL^2\psi''$ reads, with primes denoting differention with respect to ξ:

$$PL^2\psi'' = PL^2\left(2 - 12\xi + 12\xi^2\right) \tag{16}$$

Whereas the expression for $(D\psi'')''$ is

$$(D\psi'')'' = 2(2b_2 - 12b_1 + 12b_0) + 6(-12b_2 + 12b_1)\xi + 144b_2\xi^2 \tag{17}$$

We demand the sum of the expressions in Eq. (16) and (17) to vanish for any ξ. This requirement leads to the following three equations :

$$2(2b_2 - 12b_1 + 12b_0) + 2PL^2 = 0 \tag{18}$$

$$6(-12b_2 + 12b_1) - 12PL^2 = 0 \tag{19}$$

$$144b_2 + 12PL^2 = 0 \tag{20}$$

From the latter equation (20) we get the buckling load

$$P = -12\frac{b_2}{L^2} \tag{21}$$

which, remarkably, coincides with Eq. (10). Eqs. (18) and (19) lead to

$$b_0 = -b_2/6, \quad b_1 = -b_2 \tag{22}$$

Hence we obtain the sought variation of the stiffness

$$D(\xi) = \left(\frac{1}{6} + \xi - \xi^2\right)|b_2| \tag{23}$$

which takes a positive value throughout the column's axis, $\xi \in [0 ;1]$.

5 Column Simply Supported at One End and Clamped at the Other

The boundary conditions reads

$$w = 0, \quad D(\xi)\frac{d^2 w}{d\xi^2} = 0, \quad at \quad \xi = 0,$$
$$w = 0, \quad \frac{dw}{d\xi} = 0, \quad at \quad \xi = 1. \tag{24}$$

The boundary conditions are satisfied by the following polynomial function:

$$\psi(\xi) = \xi - 3\xi^3 + 2\xi^4 \tag{25}$$

Substitution of this expression into the governing differential equation, in conjunction with postulated expression for the stiffness results in

$$\left[2(-18b_1 + 24b_0) + 6(-18b_2 + 24b_1)\xi + 288b_2\xi^2\right] + PL^2\left(-18\xi + 24\xi^2\right) = 0 \tag{26}$$

Since Eq. (26) is valid for any ξ, we get following three equations :

$$-18b_1 + 24b_0 = 0\,, \text{ for } \xi^0 \tag{27}$$

$$6(-18b_2 + 24b_1) - 18PL^2 = 0\,, \text{ for } \xi^1 \tag{28}$$

$$288b_2 + 24PL^2 = 0\,, \text{ for } \xi^2 \tag{29}$$

We arrive at three equations for four unknowns : b_0, b_1, b_2 and P. We choose one of the parameters to be arbitrary, namely b_2. The Eq. (29) yields same the buckling load as in Eq. (10) and (21),

$$P = -12\frac{b_2}{L^2} \tag{30}$$

Eqs. (27) – (29) yield the following interrelation between the coefficients describing the stiffness variation :

$$b_0 = -\frac{9}{16}b_2, \quad b_1 = -\frac{3}{4}b_2 \tag{31}$$

Substituting into Eq. (3) results in the variation of the stiffness :

$$D(\xi) = \left(\frac{9}{16} + \frac{3}{4}\xi - \xi^2\right)|b_2| \tag{32}$$

which is a positive function within the length of the column.

6 Some Pertinent Comments on Present Closed – Form Solution

The following conclusions appear to be relevant :

(a) Inverse buckling problems with postulated polynomial buckling modes, as given in Eq. (2), Eq. (15) or Eq. (25) for corresponding boundary conditions have closed – form solutions ; namely, the variations of stiffness corresponding to above mode shapes are given in equations (13), (23) and (32), respectively.

(b) For three sets of boundary conditions, the fundamental buckling load is given by the same expression (see Eqs. (10), (21) and (30)). This conclusion may appear to be a paradoxial one at the first glance. To resolve it, let us consider the corresponding case of homogeneous and uniform columns. The fundamental buckling loads are :

$$P_{S-S} = \frac{\pi^2 D_1}{L^2}, \quad P_{C-C} = \frac{4\pi^2 D_2}{L^2}, \quad P_{S-C} = \frac{4.493^2 D_3}{L^2} \approx \frac{2\pi^2 D_3}{L^2} \tag{33}$$

where D_1 is the stiffness of the column that is simply supported at its both ends ; D_2 corresponds to the column that is clamped at its both ends ; D_3 is associated with the column that is simply supported at $x = 0$ and clamped $x = L$. These three columns possess the same cross – sections and have the same lengths. Now, if

$$D_1 = 4D_2 \approx 2D_3 \tag{34}$$

then all three columns have coalescing buckling loads. This implies that *only* the columns with different stiffnesses, if they are under different boundary conditions, but have the same lengths and cross – sections, *may* share the same fundamental buckling load. The same phenomenon takes place in case of our study : The columns with three different sets of boundary conditions may share the same fundamental buckling load, since their moduli of elasticity are different.

Still, it appears to be intriguing that the search of the solution of inverse buckling problem in the class of polynomial functions leads to the coincidence of buckling loads for simply supported – simply supported, clamped – clamped and simply supported – clamped boundary conditions. Here we reported the case when the fundamental buckling loads are shared by columns under different boundary conditions (see also Gottlieb, 1989, and Gladwell and Morassi, 1995, for vibration problems).

(c) To compare the results for the buckling loads, let us calculate the average stiffness in each of three cases. Average stiffness is defined as :

$$D_{av} = \int_0^1 D(\xi)d\xi \tag{35}$$

Thus, for simply supported column

$$D_{av,S-S} = \frac{7}{6}|b_2| \tag{36}$$

For clamped column

$$D_{av,C-C} = \frac{1}{3}|b_2| \tag{37}$$

For column that is simply supported at one end and clamped at the other

$$D_{av,S-C} = \frac{29}{48}|b_2| \tag{38}$$

Thus the buckling loads can be put in the following forms, by first expressing $|b_2|$ via D_{av} in Eq. (36)-(38) :

$$P_{S-S} = \frac{12|b_2|}{L^2} = \frac{72}{7}\frac{D_{av}}{L^2} \tag{39}$$

$$P_{C-C} = \frac{12|b_2|}{L^2} = 36\frac{D_{av}}{L^2} \tag{40}$$

$$P_{S-C} = \frac{12|b_2|}{L^2} = \frac{576}{29}\frac{D_{av}}{L^2} \tag{41}$$

If the average stiffnesses of these columns are chosen to be the same, then the buckling loads of the inhomogeneous columns are in the proportion

$$\frac{72}{7} : 36 : \frac{576}{29} \tag{42}$$

or 1 : 3.5 : 1.93, versus the corresponding proportion 1 : 4 : (≈ 2) for the uniform columns.

For uniform columns the polynomial expressions of the buckling mode are usually utilized to facilitate the approximate solutions, via the Bubnov – Galerkin or Rayleigh – Ritz methods. For example, Chajes (1974) uses the function $w(x) = xL^3 - 3x^3L + 2x^4$, as a *comparison function* in the Bubnov – Galerkin method for the column that is simply supported at $x = 0$ and clamped at $x = L$. It is interesting that the same polynomial function (coincident with Eq. (25)) turned out to be an *exact* buckling mode of the *inhomogeneous* column.

(d) Likewise, in his book of problems, Volmir (1984) poses a question of using the Rayleigh's method as well as the Bubnov – Galerkin method for the *approximate* estimation of the buckling load of an uniform column, simply supported at its both ends, by utilizing the trial function $w(x) = xL^3 - 2x^3L + x^4$. In our case the same buckling mode (coincident with Eq. (2)) serves as an *exact* expression of the inhomogeneous column.

(e) The question arises on the generality of the proposed method. In order the method to be acceptable it should lead to the positive buckling load, corresponding to the case of a compressive force, and modulus of elasticity that is a positive function. The formulation of some general conditions upon fulfillment of which the problem is amenable to a solution is of interest.

(f) The buckling loads turn out to depend upon only a single coefficient b_2, once the function of modulus variation is obtained. By suitable choice of b_2, the buckling load can be made arbitrary large. This conclusion is valid within the context of elastic buckling that was presupposed in this study.

(g) The solution of the buckling problems of uniform columns leads to irrational values of the buckling loads ; for example, the buckling load of the simply supported *uniform* column is written in terms of π^2 (see Eq. (33)). So is the buckling load, found by Euler (1759) in the case of the variable stiffness column, where the buckling load is $\pi^2(D_0/L^2)a^2(a+b)^2$ for the column with the stiffness $D(\xi)=D_0(a+b\xi)^4$. Here, in the case of *inhomogeneous* and/or *nonuniform* column, the solution turns out to be in terms of rational numbers. Other solutions of this kind are reported by Elishakoff and Rollot (1999). It must be stressed that this conclusion is true for special cases herein discussed, and it may be inapplicable for other cases.

(h) In this study we assume $D(\xi)$ to be the polynomial which fits the solution. The natural question arises : Can we use another functions such as *sin* or *cos*, or some other special functions instead of polynomials ? It appears that this question ought to investigated on the future studies.

7 Can A Fifth Order Polinomial Serve as Both Buckling and Vibration Mode of an Inhomogeneous Structure ?

Vibration of homogeneous uniform beams in presence of axial loads is a classical topic included in almost any vibration textbook. For the beam that is simply supported at both ends the natural frequency squared turns out to be a linear function of the applied concentrated load. In non-uniform beams or in beams that are not simply supported at both ends there is no reason to expect the above linear relationship to hold in exact terms. In present study, surprisingly, such a phenomenon of the linear relationship is established.

It appears to be instructive to list both the vibration and buckling modes of uniform beams under different boundary conditions. The vibration mode shape is denoted by $\psi_V(\xi)$, whereas the buckling mode is designated by $\psi_B(\xi)$, where ξ is a non-dimensional axial coordinate $\xi=x/L$, L being the length of the beam, x is the axial coordinate. For the beam that is simply supported at its both ends

$$\psi_V(\xi)=\psi_B(\xi)=\sin(m\pi\xi) \tag{43}$$

where m is a positive integer, denoting the serial number of the mode shape. For the beam that is clamped at its both ends

$$\begin{aligned}\psi_V(\xi)&=\sinh(\lambda_m\xi)-\sin(\lambda_m\xi)+\alpha_m[\cosh(\lambda_m\xi)-\cos(\lambda_m\xi)]\\ \alpha_m&=\frac{\sinh\lambda_m-\sin\lambda_m}{\cos\lambda_m-\cosh\lambda_m}\end{aligned} \tag{44}$$

where λ_m that satisfies the transcendental equation

$$\cos\lambda_m\cosh\lambda_m-1=0 \tag{45}$$

with roots found numerically; for example, $\lambda_1=4.73$, $\lambda_2=7.853$, $\lambda_3=10.996$, *etc*. The mode buckling shape reads

$$\psi_B(\xi)=\cos(2m\pi\xi)-1 \tag{46}$$

For the cantilever beam

$$\psi_V(\xi) = \sin(\lambda_m \xi) - \sinh(\lambda_m \xi) - \beta_m[\cos(\lambda_m \xi) - \cosh(\lambda_m \xi)]$$
$$\beta_m = \frac{\sinh\lambda_m + \sin\lambda_m}{\cos\lambda_m + \cosh\lambda_m} \tag{47}$$

where λ_m satisfies the following transcendental equation

$$\cos\lambda_m \cosh\lambda_m - 1 = 0 \tag{48}$$

with $\lambda_1 = 1.875104$, $\lambda_2 = 4.694091$, $\lambda_3 = 7.854757$, *etc*. The buckling mode reads

$$\psi_B(\xi) = \cos[(2m-1)\pi\xi] - 1 \tag{49}$$

For the beam that is clamped at left end and simply supported at the right end,

$$\psi_V(\xi) = \sinh(\lambda_m \xi) - \sin(\lambda_m \xi) + \gamma_m[\cosh(\lambda_m \xi) - \cos(\lambda_m \xi)]$$
$$\gamma_m = \frac{\sin\lambda_m - \sinh\lambda_m}{\cos\lambda_m - \cosh\lambda_m} \tag{50}$$

where λ_m satisfies the following transcendental equation

$$\tan\lambda_m = \tanh\lambda_m \tag{51}$$

with $\lambda_1 = 3.927$, $\lambda_2 = 7.069$, $\lambda_3 = 10.210$, *etc*. The buckling mode reads

$$\psi_B(\xi) = \sin(\delta_m \xi) - \delta_m \xi + \delta_m[1 - \cos(\delta_m \xi)] \tag{52}$$

where δ_m satisfies the equation

$$\tan\delta_m = \delta_m \tag{53}$$

As seen, only in the case of the uniform beam that is simply supported at both ends the vibration and buckling modes coincide, both representing a sinusoidal function.

For the inhomogeneous beams, at first glance, it may appear that there could be no compelling reason for the vibration and buckling mode shapes to coincide. Yet, we will demonstrate that the initial intuition may be wrong in some special circumstance. We postulate *ab initio* that these mode shapes coalesce. We construct beams that have this interesting property. Remarkably it turns out that such a formulation leads to a non-trivial solution, that will be reported as follows.

8 Problem Formulation

Let us first solve a simple auxiliary problem of bending of homogeneous beams. Consider first a uniform beam, simply supported at its both ends. It is subjected to the load $q(\xi) = q_0 \xi^n$. The governing differential equation reads:

$$EI\, d^4W/dx^4 = q_0 x^n \tag{54}$$

We introduce the non-dimensional coordinate

$$\xi = x/L \tag{55}$$

where L is the length. The equation (12) becomes

$$d^4W/d\xi^4 = \alpha\xi^n, \quad \alpha = q_0L^{n+4}/EI \tag{56}$$

The solution W_n of this equation that satisfies the boundary conditions reads:

$$W_n(\xi) = \frac{\alpha L^{n+4}}{(n+1)(n+2)(n+3)(n+4)}\psi(\xi)$$
$$\psi_n(\xi) = \xi^{n+4} - \frac{1}{6}(n^2+7n+12)\xi^3 + \frac{1}{6}(n^2+7n+6)\xi \tag{57}$$

For the beam that is clamped at one end and free at the other the solution of Eq. (56) subject to appropriate boundary conditions reads

$$W_n(\xi) = \frac{\alpha}{(n+1)(n+2)(n+3)(n+4)}\psi(\xi)$$
$$\psi_n(\xi) = \xi^{n+4} - \frac{1}{6}(n+2)(n+3)(n+4)\xi^3 + \frac{1}{2}(n+1)(n+3)(n+4)\xi^2 \tag{58}$$

For the beam that is clamped at its both ends, we get

$$W_n(\xi) = \frac{\alpha}{(n+1)(n+2)(n+3)(n+4)}\psi(\xi)$$
$$\psi_n(\xi) = \xi^{n+4} - (n+2)\xi^3 + (n+1)\xi^2 \tag{59}$$

Finally, for the beam that is simply supported at one end and clamped at the other, the functions $w(\xi)$ and $\psi(\xi)$ are, respectively,

$$W_n(\xi) = \frac{\alpha}{2(n+1)(n+2)(n+3)(n+4)}\psi(\xi)$$
$$\psi_n(\xi) = 2\xi^{n+4} - (n+3)\xi^3 + (n+1)\xi \tag{60}$$

We address the following problem: For a inhomogeneous vibrating beam under axially distributed loading whose exact mode shape (both in vibration and in buckling settings) is $\psi(\xi)$ when n =1, reconstruct the axial rigidity of the beam. Thus we postulate the fifth order polynomial to be the exact mode shape of a sought inhomogeneous beam. We will explore cases of reconstruction of the stiffness $D(\xi)$ for various axial variations of the inertial coefficient $R(\xi) = \rho(\xi)A$ and the distributed axial load $q(\xi)$.

9 Basic equations

The differential equation of the inhomogeneous vibrating column which is under distributed axial load reads:

$$\frac{d^2}{dx^2}\left[D(x)\frac{d^2W}{dx^2}\right] - \frac{d}{dx}\left[N(x)\frac{dW}{dx}\right] - R(x)\omega^2W(x) = 0 \tag{61}$$

where *D(x)* = stiffness, *N(x)* = axial force, $R(\xi)$ is the inertial coefficient, ω = natural frequency, *W(x)* = buckling mode, x = axial coordinate. With non-dimensional coordinate $\xi = x/L$, Eq.(61) becomes :

$$\frac{d^2}{d\xi^2}\left[D(\xi)\frac{d^2\psi}{d\xi^2}\right] - L^2\frac{d}{d\xi}\left[N(\xi)\frac{d\psi}{d\xi}\right] - \rho(\xi)A\omega^2\psi(\xi) = 0 \tag{62}$$

We consider for the inertial coefficient *R(ξ)* a *m*th order polynomial given by

$$R(\xi) = \sum_{i=0}^{m} r_i \xi^i \tag{63}$$

Considering the degree of the remaining member of the governing equation, the stiffness *D(ξ)* is taken equal to:

$$D(\xi) = \sum_{i=0}^{m+4} b_i \xi^i \tag{64}$$

and the distributed axial load reads

$$q(\xi) = q_0 \sum_{i=0}^{m+1} g_i \xi^i \tag{65}$$

where

$$g_i = q_i / q_0, \quad (i = 1,2,...,m), \quad g_0 = 1 \tag{66}$$

Since the axial load equals

$$N(\xi) = -\int_x^L q(t)dt = -L\int_\xi^1 q(t)dt, \tag{67}$$

we have

$$N(\xi) - q_0 L \sum_{i=0}^{m+2} \frac{g_i}{i+1}\left(1-\xi^{i+1}\right) \tag{68}$$

Therefore Eq. (62) becomes

$$\frac{d^2}{d\xi^2}\left[D(\xi)\frac{d^2\psi}{d\xi^2}\right] + q_0 L^3 \frac{d}{d\xi}\left[\sum_{i=0}^{m+2}\frac{g_i}{i+1}\left(1-\xi^{i+1}\right)\frac{d\psi}{d\xi}\right] - R(\xi)\omega^2\psi(\xi) = 0 \tag{69}$$

Upon substitution of the polynomial expressions of the modes shapes (57)-(60) into the differential equation (69), we get a polynomial equation of order $m + 5$ in ξ with zero as its right member. We demand, therefore, every coefficient of ξ raised to any power to be zero, we get, thus, a set of linear algebraic of $m + 6$ equations with $m + 7$ unknowns, namely, the coefficients b_i in the stiffness, the natural frequency squared Ω^2 and the axial load q_0 that is treated as a parameter.

10 Closed-Form Solution for the Simply Supported Beam

10.1 Uniform Inertial Coefficient (m = 0)

Substituting the mode shape $\psi(\xi)$ from Eq. (57) into the governing differential equation (69) we get a system of 6 linear algebraic equations with 7 unknowns, namely the coefficients b_i in the stiffness, the natural frequency squared Ω^2 and the axial load q_0 that is treated as a parameter

$$\begin{aligned}
&-40b_1 - \tfrac{7}{3}q_0L^3 = 0\\
&-120b_2 + 120b_0 - q_0L^3\left(20 + \tfrac{37}{3}g_1\right) - \tfrac{7}{3}r_0\Omega^2 = 0\\
&-240b_3 + 240b_1 + 30q_0L^3 = 0\\
&-400b_4 + 400b_2 + q_0L^3\left(20 + 30g_1\right) + \tfrac{10}{3}r_0\Omega^2 = 0\\
&600b_3 - 25q_0L^3 = 0\\
&840b_4 - 15g_1q_0L^3 - r_0\Omega^2 = 0
\end{aligned} \tag{70}$$

We have to write all unknowns in terms of a single free parameter, say, b_4 and the axial load parameter q_0. In other word, the rank of the set (70) must be less than 6. A matrix is of rank p if it contains minors of order p different from 0, while all minors of order p + 1 (if there are such) are zero (Uspensky, 1948). Accordingly, the rank of the matrix of the set (70) is less than 6 if all minors of order 6 vanish. Yet the minor obtained by extraction of the sixth column of the matrix of the set (70) equals $-5806080000000L^3$, which is non-zero. We conclude that the problem has no non-trivial solution.

10.2 Linearly Varying Inertial Coefficient (m = 1)

Substituting the mode shape $\psi(\xi)$ from Eq. (46) into the governing differential equation (58) we get a homogeneous set of 7 linear algebraic equations with 8 unknowns, namely the coefficients b_i in the stiffness, the natural frequency squared Ω^2 and the axial load q_0 that is treated as a parameter

$$\begin{aligned}
&-40b_1 - \tfrac{7}{3}q_0L^3 = 0\\
&-120b_2 + 120b_0 - q_0L^3\left(20 + \tfrac{37}{3}g_1 + \tfrac{20}{3}g_2\right) - \tfrac{7}{3}r_0\Omega^2 = 0\\
&-240b_3 + 240b_1 + q_0L^3\left(30 - \tfrac{7}{3}g_2\right) - \tfrac{7}{3}r_1\Omega^2 = 0\\
&-400b_4 + 400b_2 + q_0L^3\left(20 + 30g_1 + \tfrac{20}{3}g_2\right) + \tfrac{10}{3}r_0\Omega^2 = 0\\
&-600b_5 + 600b_3 - q_0L^3\left(25 - \tfrac{50}{3}g_2\right) + \tfrac{10}{3}r_1\Omega^2 = 0\\
&840b_4 - 15g_1q_0L^3 - r_0\Omega^2 = 0\\
&1120b_5 - \tfrac{35}{3}q_0L^3g_2 - r_1\Omega^2 = 0
\end{aligned} \tag{71}$$

We have to write all unknowns in terms of a single free parameter, say, b_5 and the axial load parameter q_0. In other word, the rank of the set (71) or its matrix must be less than 7. The rank of the matrix of the set (71) is less than 7 if all minors of order 7 vanish. We produce the minors, extracting one column of the matrix among 8; therefore we obtain a set of 8 determinantal equations for 4 unknowns, namely the parameters r_0, r_1, g_1 and g_2:

$$\begin{aligned}
&2444288 r_0 g_2 + 7999488 r_0 + 6005760 r_1 g_1 + 15353856 r_1 + 5117952 r_1 g_2 = 0\\
&r_1 = 0\\
&1419264 r_0 g_2 + 4644864 r_0 + 7520256 r_1 g_1 + 6580224 r_1 + 2193408 r_1 g_2 = 0\\
&3211264 r_1 g_2 - 2451456 r_1 = 0\\
&236544 r_0 g_2 + 774144 r_0 + 2350080 r_1 g_1 = 0\\
&1548288 r_1 g_2 + 580608 r_1 = 0\\
&19869696 g_2 + 650228096 = 0\\
&r_1 = 0
\end{aligned} \tag{72}$$

The solution of the set (72) reads

$$r_1 = 0, \quad g_2 = -{}^{36}/_{11} \tag{73}$$

The solution (73a) does not agree with the linearly varying mass density. We conclude that in this case the problem has no non-trivial solution.

10.3 Parabolically varying Inertial Coefficient (*m* = 2)

The governing differential equation (71) yields a set of 8 equations with 8 unknowns, namely the coefficients b_i in the stiffness, the natural frequency squared Ω^2 and the axial load q_0 that is treated as a parameter

$$\begin{aligned}
&-40 b_1 - {}^{7}/_{3}\, q_0 L^3 = 0\\
&-120 b_2 + 120 b_0 - q_0 L^3 \left(20 + {}^{37}/_{3}\, g_1 + {}^{20}/_{3}\, g_2 + 5 g_3\right) - {}^{7}/_{3}\, r_0 \Omega^2 = 0\\
&-240 b_3 + 240 b_1 + q_0 L^3 \left(30 - {}^{7}/_{3}\, g_2\right) - {}^{7}/_{3}\, r_1 \Omega^2 = 0\\
&-400 b_4 + 400 b_2 + q_0 L^3 \left(20 + 30 g_1 + {}^{20}/_{3}\, g_2 + {}^{8}/_{3}\, g_3\right) + \left({}^{10}/_{3}\, r_0 - {}^{7}/_{3}\, r_2\right)\Omega^2 = 0\\
&-600 b_5 + 600 b_3 - q_0 L^3 \left(25 - {}^{50}/_{3}\, g_2\right) + {}^{10}/_{3}\, r_1 \Omega^2 = 0\\
&-840 b_6 + 840 b_4 - 15 q_0 L^3 \left(g_1 - g_3\right) - \left(r_0 - {}^{10}/_{3}\, r_2\right)\Omega^2 = 0\\
&1120 b_5 - {}^{35}/_{3}\, q_0 L^3 g_2 - r_1 \Omega^2 = 0\\
&1440 b_6 - 10 q_0 L^3 g_3 - r_2 \Omega^2 = 0
\end{aligned} \tag{74}$$

In order to have a non-trivial solution, we have to write all unknowns in terms of a single free parameter, say, b_6 and the axial load parameter q_0. Consider the matrix of the set (74), where the first 7 columns represent the coefficients of the unknowns b_i while the remaining ones constitute the coefficients of the natural frequency squared Ω^2 and the axial load q_0, respectively. The matrix of the set (74) reads:

$$\begin{bmatrix} 0 & -40 & 0 & 0 & 0 & 0 & 0 & 0 & -\frac{7}{3}L^3 \\ 120 & 0 & -120 & 0 & 0 & 0 & 0 & -\frac{7}{3}r_0 & -L^3\left(20+\frac{37}{3}g_1+\frac{20}{3}g_2+5g_3\right) \\ 0 & 240 & 0 & -240 & 0 & 0 & 0 & -\frac{7}{3}r_1 & L^3\left(30-\frac{7}{3}g_2\right) \\ 0 & 0 & 400 & 0 & -400 & 0 & 0 & \left(\frac{10}{3}r_0-\frac{7}{3}r_2\right) & L^3\left(20+30g_1+\frac{20}{3}g_2+\frac{8}{3}g_3\right) \\ 0 & 0 & 0 & 600 & 0 & -600 & 0 & \frac{10}{3}r_1 & -L^3\left(25-\frac{50}{3}g_2\right) \\ 0 & 0 & 0 & 0 & 840 & 0 & 840 & -\left(r_0-\frac{10}{3}r_2\right) & -15L^3\left(g_1-g_3\right) \\ 0 & 0 & 0 & 0 & 0 & 1120 & 0 & -r_1 & -\frac{35}{3}L^3g_2 \\ 0 & 0 & 0 & 0 & 0 & 0 & 1440 & -r_2 & -10L^3g_3 \end{bmatrix} \tag{75}$$

which is a matrix with 8 rows and 9 columns. To obtain the expression of all the unknowns with respect to two parameters b_6 and q_0, we demand the rank of the matrix (75) to be less than 8. According to the definition of the rank, the rank of the matrix (75) is less than 8 if all minors of order 8 vanish. We produce the minors, extracting one column of the matrix among 9; therefore we obtain a set of 9 determinantal equations for 6 unknowns, namely the parameters r_0, r_1, r_2, g_1, g_2 and g_3. The solution of this set reads

$$r_1 = 0, \quad g_2 = -36/11 \tag{76}$$

Upon substitution of the expressions (76) into the set (74), we get the solution for the natural frequency squared

$$\Omega^2 = \left(1440b_6 - 10q_0L^3g_3\right)/r_2 \tag{77}$$

The value of q_0 that corresponds to the vanishing frequency $\Omega^2 = 0$ is identified as the buckling load $q_{0,cr}$:

$$q_{0,cr} = 144b_6 / g_3L^3 \tag{78}$$

We, thus, rewrite Eq. (77) as follows

$$\Omega^2 = 1440b_6\left(1 - q_0/q_{0,cr}\right)/r_2 \tag{79}$$

From Eq. (67), when $q_0 \equiv 0$, $\Omega^2 = 1440b_6/r_2 \equiv \Omega_0{}^2$, that is the natural frequency in *absence* of axial load q. Therefore, Eq. (67) takes the form

$$\Omega_q{}^2 = \Omega_0{}^2\left(1 - q_0/q_{0,cr}\right) \tag{80}$$

where $\Omega_q{}^2$ is the natural frequency squared in presence of the axial load. Note that the critical buckling load coincides with that of Elishakoff and Guédé (1999). When $q_0 \equiv 0$ we get Eq. (78) to be the natural frequency coefficient of the beam with axially distributed load not taken into account; this solution agrees with that determined by Elishakoff and Guédé (1999) in other manner. When q_0 tends to its critical value the natural frequency vanishes when $q_0 < q_{0,cr}$. The coefficients b_i in the stiffness are deduced from the remaining equations of the set (74)

$$
\begin{aligned}
b_0 &= \frac{(510840r_2 + 2455200r_0)b_6 - [(209g_3 - 6325g_1 + 1470)r_2 + 17050g_3r_0]q_0L^3}{138600r_2} \\
b_1 &= -7q_0L^3/120 \\
b_2 &= \frac{(255420r_2 - 712800r_0)b_6 - [(2992g_3 + 3960g_1 - 315)r_2 - 4950g_3r_0]q_0L^3}{69300r_2} \\
b_3 &= 13q_0L^3/132 \\
b_4 &= \frac{(-2376r_2 + 864r_0)b_6 + [(11g_3 + 9g_1)r_2 - 6g_3r_0]q_0L^3}{504r_2}, \quad b_5 = -3/88\, q_0L^3
\end{aligned}
\tag{81}
$$

which solve the posed problem for the parabolically varying inertial coefficient.

10.4 Cubic Inertial Coefficient ($m = 3$)

In this case, the governing differential equation yields a set of 9 equations with 10 unknowns, namely the coefficients b_i in the stiffness, the natural frequency squared Ω^2 and the axial load q_0 that is again treated as a parameter

$$
\begin{aligned}
&-40b_1 - 7/3\, q_0L^3 = 0 \\
&-120b_2 + 120b_0 - q_0L^3(20 + 37/3\, g_1 + 20/3\, g_2 + 5g_3 + 4g_4) - 7/3\, r_0\Omega^2 = 0 \\
&-240b_3 + 240b_1 + q_0L^3(30 - 7/3\, g_2) - 7/3\, r_1\Omega^2 = 0 \\
&-400b_4 + 400b_2 + q_0L^3(20 + 30g_1 + 20/3\, g_2 + 8/3\, g_3 + 4g_4) + (10/3\, r_0 - 7/3\, r_2)\Omega^2 = 0 \\
&-600b_5 + 600b_3 - q_0L^3(25 - 50/3\, g_2 + 7/3\, g_4) + (10/3\, r_1 - 7/3\, r_3)\Omega^2 = 0 \\
&-840b_6 + 840b_4 - 15q_0L^3(g_1 - g_3) - (r_0 - 10/3\, r_2)\Omega^2 = 0 \\
&-1120b_7 + 1120b_5 - q_0L^3(35/3\, g_2 - 14g_4) - (r_1 - 10/3\, r_3)\Omega^2 = 0 \\
&1440b_6 - 10q_0L^3g_3 - r_2\Omega^2 = 0 \\
&1800b_7 - 9q_0L^3g_4 - r_3\Omega^2 = 0
\end{aligned}
\tag{82}
$$

In order to have a non-trivial solution, we have to express all unknowns in terms of a single parameter, say, b_7 and the axial load parameter q_0. Therefore, we demand the rank of the matrix of (82) to be less than 9, thus, all minors of order 9 vanish. This condition produces a set of 10 determinantal equations for 8 unknowns, namely the parameters r_0, r_1, r_2, r_3, g_1, g_2, g_3 and g_4. The solution of this set reads

$$
r_1 = -74/255\, r_3, \quad g_2 = -26/55\, g_4 - 36/11 \tag{83}
$$

We substitute of the expressions (83) into the set (82) and we get the solution for the natural frequency squared

$$
\Omega^2 = (1800b_7 - 9q_0L^3g_4)/r_3 \tag{84}
$$

The value of q_0 that corresponds to the vanishing frequency $\Omega^2 = 0$ is identified as the buckling load $q_{0,cr}$:

$$q_{0,cr} = 200 b_7 / g_4 L^3 \tag{85}$$

We, thus, rewrite Eq. (84) as follows

$$\Omega^2 = 1800 b_7 \left(1 - q_0 / q_{0,cr}\right) / r_3 \tag{86}$$

From Eq. (86), when $q_0 \equiv 0$, $\Omega^2 = 1800 b_7 / r_3 \equiv \Omega_0{}^2$. Therefore, Eq. (84) takes the form

$$\Omega_q{}^2 = \Omega_0{}^2 \left(1 - q_0 / q_{0,cr}\right) \tag{87}$$

where $\Omega_q{}^2$ again denotes the natural frequency squared in presence of the axial load. Note that when q_0 tends to its critical value the natural frequency vanishes. The coefficients b_i in the stiffness are deduced from the remaining equations of the set (82)

$$\begin{aligned}
b_0 &= \{(2554200 r_2 + 12276000 r_0) b_7 + [(2744 g_4 + 13354 g_3 + 25300 g_1 - 5880) r_3 \\
&\quad - (61380 r_0 + 12771 r_2) g_4] q_0 L^3\} / 554400 r_3, \quad b_1 = -7 q_0 L^3 / 120 \\
b_2 &= -\{(7128000 r_0 - 2554200 r_2) b_7 + [(1176 g_4 + 9746 g_3 + 31680 g_1 - 2520) r_3 \\
&\quad + (12771 r_2 - 35640 r_0) g_4] q_0 L^3\} / 554400 r_3 \\
b_3 &= \tfrac{259}{51} b_7 + \left(\tfrac{13}{132} - \tfrac{35}{1683} g_4\right) q_0 L^3 \\
b_4 &= \frac{(21600 r_0 - 59400 r_2) b_7 + [(180 g_1 - 110 g_3) r_3 + (297 r_2 - 108 r_0) g_4] q_0 L^3}{10080 r_3} \\
b_5 &= -\tfrac{82}{17} b_7 + \left(\tfrac{164}{14025} g_4 - \tfrac{3}{88}\right) q_0 L^3, \quad b_6 = \left[1800 r_2 b_7 + (10 g_3 r_3 - 9 g_4 r_2) q_0 L^3\right] / 1440 r_3
\end{aligned} \tag{88}$$

which complete the solution for the inertial coefficient of cubic variation.

10.5 Quartic Inertial Coefficient ($m = 4$)

In this circumstances, the governing differential equation yields a set of 10 equations with 11 unknowns, namely the coefficients b_i in the stiffness, the natural frequency squared Ω^2 and the axial load q_0 that is treated as a parameter

$$\begin{aligned}
&-40 b_1 - \tfrac{7}{3} q_0 L^3 = 0 \\
&-120 b_2 + 120 b_0 - q_0 L^3 \left(20 + \tfrac{37}{3} g_1 + \tfrac{20}{3} g_2 + 5 g_3 + 4 g_4 + \tfrac{10}{3} g_5\right) - \tfrac{7}{3} r_0 \Omega^2 = 0 \\
&-240 b_3 + 240 b_1 + q_0 L^3 \left(30 - \tfrac{7}{3} g_2\right) - \tfrac{7}{3} r_1 \Omega^2 = 0 \\
&-400 b_4 + 400 b_2 + q_0 L^3 \left(20 + 30 g_1 + \tfrac{20}{3} g_2 + \tfrac{8}{3} g_3 + 4 g_4 + \tfrac{10}{3} g_5\right) \\
&\qquad + \left(\tfrac{10}{3} r_0 - \tfrac{7}{3} r_2\right) \Omega^2 = 0 \\
&-600 b_5 + 600 b_3 - q_0 L^3 \left(25 - \tfrac{50}{3} g_2 + \tfrac{7}{3} g_4\right) + \left(\tfrac{10}{3} r_1 - \tfrac{7}{3} r_3\right) \Omega^2 = 0 \\
&-840 b_6 + 840 b_4 - q_0 L^3 \left(15 g_1 - 15 g_3 + \tfrac{7}{3} g_5\right) - \left(r_0 - \tfrac{10}{3} r_2 + \tfrac{7}{3} r_4\right) \Omega^2 = 0
\end{aligned} \tag{89}$$

$$-1120b_7+1120b_5-q_0L^3\left(\tfrac{35}{3}g_2-14g_4\right)-\left(r_1-\tfrac{10}{3}r_3\right)\Omega^2=0$$
$$-1440b_8+1440b_6-q_0L^3\left(10g_3-\tfrac{40}{3}g_5\right)-\left(r_2-\tfrac{10}{3}r_4\right)\Omega^2=0$$
$$1800b_7-9q_0L^3g_4-r_3\Omega^2=0$$
$$2200b_8-\tfrac{25}{3}q_0L^3g_5-r_4\Omega^2=0$$

In order to have a non-trivial solution we have to write all unknowns in terms of one, say, b_7 and the parameter q_0. To obtain the expression of all the unknowns with respect to two parameters b_8 and q_0, we demand the rank of the matrix of the set (89) to be less than 10, leading to the requirement all minors of order 10 vanish. Therefore we obtain a set of 11 determinantal equations for 8 unknowns, namely the parameters r_0, r_1, r_2, r_3, g_1, g_2, g_3 and g_4. The solution of this set reads

$$r_1=-\tfrac{74}{255}r_3,\quad g_2=-\tfrac{26}{55}g_4-36/11 \tag{90}$$

Upon substitution of the expressions (90) into the set (89), we get the solution for the natural frequency squared

$$\Omega^2=\left(2200b_8-\tfrac{25}{3}q_0L^3g_5\right)/r_4 \tag{91}$$

The value of q_0 that corresponds to the vanishing frequency $\Omega^2=0$ is identified as the buckling load $q_{0,cr}$:

$$q_{0,cr}=264b_8/g_5L^3 \tag{92}$$

We, thus, rewrite Eq. (79) as follows

$$\Omega_q^{\ 2}=2200b_8\left(1-q_0/q_{0,cr}\right)/r_4 \tag{93}$$

From Eq. (93), when $q_0\equiv 0$, $\Omega^2=2200\,b_8/r_4\equiv\Omega_0^{\ 2}$. Therefore, Eq. (91) takes the familiar form

$$\Omega_q^{\ 2}=\Omega_0^{\ 2}\left(1-q_0/q_{0,cr}\right) \tag{94}$$

where $\Omega_q^{\ 2}$ is the natural frequency squared in presence of the axial load. As expected when q_0 tends to its critical value the natural frequency vanishes. The coefficients b_i in the stiffness are

$$\begin{aligned}b_0 &= \{(10071600r_4+28096200r_2+135036000r_0)b_8\\&\quad+[(45430g_5+24696g_4+120186g_3+227700g_1-52920)r_4\\&\quad-(511500r_0+106425r_2)g_5]q_0L^3\}/4989600r_4\,,\quad b_1=-7q_0L^3/120\\b_2 &= \{(10071600r_4+28096200r_2-78408000r_0)b_8\\&\quad-[(93170g_5+10584g_4+87714g_3+285120g_1)r_4\\&\quad-(297000r_0-106425r_2)g_5]q_0L^3\}/4989600r_4\end{aligned} \tag{95}$$

$$
\begin{aligned}
b_3 &= \{18803400 r_3 b_8 + [(13923 g_4 + 298350) r_4 + 71225 g_5 r_3] q_0 L^3\}/3029400 r_4 \\
b_4 &= \{(183120 r_4 - 653400 r_2 + 237600 r_0) b_8 \\
&\quad [-(938 g_5 - 990 g_3 + 1620 g_1) r_4 + (2475 r_2 - 900 r_0) g_5] q_0 L^3\}/90720 r_4 \\
b_5 &= \{-5953200 r_3 b_8 - [(12546 g_4 + 34425) r_4 - 22550 r_3 g_5] q_0 L^3\}/1009800 r_4 \\
b_6 &= \{(-10608 r_4 + 3960 r_2) b_8 + [(18 g_3 + 26 g_5) r_4 - 15 r_2 g_5] q_0 L^3\}/2592 \\
b_7 &= [6600 r_3 b_8 + (27 g_4 r_4 - 25 r_3 g_5) q_0 L^3]/5400 r_4
\end{aligned}
$$

presenting the solution of the posed problem for quartic $R(\xi)$.

10.6 Quintic Inertial Coefficient ($m = 5$)

The governing differential equation yields a set of 11 equations with 12 unknowns, namely the coefficients b_i in the stiffness, the natural frequency squared Ω^2 and the axial load q_0 that is treated as a parameter

$$
\begin{aligned}
& -40 b_1 - \tfrac{7}{3} q_0 L^3 = 0 \\
& -120 b_2 + 120 b_0 - q_0 L^3 \left(20 + \tfrac{37}{3} g_1 + \tfrac{20}{3} g_2 + 5 g_3 + 4 g_4 + \tfrac{10}{3} g_5 + \tfrac{20}{7} g_6\right) \\
& \qquad - \tfrac{7}{3} r_0 \Omega^2 = 0 \\
& -240 b_3 + 240 b_1 + q_0 L^3 \left(30 - \tfrac{7}{3} g_2\right) - \tfrac{7}{3} r_1 \Omega^2 = 0 \\
& -400 b_4 + 400 b_2 + q_0 L^3 \left(20 + 30 g_1 + \tfrac{20}{3} g_2 + \tfrac{8}{3} g_3 + 4 g_4 + \tfrac{10}{3} g_5 + \tfrac{20}{7} g_6\right) \\
& \qquad + \left(\tfrac{10}{3} r_0 - \tfrac{7}{3} r_2\right) \Omega^2 = 0 \\
& -600 b_5 + 600 b_3 - q_0 L^3 \left(25 - \tfrac{50}{3} g_2 + \tfrac{7}{3} g_4\right) + \left(\tfrac{10}{3} r_1 - \tfrac{7}{3} r_3\right) \Omega^2 = 0 \\
& -840 b_6 + 840 b_4 - q_0 L^3 \left(15 g_1 - 15 g_3 - \tfrac{7}{3} g_5\right) - \left(r_0 - \tfrac{10}{3} r_2 + \tfrac{7}{3} r_4\right) \Omega^2 = 0 \\
& -1120 b_7 + 1120 b_5 - q_0 L^3 \left(\tfrac{35}{3} g_2 - 14 g_4 + \tfrac{7}{3} g_6\right) - \left(r_1 - \tfrac{10}{3} r_3 + \tfrac{7}{3} r_5\right) \Omega^2 = 0 \\
& -1440 b_8 + 1440 b_6 - q_0 L^3 \left(10 g_3 - \tfrac{40}{3} g_5\right) - \left(r_2 - \tfrac{10}{3} r_4\right) \Omega^2 = 0 \\
& -1800 b_9 + 1800 b_7 - q_0 L^3 \left(9 g_4 - \tfrac{90}{7} g_6\right) - \left(r_3 - \tfrac{10}{3} r_5\right) \Omega^2 = 0 \\
& 2200 b_8 - \tfrac{25}{3} q_0 L^3 g_5 - r_4 \Omega^2 = 0 \\
& 2640 b_9 - \tfrac{55}{7} q_0 L^3 g_6 - r_5 \Omega^2 = 0
\end{aligned}
\tag{96}
$$

We have to write all unknowns in terms of b_8 and the axial load parameter q_0. To obtain the expression of all the unknowns with respect to two parameters b_9 and q_0, the rank of the matrix of the set (96) must be less than 10, leading to all minors of order 10 to vanish. Hence, we obtain a set of 11 determinantal equations for 10 unknowns, namely the parameters r_0, r_1, r_2, r_3, g_1, g_2, g_3 and g_4. The solution of this set reads

$$
r_1 = -\tfrac{203}{1683} r_5 - \tfrac{74}{255} r_3, \quad g_2 = -\tfrac{3}{11} g_6 - \tfrac{26}{55} g_4 - 36/11 \tag{97}
$$

Upon substitution of the expressions (97) into the set (96), we get the solution for the natural frequency squared

$$\Omega^2 = \left(2680b_9 - {}^{55}\!/_7\, q_0 L^3 g_6\right)/r_5 \tag{98}$$

The buckling load $q_{0,cr}$ is obtained by setting $\Omega^2 = 0$:

$$q_{0,cr} = 336b_9 / g_6 L^3 \tag{99}$$

We, thus, rewrite Eq. (98) as follows

$$\Omega^2 = 2680b_9\left(1 - q_0 / q_{0,cr}\right)/r_5 \tag{100}$$

From Eq. (88), when $q_0 \equiv 0$, $\Omega^2 = 2680\, b_9 / r_5 \equiv \Omega_0^{\ 2}$. Therefore, Eq. (98) takes the form

$$\Omega_q^{\ 2} = \Omega_0^{\ 2}\left(1 - q_0 / q_{0,cr}\right) \tag{101}$$

where $\Omega_q^{\ 2}$ is the natural frequency squared in presence of the axial load. One can see that when q_0 tends to its critical value the natural frequency vanishes, as it should be. The coefficients b_i in the stiffness are not reproduced here. For details one can consult the paper by Guede and Elishakoff (2001 b).

11 Closed-Form Solution for the Clamped - Free Beam

11.1 Uniform Inertial Coefficient ($m = 0$)

In the case of uniform inertial coefficient, we get a homogeneous set of 6 linear algebraic equations with 7 unknowns, namely the coefficients b_i in the stiffness, the natural frequency squared Ω^2 and the axial load q_0 that is treated as a parameter

$$\begin{aligned}
&80b_2 + 120b_1 + q_0 L^3(40 + 20g_1) = 0\\
&240b_3 - 360b_2 + 120b_0 - q_0 L^3(140 + 30g_1) = 0\\
&480b_4 - 720b_3 + 240b_1 + q_0 L^3(90 - 60g_1) - 20r_0\Omega^2 = 0\\
&-1200b_4 + 400b_2 + q_0 L^3(20 + 70g_1) + 10r_0\Omega^2 = 0\\
&600b_3 - 25q_0 L^3 = 0\\
&840b_4 - 15g_1 q_0 L^3 - r_0\Omega^2 = 0
\end{aligned} \tag{102}$$

Again we want to write all unknowns in terms of a single free parameter, say, b_4 and the axial load parameter q_0. In other word, the rank of the set (102) or its matrix must be less than 6. This implies that all minors of order 6 vanish, leading to the following 7 equations with 2 unknowns r_0 and g_1

$$\begin{aligned}
&-18247768r_0g_1+37297152r_0=0\\
&6967296r_0g_1+14432256r_0=0\\
&6137856r_0g_1+11529216r_0=0\\
&r_0=0\\
&1078272r_0g_1+456192r_0=0\\
&8957952g_1-38320128=0\\
&r_0=0
\end{aligned} \tag{103}$$

The solution of this set reads

$$r_0=0, \quad g_1={}^{77}\!/_{18} \tag{104}$$

which leads to a trivial solution for the problem since the inertial coefficient vanishes.

11.2 Linearly varying Inertial Coefficient ($m = 1$)

In this case, we obtain a set of 7 equations with 8 unknowns, namely the coefficients b_i in the stiffness, the natural frequency squared Ω^2 and the axial load q_0 that is treated as a parameter

$$\begin{aligned}
&80b_2-120b_1+q_0L^3\left(40+20g_1+{}^{40}\!/_3\,g_2\right)=0\\
&240b_3-360b_2+120b_0-q_0L^3\left(140+30g_1+20g_2\right)=0\\
&480b_4-720b_3+240b_1+q_0L^3\left(90-60g_1\right)-20r_0\Omega^2=0\\
&800b_5-1200b_4+400b_2+q_0L^3\left(20+70g_1-{}^{140}\!/_3\,g_2\right)+\Omega^2\left(10r_0-20r_1\right)=0\\
&-1800b_5+600b_3-q_0L^3\left(25-50g_2\right)+10r_1\Omega^2=0\\
&840b_4-15q_0L^3g_1-r_0\Omega^2=0\\
&1120b_5-{}^{35}\!/_3\,q_0L^3g_2-r_1\Omega^2=0
\end{aligned} \tag{105}$$

We have to express the unknowns in terms of parameters b_5 and Ω^2. Therefore, we demand the rank of the set (105) to be less than 7, leading to a system of 8 determinantal equations with 4 unknowns, namely r_0, r_1, g_1 and g_2,

$$\begin{aligned}
&2487711744r_0g_2+1233211392r_1g_2+4177281024r_0+2011226112r_1\\
&\qquad-204374016r_0g_1-1070972928r_1g_1=0\\
&-931295232r_0g_2+181149696r_1g_2-1616412672r_0+355332096r_1\\
&\qquad-780337152r_0g_1+818159616r_1g_1=0\\
&158312448r_0g_2+692084736r_1g_2-1291272192r_0+2358429696r_1\\
&\qquad-687439872r_0g_1+218474496r_1g_1=0\\
&-140341248r_1g_1+603832332r_1g_2+841300992r_1+387072000r_0g_2\\
&\qquad-309657600r_0=0
\end{aligned} \tag{106}$$

$$30772224 r_0 g_2 + 51093504 r_0 + 120766464 r_0 g_1 - 103265280 r_1 g_1 = 0$$
$$8957952 r_1 g_1 + 37158912 r_1 g_2 - 38320128 r_1 - 77414400 r_0 g_2 = 0$$
$$2584866816 g_2 - 1003290624 g_1 + 4291854336 = 0$$
$$578285568 r_1 - 743178240 r_0 = 0$$

The solution of the set (106) reads

$$r_0 = \tfrac{249}{320} r_1, \quad g_1 = \tfrac{77}{18} + \tfrac{371}{144} g_2 \tag{107}$$

Upon substitution of the expressions (107) into the set (106), we get the solution for the natural frequency squared

$$\Omega^2 = \left(1120 b_5 - {}^{35}\!/\!_3\, q_0 L^3 g_2\right) / r_1 \tag{108}$$

The value of q_0 that corresponds to the vanishing frequency $\Omega^2 = 0$ leads to the buckling load $q_{0,cr}$:

$$q_{0,cr} = 96 b_5 / g_2 L^3 \tag{109}$$

We rewrite Eq. (108) as follows

$$\Omega^2 = 1120 b_5 \left(1 - q_0 / q_{0,cr}\right) / r_1 \tag{110}$$

From Eq. (110), when $q_0 \equiv 0$, $\Omega^2 = 1120\, b_5 / r_1 \equiv \Omega_0{}^2$. Therefore, Eq. (94) takes the form

$$\Omega_q{}^2 = \Omega_0{}^2 \left(1 - q_0 / q_{0,cr}\right) \tag{111}$$

where $\Omega_q{}^2$ is the natural frequency squared in presence of the axial load. Note that when q_0 tends to its critical value the natural frequency vanishes. The coefficients b_i in the stiffness are

$$\begin{aligned}
b_0 &= \frac{16477}{120} b_5 + \left(\frac{4}{9} - \frac{1493}{1280} g_2\right) q_0 L^3, \quad b_1 = \frac{471}{20} b_5 + \left(\frac{2}{3} + \frac{289}{1920} g_2\right) q_0 L^3 \\
b_2 &= \frac{1413}{40} b_5 - \left(\frac{41}{72} + \frac{6739}{11520} g_2\right) q_0 L^3, \quad b_3 = -\frac{47}{3} b_5 + \left(\frac{1}{24} + \frac{1}{9} g_2\right) q_0 L^3 \\
b_4 &= \frac{83}{80} b_5 + \left(\frac{11}{144} + \frac{811}{23040} g_2\right) q_0 L^3
\end{aligned} \tag{112}$$

11.3 Parabolically Varying Inertial ($m = 2$)

In this case, we obtain a set of 8 equations with 8 unknowns, namely the coefficients b_i in the stiffness, the natural frequency squared Ω^2 and the axial load q_0 that is treated as a parameter

$$
\begin{aligned}
&80b_2 - 120b_1 + q_0L^3\left(40 + 20g_1 + {}^{40}\!/_3\, g_2 + 10g_3\right) = 0\\
&240b_3 - 360b_2 + 120b_0 - q_0L^3\left(140 + 30g_1 + 20g_2 + 15g_3\right) = 0\\
&480b_4 - 720b_3 + 240b_1 + q_0L^3\left(90 - 60g_1\right) - 20\Omega^2 r_0 = 0\\
&800b_5 - 1200b_4 + 400b_2 + q_0L^3\left(20 + 70g_1 - {}^{140}\!/_3\, g_2 + 5g_3\right) + \Omega^2\left(10r_0 - 20r_1\right) = 0\\
&1200b_6 - 1800b_5 + 600b_3 - q_0L^3\left(25 - 50g_2 - 50g_3\right) + \Omega^2\left(10r_1 - 20r_2\right) = 0\\
&-2520b_6 + 840b_4 - q_0L^3\left(15g_1 - 45g_3\right) - \Omega^2\left(r_0 + 10r_2\right) = 0\\
&1120b_5 - {}^{35}\!/_3\, g_2 - r_1\Omega^2 = 0\\
&1440b_6 - 10q_0L^3 g_3 - r_2\Omega^2 = 0
\end{aligned}
\tag{113}
$$

We have to express the unknowns in terms of parameters b_6 and Ω^2. Therefore, we demand the rank of the set (113) to be less than 8, leading to a system of 9 determinantal equations with 6 unknowns, namely the parameters r_0, r_1, r_2, g_1, g_2 and g_3. The solution of this set reads

$$
r_0 = \tfrac{249}{320} r_1 - \tfrac{227}{160} r_2, \quad g_1 = \tfrac{77}{18} + \tfrac{371}{144} g_2 - \tfrac{37}{18} g_3 \tag{114}
$$

Upon substitution of the expressions (114) into the set (113), we get the solution for the natural frequency squared

$$
\Omega^2 = \left(1440b_6 - 10q_0L^3 g_3\right)/r_2 \tag{115}
$$

The value of q_0 that corresponds to the vanishing frequency $\Omega^2 = 0$ is the buckling load $q_{0,cr}$

$$
q_{0,cr} = 144b_6 / g_3L^3 \tag{116}
$$

We, thus, rewrite Eq. (114) as follows

$$
\Omega^2 = 1440b_6\left(1 - q_0 / q_{0,cr}\right)/r_2 \tag{117}
$$

From Eq. (117), when $q_0 \equiv 0$, $\Omega^2 = 1440\, b_6/r_2 \equiv \Omega_0{}^2$. Therefore, Eq. (115) takes the form

$$
\Omega_q{}^2 = \Omega_0{}^2\left(1 - q_0 / q_{0,cr}\right) \tag{118}
$$

where $\Omega_q{}^2$ is the natural frequency squared in presence of the axial load. When q_0 tends to its critical value the natural frequency vanishes. The coefficients b_i in the stiffness are

$$
\begin{aligned}
b_0 = &\{(7118064r_1 - 3546144r_2)b_6\\
&+ [(17920 + 10640g_2 + 20146g_3)r_2 - 49431r_1g_3]q_0L^3\}/40320r_2
\end{aligned}
$$

$$
b_1 = \frac{(203472r_1 + 6048r_2)b_6 + [(4480 + 2660g_2 - 1162g_3)r_2 - 1413g_3r_1]q_0L^3}{6720r_2}
$$

$$b_2 = \frac{(54432r_2 + 1831248r_1)b_6 - [(22960 + 8750g_2 - 5222g_3)r_2 + 12717g_3r_1]q_0L^3}{40320r_2}$$

$$b_3 = \{(30912r_2 - 13536r_1)b_6 + [(28 - 35g_2 - 168g_3)r_2 + 94g_3r_1]q_0L^3\}/672r_2$$

$$b_4 = \frac{(107568r_1 - 1336608r_2)b_6 + [(6160 + 3710g_2 + 3682g_3)r_2 - 747g_3r_1]q_0L^3}{80640r_2}$$

$$b_5 = [864r_1b_6 + (7g_2r_2 - 6g_3r_1)q_0L^3]/672r_2$$

11.4 Cubic Inertial Coefficient ($m = 3$)

The governing differential equation yields a set of 9 equations with 10 unknowns, namely the coefficients b_i in the stiffness, the natural frequency squared Ω^2 and the axial load q_0 that is treated as a parameter

$$\begin{aligned}
&80b_2 - 120b_1 + q_0L^3(40 + 20g_1 + {}^{40}\!/\!_3\, g_2 + 10g_3 + 8g_4) = 0\\
&240b_3 - 360b_2 + 120b_0 - q_0L^3(140 + 30g_1 + 20g_2 + 15g_3 + 12g_4) = 0\\
&480b_4 - 720b_3 + 240b_1 + q_0L^3(90 - 60g_1) - 20\Omega^2 r_0 = 0\\
&800b_5 - 1200b_4 + 400b_2 + q_0L^3(20 + 70g_1 - {}^{140}\!/\!_3\, g_2 + 5g_3 + 4g_4)\\
&\qquad + \Omega^2(10r_0 - 20r_1) = 0\\
&1200b_6 - 1800b_5 + 600b_3 - q_0L^3(25 - 50g_2 - 50g_3) + \Omega^2(10r_1 - 20r_2) = 0\\
&-2520b_6 + 840b_4 - q_0L^3(15g_1 - 45g_3 + 48g_4) - \Omega^2(r_0 - 10r_2 + 20r_3) = 0\\
&-3360b_7 + 1120b_5 - q_0L^3({}^{35}\!/\!_3\, g_2 - 42g_4) - \Omega^2(r_1 - 10r_3) = 0\\
&1440b_6 - 10g_3 - r_2\Omega^2 = 0\\
&1800b_7 - 9q_0L^3g_4 - r_3\Omega^2 = 0
\end{aligned} \tag{120}$$

We have to express the unknowns in terms of parameters b_5 and Ω^2. Therefore, we require the rank of the set (120) to be less than 9. This leads to a system of 10 determinantal equations with 8 unknowns, namely the parameters r_0, r_1, r_2, r_3, g_1, g_2, g_3 and g_4. The solution of this set reads

$$r_0 = \tfrac{249}{320}r_1 - \tfrac{227}{160}r_2 + \tfrac{1393}{800}r_3, \quad g_1 = \tfrac{77}{18} + \tfrac{371}{144}g_2 - \tfrac{37}{18}g_3 + \tfrac{449}{120}g_4 \tag{121}$$

Upon substitution of the expressions (121) into the set (120), we get the solution for the natural frequency squared

$$\Omega^2 = (1800b_7 - 9q_0L^3g_4)/r_3 \tag{122}$$

The value of q_0 that corresponds to the vanishing frequency $\Omega^2 = 0$ is identified as the buckling load $q_{0,cr}$:

$$q_{0,cr} = 200b_7 / g_4L^3 \tag{123}$$

We rewrite Eq. (122) as follows

$$\Omega^2 = 1800 b_7 \left(1 - q_0 / q_{0,cr}\right) / r_3 \tag{124}$$

From Eq. (124), when $q_0 \equiv 0$, $\Omega^2 = 1800\, b_7 / r_3 \equiv \Omega_0^{\ 2}$. Therefore, Eq. (122) takes the form

$$\Omega_q^{\ 2} = \Omega_0^{\ 2}\left(1 - q_0 / q_{0,cr}\right) \tag{125}$$

where $\Omega_q^{\ 2}$ is the natural frequency squared in presence of the axial load. When q_0 tends to its critical value the natural frequency vanishes. The coefficients b_i in the stiffness are

$$\begin{aligned}
b_0 &= \{(444879000 r_1 - 221634000 r_2 + 651337200 r_3) b_7 + [(896000 + 532000 g_2 \\
&\quad - 224000 g_3 - 2577966 g_4) r_3 + (1108170 r_2 - 2224395 r_1) g_4] q_0 L^3\} / 2016000 r_3 \\
b_1 &= \{(12717000 r_1 + 378000 r_2 + 18267600 r_3) b_7 + [(224000 + 133000 g_2 - 56000 g_3 \\
&\quad + 78342 g_4) r_3 - (1890 r_2 + 63585 r_1) g_4] q_0 L^3\} / 336000 r_3 \\
b_2 &= \{(114453000 r_1 + 3402000 r_2 + 164408400 r_3) b_7 - [(1148000 + 437500 g_2 \\
&\quad + 280000 g_3 - 1382322 g_4) r_3 + (572265 r_1 + 17010 r_2) g_4] q_0 L^3\} / 2016000 r_3 \\
b_3 &= \{(-253800 r_1 + 579600 r_2 - 395280 r_3) b_7 + [(420 - 525 g_2 + 700 g_3 + 1296 g_4) r_3 \\
&\quad + (1269 r_1 - 2898 r_2) g_4] q_0 L^3\} / 10080 r_3 \\
b_4 &= \{(6723000 r_1 - 83538000 r_2 + 179780400 r_3) b_7 + [(308000 + 185500 g_2 \\
&\quad - 280000 g_3 - 439422 g_4) r_3 + (417690 r_2 - 33615 r_1) g_4] q_0 L^3\} / 4032000 r_3 \\
b_5 &= \{(5400 r_1 - 43920 r_3) b_7 + [(144 + 35 g_2) r_3 - 27 g_4 r_1] q_0 L^3\} / 3360 r_3 \\
b_6 &= [1800 r_2 b_7 + (10 g_3 r_3 - 9 g_4 r_2) q_0 L^3] / 1440 r_3
\end{aligned} \tag{126}$$

11.5 Quartic Inertial Coefficient ($m = 4$)

The governing differential equation yields a set of 10 equations with 11 unknowns, namely the coefficients b_i in the stiffness, the natural frequency squared Ω^2 and the axial load q_0 that is treated as a parameter

$$\begin{aligned}
&80 b_2 - 120 b_1 + q_0 L^3 (40 + 20 g_1 + {}^{40}\!/\!_3\, g_2 + 10 g_3 + 8 g_4 + {}^{20}\!/\!_3\, g_5) = 0 \\
&240 b_3 - 360 b_2 + 120 b_0 - q_0 L^3 (140 + 30 g_1 + 20 g_2 + 15 g_3 + 12 g_4 - 10 g_5) = 0 \\
&480 b_4 - 720 b_3 + 240 b_1 + q_0 L^3 (90 - 60 g_1) - 20 \Omega^2 r_0 = 0 \\
&800 b_5 - 1200 b_4 + 400 b_2 + q_0 L^3 (20 + 70 g_1 - {}^{140}\!/\!_3\, g_2 + 5 g_3 + 4 g_4 + {}^{10}\!/\!_3\, g_5) \\
&\qquad + \Omega^2 (10 r_0 - 20 r_1) = 0 \\
&1200 b_6 - 1800 b_5 + 600 b_3 - q_0 L^3 (25 - 50 g_2 - 50 g_3) + \Omega^2 (10 r_1 - 20 r_2) = 0 \\
&1680 b_7 - 2520 b_6 + 840 b_4 - q_0 L^3 (15 g_1 - 45 g_3 + 48 g_4) - \Omega^2 (r_0 - 10 r_2 + 20 r_3) = 0
\end{aligned} \tag{127}$$

$$2240b_8 - 3360b_7 + 1120b_5 - q_0L^3\left(\tfrac{35}{3}g_2 - 42g_4 + \tfrac{140}{3}g_5\right) - \Omega^2\left(r_1 - 10r_3 + 20r_4\right) = 0$$
$$-4320b_8 + 1440b_6 - q_0L^3\left(10g_3 - 40g_5\right) - \Omega^2\left(r_2 - 10r_4\right) = 0$$
$$1800b_7 - 9q_0L^3g_4 - r_3\Omega^2 = 0$$
$$2200b_8 - \tfrac{25}{3}q_0L^3g_5 - r_4\Omega^2 = 0$$

We have to express the unknowns in terms of parameters b_5 and Ω^2. Therefore, we demand the rank of the set (127) to be less than 10, leading to a system of 11 determinantal equations with 10 unknowns, namely the parameters r_0, r_1, r_2, r_3, r_4, g_1, g_2, g_3, g_4 and g_5. The solution for this set reads

$$r_0 = \tfrac{249}{320}r_1 - \tfrac{227}{160}r_2 + \tfrac{1393}{800}r_3 - \tfrac{3183}{1100}r_4, \quad g_1 = \tfrac{77}{18} + \tfrac{371}{144}g_2 - \tfrac{37}{18}g_3 + \tfrac{449}{120}g_4 - \tfrac{1841}{396}g_5 \tag{128}$$

Upon substitution of the expressions (128) into the set (127), we get the solution for the natural frequency squared

$$\Omega^2 = \left(2200b_8 - \tfrac{25}{3}q_0L^3g_5\right)/r_4 \tag{129}$$

The value of q_0 that corresponds to the vanishing frequency $\Omega^2 = 0$ is identified as the buckling load $q_{0,cr}$:

$$q_{0,cr} = 264b_8 / g_5L^3 \tag{130}$$

We rewrite Eq. (129) as follows

$$\Omega^2 = 2200b_8\left(1 - q_0/q_{0,cr}\right)/r_4 \tag{131}$$

From Eq. (131), when $q_0 \equiv 0$, $\Omega^2 = 2200\,b_8/r_4 \equiv \Omega_0^{\,2}$. Therefore, Eq. (129) takes the form

$$\Omega_q^{\,2} = \Omega_0^{\,2}\left(1 - q_0/q_{0,cr}\right) \tag{132}$$

where $\Omega_q^{\,2}$ is the natural frequency squared in presence of the axial load. When q_0 tends to its critical value the natural frequency vanishes. The coefficients b_i in the stiffness reads

$$\begin{aligned}
b_0 &= \{(10071600r_4 + 28096200r_2 + 135036000r_0)b_8 \\
&\quad + [(45430g_5 + 24696g_4 + 120186g_3 + 227700g_1 - 52920r_4)r_4 \\
&\quad - (511500r_0 + 106425r_2)g_5]q_0L^3\}/4989600r_4, \quad b_1 = -7q_0L^3/120 \\
b_2 &= \{(10071600r_4 + 28096200r_2 - 78408000r_0)b_8 \\
&\quad - [(93170g_5 + 10584g_4 + 87714g_3 + 285120g_1)r_4 \\
&\quad - (297000r_0 - 106425r_2)g_5]q_0L^3\}/4989600r_4 \\
b_3 &= \{18803400r_3b_8 + [(13923g_4 + 298350)r_4 + 71225g_5r_3]q_0L^3\}/3029400r_4 \\
\\
b_6 &= \{(-10608r_4 + 3960r_2)b_8 + [(18g_3 + 26g_5)r_4 - 15r_2g_5]q_0L^3\}/2592 \\
b_7 &= [6600r_3b_8 + (27g_4r_4 - 25r_3g_5)q_0L^3]/5400r_4
\end{aligned} \tag{133}$$

$$b_4 = \{(183120r_4 - 653400r_2 + 237600r_0)b_8$$
$$[-(938g_5 - 990g_3 + 1620g_1)r_4 + (2475r_2 - 900r_0)g_5]q_0L^3\}/90720r_4$$
$$b_5 = \{-5953200r_3b_8 - [(12546g_4 + 34425)r_4 - 22550r_3g_5]q_0L^3\}/1009800r_4$$

11.6 Quintic Inertial Coefficient (m = 5)

The governing differential equation yields a set of 11 equations with 12 unknowns, namely the coefficients b_i in the stiffness, the natural frequency squared Ω^2 and the axial load q_0 that is treated as a parameter

$$\begin{aligned}
&80b_2 - 120b_1 + q_0L^3(40 + 20g_1 + {}^{40}\!/_{3}\,g_2 + 10g_3 + 8g_4 + {}^{20}\!/_{3}\,g_5 + {}^{40}\!/_{7}\,g_6) = 0\\
&240b_3 - 360b_2 + 120b_0 - q_0L^3(140 + 30g_1 + 20g_2 + 15g_3 + 12g_4 - 10g_5 + {}^{60}\!/_{7}\,g_6) = 0\\
&480b_4 - 720b_3 + 240b_1 + q_0L^3(90 - 60g_1) - 20\Omega^2r_0 = 0\\
&800b_5 - 1200b_4 + 400b_2 + q_0L^3(20 + 70g_1 - {}^{140}\!/_{3}\,g_2 + 5g_3 + 4g_4 + {}^{10}\!/_{3}\,g_5 + {}^{20}\!/_{7}\,g_6)\\
&\quad + \Omega^2(10r_0 - 20r_1) = 0\\
&1200b_6 - 1800b_5 + 600b_3 - q_0L^3(25 - 50g_2 - 50g_3) + \Omega^2(10r_1 - 20r_2) = 0\\
&1680b_7 - 2520b_6 + 840b_4 - q_0L^3(15g_1 - 45g_3 + 48g_4) - \Omega^2(r_0 - 10r_2 + 20r_3) = 0\\
&2240b_8 - 3360b_7 + 1120b_5 - q_0L^3({}^{35}\!/_{3}\,g_2 - 42g_4 + {}^{140}\!/_{3}\,g_5) - \Omega^2(r_1 - 10r_3 + 20r_4) = 0\\
&3360b_7 - 4320b_8 + 1440b_6 - q_0L^3(10g_3 - 40g_5 + {}^{270}\!/_{7}\,g_6) - \Omega^2(r_2 - 10r_4 + 20r_5) = 0\\
&-5400b_9 + 1800b_7 - q_0L^3(9g_4 - {}^{270}\!/_{7}\,g_6) - \Omega^2(r_3 - 10r_5) = 0\\
&2200b_8 - {}^{25}\!/_{3}\,q_0L^3g_5 - r_4\Omega^2 = 0\\
&2640b_9 - {}^{55}\!/_{7}\,q_0L^3g_6 - r_5\Omega^2 = 0
\end{aligned} \tag{134}$$

We have to express the unknowns in terms of parameters b_5 and Ω^2. Therefore, we demand the rank of the set (134) to be less than 11, leading to a system of 12 determinantal equations with 12 unknowns, namely the parameters r_0, r_1, r_2, r_3, r_4, r_5, g_1, g_2, g_3, g_4, g_5 and g_6. The solution of this set reads

$$\begin{aligned}
r_0 &= \tfrac{249}{320}r_1 - \tfrac{227}{160}r_2 + \tfrac{1393}{800}r_3 - \tfrac{3183}{1100}r_4 + \tfrac{7553}{1760}r_5\\
g_1 &= \tfrac{77}{18} + \tfrac{371}{144}g_2 - \tfrac{37}{18}g_3 + \tfrac{449}{120}g_4 - \tfrac{1841}{396}g_5 + \tfrac{557}{72}g_6
\end{aligned} \tag{135}$$

Upon substitution of the expressions (135) into the set (134), we get the solution for the natural frequency squared

$$\Omega^2 = (2680b_9 - {}^{55}\!/_{7}\,q_0L^3g_6)/r_5 \tag{136}$$

The value of q_0 that corresponds to the vanishing frequency $\Omega^2 = 0$ is identified as the buckling load $q_{0,cr}$:

$$q_{0,cr} = 336b_9 / g_6L^3 \tag{137}$$

We rewrite Eq. (136) as follows

$$\Omega^2 = 2680 b_9 \left(1 - q_0 / q_{0,cr}\right) / r_5 \tag{138}$$

From Eq. (138), when $q_0 \equiv 0$, $\Omega^2 = 2680\, b_9 / r_5 \equiv \Omega_0^{\,2}$. Therefore, Eq. (136) takes the form

$$\Omega_q^{\,2} = \Omega_0^{\,2} \left(1 - q_0 / q_{0,cr}\right) \tag{139}$$

where $\Omega_q^{\,2}$ is the natural frequency squared in presence of the axial load. When q_0 tends to its critical value the natural frequency vanishes. The stiffness coefficients b_i read

$$\begin{aligned}
b_0 &= \{(10071600 r_4 + 28096200 r_2 + 135036000 r_0) b_8 \\
&\quad + [(45430 g_5 + 24696 g_4 + 120186 g_3 + 227700 g_1 - 52920 r_4) r_4 \\
&\quad - (511500 r_0 + 106425 r_2) g_5] q_0 L^3 \} / 4989600 r_4, \quad b_1 = -7 q_0 L^3 / 120 \\
b_2 &= \{(10071600 r_4 + 28096200 r_2 - 78408000 r_0) b_8 \\
&\quad - [(93170 g_5 + 10584 g_4 + 87714 g_3 + 285120 g_1) r_4 \\
&\quad - (297000 r_0 - 106425 r_2) g_5] q_0 L^3 \} / 4989600 r_4 \\
b_3 &= \{18803400 r_3 b_8 + [(13923 g_4 + 298350) r_4 + 71225 g_5 r_3] q_0 L^3 \} / 3029400 r_4 \\
b_4 &= \{(183120 r_4 - 653400 r_2 + 237600 r_0) b_8 \\
&\quad [-(938 g_5 - 990 g_3 + 1620 g_1) r_4 + (2475 r_2 - 900 r_0) g_5] q_0 L^3 \} / 90720 r_4 \\
b_5 &= \{-5953200 r_3 b_8 - [(12546 g_4 + 34425) r_4 - 22550 r_3 g_5] q_0 L^3 \} / 1009800 r_4 \\
b_6 &= \{(-10608 r_4 + 3960 r_2) b_8 + [(18 g_3 + 26 g_5) r_4 - 15 r_2 g_5] q_0 L^3 \} / 2592 \\
b_7 &= [6600 r_3 b_8 + (27 g_4 r_4 - 25 r_3 g_5) q_0 L^3] / 5400 r_4
\end{aligned} \tag{140}$$

12 Closed - Form Solution for the Clamped - Clamped Beam

12.1 Uniform Inertial Coefficient (*m* = 0)

In this case, we get a homogeneous set of 6 linear algebraic equations with 7 unknowns, namely the coefficients b_i in the stiffness, the natural frequency squared Ω^2 and the axial load q_0 that is treated as a parameter

$$\begin{aligned}
& 8b_2 - 36b_1 + q_0 L^3 (4 + 2g_1) = 0 \\
& 24b_3 - 108b_2 + 120b_0 - q_0 L^3 (26 + 9g_1) = 0 \\
& 48b_4 - 216b_3 + 240b_1 + q_0 L^3 (27 - 6g_1) - 2r_0 \Omega^2 = 0 \\
& -360b_4 + 400b_2 + q_0 L^3 (20 + 28g_1) + 3r_0 \Omega^2 = 0 \\
& 600b_3 - 25q_0 L^3 = 0 \\
& 840b_4 - 15g_1 q_0 L^3 - r_0 \Omega^2 = 0
\end{aligned} \tag{141}$$

We have to write all unknowns in terms of a single free parameter, say, b_4 and the axial load parameter q_0. In other word, the rank of the set (141) or its matrix must be less than 6. Thus, all minors of order 6 must vanish, leading to 7 equations with 2 unknowns r_0 and g_1

$$\begin{aligned}
&259034112 r_0 g_1 + 1135088640 r_0 = 0 \\
&80123904 r_0 g_1 + 146810880 r_0 = 0 \\
&137106432 r_0 g_1 + 334679040 r_0 = 0 \\
&r_0 = 0 \\
&41055728 r_0 g_1 + 4354560 r_0 = 0 \\
&46282752 g_1 - 365783044 = 0
\end{aligned} \tag{142}$$

The solution of the set (142) reads

$$r_0 = 0, \quad g_1 = -245/31 \tag{143}$$

which leads to a trivial solution for the problem since the inertial coefficient vanishes.

12.2 Linearly Varying Inertial Coefficient ($m = 1$)

The governing differential equation yields a set of 7 equations with 8 unknowns, namely the coefficients b_i in the stiffness, the natural frequency squared Ω^2 and the axial load q_0 that is treated as a parameter

$$\begin{aligned}
&8b_2 - 36b_1 + q_0 L^3 (4 + 2g_1 + \tfrac{4}{3} g_2) = 0 \\
&24b_3 - 108b_2 + 120b_0 - q_0 L^3 (26 + 9g_1 + 6g_2) = 0 \\
&48b_4 - 216b_3 + 240b_1 + q_0 L^3 (27 - 6g_1) - 2r_0\Omega^2 = 0 \\
&80b_5 - 360b_4 + 400b_2 + q_0 L^3 (20 + 28g_1 + \tfrac{4}{3} g_2) + \Omega^2 (3r_0 - 2r_1) = 0 \\
&-540b_5 + 600b_3 - q_0 L^3 (25 - 15g_2) + 3r_1\Omega^2 = 0 \\
&840b_4 - 15q_0 L^3 g_1 - r_0\Omega^2 = 0 \\
&1120b_5 - \tfrac{35}{3} q_0 L^3 g_2 - r_1\Omega^2 = 0
\end{aligned} \tag{144}$$

We want to express the unknowns in terms of parameters b_5 and Ω^2. Therefore, we demand the rank of the set (144) to be less than 7, leading to a system of 8 determinantal equations with 4 unknowns, namely r_0, r_1, g_1 and g_2

$$\begin{aligned}
&399940208640r_0g_2+57267689472r_1g_2+1271299276800r_0\\
&\qquad+267753185280r_1+290118205440r_0g_1-31922408448r_1g_1=0\\
&-63204986880r_0g_2+28154843136r_1g_2-164428185600r_0+69603287040r_1\\
&\qquad-89738772480r_0g_1+43898443776r_1g_1=0\\
&-87166679040r_0g_2+62465292288r_1g_2-374840524800r_0+254271467520r_1\\
&\qquad-153559203840r_0g_1+86200132608r_1g_1=0\\
&2175289344r_1g_1+3128315904r_1g_2+21920855040r_1+3483648000r_0g_2\\
&\qquad-9289728000r_0=0\\
&139055616r_0g_2+487710720r_0+459841536r_0g_1-202673664r_1g_1=0\\
&-46282752r_1g_1+13934592r_1g_2-365783040r_1-232243200r_0g_2=0\\
&11680671744g_2+5183668224g_1+40967700480=0\\
&11349725184r_1-22295347200r_0=0
\end{aligned} \tag{145}$$

The solution of the set (145) is

$$r_0=\tfrac{1629}{3200}r_1,\quad g_1=-\tfrac{245}{31}-\tfrac{3353}{1488}g_2 \tag{146}$$

Upon substitution of the expressions (146) into the set (144), we get the solution for the natural frequency squared

$$\Omega^2=\left(1120b_5-{}^{35}\!/\!_3\,q_0L^3g_2\right)/r_1 \tag{147}$$

The value of q_0 that corresponds to the vanishing frequency $\Omega^2=0$ is identified as the buckling load $q_{0,cr}$:

$$q_{0,cr}=96b_5\,/\,g_2L^3 \tag{148}$$

We rewrite Eq. (147) as follows

$$\Omega^2=1120b_5\left(1-q_0\,/\,q_{0,cr}\right)/r_1 \tag{149}$$

From Eq. (149), when $q_0\equiv 0$, $\Omega^2=1120\,b_5/r_1\equiv\Omega_0{}^2$. Therefore, Eq. (147) takes the form

$$\Omega_q{}^2=\Omega_0{}^2\left(1-q_0\,/\,q_{0,cr}\right) \tag{150}$$

where $\Omega_q{}^2$ is the natural frequency squared in presence of the axial load. We again note the following property: When q_0 tends to its critical value the natural frequency vanishes. The coefficients b_i in the stiffness are

$$
\begin{aligned}
b_0 &= \frac{100051}{40000} b_5 - \left(\frac{3409}{74400} + \frac{4456681}{119040000} g_2 \right) q_0 L^3 \\
b_1 &= \frac{771}{2000} b_5 - \left(\frac{303}{1240} + \frac{131667}{1984000} g_2 \right) q_0 L^3 \\
b_2 &= \frac{6939}{4000} b_5 + \left(\frac{933}{2480} + \frac{388997}{3968000} g_2 \right) q_0 L^3 \\
b_3 &= -\frac{47}{10} b_5 + \left(\frac{1}{24} + \frac{1}{30} g_2 \right) q_0 L^3, \quad b_4 = \frac{543}{800} b_5 - \left(\frac{35}{248} + \frac{112633}{2380800} g_2 \right) q_0 L^3
\end{aligned} \tag{151}
$$

12.3 Parabolically Varying Inertial ($m = 2$)

The governing differential equation yields a set of 8 equations with 8 unknowns, namely the coefficients 7 b_i of the stiffness, the natural frequency squared Ω^2, while the axial load q_0 is treated as a parameter

$$
\begin{aligned}
&8b_2 - 36b_1 + q_0 L^3 (4 + 2g_1 + 4/3 g_2 + g_3) = 0 \\
&24b_3 - 108b_2 + 120b_0 - q_0 L^3 (26 + 9g_1 + 6g_2 + 9/2 g_3) = 0 \\
&48b_4 - 216b_3 + 240b_1 + q_0 L^3 (27 - 6g_1) - 2r_0 \Omega^2 = 0 \\
&80b_5 - 360b_4 + 400b_2 + q_0 L^3 (20 + 28g_1 + 4/3 g_2 + 5g_3) + \Omega^2 (3r_0 - 2r_1) = 0 \\
&120b_6 - 540b_5 + 600b_3 - q_0 L^3 (25 - 15g_2 + 5g_3) + \Omega^2 (3r_1 - 2r_2) = 0 \\
&-756b_6 + 840b_4 - q_0 L^3 (15g_1 - 27/2 g_3) - \Omega^2 (r_0 - 3r_2) = 0 \\
&1120b_5 - 35/3 q_0 L^3 g_2 - r_1 \Omega^2 = 0 \\
&1440b_6 - 10q_0 L^3 g_3 - r_2 \Omega^2 = 0
\end{aligned} \tag{152}
$$

We have to express the unknowns in terms of parameters b_6 and Ω^2. Therefore, we demand the rank of the set (152) to be less than 8, leading to a system of 9 determinantal equations with 6 unknowns, namely the parameters r_0, r_1, r_2, g_1, g_2 and g_3. The solution of this set reads

$$
r_0 = \tfrac{1629}{3200} r_1 - \tfrac{681}{1600} r_2, \quad g_1 = -\tfrac{245}{31} - \tfrac{3353}{1488} g_2 - \tfrac{83}{124} g_3 \tag{153}
$$

Upon substitution of the expressions (153) into the set (152), we get the solution for the natural frequency squared

$$
\Omega^2 = \left(1440b_6 - 10q_0 L^3 g_3\right) / r_2 \tag{154}
$$

The value of q_0 that corresponds to the vanishing frequency $\Omega^2 = 0$ is identified as the buckling load $q_{0,cr}$:

$$
q_{0,cr} = 144b_6 / g_3 L^3 \tag{155}
$$

We rewrite Eq. (154) as follows

$$
\Omega^2 = 1440b_6 \left(1 - q_0 / q_{0,cr}\right) / r_2 \tag{156}
$$

From Eq. (156), when $q_0 \equiv 0$, $\Omega^2 = 1440\, b_6 / r_2 \equiv \Omega_0^{\ 2}$. Therefore, Eq. (154) takes the form

$$\Omega_q^{\ 2} = \Omega_0^{\ 2}\left(1 - q_0 / q_{0,cr}\right) \tag{157}$$

where $\Omega_q^{\ 2}$ is the natural frequency squared in presence of the axial load. When q_0 tends to its critical value $q_{0,cr}$ the natural frequency vanishes. The coefficients b_i in the stiffness are

$$\begin{aligned}
b_0 &= \{(7118064 r_1 - 3546144 r_2) b_6 \\
&\quad + [(17920 + 10640 g_2 + 20146 g_3) r_2 - 49431 r_1 g_3] q_0 L^3 \} / 40320 r_2 \\
b_1 &= \frac{(203472 r_1 + 6048 r_2) b_6 + [(4480 + 2660 g_2 - 1162 g_3) r_2 - 1413 g_3 r_1] q_0 L^3}{6720 r_2} \\
b_2 &= \frac{(54432 r_2 + 1831248 r_1) b_6 - [(22960 + 8750 g_2 - 5222 g_3) r_2 + 12717 g_3 r_1] q_0 L^3}{40320 r_2} \\
b_3 &= \{(30912 r_2 - 13536 r_1) b_6 + [(28 - 35 g_2 - 168 g_3) r_2 + 94 g_3 r_1] q_0 L^3 \} / 672 r_2 \\
b_4 &= \frac{(107568 r_1 - 1336608 r_2) b_6 + [(6160 + 3710 g_2 + 3682 g_3) r_2 - 747 g_3 r_1] q_0 L^3}{80640 r_2} \\
b_5 &= [864 r_1 b_6 + (7 g_2 r_2 - 6 g_3 r_1) q_0 L^3] / 672 r_2
\end{aligned} \tag{158}$$

12.4 Cubic Inertial Coefficient (*m* = 3)

The governing differential equation yields a set of 9 equations with 10 unknowns, namely the coefficients b_i in the stiffness, the natural frequency squared Ω^2 and the axial load q_0 that is treated as a parameter

$$\begin{aligned}
&8b_2 - 36b_1 + q_0 L^3 \left(4 + 2g_1 + \tfrac{4}{3} g_2 + g_3 + \tfrac{4}{5} g_4\right) = 0 \\
&24b_3 - 108b_2 + 120b_0 - q_0 L^3 \left(26 + 9g_1 + 6g_2 + \tfrac{9}{2} g_3 + \tfrac{18}{5} g_4\right) = 0 \\
&48b_4 - 216b_3 + 240b_1 + q_0 L^3 \left(27 - 6g_1\right) - 2r_0 \Omega^2 = 0 \\
&80b_5 - 360b_4 + 400b_2 + q_0 L^3 \left(20 + 28g_1 + \tfrac{4}{3} g_2 + 5g_3 + 4g_4\right) + \Omega^2 \left(3r_0 - 2r_1\right) = 0 \\
&120b_6 - 540b_5 + 600b_3 - q_0 L^3 \left(25 - 15g_2 + 5g_3\right) + \Omega^2 \left(3r_1 - 2r_2\right) = 0 \\
&168b_7 - 756b_6 + 840b_4 - q_0 L^3 \left(15g_1 - \tfrac{27}{2} g_3 + \tfrac{24}{5} g_4\right) - \Omega^2 \left(r_0 - 3r_2 + 2r_3\right) = 0 \\
&-1008b_7 + 1120b_5 - q_0 L^3 \left(\tfrac{35}{3} g_2 - \tfrac{63}{5} g_4\right) - \Omega^2 \left(r_1 - 3r_3\right) = 0 \\
&1440b_6 - 10q_0 L^3 g_3 - r_2 \Omega^2 = 0 \\
&1800b_7 - 9q_0 L^3 g_4 - r_3 \Omega^2 = 0
\end{aligned} \tag{159}$$

We have to express the unknowns in terms of parameters b_5 and Ω^2. Therefore, we demand the rank of the set (159) to be less than 9, leading to a system of 10 determinantal equations with 8 unknowns, namely the parameters r_0, r_1, r_2, r_3, g_1, g_2, g_3 and g_4. The solution of this set reads

$$r_0 = \tfrac{1629}{3200} r_1 - \tfrac{681}{1600} r_2 + \tfrac{23263}{80000} r_3, \quad g_1 = -\tfrac{245}{31} - \tfrac{3353}{1488} g_2 - \tfrac{83}{124} g_3 - \tfrac{77429}{62000} g_4 \tag{160}$$

Upon substitution of the expressions (160) into the set (159), we get the solution for the natural frequency squared

$$\Omega^2 = \left(1800b_7 - 9q_0L^3g_4\right)/r_3 \tag{161}$$

The value of q_0 that corresponds to the vanishing frequency $\Omega^2 = 0$ is the buckling load $q_{0,cr}$:

$$q_{0,cr} = 200b_7 / g_4L^3 \tag{168}$$

We rewrite Eq. (161) as follows

$$\Omega^2 = 1800b_7\left(1 - q_0 / q_{0,cr}\right)/r_3 \tag{163}$$

From Eq. (163), when $q_0 \equiv 0$, $\Omega^2 = 1800\,b_7/r_3 \equiv \Omega_0{}^2$. Therefore, Eq. (161) takes the form

$$\Omega_q{}^2 = \Omega_0{}^2\left(1 - q_0 / q_{0,cr}\right) \tag{164}$$

where $\Omega_q{}^2$ is the natural frequency squared in presence of the axial load. Note that when q_0 tends to its critical value the natural frequency vanishes. The coefficients b_i in the stiffness are

$$\begin{aligned}
b_0 &= \{(444879000r_1 - 221634000r_2 + 651337200r_3)b_7 + [(896000 + 532000g_2 \\
&\quad - 224000g_3 - 2577966g_4)r_3 + (1108170r_2 - 2224395r_1)g_4]q_0L^3\}/2016000r_3 \\
b_1 &= \{(12717000r_1 + 378000r_2 + 18267600r_3)b_7 + [(224000 + 133000g_2 - 56000g_3 \\
&\quad + 78342g_4)r_3 - (1890r_2 + 63585r_1)g_4]q_0L^3\}/336000r_3 \\
b_2 &= \{(114453000r_1 + 3402000r_2 + 164408400r_3)b_7 - [(1148000 + 437500g_2 \\
&\quad + 280000g_3 - 1382322g_4)r_3 + (572265r_1 + 17010r_2)g_4]q_0L^3\}/2016000r_3 \\
b_3 &= \{(-253800r_1 + 579600r_2 - 395280r_3)b_7 + [(420 - 525g_2 + 700g_3 + 1296g_4)r_3 \\
&\quad + (1269r_1 - 2898r_2)g_4]q_0L^3\}/10080r_3 \\
b_4 &= \{(6723000r_1 - 83538000r_2 + 179780400r_3)b_7 + [(308000 + 185500g_2 \\
&\quad - 280000g_3 - 439422g_4)r_3 + (417690r_2 - 33615r_1)g_4]q_0L^3\}/4032000r_3 \\
b_5 &= \{(5400r_1 - 43920r_3)b_7 + [(144 + 35g_2)r_3 - 27g_4r_1]q_0L^3\}/3360r_3 \\
b_6 &= [1800r_2b_7 + (10g_3r_3 - 9g_4r_2)q_0L^3]/1440r_3
\end{aligned} \tag{165}$$

12.5 Quartic Inertial Coefficient (*m* = 4)

The governing differential equation yields a set of 10 equations with 11 unknowns, namely the coefficients b_i in the stiffness, the natural frequency squared Ω^2 and the axial load q_0 that is treated as a parameter

$$
\begin{aligned}
&8b_2 - 36b_1 + q_0L^3\left(4 + 2g_1 + {}^4\!/_3\, g_2 + g_3 + {}^4\!/_5\, g_4 + {}^2\!/_3\, g_5\right) = 0\\
&24b_3 - 108b_2 + 120b_0 - q_0L^3\left(26 + 9g_1 + 6g_2 + {}^9\!/_2\, g_3 + {}^{18}\!/_5\, g_4 + 3g_5\right) = 0\\
&48b_4 - 216b_3 + 240b_1 + q_0L^3\left(27 - 6g_1\right) - 2r_0\Omega^2 = 0\\
&80b_5 - 360b_4 + 400b_2 + q_0L^3\left(20 + 28g_1 + {}^4\!/_3\, g_2 + 5g_3 + 4g_4 + {}^{10}\!/_3\, g_5\right)\\
&\qquad + \Omega^2\left(3r_0 - 2r_1\right) = 0 \qquad (166)\\
&120b_6 - 540b_5 + 600b_3 - q_0L^3\left(25 - 15g_2 + 5g_3\right) + \Omega^2\left(3r_1 - 2r_2\right) = 0\\
&168b_7 - 756b_6 + 840b_4 - q_0L^3\left(15g_1 - {}^{27}\!/_2\, g_3 + {}^{24}\!/_5\, g_4\right) - \Omega^2\left(r_0 - 3r_2 + 2r_3\right) = 0\\
&224b_8 - 1008b_7 + 1120b_5 - q_0L^3\left({}^{35}\!/_3\, g_2 - {}^{63}\!/_5\, g_4 + {}^{14}\!/_3\, g_5\right) - \Omega^2\left(r_1 - 3r_3 + 2r_4\right) = 0\\
&-1296b_6 + 1440b_6 - q_0L^3\left(10g_3 - 12g_5\right) - \Omega^2\left(r_2 - 3r_4\right) = 0\\
&1800b_7 - 9q_0L^3g_4 - r_3\Omega^2 = 0\\
&2200b_8 - {}^{25}\!/_3\, q_0L^3g_5 - r_4\Omega^2 = 0
\end{aligned}
$$

We have to express the unknowns in terms of parameters b_5 and Ω^2. Therefore, we demand the rank of the set (166) to be less than 10, leading to a system of 11 determinantal equations with 10 unknown, namely the parameters r_0, r_1, r_2, r_3, r_4, g_1, g_2, g_3, g_4 and g_5. The solution for this set reads

$$
\begin{aligned}
r_0 &= \tfrac{1629}{3200} r_1 - \tfrac{681}{1600} r_2 + \tfrac{23263}{80000} r_3 - \tfrac{1698}{6875} r_4\\
g_1 &= -\tfrac{245}{31} - \tfrac{3353}{1488} g_2 - \tfrac{83}{124} g_3 - \tfrac{77429}{62000} g_4 - \tfrac{20503}{409240} g_5
\end{aligned}
\qquad (167)
$$

We get the solution for the natural frequency squared

$$\Omega^2 = \left(2200b_8 - {}^{25}\!/_3\, q_0L^3g_5\right)/r_4 \qquad (168)$$

The value of q_0 that corresponds to the vanishing frequency $\Omega^2 = 0$ is identified as the buckling load $q_{0,cr}$:

$$q_{0,cr} = 264b_8 / g_5L^3 \qquad (169)$$

We put Eq. (168) as follows

$$\Omega^2 = 2200b_8\left(1 - q_0 / q_{0,cr}\right)/r_4 \qquad (170)$$

From Eq. (170), when $q_0 \equiv 0$, $\Omega^2 = 2200\, b_8/r_4 \equiv \Omega_0{}^2$. Therefore, Eq. (168) takes the form

$$\Omega_q{}^2 = \Omega_0{}^2\left(1 - q_0 / q_{0,cr}\right) \qquad (171)$$

where $\Omega_q{}^2$ is the natural frequency squared in presence of the axial load. Note that when q_0 tends to its critical value the natural frequency vanishes. The coefficients b_i in the stiffness are

$$
\begin{aligned}
b_0 &= \{(10071600r_4 + 28096200r_2 + 135036000r_0)b_8 \\
&\quad + [(45430g_5 + 24696g_4 + 120186g_3 + 227700g_1 - 52920r_4)r_4 \\
&\quad - (511500r_0 + 106425r_2)g_5]q_0L^3\}/4989600r_4, \quad b_1 = -7q_0L^3/120 \\
b_2 &= \{(10071600r_4 + 28096200r_2 - 78408000r_0)b_8 \\
&\quad - [(93170g_5 + 10584g_4 + 87714g_3 + 285120g_1)r_4 \\
&\quad - (297000r_0 - 106425r_2)g_5]q_0L^3\}/4989600r_4 \\
b_3 &= \{18803400r_3b_8 + [(13923g_4 + 298350)r_4 + 71225g_5r_3]q_0L^3\}/3029400r_4 \\
b_4 &= \{(183120r_4 - 653400r_2 + 237600r_0)b_8 \\
&\quad [-(938g_5 - 990g_3 + 1620g_1)r_4 + (2475r_2 - 900r_0)g_5]q_0L^3\}/90720r_4 \\
b_5 &= \{-5953200r_3b_8 - [(12546g_4 + 34425)r_4 - 22550r_3g_5]q_0L^3\}/1009800r_4 \\
b_6 &= \{(-10608r_4 + 3960r_2)b_8 + [(18g_3 + 26g_5)r_4 - 15r_2g_5]q_0L^3\}/2592 \\
b_7 &= [6600r_3b_8 + (27g_4r_4 - 25r_3g_5)q_0L^3]/5400r_4
\end{aligned}
\tag{172}
$$

12.6 Quintic Inertial Coefficient ($m = 5$)

The governing differential equation yields a set of 11 equations with 12 unknowns, namely the coefficients b_i in the stiffness, the natural frequency squared Ω^2 and the axial load q_0 that is treated as a parameter

$$
\begin{aligned}
& 8b_2 - 36b_1 + q_0L^3(4 + 2g_1 + 4/3\,g_2 + g_3 + 4/5\,g_4 + 2/3\,g_5 + 4/7\,g_6) = 0 \\
& 24b_3 - 108b_2 + 120b_0 - q_0L^3(26 + 9g_1 + 6g_2 + 9/2\,g_3 + 18/5\,g_4 + 3g_5 + 18/7\,g_6) = 0 \\
& 48b_4 - 216b_3 + 240b_1 + q_0L^3(27 - 6g_1) - 2r_0\Omega^2 = 0 \\
& 80b_5 - 360b_4 + 400b_2 + q_0L^3(20 + 28g_1 + 4/3\,g_2 + 5g_3 + 4g_4 + 10/3\,g_5 + 20/7\,g_6) \\
& \quad + \Omega^2(3r_0 - 2r_1) = 0 \\
& 120b_6 - 540b_5 + 600b_3 - q_0L^3(25 - 15g_2 + 5g_3) + \Omega^2(3r_1 - 2r_2) = 0 \\
& 168b_7 - 756b_6 + 840b_4 - q_0L^3(15g_1 - 27/2\,g_3 + 24/5\,g_4) - \Omega^2(r_0 - 3r_2 + 2r_3) = 0 \\
& 224b_8 - 1008b_7 + 1120b_5 - q_0L^3(35/3\,g_2 - 63/5\,g_4 + 14/3\,g_5) - \Omega^2(r_1 - 3r_3 + 2r_4) = 0 \\
& 288b_9 - 1296b_6 + 1440b_6 - q_0L^3(10g_3 - 12g_5 + 32/7\,g_6) - \Omega^2(r_2 - 3r_4 + 2r_5) = 0 \\
& -1620b_9 + 1800b_7 - q_0L^3(9g_4 - 81/7\,g_6) - \Omega^2(r_3 - 3r_5) = 0 \\
& 2200b_8 - 25/3\,q_0L^3g_5 - r_4\Omega^2 = 0 \\
& 2640b_9 - 55/7\,q_0L^3g_6 - r_5\Omega^2 = 0
\end{aligned}
\tag{173}
$$

We have to express the unknowns in terms of parameters b_5 and Ω^2. Therefore, we demand the rank of the set (173) to be less than 11, leading to a system of 12 determinantal equations with 12 unknowns, namely the parameters r_0, r_1, r_2, r_3, r_4, g_1, g_2, g_3, g_4 and g_5. The solution of this set reads

$$r_0 = \tfrac{1629}{3200} r_1 - \tfrac{681}{1600} r_2 + \tfrac{23263}{80000} r_3 - \tfrac{1698}{6875} r_4 + \tfrac{67809}{352000} r_5$$
$$g_1 = -\tfrac{245}{31} - \tfrac{3353}{1488} g_2 - \tfrac{83}{124} g_3 - \tfrac{77429}{62000} g_4 - \tfrac{20503}{409240} g_5 - \tfrac{105629}{124000} g_6 \tag{174}$$

Upon substitution of the expressions (174) into the set (173), we get the solution for the natural frequency squared

$$\Omega^2 = \left(2680 b_9 - \tfrac{55}{7} q_0 L^3 g_6\right) / r_5 \tag{175}$$

The value of q_0 that corresponds to the vanishing frequency $\Omega^2 = 0$ is identified as the buckling load $q_{0,cr}$:

$$q_{0,cr} = 336 b_9 / g_6 L^3 \tag{176}$$

We express Eq. (175) as follows

$$\Omega^2 = 2680 b_9 \left(1 - q_0 / q_{0,cr}\right) / r_5 \tag{177}$$

From Eq. (177), when $q_0 \equiv 0$, $\Omega^2 = 2680\, b_9 / r_5 \equiv \Omega_0{}^2$. Therefore, Eq. (175) takes the form

$$\Omega_q{}^2 = \Omega_0{}^2 \left(1 - q_0 / q_{0,cr}\right) \tag{178}$$

where $\Omega_q{}^2$ is the natural frequency squared in presence of the axial load. Note that when q0 tends to its critical value the natural frequency vanishes. The coefficients bi in the stiffness are

$$\begin{aligned}
b_0 = {} & \{(10071600 r_4 + 28096200 r_2 + 135036000 r_0) b_8 \\
& + [(45430 g_5 + 24696 g_4 + 120186 g_3 + 227700 g_1 - 52920 r_4) r_4 \\
& - (511500 r_0 + 106425 r_2) g_5] q_0 L^3 \} / 4989600 r_4 , \quad b_1 = -7 q_0 L^3 / 120 \\
b_2 = {} & \{(10071600 r_4 + 28096200 r_2 - 78408000 r_0) b_8 \\
& - [(93170 g_5 + 10584 g_4 + 87714 g_3 + 285120 g_1) r_4 \\
& - (297000 r_0 - 106425 r_2) g_5] q_0 L^3 \} / 4989600 r_4 \\
b_3 = {} & \{18803400 r_3 b_8 + [(13923 g_4 + 298350) r_4 + 71225 g_5 r_3] q_0 L^3 \} / 3029400 r_4 \\
b_4 = {} & \{(183120 r_4 - 653400 r_2 + 237600 r_0) b_8 \\
& [-(938 g_5 - 990 g_3 + 1620 g_1) r_4 + (2475 r_2 - 900 r_0) g_5] q_0 L^3 \} / 90720 r_4 \\
b_5 = {} & \{-5953200 r_3 b_8 - [(12546 g_4 + 34425) r_4 - 22550 r_3 g_5] q_0 L^3 \} / 1009800 r_4 \\
b_6 = {} & \{(-10608 r_4 + 3960 r_2) b_8 + [(18 g_3 + 26 g_5) r_4 - 15 r_2 g_5] q_0 L^3 \} / 2592 \\
b_7 = {} & [6600 r_3 b_8 + (27 g_4 r_4 - 25 r_3 g_5) q_0 L^3] / 5400 r_4
\end{aligned} \tag{179}$$

13 Closed-Form Solution for the Beam that is Simply Supported at One End and Clamped at the Other

13.1 Uniform Inertial Coefficient (*m* = 0)

We get a homogeneous set of 6 linear algebraic equations with 7 unknowns, namely the coefficients b_i in the stiffness, the natural frequency squared Ω^2 and the axial load q_0 that is treated as a parameter

$$\begin{aligned}
&-48b_1 - 2q_0L^3 = 0\\
&-144b_2 + 240b_0 - q_0L^3(24 + 14g_1) - 2r_0\Omega^2 = 0\\
&-288b_3 + 480b_1 + 36q_0L^3 = 0\\
&-480b_4 + 800b_2 + q_0L^3(40 + 44g_1) + 4r_0\Omega^2 = 0\\
&1200b_3 - 50q_0L^3 = 0\\
&1680b_4 - 30g_1q_0L^3 - 2r_0\Omega^2 = 0
\end{aligned} \tag{180}$$

We have to write all unknowns in terms of a single free parameter, say, b_4 and the axial load parameter q_0. In other words, the rank of the set (180) or its matrix must be less than 6. Thus, all minors of order 6 must vanish. Yet the minor obtained by extraction of the sixth column of the matrix of the set (180) equals $-74317824000000L^3$, which is non-zero. We conclude that the problem has no non-trivial solution.

13.2 Linearly Varying Inertial Coefficient (*m* = 1)

We have a homogeneous set of 7 linear algebraic equations with 8 unknowns, namely the coefficients b_i in the stiffness, the natural frequency squared Ω^2 and the axial load q_0 that is treated as a parameter

$$\begin{aligned}
&-48b_1 - 2q_0L^3 = 0\\
&-144b_2 + 240b_0 - q_0L^3(24 + 14g_1 + 8g_2) - 2r_0\Omega^2 = 0\\
&-288b_3 + 480b_1 + q_0L^3(36 - 2g_2) - 2r_1\Omega^2 = 0\\
&-480b_4 + 800b_2 + q_0L^3(40 + 44g_1 + 40/3\, g_2) + 4r_0\Omega^2 = 0\\
&-720b_5 + 1200b_3 - q_0L^3(50 - 20g_2) + 4r_1\Omega^2 = 0\\
&1680b_4 - 30g_1q_0L^3 - 2r_0\Omega^2 = 0\\
&2240b_5 - 70/3\, q_0L^3g_2 - 2r_1\Omega^2 = 0
\end{aligned} \tag{181}$$

We demand all unknowns to be expressed in terms of a single free parameter, say, b_5 and the axial load parameter q_0, leading to the condition the rank of the set (181) or its matrix must be less than 7 or all minors of order 7 must vanish. We produce the minors by extracting one column of the matrix among 8; therefore, we obtain a set of 8 determinantal equations for 4 unknowns, namely the parameters r_0, r_1, g_1 and g_2:

$$\begin{aligned}
&2397988454400 r_0 g_2 + 9591953817600 r_0 + 15786861133824 r_1 g_1 \\
&\qquad + 34792632483840 r_1 + 11597544161280 r_1 g_2 = 0 \\
&r_1 = 0 \\
&178362777600 r_0 g_2 + 713451110400 r_0 + 2201166544896 r_1 g_1 \\
&\qquad + 828396011520 r_1 + 2485188034560 r_1 g_2 = 0 \\
&6341787648 r_1 g_2 - 16052649984 r_1 = 0 \\
&495452160 r_0 g_2 + 1981808640 r_0 + 8875671552 r_1 g_1 = 0 \\
&5549064192 r_1 g_2 + 1486356480 r_1 = 0 \\
&4161798144 g_2 + 16647192576 = 0 \\
&r_1 = 0
\end{aligned} \tag{182}$$

The solution of the set (182) reads

$$r_1 = 0, \quad g_2 = -4 \tag{183}$$

Yet, $r_1 = 0$ implies that the mass density is constant. Thus, for $r_1 \neq 0$ no solution exists.

13.3 Parabolically Varying Inertial Coefficient ($m = 2$)

The governing differential equation yields a set of 8 equations with 9 unknowns, namely the coefficients b_i in the stiffness, the natural frequency squared Ω^2 and the axial load q_0

$$\begin{aligned}
&-48 b_1 - 2 q_0 L^3 = 0 \\
&-144 b_2 + 240 b_0 - q_0 L^3 (24 + 14 g_1 + 8 g_2 + 6 g_3) - 2 r_0 \Omega^2 = 0 \\
&-288 b_3 + 480 b_1 + q_0 L^3 (36 - 2 g_2) - 2 r_1 \Omega^2 = 0 \\
&-480 b_4 + 800 b_2 + q_0 L^3 (40 + 44 g_1 + {}^{40}\!/_3\, g_2 + 8 g_3) + (4 r_0 - 2 r_2)\Omega^2 = 0 \\
&-720 b_5 + 1200 b_3 - q_0 L^3 (50 - 20 g_2) + 4 r_1 \Omega^2 = 0 \\
&-1008 b_6 + 1680 b_4 - q_0 L^3 (30 g_1 - 18 g_3) - (2 r_0 - 4 r_2)\Omega^2 = 0 \\
&2240 b_5 - {}^{70}\!/_3\, q_0 L^3 g_2 - 2 r_1 \Omega^2 = 0 \\
&2880 b_6 - 20 q_0 L^3 g_3 - 2 r_2 \Omega^2 = 0
\end{aligned} \tag{184}$$

We have to express the unknowns in terms of parameters b_6 and Ω^2. Therefore, we demand the rank of the set (184) to be less than 8, leading to a system of 9 determinantal equations with 6 unknowns, namely the parameters r_0, r_1, r_2, g_1, g_2 and g_3. The solution of this set reads

$$r_1 = 0, \quad g_2 = -4 \tag{185}$$

Upon substitution of the expressions (185) into the set (184), we get the solution for the natural frequency squared

$$\Omega^2 = \left(1440b_6 - 10q_0L^3g_3\right)/r_2 \tag{186}$$

The value of q_0 that corresponds to the vanishing frequency $\Omega^2 = 0$ is identified as the buckling load $q_{0,cr}$:

$$q_{0,cr} = 144b_6 / g_3L^3 \tag{187}$$

We express Eq. (186) as follows

$$\Omega^2 = 1440b_6\left(1 - q_0 / q_{0,cr}\right)/r_2 \tag{188}$$

From Eq. (188), when $q_0 \equiv 0$, $\Omega^2 = 1440\,b_6/r_2 \equiv \Omega_0{}^2$. Therefore, Eq. (145) takes the form

$$\Omega_q{}^2 = \Omega_0{}^2\left(1 - q_0 / q_{0,cr}\right) \tag{189}$$

where $\Omega_q{}^2$ is the natural frequency squared in presence of the axial load. Note that when q_0 tends to its critical value the natural frequency vanishes. The coefficients b_i in the stiffness are

$$\begin{aligned}
b_0 &= \frac{(510840r_2 + 2455200r_0)b_6 - [(209g_3 - 6325g_1 + 1470)r_2 + 17050g_3r_0]q_0L^3}{138600r_2} \\
b_1 &= -7q_0L^3/120 \\
b_2 &= \frac{(255420r_2 - 712800r_0)b_6 - [(2992g_3 + 3960g_1 - 315)r_2 - 4950g_3r_0]q_0L^3}{69300r_2} \\
b_3 &= 13q_0L^3/132 \\
b_4 &= \frac{(-2376r_2 + 864r_0)b_6 + [(11g_3 + 9g_1)r_2 - 6g_3r_0]q_0L^3}{504r_2}, \quad b_5 = -3q_0L^3/88
\end{aligned} \tag{190}$$

13.4 Cubic Inertial Coefficient ($m = 3$)

The governing differential equation yields a set of 9 equations with 10 unknowns, namely the coefficients b_i in the stiffness, the natural frequency squared Ω^2 and the axial load q_0 that is treated as a parameter

$$\begin{aligned}
&-48b_1 - 2q_0L^3 = 0 \\
&-144b_2 + 240b_0 - q_0L^3\left(24 + 14g_1 + 8g_2 + 6g_3 + {}^{24}\!/_5\, g_4\right) - 2r_0\Omega^2 = 0 \\
&-288b_3 + 480b_1 + q_0L^3\left(36 - 2g_2\right) - 2r_1\Omega^2 = 0 \\
&-480b_4 + 800b_2 + q_0L^3\left(40 + 44g_1 + {}^{40}\!/_3\, g_2 + 8g_3 + 8g_4\right) + (4r_0 - 2r_2)\Omega^2 = 0 \\
&-720b_5 + 1200b_3 - q_0L^3\left(50 - 20g_2 + 2g_4\right) + \Omega^2\left(4r_1 - 2r_2\right) = 0 \\
&-1008b_6 + 1680b_4 - q_0L^3\left(30g_1 - 18g_3\right) - (2r_0 - 4r_2)\Omega^2 = 0 \\
&-1344b_7 + 2240b_5 - q_0L^3\left({}^{70}\!/_3\, g_2 + {}^{84}\!/_5\, g_4\right) - (2r_1 - 4r_3)\Omega^2 = 0 \\
&2880b_6 - 20q_0L^3g_3 - 2r_2\Omega^2 = 0 \\
&3600b_7 - 18q_0L^3g_4 - 2r_3\Omega^2 = 0
\end{aligned} \tag{191}$$

We have to express the unknowns in terms of parameters b_7 and Ω^2. Therefore, we demand the rank of the set (191) to be less than 9, leading to a system of 10 determinantal equations with 8 unknowns, namely the parameters r_0, r_1, r_2, r_3, g_1, g_2, g_3 and g_4. The solution of this set reads

$$r_1 = -\tfrac{1002}{5225} r_3, \quad g_2 = -4 - \tfrac{186}{625} g_4 \tag{192}$$

Upon substitution of the expressions (192) into the set (191), we get the solution for the natural frequency squared

$$\Omega^2 = \left(1800 b_7 - 9 q_0 L^3 g_4\right) / r_3 \tag{193}$$

The value of q_0 that corresponds to the vanishing frequency $\Omega^2 = 0$ is identified as the buckling load $q_{0,cr}$:

$$q_{0,cr} = 200 b_7 / g_4 L^3 \tag{194}$$

We put Eq. (193) as follows

$$\Omega^2 = 1800 b_7 \left(1 - q_0 / q_{0,cr}\right) / r_3 \tag{195}$$

From Eq. (193), when $q_0 \equiv 0$, $\Omega^2 = 1800\, b_7 / r_3 \equiv \Omega_0{}^2$. Therefore, Eq. (193) takes the form

$$\Omega_q{}^2 = \Omega_0{}^2 \left(1 - q_0 / q_{0,cr}\right) \tag{196}$$

where $\Omega_q{}^2$ is the natural frequency squared in presence of the axial load. Note that when q_0 tends to its critical value the natural frequency vanishes. The coefficients b_i in the stiffness are deduced from the remaining equations of the set (191)

$$\begin{aligned}
b_0 &= \{(2554200 r_2 + 12276000 r_0) b_7 + [(2744 g_4 + 13354 g_3 + 25300 g_1 - 5880) r_3 \\
&\quad - (61380 r_0 + 12771 r_2) g_4] q_0 L^3\} / 554400 r_3, \quad b_1 = -7 q_0 L^3 / 120 \\
b_2 &= -\{(7128000 r_0 - 2554200 r_2) b_7 + [(1176 g_4 + 9746 g_3 + 31680 g_1 - 2520) r_3 \\
&\quad + (12771 r_2 - 35640 r_0) g_4] q_0 L^3\} / 554400 r_3 \\
b_3 &= \tfrac{259}{51} b_7 + \left(\tfrac{13}{132} - \tfrac{35}{1683} g_4\right) q_0 L^3 \\
b_4 &= \frac{(21600 r_0 - 59400 r_2) b_7 + [(180 g_1 - 110 g_3) r_3 + (297 r_2 - 108 r_0) g_4] q_0 L^3}{10080 r_3} \\
b_5 &= -\tfrac{82}{17} b_7 + \left(\tfrac{164}{14025} g_4 - \tfrac{3}{88}\right) q_0 L^3, \quad b_6 = \left[1800 r_2 b_7 + (10 g_3 r_3 - 9 g_4 r_2) q_0 L^3\right] / 1440 r_3
\end{aligned} \tag{197}$$

13.5 Quartic Inertial Coefficient ($m = 4$)

The governing differential equation yields a set of 10 equations with 11 unknowns, namely the coefficients b_i in the stiffness, the natural frequency squared Ω^2 and the axial load q_0 that is treated as a parameter

$$
\begin{aligned}
&-48b_1 - 2q_0L^3 = 0\\
&-144b_2 + 240b_0 - q_0L^3\left(24 + 14g_1 + 8g_2 + 6g_3 + {}^{24}\!/\!_5\, g_4 + 4g_5\right) - 2r_0\Omega^2 = 0\\
&-288b_3 + 480b_1 + q_0L^3\left(36 - 2g_2\right) - 2r_1\Omega^2 = 0\\
&-480b_4 + 800b_2 + q_0L^3\left(40 + 44g_1 + {}^{40}\!/\!_3\, g_2 + 8g_3 + 8g_4 + {}^{20}\!/\!_3\, g_5\right) + \left(4r_0 - 2r_2\right)\Omega^2 = 0\\
&-720b_5 + 1200b_3 - q_0L^3\left(50 - 20g_2 + 2g_4\right) + \Omega^2\left(4r_1 - 2r_2\right) = 0\\
&-1008b_6 + 1680b_4 - q_0L^3\left(30g_1 - 18g_3 + 2g_5\right) - \left(2r_0 - 4r_2\right)\Omega^2 = 0\\
&-1344b_7 + 2240b_5 - q_0L^3\left({}^{70}\!/\!_3\, g_2 + {}^{84}\!/\!_5\, g_4\right) - \left(2r_1 - 4r_3\right)\Omega^2 = 0\\
&-1728b_8 + 2880b_5 - q_0L^3\left(20g_3 - 16g_5\right) - \left(2r_2 - 4r_4\right)\Omega^2 = 0\\
&3600b_7 - 18q_0L^3g_4 - 2r_3\Omega^2 = 0\\
&4400b_8 - {}^{50}\!/\!_3\, q_0L^3g_5 - 2r_4\Omega^2 = 0
\end{aligned} \tag{198}
$$

We have to express the unknowns in terms of parameters b_5 and Ω^2. Therefore, we demand the rank of the set (198) to be less than 10, leading to a system of 11 determinantal equations with 10 unknowns, namely the parameters r_0, r_1, r_2, r_3, r_4, g_1, g_2, g_3, g_4 and g_5. The solution for this set reads

$$
r_1 = -\tfrac{1002}{5225} r_3, \quad g_2 = -4 - \tfrac{186}{625} g_4 \tag{199}
$$

Upon substitution of the expressions (199) into the set (198), we get the solution for the natural frequency squared

$$
\Omega^2 = \left(2200b_8 - {}^{25}\!/\!_3\, q_0L^3g_5\right)/r_4 \tag{200}
$$

The value of q_0 that corresponds to the vanishing frequency $\Omega^2 = 0$ is identified as the buckling load $q_{0,cr}$:

$$
q_{0,cr} = 264b_8 / g_5L^3 \tag{201}
$$

We rewrite Eq. (200) as follows

$$
\Omega^2 = 2200b_8\left(1 - q_0 / q_{0,cr}\right)/r_4 \tag{202}
$$

From Eq. (202), when $q_0 \equiv 0$, $\Omega^2 = 2200\, b_8/r_4 \equiv \Omega_0{}^2$. Therefore, Eq. (200) takes the form

$$
\Omega_q{}^2 = \Omega_0{}^2\left(1 - q_0 / q_{0,cr}\right) \tag{203}
$$

where $\Omega_q{}^2$ is the natural frequency squared in presence of the axial load. Note that when q_0 tends to its critical value the natural frequency vanishes. The coefficients b_i in the stiffness are deduced from the remaining equations of the set (198)

$$
\begin{aligned}
b_0 &= \{(10071600r_4 + 28096200r_2 + 135036000r_0)b_8 \\
&\quad + [(45430g_5 + 24696g_4 + 120186g_3 + 227700g_1 - 52920r_4)r_4 \\
&\quad - (511500r_0 + 106425r_2)g_5]q_0L^3\}/4989600r_4, \quad b_1 = -7q_0L^3/120 \\
b_2 &= \{(10071600r_4 + 28096200r_2 - 78408000r_0)b_8 \\
&\quad - [(93170g_5 + 10584g_4 + 87714g_3 + 285120g_1)r_4 \\
&\quad - (297000r_0 - 106425r_2)g_5]q_0L^3\}/4989600r_4 \\
b_3 &= \{18803400r_3b_8 + [(13923g_4 + 298350)r_4 + 71225g_5r_3]q_0L^3\}/3029400r_4 \\
b_4 &= \{(183120r_4 - 653400r_2 + 237600r_0)b_8 \\
&\quad [-(938g_5 - 990g_3 + 1620g_1)r_4 + (2475r_2 - 900r_0)g_5]q_0L^3\}/90720r_4 \\
b_5 &= \{-5953200r_3b_8 - [(12546g_4 + 34425)r_4 - 22550r_3g_5]q_0L^3\}/1009800r_4 \\
b_6 &= \{(-10608r_4 + 3960r_2)b_8 + [(18g_3 + 26g_5)r_4 - 15r_2g_5]q_0L^3\}/2592 \\
b_7 &= [6600r_3b_8 + (27g_4r_4 - 25r_3g_5)q_0L^3]/5400r_4
\end{aligned} \tag{204}
$$

13.6 Quintic Inertial Coefficient ($m = 5$)

The governing differential equation yields a set of 11 equations with 12 unknowns, namely the coefficients b_i in the stiffness, the natural frequency squared Ω^2 and the axial load q_0 that is treated as a parameter

$$
\begin{aligned}
&-48b_1 - 2q_0L^3 = 0 \\
&-144b_2 + 240b_0 - q_0L^3\left(24 + 14g_1 + 8g_2 + 6g_3 + \tfrac{24}{5}g_4 + 4g_5 + \tfrac{24}{7}g_6\right) - 2r_0\Omega^2 = 0 \\
&-288b_3 + 480b_1 + q_0L^3(36 - 2g_2) - 2r_1\Omega^2 = 0 \\
&-480b_4 + 800b_2 + q_0L^3\left(40 + 44g_1 + \tfrac{40}{3}g_2 + 8g_3 + 8g_4 + \tfrac{20}{3}g_5 + \tfrac{40}{7}g_6\right) \\
&\quad + (4r_0 - 2r_2)\Omega^2 = 0 \\
&-720b_5 + 1200b_3 - q_0L^3(50 - 20g_2 + 2g_4) + \Omega^2(4r_1 - 2r_2) = 0 \\
&-1008b_6 + 1680b_4 - q_0L^3(30g_1 - 18g_3 + 2g_5) - (2r_0 - 4r_2 + 2r_4)\Omega^2 = 0 \\
&-1344b_7 + 2240b_5 - q_0L^3\left(\tfrac{70}{3}g_2 + \tfrac{84}{5}g_4 + 2g_6\right) - (2r_1 - 4r_3 + 2r_5)\Omega^2 = 0 \\
&-1728b_8 + 2880b_5 - q_0L^3(20g_3 - 16g_5) - (2r_2 - 4r_4)\Omega^2 = 0 \\
&-2160b_9 + 3600b_7 - q_0L^3\left(18g_4 - \tfrac{108}{7}g_6\right) - (2r_3 - 4r_5)\Omega^2 = 0 \\
&4400b_8 - \tfrac{50}{3}q_0L^3g_5 - 2r_4\Omega^2 = 0 \\
&5280b_9 - \tfrac{110}{7}q_0L^3g_6 - 2r_5\Omega^2 = 0
\end{aligned} \tag{205}
$$

We have to express the unknowns in terms of parameters b_5 and Ω^2. Therefore, we demand the rank of the set (205) to be less than 11, leading to a system of 12 determinantal equations with 12 unknowns, namely the parameters r_0, r_1, r_2, r_3, r_4, r_5, g_1, g_2, g_3, g_4, g_5 and g_6. The solution of this set reads

$$r_1 = -\tfrac{1002}{5225} r_3 - \tfrac{603}{11495} r_5, \quad g_2 = -4 - \tfrac{186}{625} g_4 - \tfrac{459}{4375} g_6 \tag{206}$$

Upon substitution of the expressions (206) into the set (205), we get the solution for the natural frequency squared

$$\Omega^2 = 2680 b_9 \left(1 - q_0 / q_{0,cr}\right) / r_5 \tag{207}$$

The value of q_0 that corresponds to the vanishing frequency $\Omega^2 = 0$ is identified as the buckling load $q_{0,cr}$:

$$q_{0,cr} = 336 b_9 / g_6 L^3 \tag{208}$$

Eq. (207) gets the following form:

$$\Omega^2 = 2680 b_9 \left(1 - q_0 / q_{0,cr}\right) / r_5 \tag{209}$$

From Eq. (209), when $q_0 \equiv 0$, $\Omega^2 = 2680\, b_9 / r_5 \equiv \Omega_0{}^2$. Therefore, Eq. (207) takes the form

$$\Omega_q{}^2 = \Omega_0{}^2 \left(1 - q_0 / q_{0,cr}\right) \tag{210}$$

where $\Omega_q{}^2$ is the natural frequency squared in presence of the axial load. Note that when q_0 tends to its critical value the natural frequency vanishes. The coefficients b_i in the stiffness are deduced from the remaining equations of the set (205)

$$\begin{aligned}
b_0 &= \{(10071600 r_4 + 28096200 r_2 + 135036000 r_0) b_8 \\
&\quad + [(45430 g_5 + 24696 g_4 + 120186 g_3 + 227700 g_1 - 52920 r_4) r_4 \\
&\quad - (511500 r_0 + 106425 r_2) g_5] q_0 L^3 \} / 4989600 r_4, \quad b_1 = -7 q_0 L^3 / 120 \\
b_2 &= \{(10071600 r_4 + 28096200 r_2 - 78408000 r_0) b_8 \\
&\quad - [(93170 g_5 + 10584 g_4 + 87714 g_3 + 285120 g_1) r_4 \\
&\quad - (297000 r_0 - 106425 r_2) g_5] q_0 L^3 \} / 4989600 r_4 \\
b_3 &= \{18803400 r_3 b_8 + [(13923 g_4 + 298350) r_4 + 71225 g_5 r_3] q_0 L^3 \} / 3029400 r_4 \\
b_4 &= \{(183120 r_4 - 653400 r_2 + 237600 r_0) b_8 \\
&\quad [-(938 g_5 - 990 g_3 + 1620 g_1) r_4 + (2475 r_2 - 900 r_0) g_5] q_0 L^3 \} / 90720 r_4 \\
b_5 &= \{-5953200 r_3 b_8 - [(12546 g_4 + 34425) r_4 - 22550 r_3 g_5] q_0 L^3 \} / 1009800 r_4 \\
b_6 &= \{(-10608 r_4 + 3960 r_2) b_8 + [(18 g_3 + 26 g_5) r_4 - 15 r_2 g_5] q_0 L^3 \} / 2592 \\
b_7 &= [6600 r_3 b_8 + (27 g_4 r_4 - 25 r_3 g_5) q_0 L^3] / 5400 r_4
\end{aligned} \tag{211}$$

14 Some Pertinent Comments on Higher Order Polynomial Modes

We considered a postulate involving the mode shape both in vibration and buckling to coincide with the fifth order polynomial. For each of the four considered sets of boundary conditions we studied the cases of uniform and nonuniform material densities. For the nonuniform density case the variation of the inertial coefficient was taken as linear, parabolic, cubic, quartic or quintic one. In an analogous manner one can obtain, in principle, solutions for any value of m where m is the degree of variation in the mass density.

In perfect analogy, one can postulate the sixth, seventh, etc. polynomials for the mode shapes. Thus, infinite number of closed-form solutions can be constructed. It must be noted that here we studied polynomial class of functions. It is of interest to study also variations including exponential or trigonometric functions.

Reference

Chajes A., 1974, *Principles of Structural Stability Theory*, Prentice – Hall, Englewood Cliffs, p.106.

Dinnik A. N., 1932, Design of Column of Varying Cross – Section, *Transactions of the ASME, Applied Mechanics.*

Dinnik A. N., 1955, *Prodol'nyi Izgib. Kruchenie* (Bending in Presence of Axial Forces. Torsion), Academy of USSR Publishing, pp. 96-130, (in Russian).

Elishakoff I., 2000, Both Static and Deflection and Vibration Mode of Uniform Beam Can Serve as a Buckling Mode of a Non – Uniform Column, *Journal of Sound and Vibration*, Vol. 232, 477-489.

Elishakoff I., 2001a, Inverse Buckling Problem for Inhomogeneous Columns, *International Journal of Solids and Structures*, Vol. 38, 457-464.

Elishakoff I., 2001b, A Closed – Form Solution for the Generalized Euler Problem, *Proceedings of the Royal Society of London*, Vol. 456, 2409-2417.

Elishakoff I., 2002, Euler's Problem Revisited: 222 Years Later, *Meccanica*, Vol. 36, 265-272.

Elishakoff I. and Rollot O., 1999, New Closed – Form Solutions for Buckling of a Variable Stiffness Column by Mathematica, *Journal of Sound and Vibration*, Vol. 224, 172-182.

Gladwell G. M. L. and Morassi A., 1995, On Isospectral Rods, Beams and Strings. *Inverse Problems,* 11, 533-554.

Gottlieb H. P. W., 1989, On Standard Eigenvalues of Variable – Coefficient Heat and Rod Equations, *Journal of Applied Mechanics*, 56, 146-148.

Guede Z. and Elishakoff I., 2001, Apparently First Closed Form Solutions for Inhomogeneous Vibrating Beams Under Axial Loading, *Proceedings of Royal Society of London*, Vol. 457, 623-649.

Guede Z. and Elishakoff I., 2001, A Fifth Order Polynomial That Serves as Both Buckling and Vibration Mode of an Inhomogeneous Structure, *Chaos, solutions and Fractals*, Vol. 12, 1267-1298.

Timoshenko S. P. and Gere J. M., 1961, *Theory of Elastic Stability*, Mc Graw – Hill, New York .

Uspensky J. V., *Theory of Equations*, McGrawHill, New York, 1998.

Volmir A. S. (ed), *Collection of Problems in Strength of Materials*, (Problem 7.34, p.30) "Nauka" Publishers, Moscow, (in Russian).

PART IV. THE EFFECT OF INITIAL IMPERFECTIONS ON SHELL STABILITY

J. Arbocz, J.M.A.M. Hol and J. de Vries

Delft University of Technology, The Netherlands
Faculty of Aerospace Engineering, Kluyverweg 1, 2629 HS Delft, The Netherlands

Abstract. The development of "DISDECO", the **D**elft **I**nteractive **S**hell **DE**sign **CO**de is described. The purpose of this project is to make the accumulated theoretical, numerical and practical knowledge of the last 20 years readily accessible to users interested in the analysis of buckling of sensitive structures. With this open ended, hierarchical, interactive computer code the user can access from his workstation successively programs of increasing complexity.

The initial level consists of semi-analytical solutions for the buckling load of perfect anisotropic circular cylindrical shells under axial compression, internal or external lateral pressure and torsion. Also included are modules that contain Koiter's imperfections sensitivity theory extended to anisotropic shell structures under combined loading. The nonlinear Donnell-type shell equations in terms of the radial displacement W and the Airy stress function F are used. Using a procedure that is equivalent to an approximate minimization of the second variation of the potential energy by the Rayleigh-Ritz method simple eigenvalue equations are obtained in terms of the integers m and n, representing the number of half-waves in the axial and the number of full waves in the circumferential direction, and Khot's skewedness parameter τ_K. Results of the search for the minimum eigenvalue and its imperfection sensitivity are displayed in convenient graphical form.

As a first update of DISDECO the computational modules of the initial level have been modified so as to be able to handle combined loads consisting of axial compression, internal or external pressure and clockwise or counter-clockwise torque. The applied load is assumed to have a uniform spatial distribution and is divided into a fixed and a variable part. The magnitude of the variable part is allowed to vary in proportion to a load parameter Λ. This leads to an eigenvalue problem for the critical load Λ_c.

INTRODUCTION

In modern designs, which are often obtained by one of the structural optimization codes and which are made out of high strength materials (read advanced composites), it happens frequently that the stability behavior dictates the choice of some of the critical dimensions of the structures. This implies that one has to investigate the different loading cases quite accurately by carrying out extensive numerical calculations and/or experimental verifications.

Twenty five years ago it was so that numerical results were looked upon with a certain degree of distrust and they were only accepted if supported by some other facts. Now-a-days, as the older generation of engineers (the ones who have gotten their degrees before the advent of computers) is retiring and the younger ones with extensive training in the ever-so-popular finite element techniques take over, one begins to encounter in technical discussion a new mentality; the insight of how structures behave under loading of the older generation is being more and more replaced by the nearly religious faith of the younger ones in the predictions of their favorite computer codes.

Actually what one needs is not more of one and less of the other (and the reader is free to associate his preference with one or the other), but an optimal combination of both, namely insight into structural behavior and familiarity with the appropriate numerical techniques. To provide the means for such an approach the development of DISDECO, the **D**elft **I**nteractive **S**hell **DE**sign **CO**de has been initiated. When finished this open ended, hierarchical, interactive computer code will provide for easy access to the theoretical knowledge, that has been accumulated by the many scientists who have been active in the field of shell stability, via the advanced interactive and computational facilities offered by the modern high-speed 32 bit personal workstations. Great care is being taken to present the results in a unified form so as to make it easy for the user to proceed step-by-step from the simpler approaches used by the early investigators to the more sophisticated analytical and numerical methods used presently.

Development of Level-1 of DISDECO

The central part of the whole system is the "Command and Control Processor". Its function is to control and direct the activities of the system. It starts-up and winds-down the design system, processes user input through execution of modules, creates a working environment and in general is the working partner of the user.

The link from the user to the command and control processor passes through the "Man-Machine Interface". Assuming that the user employs a terminal device or workstation which supports graphics, the man-machine interface controls the input stream from the user, analyses it, checks it for correct syntax, validates the commands and passes it in an interpretable format to the command and control processor. In a similar manner the output stream from the design and analysis system is converted to a meaningful output for the user.

The real work of DISDECO is done by a number of dedicated "Analysis Modules". Supplementary modules are required for pre- and post-processing functions, remote batch processing of analysis and general utility functions.

In DISDECO the lowest hierarchical level consists of semi-analytical solutions for the buckling load of perfect anisotropic circular cylindrical shells under axial compression, internal or external lateral pressure and torsion. Also included are modules that contain Koiter's imperfection sensitivity theory [1,2,3] extended to anisotropic shell structures.

1. DERIVATION OF THE NONLINEAR EQUILIBRIUM EQUATIONS

The nonlinear equations can be derived from the principle of stationary potential energy. A loaded shell is in equilibrium if its potential energy Π is stationary, and Π is stationary if the integrand in the expression for Π satisfies the Euler equations of the calculus of variations.

The total potential energy of an anisotropic shell subjected to axial compression, external pressure (dead load) and torsion is

$$\Pi = U_m + U_b + \Omega \tag{1.1}$$

where

U_m = membrane strain energy

$$= \frac{1}{2}\int_0^{2\pi R}\int_0^L (N_x\varepsilon_x + N_y\varepsilon_y + N_{xy}\gamma_{xy})\,dxdy \tag{1.2}$$

U_b = bending strain energy

$$\frac{1}{2}\int_0^{2\pi R}\int_0^L \left(M_x\kappa_x + M_y\kappa_y + \frac{M_{xy}+M_{yx}}{2}\kappa_{xy}\right)dxdy \tag{1.3}$$

Ω = potential energy of the applied loads

$$= -\int_0^{2\pi R}\int_0^L \{N_x|_{x=L}\,(u_{,x} - qW_{,xx}) + p_e W + N_{xy}|_{x=L}\,v_{,x}\}dxdy \tag{1.4}$$

For the notation and sign convention used see Fig. 1. The constitutive equations for an anisotropic shell can be written in matrix notation as

$$[N] = [A][\varepsilon] + [B][\kappa] \tag{1.5a}$$

$$[M] = [B][\varepsilon] + [D][\kappa] \tag{1.5b}$$

where

$$[N] = [N_x N_y N_{xy}]^T \qquad [\varepsilon] = [\varepsilon_x \varepsilon_y \gamma_{xy}]^T \tag{1.6a}$$

$$[M]=[M_x M_y \frac{M_{xy}+M_{yx}}{2}]^T \qquad [\kappa]=[\kappa_x \kappa_y \kappa_{xy}]^T \tag{1.6b}$$

and

$$[A]=\begin{bmatrix} A_{11} & A_{12} & A_{16} \\ A_{12} & A_{22} & A_{26} \\ A_{16} & A_{26} & A_{66} \end{bmatrix} \quad [B]=\begin{bmatrix} B_{11} & B_{12} & B_{16} \\ B_{12} & B_{22} & B_{26} \\ B_{16} & B_{26} & B_{66} \end{bmatrix} \quad [D]=\begin{bmatrix} D_{11} & D_{12} & D_{16} \\ D_{12} & D_{22} & D_{26} \\ D_{16} & D_{26} & D_{66} \end{bmatrix} \tag{1.7}$$

The coefficients A_{ij}, B_{ij} and D_{ij} are given in Appendix A in [12]. For the definition of stress- and moment resultants see Fig. 2.

Using a Donnell type theory the strain-displacement relations are for W and $\bar{W}$ positive inward

$$\begin{aligned} \varepsilon_x &= u_{,x} + \frac{1}{2}W_{,x}^2 + \bar{W}_{,x} W_{,x} & \kappa_x &= -W_{,xx} \\ \varepsilon_y &= v_{,y} - \frac{W}{R} + \frac{1}{2}W_{,y}^2 + \bar{W}_{,y} W_{,y} & \kappa_y &= -W_{,yy} \\ \gamma_{xy} &= u_{,y} + v_{,x} + W_{,x} W_{,y} + \bar{W}_{,x} W_{,y} + W_{,x} \bar{W}_{,y} & \kappa_{xy} &= -2W_{,xy} \end{aligned} \tag{1.8}$$

Letting now $N_x|_{x=L} = -N_o$ and $N_{xy}|_{x=L} = T_o$ then the variational statement of equilibrium for an anisotropic circular cylindrical shell under the specified external loading is

$$\delta\Pi = \delta \int_0^{2\pi R} \int_0^L F(u, u_{,x}, u_{,y}; v, v_{,x}\, v_{,y}; W, W_{,x}\, W_{,y}, W_{,xx}, W_{,xy}, W_{,yy})\, dxdy = 0 \tag{1.9}$$

where

$$\begin{aligned} F = \frac{1}{2}\{ & A_{11}\varepsilon_x^2 + 2A_{12}\varepsilon_x\varepsilon_y + 2A_{16}\varepsilon_x\gamma_{xy} + 2A_{26}\varepsilon_y\gamma_{xy} + A_{22}\varepsilon_y^2 + A_{66}\gamma_{xy}^2 \\ & +2B_{11}\varepsilon_x\kappa_x + 2B_{12}(\varepsilon_x\kappa_y + \varepsilon_y\kappa_x) + 2B_{16}(\varepsilon_x\kappa_{xy} + \kappa_x\gamma_{xy}) \\ & +2B_{26}(\varepsilon_y\kappa_{xy} + \kappa_y\gamma_{xy}) + 2B_{22}\varepsilon_y\kappa_y + 2B_{66}\gamma_{xy}\kappa_{xy} \\ & +D_{11}\kappa_x^2 + 2D_{12}\kappa_x\kappa_y + 2D_{16}\kappa_x\kappa_{xy} + 2D_{26}\kappa_y\kappa_{xy} + D_{22}\kappa_y^2 + D_{66}\kappa_{xy}^2 \\ & +2N_o(u_{,x} - qW_{,xx}) - 2p_e W - 2T_o v_{,x} \} \end{aligned} \tag{1.10}$$

Thus, in this case

$$\delta\Pi = \frac{\partial\Pi}{\partial u}\delta u + \frac{\partial\Pi}{\partial v}\delta v + \frac{\partial\Pi}{\partial W}\delta W = 0 \tag{1.11}$$

implies the 3 conditions

$$\frac{\partial\Pi}{\partial u} = \frac{\partial\Pi}{\partial v} = \frac{\partial\Pi}{\partial W} = 0 \tag{1.12}$$

Carrying out the indicated variations and integrating by parts wherever necessary one obtains the expressions

$$\frac{\partial\Pi}{\partial u} = \int_0^{2\pi R}\int_0^L \{\frac{\partial F}{\partial u} - \frac{\partial}{\partial x}\frac{\partial F}{\partial u,_x} - \frac{\partial}{\partial y}\frac{\partial F}{\partial u,_y}\}\delta u\, dxdy + \oint_c \frac{\partial F}{\partial u,_x}\delta u dy - \oint_c \frac{\partial F}{\partial u,_y}\delta u dx = 0 \tag{1.13}$$

$$\frac{\partial\Pi}{\partial v} = \int_0^{2\pi R}\int_0^L \{\frac{\partial F}{\partial v} - \frac{\partial}{\partial x}\frac{\partial F}{\partial v,_x} - \frac{\partial}{\partial y}\frac{\partial F}{\partial v,_y}\}\delta v\, dxdy + \oint_c \frac{\partial F}{\partial v,_x}\delta v dy - \oint_c \frac{\partial F}{\partial v,_y}\delta v dx = 0 \tag{1.14}$$

$$\begin{aligned}\frac{\partial\Pi}{\partial W} = &\int_0^{2\pi R}\int_0^L \{\frac{\partial F}{\partial W} - \frac{\partial}{\partial x}\frac{\partial F}{\partial W,_x} - \frac{\partial}{\partial y}\frac{\partial F}{\partial W,_y} + \frac{\partial^2}{\partial x^2}\frac{\partial F}{\partial W,_{xx}} + \frac{\partial^2}{\partial x\partial y}\frac{\partial F}{\partial W,_{xy}} + \frac{\partial^2}{\partial y^2}\frac{\partial F}{\partial W,_{yy}}\} dxdy \\ &+ \oint_c \{(\frac{\partial F}{\partial W,_x} - \frac{\partial}{\partial x}\frac{\partial F}{\partial W,_{xx}} - \frac{\partial}{\partial y}\frac{\partial F}{\partial W,_{xy}}) dW + \frac{\partial F}{\partial W,_{xx}}\frac{\partial}{\partial x}\delta W\} dy \\ &+ \oint_c \{-(\frac{\partial F}{\partial W,_y} - \frac{\partial}{\partial y}\frac{\partial F}{\partial W,_{yy}})\delta W - \frac{\partial F}{\partial W,_{xy}}\frac{\partial}{\partial x}\delta W - \frac{\partial F}{\partial W,_{yy}}\frac{\partial}{\partial y}\delta W\} dx = 0\end{aligned} \tag{1.15}$$

For this to be true for arbitrary nonzero variations $\delta u, \delta v$ and δW one must have

$$\frac{\partial F}{\partial u} - \frac{\partial}{\partial u}\frac{\partial F}{\partial u,_x} - \frac{\partial}{\partial y}\frac{\partial F}{\partial u,_y} = 0 \tag{1.16}$$

and at $x = 0, L$ either $\frac{\partial F}{\partial u,_x} = 0$ or $\delta u = 0$ (1.17)

$$\frac{\partial F}{\partial v} - \frac{\partial}{\partial x}\frac{\partial F}{\partial v,_x} - \frac{\partial}{\partial y}\frac{\partial F}{\partial v,_y} = 0 \tag{1.18}$$

$$\text{and at } x=0,L \qquad \text{either } \frac{\partial F}{\partial v_{,x}}=0 \qquad \text{or } \delta v=0 \tag{1.19}$$

$$\frac{\partial F}{\partial W}-\frac{\partial}{\partial x}\frac{\partial F}{\partial W_{,x}}-\frac{\partial}{\partial y}\frac{\partial F}{\partial W_{,y}}+\frac{\partial^2}{\partial x^2}\frac{\partial F}{\partial W_{,xx}}+\frac{\partial^2}{\partial x\partial y}\frac{\partial F}{\partial W_{,xy}}+\frac{\partial^2}{\partial y^2}\frac{\partial F}{\partial W_{,yy}}=0 \tag{1.20}$$

$$\text{and at } x=0,L \qquad \text{either } \frac{\partial F}{\partial W_{,x}}-\frac{\partial}{\partial x}\frac{\partial F}{\partial W_{,xx}}-\frac{\partial}{\partial y}\frac{\partial F}{\partial W_{,xy}}=0 \quad \text{or } \delta W=0 \tag{1.21}$$

$$\text{and} \qquad \text{either } \frac{\partial F}{\partial W_{,xx}}=0 \qquad \text{or } \frac{\partial}{\partial x}\delta W=0 \tag{1.22}$$

Notice that at x = constant dx = 0.

Carrying out the indicated differentiations, substituting and regrouping yields the following boundary value problem

$$N_{x,x}+N_{xy,y}=0 \qquad \text{at } x=0,L \qquad \text{either } N_x=-N_o \qquad \text{or } \delta u=0 \tag{1.23}$$

$$N_{xy,x}+N_{y,y}=0 \qquad \text{at } x=0,L \qquad \text{either } N_{xy}=T_o \qquad \text{or } \delta v=0 \tag{1.24}$$

$$\begin{aligned}&M_{x,xx}+(M_{xy}+M_{yx})_{,xy}+M_{y,yy}\\&\quad+\frac{1}{R}N_y+N_x(W_{,xx}+\bar{W}_{,xx})+2N_{xy}(W_{,xy}+\bar{W}_{,xy})+N_y(W_{,yy}+\bar{W}_{,yy})=-p_e\end{aligned} \tag{1.25}$$

$$\begin{aligned}\text{at } x=0,L \quad &\text{either} \quad M_{x,x}+(M_{xy}+M_{yx})_{,y}+N_x(W_{,x}+\bar{W}_{,x})+N_{xy}(W_{,y}+\bar{W}_{,y})=0\\&\text{or} \quad \delta W=0\end{aligned} \tag{1.26}$$

$$\text{and} \quad \text{either} \quad M_x=-N_o q \qquad \text{or} \quad \frac{\partial}{\partial x}\delta W=0 \tag{1.27}$$

Further, if one introduces the Airy stress function F such that

$$N_x=F_{,yy} \quad ; \quad N_y=F_{,xx} \quad ; \quad N_{xy}=-F_{,xy} \tag{1.28}$$

then the in-plane equilibrium equations (1.23) and (1.24) are identically satisfied. Inverting the first constitutive equation (1.5a) one obtains

$$[\varepsilon]=[A^*][N]+[B^*][\kappa] \tag{1.29}$$

where

$$[A^*]=[A^{-1}] \quad ; \quad [B^*]=-[A^{-1}][B] \tag{1.30}$$

Substituting the inverted first equation (1.29) into the second constitutive equation (1.5b) yields

$$[M]=[C^*][N]+[D^*][\kappa] \tag{1.31}$$

where

$$[C^*]=[B][A^{-1}] \quad ; \quad [D^*]=[D\}-[B][A^{-1}][B] \tag{1.32}$$

Notice that since [A] is symmetric, therefore also [A*] is symmetric, and since further [B] is also symmetric, therefore $[C^*]=-[B^*]^T$. Finally since [D] is symmetric, therefore also [D*] is symmetric.

Substituting now into the out-of-plane equilibrium equation (1.25) one obtains a single equation in the 2 unknowns W and F. A second equation involving the unknowns W and F is obtained by deriving a compatibility condition. This is done by eliminating u and v from the strain-displacement relations (1.8) yielding

$$\begin{aligned}\varepsilon_{x,yy}+\varepsilon_{y,xx}-\gamma_{xy,xy} &= -\frac{1}{R}W_{,xx}-\frac{1}{2}\{W_{,xx}(W_{,yy}+2\bar{W}_{,yy}) \\ &\quad -2W_{,xy}(W_{,xy}+2\bar{W}_{,xy})+W_{,yy}(W_{,xx}+2\bar{W}_{,xx})\}\end{aligned} \tag{1.33}$$

Substituting for ε_x, ε_y and γ_{xy} from Eq. (1.29) yields a second equation in the unknowns W and F. Introducing the following linear operators

$$L_{A^*}()=A^*_{22}()_{,xxxx}-2A^*_{26}()_{,xxxy}+(2A^*_{12}+A^*_{66})()_{,xxyy}-2A^*_{16}()_{,xyyy}+A^*_{11}()_{,yyyy} \tag{1.34}$$

$$\begin{aligned}L_{B^*}() &= B^*_{21}()_{,xxxx}+(2B^*_{26}-B^*_{61})()_{,xxxy}+(B^*_{11}+B^*_{22}-2B^*_{66})()_{,xxyy} \\ &\quad +(2B^*_{16}-B^*_{62})()_{,xyyy}+B^*_{12}()_{,yyyy}\end{aligned} \tag{1.35}$$

$$L_{D^*}=D^*_{11}()_{,xxxx}+4D^*_{16}()_{,xxxy}+2(D^*_{12}+2D^*_{66})()_{,xxyy}+4D^*_{26}()_{,xyyy}+D^*_{22}()_{,yyyy} \tag{1.36}$$

and the following nonlinear operator

$$L_{NL}(S,T) = S_{,xx}\,T_{,yy} - 2S_{,xy}\,T_{,xy} + S_{,yy}\,T_{,xx} \tag{1.37}$$

one can write the governing nonlinear partial differential equations as follows:

$$L_{A*}(F) - L_{B*}(W) = -\frac{1}{R}W_{,xx} - \frac{1}{2}L_{NL}(W, W + 2\bar{W}) \quad \text{- compatibility equation} \tag{1.38}$$

$$L_{B*}(F) + L_{D*}(W) = \frac{1}{R}F_{,xx} + L_{NL}(F, W + \bar{W}) + p_e \quad \text{- out-of-plane equilibrium equation} \tag{1.39}$$

These equations, together with the appropriate boundary conditions, govern the behavior of circular cylindrical shells

1. In the prebuckling stress and deformation state.
2. At the limit point or bifurcation point (if there is one).
3. In the postbuckling stress and deformation state.

2. CLASSIC LINEARIZED SMALL DEFLECTION THEORY FOR A PERFECT SHELL

For a perfect shell $\bar{W} = 0$. If we let

$$W = W^{(o)} + W^{(1)} \quad , \quad F = F^{(o)} + F^{(1)} \tag{2.1}$$

where $W^{(o)}, F^{(o)}$ represent the prebuckling solutions and $W^{(1)}, F^{(1)}$ represent small perturbations at buckling, then a direct substitution into Eqs. (1.38) and (1.39) and deletion of products of the perturbation quantities yields a set of nonlinear governing equations for the prebuckling quantities

$$L_{A*}(F^{(o)}) - L_{B*}(W^{(o)}) = -\frac{1}{R} W_{,xx}^{(o)} - \frac{1}{2} L_{NL}(W^{(o)}, W^{(o)}) \tag{2.2}$$

$$L_{B*}(F^{(o)}) + L_{D*}(W^{(o)}) = \frac{1}{R} F_{,xx}^{(o)} + L_{NL}(F^{(o)}, W^{(o)}) + p_e \tag{2.3}$$

and a set of linearized stability equations governing the perturbation quantities,

$$L_{A*}(F^{(1)}) - L_{B*}(W^{(1)}) = -\frac{1}{R} W_{,xx}^{(1)} - L_{NL}(W^{(o)}, W^{(1)}) \tag{2.4}$$

$$L_{B*}(F^{(1)}) + L_{D*}(W^{(1)}) = \frac{1}{R} F_{,xx}^{(1)} + L_{NL}(F^{(o)}, W^{(1)}) + L_{NL}(F^{(1)}, W^{(o)}) \tag{2.5}$$

2.1 Axial compression and internal pressure ($p_e = -p_i$)

If one assumes the following "membrane" solution to represent the prebuckling stress and deformation state

$$W^{(o)} = t\hat{W}_v + tW_{p_i} \tag{2.6}$$

$$F^{(o)} = \frac{Et^2}{cR}\left(-\frac{1}{2}\hat{\lambda} y^2 + \frac{1}{2}\bar{p}_i x^2\right) \tag{2.7}$$

where

$$\hat{\lambda} = \lambda - \frac{1}{2}\bar{p}_i \quad ; \quad \lambda = \frac{\sigma}{\sigma_c} = \frac{N_x}{N_c} \quad ; \quad \bar{p}_i = \frac{p_i}{\frac{t}{R}\sigma_c} \quad ; \quad \sigma_c = \frac{Et}{cR} \tag{2.8}$$

and

$$c = \sqrt{3(1-\nu^2)}$$

then the equations governing the prebuckling state are identically satisfied and the linearized stability equations reduce to

$$L_{A^*}(F^{(1)}) - L_{B^*}(W^{(1)}) = -\frac{1}{R} W_{,xx}^{(1)} \tag{2.9}$$

$$L_{B^*}(F^{(1)}) + L_{D^*}(W^{(1)}) = \frac{1}{R} F_{,xx}^{(1)} + \frac{Et^2}{cR} \bar{p}_i W_{,yy}^{(1)} - \frac{Et^2}{cR} \hat{\lambda} W_{,xx}^{(1)} \tag{2.10}$$

Assuming that the radial displacement component of the classic buckling mode can be written as

$$W^{(1)} = t \sin m\pi \frac{x}{L} \cos \frac{n}{R}(\lambda - \tau_K x) = \frac{t}{2}\{\sin({}_m x - {}_n y) + \sin({}_p x + {}_n y)\} \tag{2.11}$$

where

$$_m = m\frac{\pi}{L} + \frac{n}{R}\tau_K \qquad _p = m\frac{\pi}{L} - \frac{n}{R}\tau_K \qquad _n = \frac{n}{R} \tag{2.12}$$

and Khot's skewedness parameter τ_K [4] is introduced in order to account for the possibility of bending-twisting coupling.

Substituting into the compatibility equation yields a linear, inhomogeneous partial differential equation for $F^{(1)}$. Using the Method of Undetermined Coefficients one can obtain the following particular integral

$$F^{(1)} = \frac{Et^3}{4c} \left\{ \frac{\bar{T}_{3,m,n}}{\bar{T}_{5,m,n}} \sin({}_m x - {}_n y) + \frac{\bar{T}_{4,p,n}}{\bar{T}_{6,p,n}} \sin({}_p x + {}_n y) \right\} \tag{2.13}$$

where the coefficients $\bar{T}_{3,m,n}$, $\bar{T}_{4,p,n}$, ... are defined in Appendix B in [12]. An approximate solution of the equilibrium equation by Galerkin's procedure then yields the following eigenvalue

$$\hat{\lambda}_{mn\tau} = \lambda - \frac{1}{2}\bar{p}_i = \frac{1}{2(\alpha_m^2 + \alpha_p^2)} \left(\bar{T}_{1,m,n} + \bar{T}_{2,p,n} + \frac{\bar{T}_{3,m,n}^2}{\bar{T}_{5,m,n}} + \frac{\bar{T}_{4,p,n}^2}{\bar{T}_{6,p,n}} \right) + \frac{2\bar{p}_i \beta_n^2}{(\alpha_m^2 + \alpha_p^2)} \tag{2.14}$$

Notice that the eigenvalue $\hat{\lambda}_{mn\tau}$ depends on the wave numbers m and n and on Khot's parameter τ_K. The classic buckling load parameter λ_c is the lowest of all eigenvalues.

The quantities $\hat{W}_V$ and W_{p_i} are evaluated by enforcing the circumferential periodicity condition

$$\int_0^{2\pi R} v_{,y}\, dy = 0 \tag{2.15}$$

where for a perfect shell

$$v_{,y} = \varepsilon_y + \frac{W}{R} - \frac{1}{2} W_{,y}^{\,2} \tag{2.16}$$

$$\varepsilon_y = A^*_{12} N_x + A^*_{22} N_y + A^*_{26} N_{xy} + B^*_{21} \kappa_x + B^*_{22} \kappa_y + B^*_{26} \kappa_{xy} \tag{2.17}$$

$$W = t\hat{W}_V + tW_{p_i} + W^{(1)} \tag{2.18}$$

$$F = \frac{Et^2}{cR}\left(-\frac{1}{2}\hat{\lambda} y^2 + \frac{1}{2}\bar{p}_i x^2\right) + F^{(1)} \tag{2.19}$$

Substitution and deletion of squares of the perturbation quantities yields

$$\int_0^{2\pi R} \left\{ -A^*_{12} \frac{Et^2}{cR} \hat{\lambda} + A^*_{22} \frac{Et^2}{cR} \bar{p}_i + A^*_{12} F^{(1)}_{,yy} + A^*_{22} F^{(1)}_{,xx} - A^*_{26} F^{(1)}_{,xy} \right.$$
$$\left. - B^*_{21} W^{(1)}_{,xx} - B^*_{22} W^{(1)}_{,yy} - 2B^*_{26} W^{(1)}_{,xy} + \frac{t}{R}\hat{W}_V + \frac{t}{R} W_{p_i} + \frac{W^{(1)}}{R} \right\} dy = 0 \tag{2.20}$$

Substituting for $W^{(1)}$ and $F^{(1)}$ from Eqs. (2.11) and (2.13) and carrying out the integrals one obtains

$$\left\{ -A^*_{12} \frac{Et^2}{cR} \hat{\lambda} + A^*_{22} \frac{Et^2}{cR} \bar{p}_i + \frac{t}{R}\hat{W}_V + \frac{t}{R} W_{p_i} \right\} (2\pi R) = 0 \tag{2.21}$$

Thus if

$$\hat{W}_v = \frac{Et}{c} A_{12}^* \hat{\lambda} = \bar{A}_{12}^* \frac{\hat{\lambda}}{c} \quad ; \quad W_{p_i} = -\frac{Et}{c} A_{22}^* \bar{p}_i = -\bar{A}_{22}^* \frac{\bar{p}_i}{c} \tag{2.22}$$

then the circumferential periodicity condition is identically satisfied.

2.2 External pressure and axial compression

If one assumes the following "membrane" solution to represent the prebuckling stress- and deformation state

$$W^{(o)} = tW_v + tW_{p_e} \tag{2.23}$$

$$F^{(o)} = \frac{Et^2}{cR}\left(-\frac{1}{2}\bar{p}_e x^2 - \frac{1}{2}\lambda y^2\right) \tag{2.24}$$

where

$$\bar{p}_e = \frac{p_e}{\frac{t}{R}\sigma_c} \tag{2.25}$$

then the equations governing the prebuckling state are identically satisfied and the linearized stability equations reduce to

$$L_{A^*}(F^{(1)}) - L_{B^*}(W^{(1)}) = -\frac{1}{R} W_{,xx}^{(1)} \tag{2.26}$$

$$L_{B^*}(F^{(1)}) + L_{D^*}(W^{(1)}) = \frac{1}{R} F_{,xx}^{(1)} - \frac{Et^2}{cR}\bar{p}_e W_{,yy}^{(1)} - \frac{Et^2}{cR}\lambda W_{,xx}^{(1)} \tag{2.27}$$

By assuming that the radial displacement component of the classic buckling mode can be written as

$$W^{(1)} = t \sin m\pi \frac{x}{L} \cos \frac{n}{R}(y - \tau_K x) \tag{2.28}$$

an approximate solution of the stability equations can be obtained as in the preceding case by Galerkin's method yielding the following eigenvalue

$$\lambda_{mn\tau} = \frac{1}{2(\alpha_m^2 + \alpha_p^2)}(\overline{T}_{1,m,n} + \overline{T}_{2,p,n} + \frac{\overline{T}_{3,m,n}^2}{\overline{T}_{5,m,n}} + \frac{\overline{T}_{4,p,n}^2}{\overline{T}_{6,p,n}}) - \frac{2\overline{p}_e \beta_n^2}{(\alpha_m^2 + \alpha_p^2)} \tag{2.29}$$

Introducing the nondimensional ratio $\hat{R} = \lambda / \overline{p}_e$, then the eigenvalue can be written as

$$\overline{P}_{mn\tau} = \frac{1}{2}\frac{1}{\hat{R}(\alpha_m^2 + \alpha_p^2) + 2\beta_n^2}(\overline{T}_{1,m,n} + \overline{T}_{2,p,n} + \frac{\overline{T}_{3,m,n}^2}{\overline{T}_{5,m,n}} + \frac{\overline{T}_{4,p,n}^2}{\overline{T}_{6,p,n}}) \tag{2.30}$$

Notice that the eigenvalue $\overline{P}_{mn\tau}$ depends on the wave numbers m and n and on Khot's parameter τ_K. The classic buckling load parameter $\overline{p}_c$ is the lowest of all eigenvalues.

The quantities W_v and W_{p_e} are evaluated by enforcing the circumferential periodicity condition yielding

$$W_v = \overline{A}_{12}^* \frac{\lambda}{c} \quad ; \quad W_{p_e} = \overline{A}_{22}^* \frac{\overline{p}_e}{c} \tag{2.31}$$

2.3 Torsion and axial compression

If one assumes the following "membrane" solution to represent the prebuckling stress- and deformation state

$$W^{(o)} = tW_v + tW_t \tag{2.32}$$

$$F^{(o)} = \frac{Et^2}{cR}(-\overline{\tau}xy - \frac{1}{2}\lambda y^2) \tag{2.33}$$

where

$$\overline{\tau} = \frac{\tau_{xy}}{\sigma_c} = \frac{N_{xy}}{N_c} \quad ; \quad N_{xy} = \frac{T_o}{2\pi R^2} \tag{2.34}$$

and T_o is the applied torque, then the equations governing the prebuckling state are identically satisfied and the linearized stability equations reduce to

$$L_{A*}(F^{(1)}) - L_{B*}(W^{(1)}) = -\frac{1}{R}W_{,xx}^{(1)} \tag{2.35}$$

$$L_{B*}(F^{(1)}) + L_{D*}(W^{(1)}) = \frac{1}{R} F_{,xx}^{(1)} + 2\frac{Et^2}{cR} \tau W_{,xy}^{(1)} - \frac{Et^2}{cR} \lambda W_{,xx}^{(1)} \tag{2.36}$$

Under torsional loading the buckling deformation of a cylindrical shell consists of a number of circumferential waves that spiral around the cylinder from one end to the other. Such waves can be represented by the following deflection function

$$W^{(1)} = t \sin m\pi \frac{x}{L} \cos \frac{n}{R}(y - \tau_K x) \tag{2.37}$$

Proceeding as in the previous 2 cases one can obtain an approximate solution of the stability equations yielding the following eigenvalue

$$\bar{\tau}_{mn\tau} = \frac{1}{4(\alpha_m - \alpha_p)\beta_n}(\bar{T}_{1,m,n} + \bar{T}_{2,p,n} + \frac{\bar{T}^2_{3,m,n}}{\bar{T}_{5,m,n}} + \frac{\bar{T}^2_{4,p,n}}{\bar{T}_{6,p,n}}) - \frac{(\alpha_m^2 + \alpha_p^2)\lambda}{2(\alpha_m - \alpha_p)\beta_n} \tag{2.38}$$

Notice that the eigenvalue $\bar{\tau}_{mn\tau}$ depends on the wave numbers m and n and on Khot's parameter τ_K. The classic buckling load $\bar{\tau}_c$ is the lowest of all eigenvalues.
Notice also that if $\tau_K = 0$ then $\alpha_m = \alpha_p$. Thus the above formula is only valid if τ_K is not equal to 0.
The quantities W_v and W_t can be evaluated by enforcing the circumferential periodicity condition yielding

$$W_v = \bar{A}_{12}^* \frac{\lambda}{c} \quad ; \quad W_t = -\bar{A}_{26}^* \frac{\bar{\tau}}{c} \tag{2.39}$$

For torsional loading the buckling patterns are always skewed. Thus τ_K is always different from zero.
For counter-clockwise torsional loading $\bar{\tau}_c > 0$, hence the corresponding value of Khot's skewedness parameter will be positive (that is, $\tau_K > 0$). On the other hand, for clockwise torsional loading $\bar{\tau}_c < 0$ and the corresponding value of Khot's skewedness parameter τ_K is also negative (that is $\tau_K < 0$).

2.4 Combined Loading

If besides axial compression both external (or internal) pressure and torsion are applied to a cylindrical shell, then the prebuckling stress and deformation can be represented by the following "membrane" solution

$$W^{(o)} = tW_v + tW_p + tW_t \tag{2.40}$$

$$F^{(o)} = \frac{Et^2}{cR}\left(-\frac{1}{2}\lambda y^2 - \frac{1}{2}\bar{p}_e x^2 - \bar{\tau} xy\right) \tag{2.41}$$

where

$$\lambda = \frac{\sigma}{\sigma_c} \quad ; \quad \bar{p}_e = \frac{p_e}{\frac{t}{R}\sigma_c} \quad ; \quad \bar{\tau} = \frac{\tau_{xy}}{\sigma_c} \tag{2.42}$$

and

$$\sigma = \frac{P}{2\pi R t} \quad ; \quad \tau_{xy} = \frac{T_o}{2\pi R^2 t} \tag{2.43}$$

Notice that here P is the applied axial load, p_e is the applied external lateral pressure and T_o is the applied torque. Internal pressure is represented by a negative value of the external pressure, thus $p_i = -p_e$. One must also remember that when internal pressure is used, the normalized axial load parameter λ must be replaced by $\hat{\lambda} = \lambda - \frac{1}{2}\bar{p}_i$ in order to account for the axial component of the specified internal pressure p_i. The values of the radial displacements W_v, W_p and W_t have been defined earlier (see Eqs. 2.22, 2.31 and 2.39).

The membrane solutions of Eqs. (2.40) and (2.41) satisfy the equations governing the prebuckling state identically and reduce the linearized stability equations to

$$L_{A*}(F^{(1)}) - L_{B*}(W^{(1)}) = -\frac{1}{R} W_{,xx}^{(1)} \tag{2.44}$$

$$L_{B*}(F^{(1)}) + L_{D*}(W^{(1)}) = \frac{1}{R} F_{,xx}^{(1)} - \frac{Et^2}{cR}\lambda W_{,xx}^{(1)} + \frac{2Et^2}{cR}\bar{\tau} W_{,xy}^{(1)} - \frac{Et^2}{cR}\bar{p}_e W_{,yy}^{(1)} \tag{2.45}$$

Assuming the radial displacement component of the classic buckling mode to be

$$W^{(1)} = t \sin m\pi \frac{x}{L} \cos \frac{n}{R}(y - \tau_K x) \tag{2.46}$$

one can obtain an approximate solution of the stability equations by Galerkin's method as described earlier yielding the following characteristic equation

$$\lambda\, 2(\alpha_m^2 + \alpha_p^2) + \bar{p}_e 4\beta_n^2 + \bar{\tau}\, 4(\alpha_m - \alpha_p)\beta_n = \bar{T}_{1,m,n} + \bar{T}_{2,p,n} + \frac{\bar{T}_{3,m,n}^2}{\bar{T}_{5,m,n}} + \frac{\bar{T}_{4,p,n}^2}{\bar{T}_{6,p,n}} \tag{2.47}$$

If one chooses the variable load Λ to be the axial compression λ, then its critical value is given by

$$\lambda_c = \frac{1}{(\alpha_m^2 + \alpha_p^2)} \left\{ \frac{1}{2}(\bar{T}_{1,m,n} + \bar{T}_{2,p,n} + \frac{\bar{T}_{3,m,n}^2}{\bar{T}_{5,m,n}} + \frac{\bar{T}_{4,p,n}^2}{\bar{T}_{6,p,n}}) - 2\bar{p}_e\beta_n^2 - 2\bar{\tau}(\alpha_m - \alpha_p)\beta_n \right\} \quad (2.48)$$

where $\bar{p}_e$ and $\bar{\tau}$ are specified fixed loads. Notice that the eigenvalue $\lambda_{mn\tau}$ depends on the wave numbers m and n and on Khot's parameter τ_K. The classic buckling load parameter $\Lambda_c = \lambda_c$ is the lowest of all eigenvalues $\lambda_{mn\tau}$.

If instead of external pressure $\bar{p}_e$, a certain internal pressure $\bar{p}_i$ is specified, then Eq. (2.48) becomes

$$\hat{\lambda}_c = \hat{\lambda}_{mn\tau} = \lambda - \frac{1}{2}\bar{p}_i$$

$$= \frac{1}{\alpha_m^2 + \alpha_p^2} \left\{ \frac{1}{2}(\bar{T}_{1,m,n} + \bar{T}_{2,p,n} + \frac{\bar{T}_{3,m,n}^2}{\bar{T}_{5,m,n}} + \frac{\bar{T}_{4,p,n}^2}{\bar{T}_{6,p,n}}) + 2\bar{p}_i\beta_n^2 - 2\bar{\tau}(\alpha_m - \alpha_p)\beta_n \right\} \quad (2.49)$$

On the other hand, if one chooses the variable load Λ to be the external lateral pressure $\bar{p}_e$ then its critical value is given by

$$\Lambda_c = \bar{p}_{mn\tau}$$

$$= \frac{1}{2\beta_n^2} \left\{ \frac{1}{2}(\bar{T}_{1,m,n} + \bar{T}_{2,p,n} + \frac{\bar{T}_{3,m,n}^2}{\bar{T}_{5,m,n}} + \frac{\bar{T}_{4,p,n}^2}{\bar{T}_{6,p,n}}) - \lambda(\alpha_m^2 + \alpha_p^2) - 2\bar{\tau}(\alpha_m - \alpha_p)\beta_n \right\} \quad (2.50)$$

where now λ and $\bar{\tau}$ have specified fixed values. Notice that once again the eigenvalue $\bar{p}_{mn\tau}$ depends on the wave numbers m and n on Khot's parameter τ_K, and that the value of the classic buckling load parameter $\Lambda_c = \bar{p}_c$ is the lowest of all eigenvalues $\bar{p}_{mn\tau}$.

Often it is convenient to introduce the relationship

$$\lambda = \hat{R}\,\bar{p}_e \quad (2.51)$$

where $\hat{R}$ is a nondimensional constant. A case of particular interest is represented by the ratio $\hat{R} = 0.5$. For this value

$$P = \pi R^2 p_e \tag{2.52}$$

and the cylinder is seen to be subjected to a uniform external pressure p_e applied both to its lateral surface and to its ends. Such loading is called hydrostatic pressure loading and its critical value is given by

$$\Lambda_c = \bar{p}_{mn\tau}$$

$$= \frac{1}{\hat{R}(\alpha_m^2 + \alpha_p^2) + 2\beta_n^2} \left\{ \frac{1}{2}\left(\bar{T}_{1,m,n} + \bar{T}_{2,p,n} + \frac{\bar{T}_{3,m,n}^2}{\bar{T}_{5,m,n}} + \frac{\bar{T}_{4,p,n}^2}{\bar{T}_{6,p,n}}\right) - 2\bar{\tau}(\alpha_m - \alpha_p)\beta_n \right\} \tag{2.53}$$

where $\bar{\tau}$ has a specified fixed value. Once again the eigenvalue $\bar{p}_{mn\tau}$ depends on the wave numbers m and n and on Khot's parameter τ_K. The value of the classic buckling load parameter $\Lambda_c = \bar{p}_c$ is the lowest of all eigenvalues $\bar{p}_{mn\tau}$.

If, finally, one chooses the variable load Λ to be the normalized torsion parameter $\bar{\tau}$ then its critical value is given by

$$\Lambda_c = \bar{\tau}_{mn\tau}$$

$$= \frac{1}{2(\alpha_m - \alpha_p)\beta_n} \left\{ \frac{1}{2}\left(\bar{T}_{1,m,n} + \bar{T}_{2,p,n} + \frac{\bar{T}_{3,m,n}^2}{\bar{T}_{5,m,n}} + \frac{\bar{T}_{4,p,n}^2}{\bar{T}_{6,p,n}}\right) - \lambda(\alpha_m^2 + \alpha_p^2) - 2\,\bar{p}_e\beta_n^2 \right\} \tag{2.54}$$

where λ and $\bar{p}_e$ have specified fixed values. As before, the eigenvalue $\bar{\tau}_{mn\tau}$ depends on the wave numbers m and n and on Khot's parameter τ_K. The value of the classic buckling load parameter $\Lambda_c = \bar{\tau}_c$ is the lowest of all eigenvalues $\bar{\tau}_{mn\tau}$.

3. EFFECT OF AXISYMMETRIC IMPERFECTION

If we let

$$W = W^{(o)} + W^{(1)} \quad , \quad F = F^{(o)} + F^{(1)} \tag{3.1}$$

where $W^{(o)}, F^{(o)}$ represent the prebuckling solutions and $W^{(1)}, F^{(1)}$ represent small perturbations at buckling, then a direct substitution into Eqs. (1.38) and (1.39) and deletion of products of the perturbation quantities yield a set of nonlinear governing equations for the prebuckling quantities

$$L_{A^*}(F^{(o)}) - L_{B^*}(W^{(o)}) = -\frac{1}{R}W,_{xx}^{(o)} - \frac{1}{2}L_{NL}(W^{(o)} + 2\bar{W}, W^{(o)}) \tag{3.2}$$

$$L_{B^*}(F^{(o)}) + L_{D^*}(W^{(o)}) = \frac{1}{R}F,_{xx}^{(o)} + L_{NL}(F^{(o)}, W^{(o)} + \bar{W}) + p_e \tag{3.3}$$

and a set of linearized equations governing the perturbation quantities

$$L_{A^*}(F^{(1)}) - L_{B^*}(W^{(1)}) = -\frac{1}{R}W,_{xx}^{(1)} - L_{NL}(W^{(o)} + \bar{W}, W^{(1)}) \tag{3.4}$$

$$L_{B^*}(F^{(1)}) + L_{D^*}(W^{(1)}) = \frac{1}{R}F,_{xx}^{(1)} + L_{NL}(F^{(o)}, W^{(1)}) + L_{NL}(F^{(1)}, W^{(o)} + \bar{W}) \tag{3.5}$$

3.1 Axial compression and internal pressure ($p_e = -p_i$)

If the initial imperfection is axisymmetric, e.g.,

$$\bar{W} = t\xi_1 \cos i\pi\frac{x}{L} \tag{3.6}$$

then the prebuckling solution will also be axisymmetric, namely

$$W^{(o)} = t\,\hat{W}_v + tW_{p_i} + w_o(x) \tag{3.7}$$

$$F^{(o)} = \frac{Et^2}{cR}\left(-\frac{1}{2}\hat{\lambda}\, y^2 + \frac{1}{2}\bar{p}_i x^2\right) + f_o(x) \tag{3.8}$$

Substitution into the equations governing the prebuckling state (Eqs. 3.2 - 3.3) yields

$$A^*_{22} f_{o,xxxx} - B^*_{21} w_{o,xxxx} = -\frac{1}{R} w_{o,xx} \tag{3.9}$$

$$B^*_{21} f_{o,xxxx} + D^*_{11} w_{o,xxxx} = \frac{1}{R} f_{o,xx} - \frac{Et^2}{cR}\hat{\lambda}\, w_{o,xx} + \frac{Et^3}{cR}\left(i\frac{\pi}{L}\right)^2 \bar{\xi}_1 \hat{\lambda} \cos i\pi\frac{x}{L} \tag{3.10}$$

Neglecting the effect of boundary conditions the (particular) solution of these equations is

$$w_o(x) = t\frac{\hat{\lambda}}{\lambda_{c_i} - \hat{\lambda}} \bar{\xi}_1 \cos i\pi\frac{x}{L} \tag{3.11}$$

$$f_o(x) = \frac{Et^3}{c} \frac{\hat{\lambda}}{\lambda_{c_i} - \hat{\lambda}} \frac{(1+\bar{B}^*_{21}\alpha_i^2)}{2\alpha_i^2 \bar{A}^*_{22}} \bar{\xi}_1 \cos i\pi\frac{x}{L} \tag{3.12}$$

where

$$\lambda_{c_i} = \frac{1}{2}\left\{\alpha_i^2 \bar{D}^*_{11} + \frac{\left(1+\bar{B}^*_{21}\alpha_i^2\right)^2}{\alpha_i^2 \bar{A}^*_{22}}\right\} \tag{3.13}$$

is the classic axisymmetric buckling load for axial compression only.
For the axisymmetric imperfection and prebuckling state the linearized stability equations reduce to

$$L_{A^*}(F^{(1)}) - L_{B^*}(W^{(1)}) = -\frac{1}{R} W^{(1)}_{,xx} + \frac{2c}{R}\alpha_i^2(\bar{\xi}_1 + \bar{A})\cos i\pi\frac{x}{L} W^{(1)}_{,yy} \tag{3.14}$$

$$\begin{aligned} L_{B^*}(F^{(1)}) + L_{D^*}(W^{(1)}) = {} & \frac{1}{R} F^{(1)}_{,xx} + \frac{Et^2}{cR}\left(\bar{p}_i - 2c\,\alpha_i^2 \bar{B}\cos i\pi\frac{x}{L}\right) W^{(1)}_{,yy} \\ & - \frac{Et^2}{cR}\hat{\lambda}\, W^{(1)}_{,xx} - \frac{2c}{R}\alpha_i^2(\bar{\xi}_1 + \bar{A})\cos i\pi\frac{x}{L} F^{(1)}_{,yy} \end{aligned} \tag{3.15}$$

where

$$\bar{A} = \frac{\hat{\lambda}}{\lambda_{c_i} - \hat{\lambda}} \bar{\xi}_1 \quad ; \quad \bar{B} = \frac{\hat{\lambda}}{\lambda_{c_i} - \hat{\lambda}} \frac{(1 + \bar{B}^*_{21}\alpha_i^2)}{2\alpha_i^2 \bar{A}^*_{22}} \bar{\xi}_1 \tag{3.16}$$

Assuming that the radial displacement component of the classic buckling mode can be written as

$$W^{(1)} = t\hat{\xi} \sin m\pi \frac{x}{L} \cos \frac{n}{R}(y - \tau_K x) \tag{3.17}$$

then substitution into the compatibility equation (3.14) yields a linear, inhomogeneous partial differential equation for $F^{(1)}$. By the Method of Undetermined Coefficients one can obtain the following particular integral

$$\begin{aligned} F^{(1)} = {} & \hat{A}_1 \sin({}_{m}x - {}_{n}y) + \hat{A}_2 \sin {}_{i+m}x \cos {}_{n}y + \hat{A}_3 \sin {}_{i-m}x \cos {}_{n}y \\ & + \hat{A}_4 \cos {}_{i+m}x \sin {}_{n}y + \hat{A}_5 \cos {}_{i-m}x \sin {}_{n}y + \hat{B}_1 \sin({}_{p}x + {}_{n}y) \\ & + \hat{B}_2 \sin {}_{i+p}x \cos {}_{n}y + \hat{B}_3 \sin {}_{i-p}x \cos {}_{n}y + \hat{B}_4 \cos {}_{i+p}x \sin {}_{n}y \\ & + \hat{B}_5 \cos {}_{i-p}x \sin {}_{n}y \end{aligned} \tag{3.18}$$

where

$$\begin{aligned} {}_{i+m} &= (i+m)\frac{\pi}{L} + \frac{n}{R}\tau_K \quad ; \quad {}_{i+p} = (i+p)\frac{\pi}{L} - \frac{n}{R}\tau_K \\ {}_{i-m} &= (i-m)\frac{\pi}{L} - \frac{n}{R}\tau_K \quad ; \quad {}_{i-p} = (i-p)\frac{\pi}{L} + \frac{n}{R}\tau_K \end{aligned} \tag{3.19}$$

and the constants $\hat{A}_1, \hat{A}_2, \ldots, \hat{B}_1, \ldots, \hat{B}_5$ are defined in Appendix C in [12].
An approximate solution of the equilibrium equation (3.15) by Galerkin's procedure then yields the following cubic polynomial for the eigenvalue $\hat{\lambda}$:

$$\hat{\lambda}^3 - \left\{ \hat{\lambda}_{mn\tau} + 2\lambda_{c_i} - \hat{C}_1 \bar{\xi}\delta_{i=2m} \right\} \hat{\lambda}^2 \tag{3.20}$$

$$+ \left\{ 2\hat{\lambda}_{mn\tau} + \lambda_{c_i} + (\hat{C}_2 - \hat{C}_1)\bar{\xi}_1 \delta_{i=2m} \right\} \lambda_{c_i} \hat{\lambda} - \left\{ \hat{\lambda}_{mn\tau} + \hat{C}_2 \bar{\xi}_1 \delta_{i=2m} + (\hat{C}_3 - \hat{C}_4 \delta_{im} \delta_{\tau o}) \bar{\xi}^2 \right\} \lambda_{c_i}^2 = 0$$

where

$$\begin{aligned} \delta_{i=2m} &= 1 \text{ if } i = 2m \\ &= 0 \text{ otherwise} \end{aligned} \quad ; \quad \begin{aligned} \delta_{im} &= 1 \text{ if } i = m \\ &= 0 \text{ otherwise} \end{aligned} \quad ; \quad \begin{aligned} \delta_{\tau o} &= 1 \text{ if } \tau_K = 0 \\ &= 0 \text{ otherwise} \end{aligned} \tag{3.21}$$

and the constants $\hat{C}_1, \ldots, \hat{C}_4$ are defined in Appendix C in [12].

In using Eq. (3.20) one must remember that because of the presence of the Kronecker delta $\delta_{i=2m}$ the critical buckling load $\hat{\lambda}_c$ is noticeably influenced by the given imperfection (see Eq. (3.6)) only if the condition $i = 2m$ is satisfied. The circumferential wave number n must be chosen so as to obtain the minimum value of $\hat{\lambda}_c$ for a given imperfection amplitude $\bar{\xi}$, from Eq. (3.20).

3.2 External pressure and axial compression

If the initial imperfection is axisymmetric, e.g.

$$\bar{W} = t\bar{\xi}_1 \cos i\pi \frac{x}{L} \tag{3.22}$$

then the prebuckling solution will also be axixymmetric, namely

$$W^{(o)} = tW_c + w_o(x) \tag{3.23}$$

$$F^{(o)} = \frac{Et^2}{cR}(-\frac{1}{2}\bar{p}_e)(x^2 + \hat{R}\, y^2) + f_o(x) \tag{3.24}$$

where

$$W_c = W_v + W_{p_e} = \frac{\bar{p}_e}{c}(\bar{A}_{22}^* + \hat{R}\,\bar{A}_{12}^*) \quad ; \quad \hat{R} = \frac{\lambda}{\bar{p}_e}$$

A substitution into the equations governing the prebuckling state (Eqs. 3.2 and 3.3) yields

$$A_{22}^{*} f_{o,xxxx} - B_{21}^{*} w_{o,xxxx} = -\frac{1}{R} w_{o,xx} \tag{3.25}$$

$$B_{21}^{*} f_{o,xxxx} + D_{11}^{*} w_{o,xxxx} = \frac{1}{R} f_{o,xx} - \frac{Et^2}{cR} \hat{R}\,\bar{p}_e w_{o,xx} + \frac{Et^3}{cR} \left(i\frac{\pi}{L}\right)^2 \bar{\xi}_1 \hat{R}\,\bar{p}_e \cos i\pi\frac{x}{L} \tag{3.26}$$

Neglecting the effect of boundary conditions, the (particular) solution of these equations is

$$w_o(x) = t\frac{\hat{R}\,\bar{p}_e}{\lambda_{c_i} - \hat{R}\,\bar{p}_e} \bar{\xi}_1 \cos i\pi\frac{x}{L} \tag{3.27}$$

$$f_o(x) = \frac{Et^3}{c} \frac{\hat{R}\,\bar{p}_e}{\lambda_{c_i} - \hat{R}\,\bar{p}_e} \frac{(1+\bar{B}_{21}^{*}\alpha_i^2)}{2\alpha_i^2 \bar{A}_{22}^{*}} \bar{\xi}_1 \cos i\pi\frac{x}{L} \tag{3.28}$$

where

$$\lambda_{c_i} = \frac{1}{2}\left\{\alpha_i^2 \bar{D}_{11}^{*} + \frac{(1+\bar{B}_{21}^{*}\alpha_i^2)^2}{\alpha_i^2 \bar{A}_{22}^{*}}\right\}$$

is the classic axisymmetric buckling load for axial compression only.

For the axisymmetric imperfection and prebuckling state the linearized stability equations reduce to

$$L_{A^*}(F^{(1)}) - L_{B^*}(W^{(1)}) = -\frac{1}{R} W_{,xx}^{(1)} + \frac{2c}{R}\alpha_i^2 (\bar{\xi}_1 + A)\cos i\pi\frac{x}{L} W_{,yy}^{(1)} \tag{3.29}$$

$$\begin{aligned} L_{B^*}(F^{(1)}) + L_{D^*}(W^{(1)}) = {} & \frac{1}{R} F_{,xx}^{(1)} - \frac{Et^2}{cR}(\bar{p}_e + 2c\alpha_i^2 B\cos i\pi\frac{x}{L}) W_{,yy}^{(1)} \\ & - \frac{Et^2}{cR}\hat{R}\,\bar{p}_e W_{,xx}^{(1)} - \frac{2c}{R}\alpha_i^2 (\bar{\xi}_1 + A)\cos i\pi\frac{x}{L} F_{,yy}^{(1)} \end{aligned} \tag{3.30}$$

where

$$A = \frac{\hat{R}\bar{p}_e}{\lambda_{c_i} - \hat{R}\bar{p}_e}\bar{\xi}_1 \quad ; \quad B = \frac{\hat{R}\bar{p}_e}{\lambda_{c_i} - \hat{R}\bar{p}_e}\frac{(1+\bar{B}^*_{21}\alpha_i^2)}{2\alpha_i^2\,\bar{A}^*_{22}}\bar{\xi}_1 \tag{3.31}$$

By assuming that the radial displacement component of the classic buckling mode can be written as

$$W^{(1)} = t\hat{\xi}\sin m\pi\frac{x}{L}\cos\frac{n}{R}(y - \tau_K x) \tag{3.32}$$

an approximate solution of the stability equations can be obtained as in the preceding case yielding the following cubic polynomial for the eigenvalue $\bar{p}_e$:

$$\begin{aligned}&\hat{R}^2\,\bar{p}_e^3 - \{\bar{p}_{mn\tau}\,\hat{R}^2 + 2\lambda_{c_i}\,\hat{R} - \hat{D}_1\,\bar{\xi}_1\,\hat{R}^2\,\delta_{i=2m}\}\,\bar{p}_e^2\\ &\quad + \{2\bar{p}_{mn\tau}\,\hat{R} + \lambda_{c_i} + (\hat{D}_2 - \hat{D}_1)\bar{\xi}_1\,\hat{R}\,\delta_{i=2m}\}\lambda_{c_i}\bar{p}_e\\ &\quad - \{\bar{p}_{mn\tau} + \hat{D}_2\,\bar{\xi}_1\delta_{i=2m} + (\hat{D}_3 - \hat{D}_4\,\delta_{im}\delta_{\tau o})\bar{\xi}_1^2\}\lambda_{c_i}^2 = 0\end{aligned} \tag{3.33}$$

Thus for $\hat{R} = 0$ → **external pressure case**

$$\bar{p}_e = \bar{p}_{mn\tau} + \hat{D}_2\,\bar{\xi}_1\delta_{i=2m} + (\hat{D}_3 - \hat{D}_4\,\delta_{im}\delta_{\tau o})\,\bar{\xi}_1^2 \tag{3.34}$$

whereas for $\hat{R} > 0$ → **combined loading case**

$$\begin{aligned}&\bar{p}_e^3 - \{\bar{p}_{mn\tau} + 2\lambda_{c_i}/\hat{R} - \hat{D}_1\,\bar{\xi}_1\,\delta_{i=2m}\}\,\bar{p}_e^2\\ &\quad + (1/\hat{R})\{2\bar{p}_{mn\tau} + \lambda_{c_i}/\hat{R} + (\hat{D}_2 - \hat{D}_1)\bar{\xi}_1\,\delta_{i=2m}\}\lambda_{c_i}\bar{p}_e\\ &\quad - (1/\hat{R}^2)\{\bar{p}_{mn\tau} + \hat{D}_2\,\bar{\xi}_1\delta_{i=2m} + (\hat{D}_3 - \hat{D}_4\,\delta_{im}\delta_{\tau o})\bar{\xi}_i^2\}\lambda_{c_i}^2 = 0\end{aligned} \tag{3.35}$$

The constants $\hat{D}_1, \ldots\ldots, \hat{D}_4$ are listed in Apppendix C in [12].

In using Eqs. (3.34) or (3.35) one must remember that because of the presence of the Kronecker delta $\delta_{i=2m}$ the critical buckling pressure $\bar{p}_c$ is noticeably influenced by the given imperfection (see Eq. (3.22)) only if the condition $i=2m$ is satisfied. The circumferential wave number n must be chosen so as to obtain the minimum value of $\bar{p}_c$ for a given imperfection amplitude $\bar{\xi}_1$ from Eq. (3.34) or Eq. (3.35).

3.3 Torsion and Axial Compression

If the initial imperfection is axisymmetric, e.g.

$$\bar{W} = t\bar{\xi}_1 \cos i\pi\frac{x}{L} \tag{3.36}$$

then the prebuckling solution can be represented by

$$W^{(o)} = tW_v + tW_t + w_o(x) \tag{3.37}$$

$$F^{(o)} = \frac{Et^2}{cR}(-\frac{1}{2}\lambda y^2 - \bar{\tau}xy) + f_o(x) \tag{3.38}$$

Substituting into the equations governing the prebuckling state (Eqs. 3.2-3.3) yields

$$A_{22}^* f_{o,xxxx} - B_{21}^* w_{o,xxxx} = -\frac{1}{R} w_{o,xx} \tag{3.39}$$

$$B_{21}^* f_{o,xxxx} + D_{11}^* w_{o,xxxx} = \frac{1}{R} f_{o,xx} - \frac{Et^2}{cR}\lambda w_{o,xx} + \frac{Et^3}{cR}(i\frac{\pi}{L})^2 \bar{\xi}_1 \lambda \cos i\pi\frac{x}{L} \tag{3.40}$$

Neglecting the effect of boundary conditions, the particular solution of these equations is

$$w_o(x) = t\frac{\lambda}{\lambda_{c_i} - \lambda}\bar{\xi}_1 \cos i\pi\frac{x}{L} \tag{3.41}$$

$$f_o(x) = \frac{Et^3}{c}\frac{\lambda}{\lambda_{c_i} - \lambda}\frac{(1+\bar{B}_{21}^*\alpha_i^2)}{2\alpha_i^2\bar{A}_{22}^*}\bar{\xi}_1 \cos i\pi\frac{x}{L} \tag{3.42}$$

where

$$\lambda_{c_i} = \frac{1}{2}\left\{\alpha_i^2 \bar{D}_{11}^* + \frac{(1+\bar{B}_{21}^*\alpha_i^2)^2}{2\alpha_i^2\bar{A}_{22}^*}\right\} \tag{3.43}$$

is the classic axisymmetric buckling load for axial compression only.

For the axisymmetric imperfection and prebuckling state the linearized stability equations reduce to

$$L_{A*}(F^{(1)}) - L_{B*}(W^{(1)}) = -\frac{1}{R}W_{,xx}^{(1)} + \frac{2c}{R}\alpha_i^2(\bar{\xi}_1 + \bar{A})\cos i\pi\frac{x}{L}W_{,yy}^{(1)} \tag{3.44}$$

$$\begin{aligned} L_{B*}(F^{(1)}) + L_{D*}(W^{(1)}) = {} & \frac{1}{R}F_{,xx}^{(1)} - \frac{Et^2}{cR}\lambda W_{,xx}^{(1)} + 2\frac{Et^2}{cR}\bar{\tau}\, W_{,xy}^{(1)} \\ & -2\frac{Et^2}{cR}\alpha_i^2\bar{B}\cos i\pi\frac{x}{L}W_{,yy}^{(1)} - \frac{2c}{R}\alpha_i^2(\bar{\xi}_1 + \bar{A})\cos i\pi\frac{x}{L}F_{,yy}^{(1)} \end{aligned} \tag{3.45}$$

where

$$\bar{A} = \frac{\lambda}{\lambda_{c_i} - \lambda}\bar{\xi}_1 \quad ; \quad \bar{B} = \frac{\lambda}{\lambda_{c_i} - \lambda}\frac{(1 + \bar{B}_{21}^*\alpha_i^{2})}{2\alpha_i^2\,\bar{A}_{22}^*}\bar{\xi}_1 \tag{3.46}$$

Assuming that the radial displacement component of the classic buckling mode can be written as

$$W^{(1)} = t\hat{\xi}\sin m\pi\frac{x}{L}\cos\frac{n}{R}(y - \tau_K x) \tag{3.47}$$

then substitution into the compatibility equation (3.44) yields a linear, inhomogeneous partial differential equation for $F^{(1)}$. By the Method of Undetermined Coefficients one can easily obtain a particular integral of the form given by Eq. (3.18). An approximate solution of the equilibrium equation (3.45) by Galerkin's procedure then yields the following characteristic equation for the eigenvalue $\bar{\tau}$:

$$\begin{aligned} & \left\{\frac{1}{2}\left(\bar{T}_{1,m,n} + \bar{T}_{2,p,n} + \frac{\bar{T}_{3,m,n}^2}{\bar{T}_{5,m,n}} + \frac{\bar{T}_{4,p,n}^2}{\bar{T}_{6,p,n}}\right) - \lambda(\alpha_m^2 + \alpha_p^2) - 2\bar{\tau}(\alpha_m - \alpha_p)\beta_n\right\}(\lambda_{c_i} - \lambda)^2 \\ & + (\mathbf{A}\lambda_{c_i} + \mathbf{B}\lambda)\delta_{i=2m}\bar{\xi}_1(\lambda_{c_i} - \lambda) + (\mathbf{C} - \mathbf{D}\delta_{im}\delta_{\tau o})\lambda_{c_i}^2\bar{\xi}_1^2 = 0 \end{aligned} \tag{3.48}$$

where

$$\delta_{i=2m} \begin{cases} = 1 \text{ if } i = 2m \\ = 0 \text{ otherwise} \end{cases} ; \quad \delta_{im} \begin{cases} = 1 \text{ if } i = m \\ = 0 \text{ otherwise} \end{cases} ; \quad \delta_{\tau o} \begin{cases} = 1 \text{ if } \tau_K = 0 \\ = 0 \text{ otherwise} \end{cases} \tag{3.49}$$

and the constants **A**, **B**, **C** and **D** are defined in Appendix C.
Solving for $\bar{\tau}$ one obtains

$$\overline{\tau} = \overline{\tau}_{mn\tau} + (\hat{\mathbf{A}}\lambda_{c_i} + \hat{\mathbf{B}}\lambda)\delta_{i=2m}\frac{\overline{\xi}_1}{\lambda_{c_i} - \lambda} + (\hat{\mathbf{C}} - \hat{\mathbf{D}}\delta_{im}\delta_{\tau o})\lambda_{c_i}^2 \frac{\overline{\xi}_1^2}{(\lambda_{c_i} - \lambda)^2} \tag{3.50}$$

where $\overline{\tau}_{mn\tau}$ is given by Eq. (2.38).

The constants $\hat{\mathbf{A}}$, $\hat{\mathbf{B}}$, $\hat{\mathbf{C}}$ and $\hat{\mathbf{D}}$ are listed in Appendix C in [12].

In using Eq. (3.50) one must once again remember that because of the presence of the Kronecker delta $\delta_{i=2m}$ the critical buckling torsion parameter $\overline{\tau}_c$ is noticeably influenced by the given imperfection (see Eq. (3.36)) only of the condition $i = 2m$ is satisfied. The circumferential wave number n must be chosen so as to obtain the minimum value of $\overline{\tau}_c$ for a given imperfection amplitude $\overline{\xi}_1$ from Eq. (3.50).

3.4 Combined Loading

If the initial imperfection is axisymmetric

$$\overline{W} = t\overline{\xi}_1 \cos i\pi\frac{x}{L} \tag{3.51}$$

and besides axial compression both external (or internal) pressure and torsion are applied to a cylindrical shell, then the prebuckling stress and deformation state can be represented by the following "membrane" solution

$$w^{(o)} = tW_v + tW_p + tW_t + w_o(x) \tag{3.52}$$

$$F^{(o)} = \frac{Et}{cR}(-\frac{1}{2}\lambda y^2 - \frac{1}{2}\overline{p}_e x^2 - \overline{\tau}xy) + f_o(x) \tag{3.53}$$

Substituting these solutions into the equations governing the prebuckling state (Eqs. 3.2 and 3.3) yields

$$A_{22}^* f_{o,xxxx} - B_{21}^* w_{o,xxxx} = -\frac{1}{R} w_{o,xx} \tag{3.54}$$

$$B_{21}^* f_{o,xxxx} + D_{11}^* w_{o,xxxx} = \frac{1}{R} f_{o,xx} - \frac{Et^2}{cR}\lambda w_{o,xx} + \frac{Et^2}{cR}(i\frac{\pi}{L})^2 \overline{\xi}_1 \lambda \cos i\pi\frac{x}{L} \tag{3.55}$$

the same set of equations as obtained earlier (see Eqs. (3.9) and (3.10), or Eqs. (3.39) and (3.40), respectively). Neglecting the effect of boundary conditions, the particular solution of these equations is

$$w_o(x) = t\frac{\lambda}{\lambda_{c_i} - \lambda}\bar{\xi}_1 \cos i\pi\frac{x}{L} \tag{3.56}$$

$$f_o(x) = \frac{Et^3}{c}\frac{\lambda}{\lambda_{c_i} - \lambda}\frac{(1+\bar{B}^*_{21}\alpha_i^2)}{2\alpha_i^2\bar{A}^*_{22}}\bar{\xi}_1 \cos i\pi\frac{x}{L} \tag{3.57}$$

For the axisymmetric imperfection and the above prebuckling state the linearized stability equations reduce to

$$L_{A^*}(F^{(1)}) - L_{B^*}(W^{(1)}) = -\frac{1}{R}W^{(1)}_{,xx} + \frac{2c}{R}\alpha_i^2(\bar{\xi}_1 + \bar{A})\cos i\pi\frac{x}{L}W^{(1)}_{,yy} \tag{3.58}$$

$$\begin{aligned} L_{B^*}(F^{(1)}) + L_{D^*}(W^{(1)}) = {} & \frac{1}{R}F^{(1)}_{,xx} - \frac{Et^2}{cR}\lambda W^{(1)}_{,xx} + 2\frac{Et^2}{cR}\bar{\tau}\, W^{(1)}_{,xy} \\ & - \frac{Et^2}{cR}(\bar{p}_e + 2c\alpha_i^2\bar{B}\cos i\pi\frac{x}{L})W^{(1)}_{,yy} - \frac{2c}{R}\alpha_i^2(\bar{\xi}_1 + \bar{A})\cos i\pi\frac{x}{L}F^{(1)}_{,yy} \end{aligned} \tag{3.59}$$

Assuming that the radial displacement component of the classic buckling mode can be written as

$$W^{(1)} = t\hat{\xi}\sin m\pi\frac{x}{L}\cos\frac{n}{R}(y - \tau_K x) \tag{3.60}$$

then an approximate solution of the stability equations can be obtained as described in the preceding cases yielding the following characteristic equation

$$\begin{aligned} & \left\{\frac{1}{2}(\bar{T}_{1,m,n} + \bar{T}_{2,p,n} + \frac{\bar{T}^2_{3,m,n}}{\bar{T}_{5,m,n}} + \frac{\bar{T}^2_{4,p,n}}{\bar{T}_{6,p,n}}) - \lambda(\alpha_m^2 + \alpha_p^2) - 2\bar{p}_e\beta_n^2 - 2\bar{\tau}(\alpha_m - \alpha_p)\beta_n\right\}(\lambda_{c_i} - \lambda)^2 \\ & \qquad + (\mathbf{A}\lambda_{c_i} + \mathbf{B}\lambda)\delta_{i=2m}\bar{\xi}_1(\lambda_{c_i} - \lambda) + (\mathbf{C} - \mathbf{D}\delta_{im}\delta_{\tau o})\lambda^2_{c_i}\bar{\xi}_1^2 = 0 \end{aligned} \tag{3.61}$$

If one chooses the variable load Λ to be the axial compression λ then Eq. (3.61) can be written in the following form

$$\left\{[\frac{1}{2(\alpha_m^2+\alpha_p^2)}(\overline{T}_{1,m,n}+\overline{T}_{2,p,n}+\frac{\overline{T}_{3,m,n}^2}{\overline{T}_{5,m,n}}+\frac{\overline{T}_{4,p,n}^2}{\overline{T}_{6,p,n}})-\frac{2\overline{p}_e\beta_n^2}{\alpha_m^2+\alpha_p^2}-\frac{2\overline{\tau}(\alpha_m^2-\alpha_p^2)\beta_n}{\alpha_m^2+\alpha_p^2}]-\lambda\right\}(\lambda_{c_i}-\lambda)^2$$

$$+\frac{1}{(\alpha_m^2+\alpha_p^2)}(\mathbf{A}\lambda_{c_i}+\mathbf{B}\lambda)\delta_{i=2m}\overline{\xi}_1(\lambda_{c_i}-\lambda)+\frac{1}{(\alpha_m^2+\alpha_p^2)}(\mathbf{C}-\mathbf{D}\delta_{im}\delta_{\tau o})\lambda_{c_i}^2\overline{\xi}_1^2=0 \tag{3.62}$$

Multiplying it out and regrouping yields the following cubic polynomial for the eigenvalue λ:

$$\lambda^3-\{\lambda_{mn\tau}+2\lambda_{c_i}-\mathbf{B}^*\overline{\xi}_1\,\delta_{i=2m}\}\lambda^2 \tag{3.63}$$

$$+\{2\lambda_{mn\tau}+\lambda_{c_i}+(\mathbf{A}^*-\mathbf{B}^*)\overline{\xi}_1\,\delta_{i=2m}\}\lambda_{c_i}\lambda-\left\{\lambda_{mn\tau}+\mathbf{A}^*\,\overline{\xi}_1\delta_{i=2m}+(\mathbf{C}^*-\mathbf{D}^*\delta_{im}\delta_{\tau o})\overline{\xi}_1^2\right\}\lambda_{c_i}^2=0$$

where

$$\lambda_{mn\tau}=\frac{1}{\Delta_\lambda}\left\{\frac{1}{2}(\overline{T}_{1,m,n}+\overline{T}_{2,p,n}+\frac{\overline{T}_{3,m,n}^2}{\overline{T}_{5,m,n}}+\frac{\overline{T}_{4,p,n}^2}{\overline{T}_{6,p,n}})-2\overline{p}_e\beta_n^2-2\overline{\tau}(\alpha_m-\alpha_p)\beta_n\right\}$$

$$\Delta_\lambda=\alpha_m^2+\alpha_p^2 \tag{3.64}$$

$$\mathbf{A}^*=\frac{\mathbf{A}}{\Delta_\lambda}\quad;\quad\mathbf{B}^*=\frac{\mathbf{B}}{\Delta_\lambda}\quad;\quad\mathbf{C}^*=\frac{\mathbf{C}}{\Delta_\lambda}\quad;\quad\mathbf{D}^*=\frac{\mathbf{D}}{\Delta_\lambda}$$

and the constants **A**, **B**, **C** and **D** are defined in Appendix C in [12]. Notice that $\overline{p}_e$ and $\overline{\tau}$ are specified fixed loads. In using Eq. (3.63) one must remember that according to Koiter's work [1] the critical buckling load λ is only influenced by the given initial imperfection (see Eq. (3.51)) if the condition i=2m is satisfied. The circumferential full wave number n must be chosen so as to obtain the minimum value of λ for a given imperfection amplitude $\overline{\xi}_1$ from Eq. (3.63).

If instead of external pressure $\overline{p}_e$, a certain internal pressure $\overline{p}_i$ is specified, then Eq. (3.63) becomes

$$\lambda^3 - \{\lambda_{mn\tau} + 2\lambda_{c_i} - \mathbf{B}^* \overline{\xi}_1 \, \delta_{i=2m}\} \hat{\lambda}^2 \tag{3.65}$$

$$+\left\{2\hat{\lambda}_{mn\tau} + \lambda_{c_i} + (\mathbf{A}^* - \mathbf{B}^*)\overline{\xi}_1 \, \delta_{i=2m}\right\}\lambda_{c_i}\lambda - \left\{\hat{\lambda}_{mn\tau} + \mathbf{A}^* \, \overline{\xi}_1 \delta_{i=2m} + (\mathbf{C}^* - \mathbf{D}^* \delta_{im}\delta_{\tau o})\overline{\xi}_1^2\right\}\lambda_{c_i}^2 = 0$$

where

$$\hat{\lambda}_{mn\tau} = \lambda - \frac{1}{2}\overline{p}_i = \frac{1}{\Delta_\lambda}\left\{\frac{1}{2}(\overline{T}_{1,m,n} + \overline{T}_{2,p,n} + \frac{\overline{T}_{3,m,n}^2}{\overline{T}_{5,m,n}} + \frac{\overline{T}_{4,p,n}^2}{\overline{T}_{6,p,n}}) + 2\overline{p}_i\beta_n^2 - 2\overline{\tau}(\alpha_m - \alpha_p)\beta_n\right\}$$

$$\Delta_\lambda = \alpha_m^2 + \alpha_p^2 \tag{3.66}$$

$$\mathbf{A}^* = \frac{\mathbf{A}}{\Delta_\lambda} \; ; \; \mathbf{B}^* = \frac{\mathbf{B}}{\Delta_\lambda} \; ; \; \mathbf{C}^* = \frac{\mathbf{C}}{\Delta_\lambda} \; ; \; \mathbf{D}^* = \frac{\mathbf{D}}{\Delta_\lambda}$$

and the constants **A**, **B**, **C** and **D** are defined in Appendix C in [12]. Notice that $\overline{p}_i$ and $\overline{\tau}$ are specified loads.

If. on the other hand, one chooses the variable load Λ to be the external pressure $\overline{p}_e$, then Eq. (3.61) can be written in the following form

$$\left\{[\frac{1}{4\beta_n^2}(\overline{T}_{1,m,n} + \overline{T}_{2,p,n} + \frac{\overline{T}_{3,m,n}^2}{\overline{T}_{5,m,n}} + \frac{\overline{T}_{4,p,n}^2}{\overline{T}_{6,p,n}}) - \frac{\lambda(\alpha_m^2 + \alpha_p^2)}{2\beta_n^2} - \frac{\overline{\tau}(\alpha_m - \alpha_p)}{\beta_n}] - \overline{p}_e\right\}(\lambda_{c_i} - \lambda)^2$$

$$+\frac{1}{2\beta_n^2}(\mathbf{A}\lambda_{c_i} + \mathbf{B}\lambda)\delta_{i=2m}\overline{\xi}_1(\lambda_{c_i} - \lambda) + \frac{1}{2\beta_n^2}(\mathbf{C} - \mathbf{D}\delta_{im}\delta_{\tau o})\lambda_{c_i}^2\overline{\xi}_1^2 = 0 \tag{3.67}$$

Multiplying it out and solving for $\overline{p}_e$ one obtains

$$\overline{p}_e = \overline{p}_{mn\tau} + \frac{1}{\lambda_{c_i} - \lambda}(\mathbf{A}^*\lambda_{c_i} + \mathbf{B}^*\lambda)\overline{\xi}_1\delta_{i=2m} + \frac{1}{(\lambda_{c_i} - \lambda)^2}(\mathbf{C}^* - \mathbf{D}^*\delta_{im}\delta_{\tau o})\lambda_{c_i}^2\overline{\xi}_1^2 \tag{3.68}$$

where

$$\bar{p}_{mn\tau} = \frac{1}{\Delta_p}\left\{\frac{1}{2}(\bar{T}_{1,m,n} + \bar{T}_{2,p,n} + \frac{\bar{T}^2_{3,m,n}}{\bar{T}_{5,m,n}} + \frac{\bar{T}^2_{4,p,n}}{\bar{T}_{6,p,n}}) - \lambda(\alpha_m^2 + \alpha_p^2) - 2\bar{\tau}(\alpha_m - \alpha_p)\beta_n\right\}$$

$$\Delta_p = 2\beta_n^2 \tag{3.69}$$

$$\mathbf{A}^* = \frac{\mathbf{A}}{\Delta_p} \;\; ; \;\; \mathbf{B}^* = \frac{\mathbf{B}}{\Delta_p} \;\; ; \;\; \mathbf{C}^* = \frac{\mathbf{C}}{\Delta_p} \;\; ; \;\; \mathbf{D}^* = \frac{\mathbf{D}}{\Delta_p}$$

and the constants **A**, **B**, **C** and **D** are listed in Appendix C in [12]. Notice that λ and $\bar{\tau}$ are specified fixed loads.

As mentioned before, it is often convenient to introduce the relationship

$$\lambda = \hat{R}\,\bar{p}_e \tag{3.70}$$

where $\hat{R}$ is a nondimensional constant. In this case both $\bar{p}_e$ and λ are variable loads and Eq. (3.61) can now be written in the following form

$$\left\{\frac{1}{2[\hat{R}(\alpha_m^2 + \alpha_p^2) + 2\beta_n^2]}[\bar{T}_{1,m,n} + \bar{T}_{2,p,n} + \frac{\bar{T}^2_{3,m,n}}{\bar{T}_{5,m,n}} + \frac{\bar{T}^2_{4,p,n}}{\bar{T}_{6,p,n}}) - 4\bar{\tau}(\alpha_m - \alpha_p)\beta_n] - \bar{p}_e\right\}(\lambda_{c_i} - \lambda)^2$$

$$+ \frac{1}{2[\hat{R}(\alpha_m^2 + \alpha_n^2) + 2\beta_n^2}(\mathbf{A}\lambda_{c_i} + \mathbf{B}\lambda)\delta_{i=2m}\bar{\xi}_1(\lambda_{c_i} - \lambda) \tag{3.71}$$

$$+ \frac{1}{2[\hat{R}(\alpha_m^2 + \alpha_p^2) + 2\beta_n^2}(\mathbf{C} - \mathbf{D}\delta_{im}\delta_{\tau o})\lambda_{c_i}^2\bar{\xi}_1^2 = 0$$

Multiplying it out and regrouping yields the following cubic polynomial for the eigenvalue $\bar{p}_e$

$$\bar{p}_e^3 - \{\bar{p}_{mn\tau} + (2\lambda_{c_i}/\hat{R}) - \mathbf{B}^*\bar{\xi}_i\,\delta_{i=2m}\}\bar{p}_e^2$$

$$+\{2\bar{p}_{mn\tau} + (2\lambda_{c_i}/\hat{R}) + (\mathbf{A}^* - \mathbf{B}^*)\bar{\xi}_i\,\delta_{i=2m}\}2\lambda_{c_i}/\hat{R})\bar{p}_e \tag{3.72}$$

$$-\{\bar{p}_{mn\tau} + \mathbf{A}^*\bar{\xi}_1\delta_{i=2m} + (\mathbf{C}^* - \mathbf{D}^*\delta_{im}\delta_{\tau o})\bar{\xi}_1^2\}(2\lambda_{c_i}/\hat{R})^2 = 0$$

where

$$\bar{p}_{mn\tau} = \frac{1}{\Delta_c}\left\{\frac{1}{2}(\bar{T}_{1,m,n} + \bar{T}_{2,p,n} + \frac{\bar{T}_{3,m,n}^2}{\bar{T}_{5,m,n}} + \frac{\bar{T}_{4,p,n}^2}{\bar{T}_{6,p,n}}) - 2\bar{\tau}(\alpha_m - \alpha_p)\beta_n\right\}$$

$$\Delta_c = \hat{R}(\alpha_m^2 + \alpha_p^2) + 2\beta_n^2 \tag{3.73}$$

$$\mathbf{A}^* = \frac{\mathbf{A}}{\Delta_c}\;;\;\mathbf{B}^* = \frac{\mathbf{B}}{\Delta_c}\;;\;\mathbf{C}^* = \frac{\mathbf{C}}{\Delta_c}\;;\;\mathbf{D}^* = \frac{\mathbf{D}}{\Delta_c}$$

and the constants **A**, **B**, **C** and **D** are listed in Appendix C in [12]. Notice that $\bar{\tau}$ is a specified fixed load.

Finally, if one chooses the variable load Λ to be the normalized torsion parameter $\bar{\tau}$ then Eq. (3.61) can be written as

$$\left\{\frac{1}{2(\alpha_m - \alpha_p)\beta_n}[\frac{1}{2}(\bar{T}_{1,m,n} + \bar{T}_{2,p,n} + \frac{\bar{T}_{3,m,n}^2}{\bar{T}_{5,m,n}} + \frac{\bar{T}_{4,p,n}^2}{\bar{T}_{6,p,n}}) - \lambda(\alpha_m^2 + \alpha_p^2) - 2\bar{p}_e\beta_n^2] - \bar{\tau}\right\}(\lambda_{c_i} - \lambda)$$

$$+\frac{1}{2(\alpha_m - \alpha_p)\beta_n}(\mathbf{A}\lambda_{c_i} + \mathbf{B}\lambda)\delta_{i=2m}\bar{\xi}_1(\lambda_{c_i} - \lambda) \tag{3.74}$$

$$+\frac{1}{2(\alpha_m - \alpha_p)\beta_n}(\mathbf{C} - \mathbf{D}\delta_{im}\delta_{\tau o})\lambda_{c_i}^2\bar{\xi}_1^2 = 0$$

Multiplying it out and solving for $\bar{\tau}$ one obtains

$$\bar{\tau} = \bar{\tau}_{mn\tau} + \frac{1}{\lambda_{c_i} - \lambda}(\mathbf{A}^*\lambda_{c_i} + \mathbf{B}^*\lambda)\bar{\xi}_1\delta_{i=2m} + \frac{1}{(\lambda_{c_i} - \lambda)^2}(\mathbf{C}^* - \mathbf{D}^*\delta_{im}\delta_{\tau o})\lambda_{c_i}^2\bar{\xi}_1^2 \tag{3.75}$$

where

$$\overline{\tau}_{mn\tau} = \frac{1}{\Delta_\tau}\left\{\frac{1}{2}\left(\overline{T}_{1,m,n} + \overline{T}_{2,p,n} + \frac{\overline{T}^2_{3,m,n}}{\overline{T}_{5,m,n}} + \frac{\overline{T}^2_{4,p,n}}{\overline{T}_{6,p,n}}\right) - \lambda(\alpha_m^2 + \alpha_p^2) - 2\overline{p}_e\,\beta_n\right\}$$

$$\Delta_\tau = 2(\alpha_m - \alpha_p)\beta_n \tag{3.76}$$

$$\mathbf{A}^* = \frac{\mathbf{A}}{\Delta_\tau} \;;\; \mathbf{B}^* = \frac{\mathbf{B}}{\Delta_\tau} \;;\; \mathbf{C}^* = \frac{\mathbf{C}}{\Delta_\tau} \;;\; \mathbf{D}^* = \frac{\mathbf{D}}{\Delta_\tau}$$

and the constants **A**, **B**, **C** and **D** are listed in Appendix C in [12]. Notice that λ and $\overline{p}_e$ are specified fixed loads.

Finally, it must be remembered that when one is using the buckling equations (Eqs. (3.63), or (3.65), or (3.68), or (3.72), or (3.75)) because of the presence of the Kronecker delta $\delta_{i=2m}$ the critical value of the respective load λ_c is noticeably influenced by the given imperfection (see Eq. (3.51)) only if the condition i=2m is satisfied. Further, the circumferential wave number n must be chosen so as to obtain the minimum value of Λ_c for a given imperfection amplitude $\overline{\xi}_1$ from the respective buckling equation.

4. EFFECT OF ASYMMETRIC IMPERFECTION (b-FACTOR METHOD)

Koiter [1] has shown that the imperfection sensitivity of shell structures is closely related to their initial postbuckling behavior. That is, one is interested in the variation of Λ with ξ in the vicinity of the bifurcation point $\Lambda = \Lambda_c$, where Λ is the loading parameter and ξ is the amplitude of the buckling mode normalized with respect to the wall thickness t. If the shell structure possesses a unique buckling mode associated with the lowest buckling load, then its buckling and initial postbuckling behavior can be expressed as

$$\Lambda = \Lambda_c + a\xi + b\xi^2 + \ldots \tag{4.1}$$

Figure 3a illustrates the case of asymmetric imperfection for $a < 0$, whereas Figures 3b and 3c show cases of symmetric bifurcation with $a = 0$ and $b > 0$ and $b < 0$, respectively.
As can be seen from Fig. 3 the shape of the postbuckling equilibrium path plays a central role in determining the influence of the initial imperfections. When the initial portion of the postbuckling path has a positive curvature (see Fig. 3b) the structure can develop considerable postbuckling strength, and loss of stability of the primary path does not result in structural collapse. However, when the initial portion of the postbuckling path has a negative curvature (see Fig. 3c), then in most cases buckling will occur violently and the magnitude of the critical load is subject to the degrading influence of initial imperfections.

Thus, in order to describe the expected behavior of real (read imperfect) shell structures, one is interested in the variation of $\Lambda(\xi, \bar{\xi})$ with ξ for an imperfect shell ($\bar{\xi} \neq 0$) in the vicinity of $\Lambda = \Lambda_c$, so that the expressions used for perfect shells are still useful. In the following a small initial stress free imperfection $\bar{W} = \bar{\xi}\hat{W}$ will be assumed, where $\hat{W}$ is a suitably normalized continuously differentiable radial displacement field and $\bar{\xi}$ is its amplitude. If only the lowest order terms in $\bar{\xi}$ are retained then an asymptotic expansion of the following form

$$(\Lambda - \Lambda_c)\xi = a\Lambda_c\xi^2 + b\Lambda_c\xi^3 + \ldots - \alpha\Lambda_c\bar{\xi} - \beta(\Lambda - \Lambda_c)\bar{\xi} - \ldots \tag{4.2}$$

becomes appropriate [4]. Notice that

$$\lim_{\xi \to 0} \{\lim_{\bar{\xi} \to 0} \Lambda\} = \Lambda_c \quad \text{but for } \bar{\xi} \neq 0 \quad \lim_{\xi \to 0} \Lambda = 0 \tag{4.3}$$

The relation between the limit loads Λ_s of the imperfect structure and the bifurcation load Λ_c of the perfect structure can now be found from Eq. (4.2) and the condition

$$\left.\frac{d\Lambda}{d\xi}\right|_{\Lambda=\Lambda_s} = 0 \tag{4.4}$$

For $a = 0$, one obtains after some algebraic manipulation the formula

$$(1-\frac{\Lambda_s}{\Lambda_c})^{3/2} = \frac{3}{2}\sqrt{-3\alpha^2 b}\,|\,\xi\,|\{1-\frac{\beta}{\alpha}(1-\frac{\Lambda_s}{\Lambda_c})\} \tag{4.5}$$

Using the following perturbation expansion

$$W = W^{(o)}(\Lambda) + \xi W^{(1)} + \xi^2 W^{(2)} + \ldots \tag{4.6}$$

$$F = F^{(o)}(\Lambda) + \xi F^{(1)} + \xi^2 F^{(2)} + \ldots \tag{4.7}$$

and membrane prebuckling solutions, formulas for the postbuckling coefficients “a” and “b” have been derived by Budiansky and Hutchinson [2] yielding

$$a = -(3/2)F^{(1)} * (W^{(1)}, W^{(1)})/(\Lambda_c \hat{\Delta}) \tag{4.8}$$

$$b = -\left\{2F^{(1)} * (W^{(1)}, W^{(2)}) + F^{(2)} * (W^{(1)}, W^{(1)})\right\}/(\Lambda_c \hat{\Delta} \tag{4.9}$$

where

$$\hat{\Delta} = F_c{}^*(W^{(1)}, W^{(1)}) \;\; ; \;\; (\;)_c = \frac{\partial}{\partial\Lambda}(\;)_c \tag{4.10}$$

and the subscript $(\;)_c$ denotes the fact that the prebuckling solution is evaluated at the bifurcation point.
Using the same perturbation expansion, the formulas for the first and second imperfection form factors α and β published by Cohen [4] become for membrane prebuckling solutions

$$\alpha = F_c * (\hat{W}, W^{(1)})/(\Lambda_c \hat{\Delta}) \quad ; \quad \beta = (1/\hat{\Delta}) F_c{}^*(\hat{W}, W^{(1)}) \tag{4.11}$$

In these formulas the following shorthand notation has been used

$$A^*(B,C) = \int_0^{2\pi R}\int_0^{L} \left\{A_{,xx} B_{,y} C_{,y} - A_{,xy} (B_{,x} C_{,y} + B_{,y} C_{,x}) + A_{,yy} B_{,x} C_{,x}\right\} dxdy \tag{4.12}$$

Thus the calculation of the postbuckling and the imperfection form coefficients involves the solution of the zeroth, the first and the second order states. The equations governing these states can be derived as follows:

The governing equations in terms of W and F for a perfect shell $(\bar{W}=0)$, made out of general anisotropic material are (with W positive inward)

$$L_{A*}(F)-L_{B*}(W)=-\frac{1}{R}W_{,xx}-\frac{1}{2}L_{NL}(W,W) \tag{1.38}$$

$$L_{B*}(F)+L_{D*}(W)=\frac{1}{R}F_{,xx}+L_{NL}(F,W)+p_e \tag{1.39}$$

A formal substitution of the asymptotic expansion for W and F (Eqs. 4.6 and 4.7) and regrouping according to powers of ξ generate the following sequence of linear equations for the functions appearing in the expansion.
For the zeroth order state:

$$L_{A*}(F^{(o)})-L_{B*}(W^{(o)})=-\frac{1}{R}W_{,xx}^{(o)}-\frac{1}{2}L_{NL}(W^{(o)},W^{(o)}) \tag{4.13}$$

$$L_{B*}(F^{(o)})+L_{D*}(W^{(o)})=\frac{1}{R}F_{,xx}^{(o)}+L_{NL}(F^{(o)},W^{(o)})+p_e \tag{4.14}$$

For the first order state:

$$L_{A*}(F^{(1)})-L_{B*}(W^{(1)})=-\frac{1}{R}W_{,xx}^{(1)}-L_{NL}(W^{(o)},W^{(o)}) \tag{4.15}$$

$$L_{B*}(F^{(1)})+L_{D*}(W^{(1)})=\frac{1}{R}F_{,xx}^{(1)}+L_{NL}(F^{(o)},W^{(1)})+L_{NL}(F^{(1)},W^{(o)}) \tag{4.16}$$

For the second order state:

$$L_{A*}(F^{(2)})-L_{B*}(W^{(2)})=-\frac{1}{R}W_{,xx}^{(2)}-L_{NL}(W^{(o)},W^{(2)})-\frac{1}{2}L_{NL}(W^{(1)},W^{(1)}) \tag{4.17}$$

$$\begin{aligned}L_{B*}(F^{(2)})+L_{D*}(W^{(2)})=&\frac{1}{R}F_{,xx}^{(2)}+L_{NL}(F^{(o)},W^{(2)})\\&+L_{NL}(F^{(2)},W^{(o)})+L_{NL}(F^{(1)},W^{(1)})\end{aligned} \tag{4.18}$$

4.1 Axial compression and internal pressure $(p_e = -p_i)$

The membrane prebuckling solution

$$W^{(o)} = tW_i \tag{4.19}$$

$$F^{(o)} = \frac{Et^2}{cR}\left(-\frac{1}{2}\hat{\lambda}\, y^2 + \frac{1}{2}\bar{p}_i\, x^2\right) \tag{4.20}$$

where

$$W_i = \hat{W}_v + W_{p_i} = \frac{1}{c}\left(\bar{A}^*_{12}\,\hat{\lambda} - \bar{A}^*_{22}\bar{p}_i\right) \quad ; \quad \hat{\lambda} = \lambda - \frac{1}{2}\bar{p}_i \tag{4.21}$$

satisfies the governing equations of the zeroth order state identically, and reduces the governing equations of the first- and second order state to

$$L_{A^*}(F^{(1)}) - L_{B^*}(W^{(1)}) = -\frac{1}{R}W,^{(1)}_{xx} \tag{4.22}$$

$$L_{B^*}(F^{(1)}) + L_{D^*}(W^{(1)}) = \frac{1}{R}F,^{(1)}_{xx} + \frac{Et^2}{cR}\bar{p}_i W,^{(1)}_{yy} - \frac{Et^2}{cR}\hat{\lambda}\, W,^{(1)}_{xx} \tag{4.23}$$

and

$$L_{A^*}(F^{(2)}) - L_{B^*}(W^{(2)}) = -\frac{1}{R}W,^{(2)}_{xx} + W,^{(1)}_{xy}W,^{(1)}_{xy} - W,^{(1)}_{xx}W,^{(1)}_{yy} \tag{4.24}$$

$$\begin{aligned} L_{B^*}(F^{(2)}) + L_{D^*}(W^{(2)}) = {} & \frac{1}{R}F,^{(2)}_{xx} + \frac{Et^2}{cR}\bar{p}_i W,^{(2)}_{yy} - \frac{Et^2}{cR}\hat{\lambda}\, W,^{(2)}_{xx} \\ & + F,^{(1)}_{xx}W,^{(1)}_{yy} - 2F,^{(1)}_{xy}W,^{(1)}_{xy} + F,^{(1)}_{yy}W,^{(1)}_{xx} \end{aligned} \tag{4.25}$$

In this case the set of equations for $W^{(1)}$ and $F^{(1)}$ represent the classic eigenvalue problem (see Eq. 2.9, 2.10), the solution of which is for $\hat{\lambda} = \hat{\lambda}_{mn\tau}$

$$W^{(1)} = t\sin m\pi\frac{x}{L}\cos\frac{n}{R}(y - \tau_K x) = \frac{t}{2}\left\{\sin(\;_m x - \;_n y) + \sin(\;_p x + \;_n y)\right\} \tag{4.26}$$

$$F^{(1)} = \frac{Et^3}{4c}\left\{\frac{\overline{T}_{3,m,n}}{\overline{\overline{T}}_{5,m,n}}\sin({}_m x - {}_n y) + \frac{\overline{T}_{4,p,n}}{\overline{\overline{T}}_{6,p,n}}\sin({}_p x + {}_n y)\right\} \tag{4.27}$$

where

$${}_m = m\frac{\pi}{L} + \frac{n}{R}\tau_K \quad ; \quad {}_p = m\frac{\pi}{L} - \frac{n}{R}\tau_K \quad ; \quad {}_n = \frac{n}{R} \tag{4.28}$$

Substituting for $W^{(1)}$ and $F^{(1)}$, the governing equations for the second order state become

$$L_{A*}(F^{(2)}) - L_{B*}(W^{(2)}) = -\frac{1}{R}W^{(2)}_{xx} - \frac{c^2}{R^2}(\alpha_m + \alpha_p)^2\beta_n^2 \sin({}_m x - {}_n y)\sin({}_p x + {}_n y) \tag{4.29}$$

$$L_{B*}(F^{(2)}) + L_{D*}(W^{(2)}) = \frac{1}{R}F,^{(2)}_{xx} + \frac{Et^2}{cR}\overline{p}_i W,^{(2)}_{yy} - \frac{Et^2}{cR}\hat{\lambda}_{mn\tau} W,^{(2)}_{xx}$$
$$+\frac{Et^2}{2R^2}c(\alpha_m + \alpha_p)^2\beta_n^2(2\overline{F})\sin({}_m x - {}_n y)\sin({}_p x + {}_n y) \tag{4.30}$$

where

$$2\overline{F} = \frac{\overline{T}_{3,m,n}}{\overline{\overline{T}}_{5,m,n}} + \frac{\overline{T}_{4,p,n}}{\overline{\overline{T}}_{6,p,n}} \tag{4.31}$$

These equations admit separable solutions of the following form

$$W^{(2)} = t\left\{\sum_{i=1,3,...}^{\infty} A_i \sin {}_i x + \frac{1}{2}\sum_{i=1,3,...}^{\infty} B_i[\sin({}_{m_i} x - {}_{2n} y) + \sin({}_{p_i} x + {}_{2n} y)]\right\} \tag{4.32}$$

$$F^{(2)} = \frac{Et^3}{2c}\left\{\sum_{i=1,3,...}^{\infty} C_i \sin {}_i x + \frac{1}{2}\sum_{i=1,3,...}^{\infty} D_i[\sin({}_{m_i} x - {}_{2n} y) + \sin({}_{p_i} x + {}_{2n} y)]\right\} \tag{4.33}$$

where

$${}_i = i\frac{\pi}{L} \quad ; \quad {}_{m_i} = i\frac{\pi}{L} + 2\frac{n}{R}\tau_K \quad ; \quad {}_{p_i} = i\frac{\pi}{L} - 2\frac{n}{R}\tau_K \quad ; \quad {}_{2n} = 2\frac{n}{R} \tag{4.34}$$

and the summation sign over i implies summation over odd integers only.

The coefficients are readily determined by Galerkin's procedure and are listed in Appendix D in [12]. Finally, the postbuckling coefficients "a" and "b" are calculated by evaluating the integrals indicated by Eqs. (4.8) and (4.10). In this case

$$a = 0 \tag{4.35}$$

and

$$b = -2\frac{c}{\pi}\frac{(\alpha_m+\alpha_p)^2\beta_n^2}{(\alpha_m^2+\alpha_p^2)}\frac{1}{\hat{\lambda}_{mn\tau}}\{\sum_{i=1,3,\ldots}(2\bar{F}B_i+D_i)(\frac{1}{i})+2\sum_{i=1,3,\ldots}(2\bar{F}A_i+C_i)(\frac{i}{4m^2-i^2})\} \tag{4.36}$$

4.2 External pressure and axial compression

The membrane prebuckling solution

$$W^{(o)} = tW_c \tag{4.37}$$

$$F^{(o)} = \frac{Et^2}{cR}(-\frac{1}{2}\bar{p}_e)(x^2+\hat{R}\,y^2) \tag{4.38}$$

where

$$W_c = W_v + W_{p_e} = \frac{\bar{p}_e}{c}(\bar{A}_{22}^* + \hat{R}\,\bar{A}_{12}^*) \quad ; \quad \hat{R} = \frac{\lambda}{\bar{p}_e} \tag{4.39}$$

satisfies the governing equations of the zeroth order state identically and reduces the governing equations of the first- and second order state to

$$L_{A^*}(F^{(1)}) - L_{B^*}(W^{(1)}) = -\frac{1}{R}W_{,xx}^{(1)} \tag{4.40}$$

$$L_{B^*}(F^{(1)}) + L_{D^*}(W^{(1)}) = \frac{1}{R}F_{,xx}^{(1)} - \frac{Et^2}{cR}\bar{p}_e W_{,yy}^{(1)} - \frac{Et^2}{cR}\hat{R}\,\bar{p}_e W_{,xx}^{(1)} \tag{4.41}$$

and

$$L_{A^*}(F^{(2)}) + L_{B^*}(W^{(2)}) = -\frac{1}{R}W_{,xx}^{(2)} + W_{,xy}^{(1)}W_{,xy}^{(1)} - W_{,xx}^{(1)}W_{,yy}^{(1)} \tag{4.42}$$

$$L_{B^*}(F^{(2)}) + L_{D^*}(W^{(2)}) = \frac{1}{R} F_{,xx}^{(2)} - \frac{Et^2}{cR} \overline{p}_e W_{,yy}^{(2)} - \frac{Et^2}{cR} \hat{R}\, \overline{p}_e W_{,xx}^{(2)}$$
$$+ F_{,xx}^{(1)} W_{,yy}^{(1)} - 2F_{,xy}^{(1)} W_{,xy}^{(1)} + F_{,yy}^{(1)} W_{,xx}^{(1)} \tag{4.43}$$

In this case the set of equations of the first order state represent the classic eigenvalue problem (see Eqs. 2.26 and 2.27), the solution of which is for $\overline{p}_e = \overline{p}_{mn\tau}$

$$W^{(1)} = t \sin m\pi \frac{x}{L} \cos \frac{n}{R}(y - \tau_K x) = \frac{t}{2}\left\{\sin({}_m x - {}_n y) + \sin({}_p x + {}_n y)\right\} \tag{4.44}$$

$$F^{(1)} = \frac{Et^3}{4c}\left\{\frac{\overline{T}_{3,m,n}}{\overline{T}_{5,m,n}} \sin({}_m x - {}_n y) + \frac{\overline{T}_{4,p,n}}{\overline{T}_{6,p,n}} \sin({}_p x + {}_n y)\right\} \tag{4.45}$$

where

$$_m = \frac{m\pi}{L} + \frac{n}{R}\tau_K \quad ; \quad _p = \frac{m\pi}{L} - \frac{n}{R}\tau_K \quad ; \quad _n = \frac{n}{R} \tag{4.46}$$

Substituting for $W^{(1)}$ and $F^{(1)}$ the governing equations for the second order state become

$$L_{A^*}(F^{(2)}) - L_{B^*}(W^{(2)}) =$$
$$-\frac{1}{R} W_{,xx}^{(2)} - \frac{c^2}{R^2}(\alpha_m + \alpha_p)^2 \beta_n^2 \sin({}_m x - {}_n y) \sin({}_p x + {}_n y) \tag{4.47}$$

$$L_{B^*}(F^{(2)}) + L_{D^*}(W^{(2)}) = \frac{1}{R} F_{,xx}^{(2)} - \frac{Et^2}{cR} \overline{p}_e W_{,yy}^{(2)} - \frac{Et^2}{cR} \hat{R}\, \overline{p}_e W_{,xx}^{(2)}$$
$$+ \frac{Et^2}{2R^2} c(\alpha_m + \alpha_p)^2 \beta_n^2 (2\overline{F}) \sin({}_m x - {}_n y) \sin({}_p x + {}_n y) \tag{4.48}$$

where

$$2\overline{F} = \frac{\overline{T}_{3,m,n}}{\overline{T}_{5,m,n}} + \frac{\overline{T}_{4,p,n}}{\overline{T}_{6,p,n}} \tag{4.49}$$

These equations admit separable solutions of the following form

$$W^{(2)} = t\left\{ \sum_{i=1,3,...}^{\infty} A_i \sin \;{}_i x + \frac{1}{2} \sum_{i=1,3,...}^{\infty} B_i [\sin(\;{}_{m_i} x - \;{}_{2n} y) + \sin(\;{}_{p_i} x + \;{}_{2n} y)] \right\} \quad (4.50)$$

$$F^{(2)} = \frac{Et^3}{2c}\left\{ \sum_{i=1,3,...}^{\infty} C_i \sin \;{}_i x + \frac{1}{2} \sum_{i=1,3,...}^{\infty} D_i [\sin(\;{}_{m_i} x - \;{}_{2n} y) + \sin(\;{}_{p_i} x + \;{}_{2n} y)] \right\} \quad (4.51)$$

where

$$_i = i\frac{\pi}{L} \;\; ; \;\; {}_{m_i} = i\frac{\pi}{L} + 2\frac{n}{R}\tau_K \;\; ; \;\; {}_{p_i} = i\frac{\pi}{L} - 2\frac{n}{R}\tau_K \;\; ; \;\; {}_{2n} = 2\frac{n}{R} \quad (4.52)$$

and the summation sign over i implies summation over odd integers only.
The coefficients are readily determined by Galerkin's procedure and are listed in Appendix D in [12]. Finally, the postbuckling coefficients "a" and "b" are calculated by evaluating the integrals indicated by Eqs. (4.8), (4.9) and (4.10) yielding

$$a = 0 \quad (4.53)$$

and

$$b = -\frac{2c}{\pi} \frac{(\alpha_m + \alpha_p)^2 \beta_n^2}{[2\beta_n^2 + \hat{R}(\alpha_m^2 + \alpha_p^2)]} \frac{1}{\bar{p}_{mn\tau}} \left\{ \sum_{i=1,3,...} (2\bar{F}B_i + D_i)(\frac{1}{i}) + 2 \sum_{i=1,3,...} (2\bar{F}A_i + C_i)(\frac{i}{4m^2 - i^2}) \right\} \quad (4.54)$$

4.3 Torsion and Axial Compression

The membrane prebuckling solution

$$W^{(o)} = tW_v + tW_t \quad (4.55)$$

$$F^{(o)} = \frac{Et^2}{cR}\left(-\frac{1}{2}\lambda y^2 - \bar{\tau} xy\right) \quad (4.56)$$

where

$$W_v = \frac{\lambda}{c}\bar{A}_{12}^* \quad ; \quad W_t = -\frac{\bar{\tau}}{c} A_{26}^* \quad (4.57)$$

satisfies the governing equations of the zeroth order state identically and reduces the governing equations of the first- and second order state to

$$L_{A^*}(F^{(1)}) - L_{B^*}(W^{(1)}) = -\frac{1}{R} W_{,xx}^{(1)} \tag{4.58}$$

$$L_{B^*}(F^{(1)}) + L_{D^*}(W^{(1)}) = \frac{1}{R} F_{,xx}^{(1)} - \frac{Et^2}{cR} \lambda W_{,xx}^{(1)} + 2\frac{Et^2}{cR} \bar{\tau} W_{,xy}^{(1)} \tag{4.59}$$

and

$$L_{A^*}(F^{(2)}) - L_{B^*}(W^{(2)}) = -\frac{1}{R} W_{,xx}^{(2)} + W_{,xy}^{(1)} W_{,xy}^{(1)} - W_{xx}^{(1)} W_{,yy}^{(1)} \tag{4.60}$$

$$\begin{aligned} L_{B^*}(F^{(2)}) + L_{D^*}(W^{(2)}) = \frac{1}{R} F_{,xx}^{(2)} - \frac{Et^2}{cR} \lambda W_{,xx}^{(2)} + 2\frac{Et^2}{cR} \bar{\tau} W_{,xy}^{(2)} \\ + F_{,xx}^{(1)} W_{,yy}^{(1)} - 2F_{,xy}^{(1)} W_{,xy}^{(1)} + F_{,yy}^{(1)} W_{,xx}^{(1)} \end{aligned} \tag{4.61}$$

In this case the set of equations for $W^{(1)}$ and $F^{(1)}$ represent the classic eigenvalue problem (see Eqs. (2.35) and (2.36)), the solution of which is for $\bar{\tau} = \bar{\tau}_{mn\tau}$

$$W^{(1)} = t \sin m\pi \frac{x}{L} \cos \frac{n}{R}(y - \tau_K x) = \frac{t}{2}\left\{ \sin({}_m x - {}_n y) + \sin({}_p x + {}_n y) \right\} \tag{4.62}$$

$$F^{(1)} = \frac{Et^3}{4c}\left\{ \frac{\bar{T}_{3,m,n}}{\bar{T}_{5,m,n}} \sin({}_m x - {}_n y) + \frac{\bar{T}_{4,p,n}}{\bar{T}_{6,p,n}} \sin({}_p x + {}_n y) \right\} \tag{4.63}$$

where

$${}_m = m\frac{\pi}{L} + \frac{n}{R}\tau_K \quad ; \quad {}_p = m\frac{\pi}{L} - \frac{n}{R}\tau_K \quad ; \quad {}_n = \frac{n}{R} \tag{4.64}$$

Substituting for $W^{(1)}$ and $F^{(1)}$, the governing equations for the 2nd order state become

$$\begin{aligned} L_{A^*}(F^{(2)}) - L_{B^*}(W^{(2)}) = \\ -\frac{1}{R} W_{,xx}^{(2)} - \frac{c^2}{R^2}(\alpha_m + \alpha_p)^2 \beta_n^2 \sin({}_m x - {}_n y) \sin({}_p x + {}_n y) \end{aligned} \tag{4.65}$$

$$L_{B*}(F^{(2)})+L_{D*}(W^{(2)})=\frac{1}{R}F_{,xx}^{(2)}-\frac{Et^2}{cR}\lambda W_{,xx}^{(2)}+2\frac{Et^2}{cR}\bar{\tau}W_{,xy}^{(2)}$$

$$+\frac{Et^2}{2R^2}c(\alpha_m+\alpha_p)^2\beta_n^2(2\bar{F})\sin(\ _m x-\ _n y)\sin(\ _p x+\ _n y) \tag{4.66}$$

where

$$2\bar{F}=\frac{\bar{T}_{3,m,n}}{\bar{T}_{5,m,n}}+\frac{\bar{T}_{4,p,n}}{\bar{T}_{6,p,n}} \tag{4.67}$$

These equations admit separable solutions of the following form

$$W^{(2)}=t\left\{\sum_{i=1,3,\ldots}^{\infty}\hat{A}_i\sin\ _i x+\frac{1}{2}\sum_{i=1,3,\ldots}^{\infty}\hat{B}_i[\sin(\ _{m_i}x-\ _{2n}y)+\sin(\ _{p_i}x+\ _{2n}y)]\right\} \tag{4.68}$$

$$F^{(2)}=\frac{Et^3}{2c}\left\{\sum_{i=1,3,\ldots}^{\infty}\hat{C}_i\sin\ _i x+\frac{1}{2}\sum_{i=1,3,\ldots}^{\infty}\hat{D}_i[\sin(\ _{m_i}x-\ _{2n}y)+\sin(\ _{p_i}x+\ _{2n}y)]\right\} \tag{4.69}$$

where

$$_i=i\frac{\pi}{L}\ ;\quad _{m_i}=i\frac{\pi}{L}+2\frac{n}{R}\tau_K\ ;\quad _{p_i}=i\frac{\pi}{L}-2\frac{n}{R}\tau_K\ ;\quad _{2n}=2\frac{n}{R} \tag{4.70}$$

and the summation sign over i implies summation over odd integers only.

The coefficients are readily determined by Galerkin's procedure and are listed in Appendix D in [12]. Finally, the postbuckling coefficients "a" and "b" are calculated by evaluating the integrals indicated by Eqs. (4.8) and (4.9) yielding

$$a=0 \tag{4.71}$$

and

$$b=-\frac{c}{\pi}\frac{(\alpha_m+\alpha_p)^2\beta_n}{\alpha_m-\alpha_p}\frac{1}{\bar{\tau}_{mn\tau}}\left\{\sum_{i=1,3,\ldots}(2\bar{F}\hat{B}_i+\hat{D}_i)(\frac{1}{i})+2\sum_{i=1,3,\ldots}(2\bar{F}\hat{A}_i+\hat{C}_i)(\frac{i}{4m^2-i^2})\right\} \tag{4.72}$$

and Khot's skewedness parameter τ_K cannot be equal to zero.

4.4 Combined Loading

The membrane prebuckling solution

$$W^{(o)} = tW_v + tW_p + tW_t \tag{4.73}$$

$$F^{(o)} = \frac{Et^2}{cR}\left(-\frac{1}{2}\lambda y^2 - \frac{1}{2}\bar{p}_e - \bar{\tau} xy\right) \tag{4.74}$$

where

$$W_v = \frac{\lambda}{c}\bar{A}_{12}^* \quad ; \quad W_p = \frac{\bar{p}_e}{c}\bar{A}_{22}^* \quad ; \quad W_t = -\frac{\bar{\tau}}{c}\bar{A}_{26}^* \tag{4.75}$$

satisfies the governing equations of the zeroth order state identically and reduces the governing equations of the first and second order state to

$$L_{A^*}(F^{(1)}) - L_{B^*}(W^{(1)}) = -\frac{1}{R}W_{,xx}^{(1)} \tag{4.76}$$

$$L_{B^*}(F^{(1)}) + L_{D^*}(W^{(1)}) = \frac{1}{R}F_{,xx}^{(1)} - \frac{Et^2}{cR}\lambda W_{,xx}^{(1)} + 2\frac{Et^2}{cR}\bar{\tau} W_{,xy}^{(1)} - \frac{Et^2}{cR}\bar{p}_e W_{,yy}^{(1)} \tag{4.77}$$

and

$$L_{A^*}(F^{(2)}) - L_{B^*}(W^{(2)}) = -\frac{1}{R}W_{,xx}^{(2)} + W_{,xy}^{(1)}W_{,xy}^{(1)} - W_{,xx}^{(1)}W_{,yy}^{(1)} \tag{4.78}$$

$$\begin{aligned} L_{B^*}(F^{(2)}) + L_{D^*}(W^{(2)}) = {} & \frac{1}{R}F_{,xx}^{(2)} - \frac{Et^2}{cR}\lambda W_{,xx}^{(2)} + 2\frac{Et^2}{cR}\bar{\tau} W_{,xy}^{(2)} - \frac{Et^2}{cR}\bar{p}_e W_{,yy}^{(2)} \\ & + F_{,xx}^{(1)}W_{,yy}^{(1)} - 2F_{,xy}^{(1)}W_{,xy}^{(1)} + F_{,yy}^{(1)}W_{,xx}^{(1)} \end{aligned} \tag{4.79}$$

Notice that the set of equations for $W^{(1)}$ and $F^{(1)}$ represent the classic eigenvalue problem for combined loading (see Eqs. (2.44) and (2.45)) which has been solved earlier in section 2.4. At the critical value Λ_c of the chosen variable load the buckling modes $W^{(1)}$ and $F^{(1)}$ are

$$W^{(1)} = t\sin m\pi\frac{x}{L}\cos\frac{n}{R}(y - \tau_K x) = \frac{t}{2}\left\{\sin({}_m x - {}_n y) + \sin({}_p x + {}_n y)\right\} \tag{4.80}$$

$$F^{(1)} = \frac{Et^3}{4c}\left\{\frac{\bar{T}_{3,m,n}}{\bar{\bar{T}}_{5,m,n}}\sin({}_{m}x - {}_{n}y) + \frac{\bar{T}_{4,p,n}}{\bar{\bar{T}}_{6,p,n}}\sin({}_{p}x + {}_{n}y)\right\} \tag{4.81}$$

where

$$ {}_{m} = m\frac{\pi}{L} + \frac{n}{R}\tau_K \quad ; \quad {}_{p} = m\frac{\pi}{L} - \frac{n}{R}\tau_K \quad ; \quad {}_{n} = \frac{n}{R} \tag{4.82}$$

Substituting for $W^{(1)}$ and $F^{(1)}$, the governing equations for the 2nd order state become

$$L_{A*}(F^{(2)}) - L_{B*}(W^{(2)}) =$$

$$ -\frac{1}{R}W_{,xx}^{(2)} - \frac{c^2}{R^2}(\alpha_m + \alpha_p)^2\beta_n^2 \sin({}_{m}x - {}_{n}y)\sin({}_{p}x + {}_{n}y) \tag{4.83}$$

$$L_{B*}(F^{(2)}) + L_{D*}(W^{(2)}) = \frac{1}{R}F_{,xx}^{(2)} - \frac{Et^2}{cR}\lambda W_{,xx}^{(2)} + 2\frac{Et^2}{cR}\bar{\tau} W_{,xy}^{(2)} - \frac{Et^2}{cR}\bar{p}_e W_{,yy}^{(2)}$$

$$ + \frac{Et^2}{2R^2}c(\alpha_m + \alpha_p)^2\beta_n^2(2\bar{F})\sin({}_{m}x - {}_{n}y)\sin({}_{p}x + {}_{n}y) \tag{4.84}$$

where (2F) is defined by Eq. (4.67).

These equations admit separable solutions of the following form

$$W^{(2)} = t\left\{\sum_{i=1,3,\ldots}^{\infty} A_i^* \sin {}_{i}x + \frac{1}{2}\sum_{i=1,3,\ldots}^{\infty} B_i^*[\sin({}_{m_i}x - {}_{2n}y) + \sin({}_{p_i}x + {}_{2n}y)]\right\} \tag{4.85}$$

$$F^{(2)} = \frac{Et^3}{2c}\left\{\sum_{i=1,3,\ldots}^{\infty} C_i^* \sin {}_{i}x + \frac{1}{2}\sum_{i=1,3,\ldots}^{\infty} D_i^*[\sin({}_{m_i}x - {}_{2n}y) + \sin({}_{p_i}x + {}_{2n}y)]\right\} \tag{4.86}$$

where

$$ {}_{i} = i\frac{\pi}{L} \quad ; \quad {}_{m_i} = i\frac{\pi}{L} + 2\frac{n}{R}\tau_K \quad ; \quad {}_{p_i} = i\frac{\pi}{L} - 2\frac{n}{R}\tau_K \quad ; \quad {}_{2n} = 2\frac{n}{R} \tag{4.87}$$

and the summation sign over i implies summation over odd integers only.

The coefficients are readily determined by Galerkin's procedure and are listed in Appendix D in [12]. Finally, the postbuckling coefficients "a" and "b" are calculated by evaluating the integrals indicated by Eqs. (4.8), (4.9) and (4.10). Notice that the evaluation of the integrals involved in Eq. (4.10)

$$\hat{\Delta} = F_c{}^*(W^{(1)}, W^{(1)}) \quad \text{where} \quad F_c = \frac{\partial}{\partial \Lambda} F_c \tag{4.10}$$

depends on the choice of the variable load Λ.

Thus if one chooses the variable load Λ to be the axial compression λ then the evaluation of Eq. (4.10) yields

$$\Lambda_c \hat{\Delta} = -\lambda_{mn\tau} \frac{Et}{2} (\pi RL) (\frac{t}{R})^2 (\alpha_m^2 + \alpha_p^2) \tag{4.88}$$

whereby $\bar{p}_e$ and $\bar{\tau}$ are specified fixed loads.

For the special case of specified internal pressure $\bar{p}_e = -\bar{p}_i$, and Eq. (4.88) becomes

$$\hat{\Lambda}_c \hat{\Delta} = -\hat{\lambda}_{mn\tau} \frac{Et}{2} (\pi RL) (\frac{t}{R})^2 (\alpha_m^2 + \alpha_p^2) \tag{4.89}$$

If one chooses the variable load Λ to be the external lateral pressure $\bar{p}_e$ then the evaluation of Eq. (4.10) yields

$$\Lambda_c \hat{\Delta} = -\bar{p}_{mn\tau} \frac{Et}{2} (\pi RL) (\frac{t}{R})^2 2\beta_n^2 \tag{4.90}$$

whereby λ and $\bar{\tau}$ are specified fixed loads.

For the special case of $\lambda = \hat{R}\,\bar{p}_e$ the evaluation of Eq. (4.10) yields

$$\Lambda_c \hat{\Delta} = -\bar{p}_{mn\tau} \frac{Et}{2} (\pi RL) (\frac{t}{R})^2 [\hat{R}(\alpha_m^2 + \alpha_p^2) + 2\beta_n^2] \tag{4.91}$$

where now $\bar{\tau}$ is a fixed load.

Finally, if one chooses the variable load Λ to be the normalized torsion parameter $\bar{\tau}$ then the evaluation of Eq. (4.10) yields

$$\Lambda_c \hat{\Delta} = -\bar{\tau}_{mn\tau} \frac{Et}{2} (\pi RL) (\frac{t}{R})^2 2\hat{\beta}_n (\alpha_m - \alpha_p) \tag{4.92}$$

whereby λ and $\overline{p}_e$ are specified fixed loads.

Summarizing the results, in all cases the first postbuckling coefficient

$$a = 0 \tag{4.93}$$

whereas the second postbuckling coefficient b assumes different values depending on the variable load Λ.

Thus the variable load is $\Lambda = \lambda$ and both $\overline{p}_e$ and $\overline{\tau}$ are fixed loads

$$b = -\frac{2c}{\pi}\frac{(\alpha_m+\alpha_p)^2\beta_n^2}{(\alpha_m^2+\alpha_p^2)}\frac{1}{\lambda_{mn\tau}}\left\{\sum_{i=1,3,...}(2\overline{F}B_i+D_i)(\frac{1}{i})+2\sum_{i=1,3,...}(2\overline{F}A_i+C_i)(\frac{i}{4m^2-i^2})\right\} \tag{4.94}$$

For the case of a specified internal pressure $(\overline{p}_e = -\overline{p}_i)$ the eigenvalue $\lambda_{mn\tau}$ must be replaced by $\hat{\lambda}_{mn\tau} = \lambda_{mn\tau} - \frac{1}{2}\overline{p}_i$.

If the variable load is $\Lambda = \overline{p}_e$ and both λ and $\overline{\tau}$ are fixed loads

$$b = -\frac{c}{\pi}(\alpha_m+\alpha_p)^2\frac{1}{\overline{p}_{mn\tau}}\left\{\sum_{i=1,3,...}(2\overline{F}B_i+D_i)(\frac{1}{i})+2\sum_{i=1,3,...}(2\overline{F}A_i+C_i)(\frac{i}{4m^2-i^2})\right\} \tag{4.95}$$

For the special case that the variable load is $\Lambda = \overline{p}_e$ and $\lambda = \hat{R}\,\overline{p}_e$ whereas $\overline{\tau}$ remains a fixed load

$$b = -\frac{2c}{\pi}\frac{(\alpha_m+\alpha_p)^2\beta_n^2}{[\hat{R}(\alpha_m^2+\alpha_p^2)+2\beta_n^2]\overline{p}_{mn\tau}}\left\{\sum_{i=1,3,...}(2\overline{F}B_i+D_i)(\frac{1}{i})+2\sum_{i=1,3,...}(2\overline{F}A_i+C_i)(\frac{i}{4m^2-i^2})\right\} \tag{4.96}$$

Finally, if the variable load is $\Lambda = \bar{\tau}$ and both λ and $\bar{p}_e$ are fixed loads

$$b = -\frac{c}{\pi}\frac{(\alpha_m + \alpha_p)^2 \beta_n}{(\alpha_m - \alpha_p)}\frac{1}{\bar{\tau}_{mn\tau}}$$

$$\left\{ \sum_{i=1,3,\ldots} (2\bar{F}B_i^* + D_i^*)(\frac{1}{i}) + 2\sum_{i=1,3,\ldots} (2\bar{F}A_i^* + C_i^*)(\frac{i}{4m^2 - i^2}) \right\} \tag{4.97}$$

whereby Khot's skewedness parameter τ_K cannot be equal to zero.

Notice that in all cases summation over odd integers is implied, whereby the evaluation of the summations is truncated once the desired error tolerance is reached.

5. GENERALIZED "LOAD-SHORTENING RELATION"

Information concerning the extent to which buckling can be expected to be gradual or sudden can be obtained from the postbuckling variation of the applied variable load Λ with the generalized displacement Δ. Notice that $\Lambda\,\Delta$ represents the decrease in potential energy of the applied variable load. Thus

$$\Lambda\,\Delta = \iint\limits_S \bar{N}_{\alpha\beta}\varepsilon_{\alpha\beta}\,dxdy \tag{5.1}$$

where $\bar{N}_{\alpha\beta}$ is the variable applied stress resultant and $\varepsilon_{\alpha\beta}$ is the corresponding linear strain. Notice the $\varepsilon_{\alpha\beta}$ can be obtained from the nonlinear stress-strain relations

$$E_{\alpha\beta} = \varepsilon_{\alpha\beta} + \frac{1}{2}W_{,\alpha}W_{,\beta} \tag{5.2}$$

5.1 Axial compression and internal pressure ($p_e = -p_i$)

In this case the initial slope of the load versus end-shortening curve evaluated at the bifurcation point (see Fig. 4) yields further information about the extent to which buckling can be expected to be gradual or sudden. By definition, the (average) axial end-shortening is

$$\varepsilon = -\frac{1}{2\pi RL}\int_0^{2\pi R}\int_0^{L}(u_{,x} - qW_{,xx})\,dxdy \tag{5.3}$$

where

$$u_{,x} = \varepsilon_x - \frac{1}{2}W_{,x}^2 =$$
$$A_{11}^*F_{,yy} + A_{12}^*F_{,xx} - A_{16}^*F_{,xy} - B_{11}^*W_{,xx} - B_{12}^*W_{,yy} - 2B_{16}^*W_{,xy} - \frac{1}{2}W_{,x}^2 \tag{5.4}$$

$q =$ load eccentricity (positive inward)

Introducing the perturbation expansion for W and F (Eqs. 4.6 and 4.7) and regrouping according to powers of ξ yields

$$u_x = A^*_{11}F^{(o)}_{,yy} + A^*_{12}F^{(o)}_{,xx} - A^*_{16}F^{(o)}_{,xy} - B^*_{11}W^{(o)}_{,xx} - B^*_{12}W^{(o)}_{,yy} - 2B^*_{16}W^{(o)}_{,xy} - \frac{1}{2}W^{(o)}_{,x}W^{(o)}_{,x}$$

$$+\xi\left\{A^*_{11}F^{(1)}_{,yy} + A^*_{12}F^{(1)}_{,xx} - A^*_{16}F^{(1)}_{,xy} - B^*_{11}W^{(1)}_{,xx} - B^*_{12}W^{(1)}_{,yy} - 2B^*_{16}W^{(1)}_{,xy} - W^{(o)}_{,x}W^{(1)}_{,x}\right\} \tag{5.5}$$

$$+\xi^2\left\{A^*_{11}F^{(2)}_{,yy} + A^*_{12}F^{(2)}_{,xx} - A^*_{16}F^{(2)}_{,xy}\right.$$

$$\left. - B^*_{11}W^{(2)}_{,xx} - B^*_{12}W^{(2)}_{,yy} - 2B^*_{16}W^{(2)}_{,xy} - W^{(o)}_{,x}W^{(2)}_{,x} - \frac{1}{2}W^{(1)}_{,x}W^{(1)}_{,x}\right\}$$

$$+O(\xi^3)$$

Expanding the prebuckling variables in the usual Taylor series one gets

$$W^{(o)} = W_c + \xi a\hat{\lambda}_c W_c + \xi^2\{b\hat{\lambda}_c W_c + \frac{1}{2}(a\hat{\lambda}_c)^2 W_c\} + O(\xi^3) \tag{5.6a}$$

$$F^{(o)} = F_c + \xi a\hat{\lambda}_c F_c + \xi^2\{b\hat{\lambda}_c F_c + \frac{1}{2}(a\hat{\lambda}_c)^2 F_c\} + O(\xi^3) \tag{5.6b}$$

where $(\ ) = \partial/\partial\hat{\lambda}$ and $(\ )_c$ indicates that the corresponding variable is evaluated at the bifurcation point.

Notice that for a membrane prebuckling state where

$$W^{(o)} = \frac{t}{c}(\bar{A}^*_{12}\hat{\lambda} - \bar{A}^*_{22}\bar{p}_i) \tag{2.6}$$

$$F^{(o)} = \frac{Et^2}{cR}(-\frac{1}{2}\hat{\lambda} + \frac{1}{2}\bar{p}_i x^2) \tag{2.7}$$

one obtains

$$W_c = \frac{t}{c}\bar{A}^*_{12} \qquad ; \qquad W_c = 0 \tag{5.7a}$$

$$F_c = -\frac{1}{2}\frac{Et^2}{cR}y^2 \qquad ; \qquad F_c = 0 \tag{5.7b}$$

Introducing the normalized axial end-shortening $\delta = \varepsilon / \varepsilon_c$ and assuming that the first-postbuckling coefficient $a = 0$, then Eq. (5.3) becomes upon substitution and regrouping

$$\delta = \varepsilon / \varepsilon_c = \delta_o + \xi \Big\{ -\frac{1}{\varepsilon_c} \frac{1}{2\pi RL} \int_0^{2\pi R} \int_0^L (A_{11}^* F_{,yy}^{(1)} + A_{12}^* F_{,xx}^{(1)} - A_{16}^* F_{,xy}^{(1)} - B_{11}^* W_{,xx}^{(1)}$$

$$- B_{12}^* W_{,yy}^{(1)} - 2B_{16}^* W_{,xy}^{(1)} - W_{c,x} W_{,x}^{(1)} - q W_{,xx}^{(1)}) dxdy \Big\}$$

$$+ \xi^2 \Big\{ -\frac{1}{\varepsilon_c} \frac{1}{2\pi RL} \int_0^{2\pi R} \int_0^L (A_{11}^* F_{,yy}^{(2)} + A_{12}^* F_{,xx}^{(2)} - A_{16}^* F_{,xy}^{(2)} - B_{11}^* W_{,xx}^{(2)} - B_{12}^* W_{,yy}^{(2)} \qquad (5.8)$$

$$- 2B_{16}^* W_{,xy}^{(2)} - W_{c,x} W_{,x}^{(2)} - q W_{,xx}^{(2)} - \frac{1}{2} W_{,x}^{(1)} W_{,x}^{(1)} + b \hat{\lambda}_c A_{11}^* F_{c,yy}) dxdy \Big\}$$

$$+ O(\xi^3)$$

and where the **end-shortening along the prebuckling path** δ_o evaluated at the bifurcation point $\hat{\lambda} = \hat{\lambda}_c = \hat{\lambda}_{mn\tau}$ is given by

$$\delta_c = -\frac{1}{\varepsilon_c} \frac{1}{2\pi RL} \int_0^{2\pi R} \int_0^L (A_{11}^* F_{c,yy} + A_{12}^* F_{c,xx} - A_{16}^* F_{c,xy} - B_{11}^* W_{c,xx}$$

$$- B_{12}^* W_{c,yy} - 2B_{16}^* W_{c,xy} - \frac{1}{2} W_{c,x}^2 - q W_{c,xx}) dxdy \qquad (5.9)$$

Further, by definition

$$\varepsilon_c = \left| \varepsilon_x^{(o)} \Big|_{\hat{\lambda} = \hat{\lambda}_c = \hat{\lambda}_{mn\tau}} \right| = \frac{1}{c} \frac{t}{R} (\bar{A}_{11}^* \hat{\lambda}_{mn\tau} - \bar{A}_{12}^* \bar{p}_i) \qquad (5.10)$$

Notice also that since the buckling modes $F^{(1)}, W^{(1)}$ and their derivatives are periodic functions in the circumferential direction, whereas the prebuckling solution and its derivatives are axisymmetric, therefore

$$\int_0^{2\pi R}\int_0^L (A_{11}^* F_{,yy}^{(1)} + A_{12}^* F_{,xx}^{(1)} - A_{16}^* F_{,xy}^{(1)} - B_{11}^* W_{,xx}^{(1)} - B_{12}^* W_{,yy}^{(1)} - 2B_{16}^* W_{,xy}^{(1)} \tag{5.11}$$

$$- W_{c,x} W_{,x}^{(1)} - qW_{,xx}^{(1)})\,dxdy = 0$$

Recalling now the asymptotic expansion

$$(\hat{\lambda} - \hat{\lambda}_c)\xi = a\hat{\lambda}_c \xi^2 + b\hat{\lambda}_c \xi_3 + \ldots - \alpha\hat{\lambda}_c \bar{\xi} - \beta(\hat{\lambda} - \hat{\lambda}_c)\bar{\xi} + \ldots \tag{4.2}$$

But for a perfect shell $\bar{\xi} = 0$ and since here $a = 0$ then

$$\xi^2 = \frac{1}{b\hat{\lambda}_c}(\hat{\lambda} - \hat{\lambda}_c) = \frac{1}{b}(\hat{\rho} - 1) \tag{5.12}$$

where

$$\hat{\rho} = \frac{\hat{\lambda}}{\hat{\lambda}_c} = \text{ a new load parameter normalized so that if } \hat{\lambda} = \hat{\lambda}_c = \hat{\lambda}_{mn\tau} \text{ then } \hat{\rho} = 1. \tag{5.13}$$

Hence the **end-shortening after buckling** can be written as

$$\delta = \delta_c + \frac{1}{b}(\hat{\rho} - 1)\Big\{ -\frac{1}{\varepsilon_c}\frac{1}{2\pi RL}\int_0^{2\pi R}\int_0^L (A_{11}^* F_{,yy}^{(2)} + A_{12}^* F_{,xx}^{(2)} - A_{16}^* F_{,xy}^{(2)} \tag{5.14}$$

$$- B_{11}^* W_{,xx}^{(2)} - B_{12}^* W_{,yy}^{(2)} - 2B_{16}^* W_{,xy}^{(2)}$$

$$- W_{c,x} W_{,x}^{(2)} - qW_{,xx}^{(2)} - \frac{1}{2} W_{,x}^{(1)} W_{,x}^{(1)} + b\hat{\lambda}_c A_{11}^* F_{c,yy})\,dxdy \Big\} + O(\xi^3)$$

Substituting for $W^{(o)}$ and $F^{(o)}$ from Eq. (4.19) and (4.20) into Eq. (5.9), one can calculate the end-shortening along the prebuckling path δ_o

$$\delta_o = \frac{\hat{\lambda}_c}{\varepsilon_c}\left(\frac{1}{c}\frac{t}{R}\right)\left(\bar{A}_{11}^* \hat{\rho} - \bar{A}_{12}^* \frac{\bar{p}_i}{\hat{\lambda}_c}\right) = \frac{1}{\bar{A}_{11}^* - \bar{A}_{12}^* \bar{p}_i / \hat{\lambda}_{mn\tau}}\left\{ \bar{A}_{11}^* \hat{\rho} - \bar{A}_{12}^* \bar{p}_i / \hat{\lambda}_{mn\tau} \right\} \tag{5.15}$$

Similarly, substituting for $W^{(1)}$ and $W^{(2)}$, $F^{(2)}$ from Eqs. (4.26), (4.32) and (4.33) into Eq. (5.14) and evaluating the integrals, one obtains the end-shortening after buckling δ as

$$\delta = \delta_c + \frac{1}{b\hat{\lambda}_{mn\tau}}(\hat{\rho}-1)\frac{1}{\bar{A}_{11}^* - \bar{A}_{12}^*\bar{p}_i/\hat{\lambda}_{mn\tau}}\left\{\bar{A}_{11}^* b\hat{\lambda}_{mn\tau} + \Sigma_2\right\} \tag{5.16}$$

where

$$\Sigma_2 = \frac{\pi t}{L}\frac{R}{L}\left\{\bar{A}_{12}^* \sum_{i=1,3,\ldots} (iD_i) - (\bar{B}_{11}^* + \frac{1}{2}\frac{t}{R}\bar{q}) \sum_{i=1,3,\ldots} (iA_i)\right\} + \frac{c^2}{8}(\alpha_m^2 + \alpha_p^2) + O(\xi^3) \tag{5.17}$$

Notice that the summation sign over i implies summation over odd integers only.
From Eqs. (5.15) and (5.16)

$$\frac{\partial}{\partial\hat{\rho}}\delta_o = \frac{\bar{A}_{11}^*}{\bar{A}_{11}^* - \bar{A}_{12}^*\bar{p}_i/\hat{\lambda}_{mn\tau}} \tag{5.18}$$

$$\frac{\partial}{\partial\hat{\rho}}\delta = \frac{1}{b\hat{\lambda}_{mn\tau}}\frac{1}{\bar{A}_{11}^* - \bar{A}_{12}^*\bar{p}_i/\hat{\lambda}_{mn\tau}}\left\{\bar{A}_{11}^* b\hat{\lambda}_{mn\tau} + \Sigma_2\right\} \tag{5.19}$$

Hence the slope of the load vs end-shortening curve along the prebuckling path becomes

$$K_c = \frac{\partial\hat{\rho}}{\partial\rho_o} = 1 - \frac{\bar{A}_{12}^*}{\bar{A}_{11}^*}\frac{\bar{p}_i}{\hat{\lambda}_{mn\tau}} \tag{5.20}$$

Whereas the slope of the load vs end/shortening curve just after buckling is

$$K_c^* = \frac{\partial\hat{\rho}}{\partial\delta} = \frac{\bar{A}_{11}^* - \bar{A}_{12}^*\bar{p}_i/\hat{\lambda}_{mn\tau}}{\bar{A}_{11}^* + (1/b\hat{\lambda}_{mn\tau})\sum_2} \tag{5.21}$$

and both quantities are evaluated at the bifurcation point $\hat{\lambda}_c = \hat{\lambda}_{mn\tau}$.

As can be seen from Fig. 4, the angle of the initial slope just after buckling θ^*, where

$$\theta^* = \mathrm{TAN}^{-1}(K_c^*) \tag{5.22}$$

is indeed helpful in determining wether the buckling will be gradual or catastrophic. Notice that by normalizing the generalized displacement Δ by the appropriate membrane strain ε_c , for membrane prebuckling (if $\bar{p}_i = 0$) the angle of the fundamental path $\theta = \mathrm{TAN}^{-1}(K_c) = 45$.

5.2 External pressure and axial compression

If external pressure is the variable load (that is $\Lambda = \bar{p}_e$) then the appropriate generalized displacement Δ is the average normal displacement defined as

$$W_{ave} = \frac{1}{2\pi RL} \int_0^{2\pi R} \int_0^{L} W \, dx dy \tag{5.23}$$

where now

$$W = W^{(o)} + \xi W^{(1)} + \xi W^{(2)} + \dots \tag{5.24a}$$

$$W^{(o)} = W_c + (\bar{p}_e - \bar{p}_c) \dot{W}_c + \frac{1}{2} (\bar{p}_e - \bar{p}_c)^2 \ddot{W}_c + \dots \tag{5.24b}$$

$$\bar{p}_e - \bar{p}_c = a\bar{p}_c \xi + b\bar{p}_c \xi^2 + \dots \tag{5.24c}$$

and $\dot{(\ )} = \partial / \partial \bar{p}_e$. The subscript c indicates that the corresponding variable is evaluated at the bifurcation point.

Notice that for a membrane prebuckling state where

$$W^{(o)} = \frac{t}{c} (\bar{A}_{12}^* \hat{R} + \bar{A}_{22}^*) \bar{p}_e \tag{5.25}$$

$$F^{(o)} = \frac{Et^2}{cR} (-\frac{1}{2} \bar{p}_e x^2 - \frac{1}{2} \hat{R} \bar{p}_e y^2) \tag{5.26}$$

and $\hat{R} = \lambda / \bar{p}_e$ one obtains

$$\dot{W}_c = \frac{t}{c} (\bar{A}_{12}^* \hat{R} + \bar{A}_{22}^*) \quad ; \quad \ddot{W}_c = 0 \tag{5.27a}$$

$$\dot{F}_c = \frac{Et^2}{cR} (-\frac{1}{2} x^2 - \frac{1}{2} \hat{R} y^2) \quad ; \quad \ddot{F}_c = 0 \tag{5.27b}$$

Introducing the nondimensional average normal displacement $\bar{W}_{ave} = W_{ave} / W_c$, and assuming that the first-postbuckling coefficient $a = 0$, then Eq. (5.23) becomes upon substituting of the appropriate perturbation expansions (Eqs. (5.24)) and regrouping according to powers of ξ

$$\bar{W}_{ave} = W_{ave} / W_c = \bar{W}_c + \xi\left\{\frac{1}{W_c}\frac{1}{2\pi RL}\int_0^{2\pi R}\int_0^{L} W^{(1)} dxdy\right\} + \xi^2\left\{\frac{1}{W_c}\frac{1}{2\pi RL}\int_0^{2\pi R}\int_0^{L}(W^{(2)} + b\bar{p}_c W_c)dxdy\right\} + O(\xi^3) \tag{5.28}$$

Notice that the nondimensional average normal displacement along the prebuckling path $\bar{W}_o$ evaluated at the bifurcation point $\bar{p}_e = \bar{p}_c = \bar{p}_{mn\tau}$ is given by

$$\bar{W}_c = \frac{1}{W_c}\frac{1}{2\pi RL}\int_0^{2\pi R}\int_0^{L} W_c\, dxdy \tag{5.29}$$

Further, by definition

$$W_c = t(W_v + W_{pe})\Big|_{\bar{p}_e=\bar{p}_c=\bar{p}_{mn\pi}} = \frac{t}{c}(A_{12}^{*}\hat{R} + \bar{A}_{22}^{*})\bar{p}_{mn\tau} \tag{5.30}$$

Notice also that since the buckling mode $W^{(1)}$ is a periodic function in the circumferential direction

$$\int_0^{2\pi R}\int_0^{L} W^{(1)} dxdy = 0 \tag{5.31}$$

Recalling now the asymptotic expansion Eq. (5.24c) one sees that for $a = 0$

$$\xi^2 = \frac{1}{b\bar{p}_c}(\bar{p}_e - \bar{p}_c) = \frac{1}{b}(\rho - 1) \tag{5.32}$$

Where now

$$\rho = \frac{\bar{p}_e}{\bar{p}_c} = \text{a new load parameter normalized so that if } \bar{p}_e = \bar{p}_c = \bar{p}_{mn\tau} \text{ then } \rho = 1 \tag{5.33}$$

Hence the nondimensional average normal displacement after buckling can be written as

$$\bar{W}_{ave} = \bar{W}_c + \frac{1}{b}(\rho - 1)\left\{\frac{1}{W_c}\frac{1}{2\pi RL}\int_0^{2\pi R}\int_0^{L}(W^{(2)} + b\bar{p}_c W_c)\,dxdy\right\} \tag{5.34}$$

Substituting for $W^{(o)}$ from Eq. (4.37) into Eq. (5.29) one can calculate the nondimensional average normal displacement along the prebuckling path $\bar{W}_o$ as

$$\bar{W}_o = \frac{\bar{p}_e}{W_c}\frac{t}{c}(\bar{A}^*_{12}\hat{R} + \bar{A}^*_{22}) = \rho \tag{5.35}$$

Similarly, substituting for $W^{(2)}$ from Eq. (4.50) into Eq. (5.35) and evaluating the integrals, one obtains the nondimensional average normal displacement after buckling $\bar{W}_{ave}$ as

$$\bar{W}_{ave} = \bar{W}_c + \frac{1}{b\bar{p}_{mn\tau}}(\rho - 1)\frac{1}{\bar{A}^*_{12}\hat{R} + \bar{A}^*_{22}}\left\{\bar{A}^*_{12}\hat{R} + \bar{A}^*_{22})b\bar{p}_{mn\tau} + \Sigma_3\right\} \tag{5.36}$$

where

$$\Sigma_3 = \frac{2c}{\pi}\sum_{i=1,3,\ldots}\left(\frac{1}{i}A_i\right) \tag{5.37}$$

Notice that the summation sign over i implies summation over odd integers only.
From Eqs. (5.35) and (5.36)

$$\frac{\partial}{\partial\rho}\bar{W}_o = 1 \tag{5.38}$$

$$\frac{\partial}{\partial\rho}\bar{W}_{ave} = \frac{1}{b\bar{p}_{mn\tau}}\frac{1}{\bar{A}^*_{12}\hat{R} + \bar{A}^*_{22}}\left\{(\bar{A}^*_{12}\hat{R} + \bar{A}^*_{22})\,b\bar{p}_{mn\tau} + \Sigma_3\right\} \tag{5.39}$$

Hence the slope of the external pressure vs average normal displacement curve along the prebuckling path becomes

$$K_c = \frac{\partial\rho}{\partial\bar{W}_o} = 1 \tag{5.40}$$

Whereas the slope of the external pressure vs average normal displacement curve just after buckling is

$$K_c^* = \frac{\partial \rho}{\partial \overline{W}_{ave}} = \frac{1}{1 + \dfrac{1}{b\overline{p}_{mn\tau}(\overline{A}_{12}^* \hat{R} + \overline{A}_{22}^*)} \Sigma_3} \tag{5.41}$$

Notice that due to the appropriate choice of the generalized displacement Δ (see Eq. (5.23)) also in this case, where the external pressure $\overline{p}_c$ is the variable load, the angle of the fundamental path $\theta = TAN^{-1}(K_c) = 45$ in the generalized load vs displacement plane.

5.3 Torsional and Axial Compression

If torsion is the variable load (that is $\Lambda = \overline{\tau}$) then the appropriate generalized displacement Δ is the apparent shear strain defined as

$$\gamma = \frac{1}{2\pi RL} \int_0^{2\pi R} \int_0^{L} v_{,x} \, dx dy \tag{5.42}$$

where

$$\begin{aligned} v_{,x} &= \gamma_{xy} - u_{,y} - W_{,x} W_{,y} \\ &= A_{16}^* F_{,yy} + A_{26}^* F_{,xx} - A_{66}^* F_{,xy} - B_{61}^* W_{,xx} - B_{62}^* W_{,yy} - 2B_{66}^* W_{,xy} - u_{,y} - W_{,x} W_{,y} \end{aligned} \tag{5.43}$$

Notice that upon substituting this expression into Eq. (5.42) the contribution of $u_{,y}$ vanishes, since because of the periodicity requirement

$$\int_0^{2\pi R} u_{,y} dy = 0 \tag{5.44}$$

Introducing the perturbation expansion for W and F (Eqs. 4.6 and 4.7) and regrouping according to powers of ξ yields

$$v_{,x} = A^*_{16}F^{(o)}_{,yy} + A^*_{26}F^{(o)}_{,xx} - A^*_{66}F^{(o)}_{,xy} - B^*_{61}W^{(o)}_{,xx} - B^*_{62}W^{(o)}_{,yy} - 2B^*_{66}W^{(o)}_{,xy} - W^{(o)}_{,x}W^{(o)}_{,y}$$

$$+\xi\Big\{A^*_{16}F^{(1)}_{,yy} + A^*_{26}F^{(1)}_{,xx} - A^*_{66}F^{(1)}_{,xy} - B^*_{61}W^{(1)}_{,xx} - B^*_{62}W^{(1)}_{,yy}$$

$$-2B^*_{66}W^{(1)}_{,xy} - W^{(o)}_{,y}, W^{(1)}_{,x} - W^{(o)}_{,x}W^{(1)}_{,y}\Big\} \tag{5.45}$$

$$+\xi^2\Big\{A^*_{16}F^{(2)}_{,yy} + A^*_{26}F^{(2)}_{,xx} - A^*_{66}F^{(2)}_{,xy} - B^*_{61}W^{(2)}_{,xx} - B^*_{62}W^{(2)}_{,yy}$$

$$-2B^*_{66}W^{(2)}_{,xy} - W^{(o)}_{,y}W^{(2)}_{,x} - W^{(o)}_{,x}W^{(2)}_{,y} - W^{(1)}_{,x}W^{(1)}_{,y}\Big\} + O(\xi^3)$$

whereby one must remember that the prebuckling displacement $W^{(o)}$ is axisymmetric, thus $W^{(o)}_{,y} = 0$.

Expanding the prebuckling variables in the usual Taylor series one gets

$$W^{(o)} = W_c + \xi a\overline{\tau}_c W_c + \xi^2\{b\overline{\tau}_c W_c + \frac{1}{2}(a\overline{\tau}_c)^2 W_c\} + O(\xi^3) \tag{5.46a}$$

$$F^{(o)} = F_c + \xi a\overline{\tau}_c F_c + \xi^2\{b\overline{\tau}_c F_c + \frac{1}{2}(a\overline{\tau}_c)^2 F_c\} + O(\xi^3) \tag{5.46b}$$

where $(\ ) = \partial/\partial\overline{\tau}$ and $(\ )_c$ indicates that the corresponding variable is evaluated at the bifurcation point.

Notice that for a membrane prebuckling state where

$$W^{(o)} = \frac{t}{c}(\overline{A}^*_{12}\lambda - \overline{A}^*_{26}\overline{\tau}) \tag{5.47}$$

$$F^{(o)} = \frac{Et^2}{cR}(-\frac{1}{2}\lambda - \overline{\tau}xy) \tag{5.48}$$

one obtains

$$W_c = -\frac{t}{c}\overline{A}^*_{26} \quad ; \quad W_c = 0 \tag{5.49a}$$

$$F_c = -\frac{Et^2}{cR}xy \qquad ; \qquad F_c = 0 \tag{5.49b}$$

Introducing the normalized apparent shearing strain $\bar{\gamma} = \gamma / \gamma_c$ and recalling that in this case the first postbuckling coefficient $a = 0$, then Eq. (5.42) becomes upon substitution and regrouping

$$\begin{aligned}
\bar{\gamma} = \gamma/\gamma_c = \bar{\gamma}_o &+ \xi\Big\{\frac{1}{\gamma_c}\frac{1}{2\pi RL}\int_0^{2\pi R}\int_0^{L}(A_{16}^* F_{,yy}^{(1)} + A_{26}^* F_{,xx}^{(1)} - A_{66}^* F_{,xy}^{(1)} - B_{61}^* W_{,xx}^{(1)} \\
&\qquad - B_{62}^* W_{,yy}^{(1)} - 2B_{66}^* W_{,xy}^{(1)} - W_{c,x} W_{,y}^{(1)})\,dxdy\Big\} \\
&+ \xi^2\Big\{\frac{1}{\gamma_c}\frac{1}{2\pi RL}\int_0^{2\pi R}\int_0^{L}(A_{16}^* F_{,yy}^{(2)} + A_{26}^* F_{,xx}^{(2)} - A_{66}^* F_{,xy}^{(2)} - B_{61}^* W_{,xx}^{(2)} \\
&\qquad - B_{62}^* W_{,yy}^{(2)} - 2B_{66}^* W_{,xy}^{(2)} - W_{c,x} W_{,y}^{(2)} - W_{,x}^{(1)} W_{,y}^{(1)})\,dxdy \\
&+ b\bar{\tau}_c \frac{1}{\gamma_c}\frac{1}{2\pi RL}\int_0^{2\pi R}\int_0^{L}(A_{16}^* F_{c,yy} + A_{26}^* F_{c,xx} - A_{66}^* F_{c,xy} - B_{61}^* W_{c,xx})\,dxdy\Big\} + O(\xi^3)
\end{aligned} \tag{5.50}$$

where by definition

$$\gamma_c = \left|\gamma_{xy}^{(o)}\right|_{\bar{\tau}=\bar{\tau}_c} = \frac{1}{c}\frac{t}{R}\left|-\bar{A}_{16}^*\lambda + \bar{A}_{66}^*\bar{\tau}_c\right| \tag{5.51}$$

The normalized apparent shearing strain along the prebuckling path $\bar{\gamma}_o = \gamma_o / \gamma_c$ evaluated at the bifurcation point $\bar{\tau} = \bar{\tau}_c = \bar{\tau}_{mn\tau}$ is given by

$$\bar{\gamma}_c = \gamma_c / \gamma_c = \frac{1}{\gamma_c}\frac{1}{2\pi RL}\int_0^{2\pi R}\int_0^{L} A_{16}^* F_{c,yy} + A_{26}^* F_{c,xx} - A_{66}^* F_{c,xy} - B_{61}^* W_{c,xx})\,dxdy \tag{5.52}$$

Notice also that since the buckling modes $F^{(1)}$, $W^{(1)}$ and their derivatives are periodic functions in the circumferential direction, whereas the prebuckling solution and its derivatives are axisymmetric, therefore

$$\int_0^{2\pi R}\int_0^L (A^*_{16}F_{,yy}^{(1)} + A^*_{26}F_{,xx}^{(1)} - A^*_{66}F_{,xy}^{(1)}$$

$$- B^*_{61}W_{,xx}^{(1)} - B^*_{62}W_{,yy}^{(1)} - 2B^*_{66}W_{,xy}^{(1)} - W_{c,x}W_{,y}^{(1)})\,dxdy = 0 \tag{5.53}$$

Recalling now the asymptotic expansion (using $\Lambda = \overline{\tau}$)

$$(\overline{\tau} - \overline{\tau}_c)\xi = a\overline{\tau}_c\xi^2 + b\overline{\tau}_c\xi^3 + \ldots - a\overline{\tau}_c\overline{\xi} - \beta(\overline{\tau} - \overline{\tau}_c)\overline{\xi}\ldots \tag{4.2}$$

where for a perfect shell $\overline{\xi} = 0$. Since in the present case $a = 0$ therefore

$$\xi^2 = \frac{1}{b\overline{\tau}_c}(\overline{\tau} - \overline{\tau}_c) = \frac{1}{b}(\rho - 1) \tag{5.54}$$

where

$$\rho = \frac{\overline{\tau}}{\overline{\tau}_c} = \text{ a new load parameter normalized so that if } \overline{\tau} = \overline{\tau}_c = \overline{\tau}_{mn\tau} \text{ then } \rho = 1$$

Thus the normalized apparent shearing strain after buckling can be written as

$$\overline{\gamma} = \overline{\gamma}_c + \frac{1}{b}(\rho - 1)\Big\{\frac{1}{\gamma_c}\frac{1}{2\pi RL}\int_0^{2\pi R}\int_0^L (A^*_{12}F_{,yy}^{(2)} + A^*_{26}F_{,xx}^{(2)} - A^*_{66}F_{,xy}^{(2)}$$

$$- B^*_{61}W_{,xx}^{(2)} - B^*_{62}W_{,yy}^{(2)} - 2B^*_{66}W_{,xy}^{(2)} - W_{c,x}W_{,y}^{(2)} - W_{,x}^{(1)}W_{,y}^{(1)})\,dxdy \tag{5.55}$$

$$+ b\overline{\tau}_c\frac{1}{\gamma_c}\frac{1}{2\pi RL}\int_0^{2\pi R}\int_0^L (A^*_{16}F_{c,yy} + A^*_{26}F_{c,xx} - A^*_{66}F_{c,xy} - B^*_{61}W_{c,xx})\,dxdy\Big\}$$

Substituting for $W^{(o)}$ and $F^{(o)}$ from Eqs. (5.47) and (5.48) one can evaluate the integrals in Eq. (5.52) obtaining the following expression for the normalized shearing strain $\overline{\gamma}_o$ along the prebuckling path

$$\overline{\gamma}_o = \frac{1}{\left|-\overline{A}^*_{16}\lambda + \overline{A}^*_{66}\overline{\tau}_c\right|}(-\overline{A}^*_{16}\lambda + \overline{A}^*_{66}\overline{\tau}) = \frac{1}{\left|-\overline{A}^*_{16}\lambda/\overline{\tau}_c + \overline{A}^*_{66}\right|}(-A^*_{16}\lambda/\overline{\tau}_c + \overline{A}^*_{66}\rho \tag{5.56}$$

Similarly, substituting for $W^{(1)}$ and $W^{(2)}$, $F^{(2)}$ from Eqs. (4.62), (4.68) and (4.69) one can evaluate the integrals in Eq. (5.55). This yields the following expression for the normalized apparent shearing strain $\bar{\gamma}$ after buckling

$$\bar{\gamma} = \bar{\gamma}_c + (\rho - 1)\frac{1}{b\bar{\tau}_c}\frac{1}{\left|-\bar{A}^*_{16}\lambda/\bar{\tau}_c + \bar{A}^*_{66}\right|}\left\{b\bar{\tau}_c\bar{A}^*_{66} + \Sigma_3\right\} \tag{5.57}$$

where

$$\Sigma_3 = \pi\frac{t}{L}\frac{R}{L}\left\{-\bar{A}^*_{26}\sum_{i=1,3,\ldots} iD_i + \bar{B}^*_{61}\sum_{i=1,3,\ldots} iA_i\right\} + \frac{c^2}{4}\beta_n(\alpha_m - \alpha_p) \tag{5.58}$$

Notice that the summation sign over i implies summation over odd integers only and that $\bar{\tau}_c = \bar{\tau}_{mn\tau}$.

From Eqs. (5.56) and (5.57)

$$\frac{\partial}{\partial p}\bar{\gamma}_o = \frac{\bar{A}^*_{66}}{\left|-\bar{A}^*_{16}\lambda/\bar{\tau}_c + \bar{A}^*_{66}\right|} \tag{5.59}$$

$$\frac{\partial}{\partial p}\bar{\gamma} = \frac{1}{b\bar{\tau}_c}\frac{1}{\left|-\bar{A}^*_{16}\lambda/\bar{\tau}_c + \bar{A}^*_{66}\right|}\left\{b\bar{\tau}_c\bar{A}^*_{66} + \Sigma_3\right\} \tag{5.60}$$

Hence the slope of the load vs normalized apparent shearing strain curve along the prebuckling path becomes

$$K_c = \frac{\partial p}{\partial \bar{\gamma}_o} = \frac{\left|-\bar{A}^*_{16}\lambda/\bar{\tau}_c + \bar{A}^*_{66}\right|}{\bar{A}^*_{66}} \tag{5.61}$$

Whereas the slope of the load vs normalized apparent shearing strain curve just after buckling is

$$K^*_c = \frac{\partial p}{\partial \bar{\gamma}} = \frac{\left|-\bar{A}^*_{16}\lambda/\bar{\tau}_c + \bar{A}^*_{66}\right|}{\bar{A}^*_{66} + \frac{1}{b\bar{\tau}_c}\Sigma_3} \tag{5.62}$$

Notice that both slopes are evaluated at the bifurcation point $\bar{\tau} = \bar{\tau}_c = \bar{\tau}_{mn\tau}$. Notice further, that due to the appropriate choice of the generalized displacement Δ (see Eq. (5.42)) also in this case, where the nondimensional torque parameter $\bar{\tau}$ is the variable load, the angle of the fundamental path $\theta = TAN^{-1}(K_c) = 45$ in the generalized load vs displacement plane, if the specified fixed loads are zero.

5.4 Combined Loading

In this case the membrane prebuckling solution is represented by

$$W^{(o)} = \frac{t}{c}(A_{12}^{*}\lambda + A_{22}^{*}\overline{p}_e - A_{26}^{*}\overline{\tau}) \tag{5.63}$$

$$F^{(o)} = \frac{Et^2}{cR}(-\frac{1}{2}\lambda y^2 - \frac{1}{2}\overline{p}_e x^2 - \overline{\tau}xy) \tag{5.64}$$

Before proceeding with the derivation one must choose the variable load Λ. Once the variable load is known the corresponding generalized displacement Δ can be assigned such that the product $\Lambda\,\Delta$ represents a decrease in potential energy of the applied variable load. There are essentially the following cases possible.

If one chooses the variable load Λ to be the axial compression λ then the other loads $\overline{p}_e$ and $\overline{\tau}$ are fixed loads and the associated generalized displacement Δ is the normalized end-shortening $\delta = \varepsilon / \varepsilon_c$ (see Eq. (5.8)) where now

$$\varepsilon_c = \left| \varepsilon_x^{(o)} \Big|_{\lambda=\lambda_c=\lambda_{mn\tau}} \right| = \frac{1}{c}\frac{t}{R}(\overline{A}_{11}^{*}\lambda_{mn\tau} + \overline{A}_{12}^{*}\overline{p}_e - \overline{A}_{16}^{*}\overline{\tau}) \tag{5.65}$$

Notice, that if one uses a specified internal pressure $\overline{p}_i = -\overline{p}_e$, then the variable load Λ is equal to $\hat{\lambda} = \lambda - \frac{1}{2}\overline{p}_i$ and ε_c becomes

$$\varepsilon_c = \left| \varepsilon_x^{(o)} \Big|_{\hat{\lambda}=\hat{\lambda}_c=\hat{\lambda}_{mn\tau}} \right| = \frac{1}{c}\frac{t}{R}(\overline{A}_{11}^{*}\hat{\lambda}_{mn\tau} - \overline{A}_{12}^{*}\overline{p}_i - \overline{A}_{16}^{*}\overline{\tau}) \tag{5.66}$$

If, on the other hand, one chooses the variable load Λ to be the external pressure $\overline{p}_e$ then the other loads λ and $\overline{\tau}$ are fixed loads and the associated generalized displacement Δ is the normalized average normal displacement $\overline{W}_{ave} = W_{ave} / W_c$ (see Eq. (5.28)) where now

$$W_c = t(W_v + W_p + W_t)\Big|_{\overline{p}_e=\overline{p}_c=\overline{p}_{mn\tau}} = \frac{1}{c}(\overline{A}_{12}^{*}\lambda + \overline{A}_{22}^{*}\overline{p}_{mn\tau} - \overline{A}_{26}^{*}\overline{\tau}) \tag{5.67}$$

For the special case that $\lambda = \hat{R}\,\overline{p}_e$ this expression becomes

$$W_c = \frac{1}{c}\left\{(\bar{A}^*_{12}\hat{R} + \bar{A}^*_{22})\bar{p}_{mn\tau} - \bar{A}^*_{26}\bar{\tau}\right\} \tag{5.68}$$

Finally, if the variable load Λ is chosen to be the normalized torsion parameter $\bar{\tau}$ then the other loads λ and $\bar{p}_e$ are fixed loads and the associated generalized displacement Δ is the normalized apparent shearing strain $\bar{\gamma} = \gamma / \gamma_c$ (see Eq. (5.50)) where now

$$\gamma_c = \left|\gamma^{(o)}_{xy}\right|_{\bar{\tau}=\bar{\tau}_c=\bar{\tau}_{mn\tau}} = \frac{1}{c}\frac{t}{R}\left|-\bar{A}^*_{16}\lambda - \bar{A}^*_{26}\bar{p}_e + \bar{A}^*_{66}\bar{\tau}_{mn\tau}\right| \tag{5.69}$$

6. NUMERICAL RESULTS FOR ISOTROPIC AND ORTHOTROPIC SHELLS

Thanks to extensive NASA sponsored research programs carried out in the sixties and in the early seventies, much is known about the buckling behavior of thin-walled shell structures. The degrading effect of initial geometric imperfections has been extensively investigated using Koiter's general elastic postbuckling theory [1]. Though it was found that besides boundary conditions [5] and nonlinear modal interaction [6] also the use of a rigorous prebuckling analysis [7] can affect the predicted imperfection sensitivity of the buckling loads, it is customary, because of the overall view it provides to rely on membrane prebuckling analysis for the initial computations. Thus in the Level-1 computational modules of DISDECO, described in this report, all computations are carried out assuming a membrane prebuckling state. The accuracy of the final predictions can then be verified by Level-2 computational modules such as ANILISA [8], which includes both a rigorous prebuckling analysis and a rigorous satisfaction of the prescribed boundary conditions.

To test the accuracy and reliability of the Level-1 computational modules of DISDECO, initially an attempt will be made to reproduce the results published by Hutchinson and Amazigo [9] in 1967. In all cases first the figure out of the original publication will be presented followed by hard copies of the graphic output of DISDECO.

6.1 Classical buckling and imperfection sensitivity of stringer stiffened cylinders under axial compression

The results of the original paper (Ref. [9], Fig. 3) are here reproduced as Fig. 5 (for light stiffening), Fig. 8 (for medium stiffening) and as Fig. 11 (for heavy stiffening). Notice that for light stiffening from the nondimensional parameters given in Fig. 5 and from the geometric properties listed in Table 1 one can easily calculate the quantities A_s, I_s and e_s. Using these values with Eq. (2.14) yields the curves for the classical normalized buckling load λ_c as a function of Batdorf's Z-parameter displayed in Fig. 6.

Once for a given Z-value the lowest buckling load λ_c and the corresponding wave numbers m and n have been determined, one can use Eq. (4.36) to evaluate the b-factors displayed in Fig. 7. The reader is reminded that according to Koiter's theory a negative b-factor is a measure of the expected imperfection sensitivity. For a given negative b-factor one can use Eq. (4.5) to calculate the decrease in buckling load as a function of the amplitude $\bar{\xi}$ of an asymmetric imperfection which has the shape of the classical buckling mode

$$\bar{W} = \bar{\xi}\,\hat{W} = \bar{\xi} t \sin m\frac{x}{L} \cos n\frac{y}{R} \tag{6.1}$$

For medium stiffening, the results obtained by DISDECO are displayed in Figures 9 and 10, whereas for heavy stiffening the DISDECO results are shown in Figures 12 and 13.

In general the agreement between the present results and those of Reference [9] is very good, whereby it must be remembered that Hutchinson and Amazigo minimized their eigenvalues by considering $\bar{n} = nL/\pi R$ as a continuous variable, whereas in DISDECO minimization is carried out using the wave numbers m and n as integers.

6.2 Classical buckling and imperfection sensitivity of isotropic cylinders under torsion

Parts of the results of the original paper (Ref. [10], Figures 3 and 4) are here reproduced as Fig. 14. Notice that the figures have been replotted using as the nondimensional torsional buckling parameter

$$\bar{\tau}_c = \frac{\tau_{xy}}{\sigma_c} \quad ; \quad \sigma_c = \frac{Et}{cR} \quad ; \quad c = \sqrt{3(1-\nu^2)}$$

instead of

$$\Lambda_c = \bar{\tau}_c \frac{L^2 t}{D} \quad ; \quad D = \frac{Et^3}{12(1-\nu^2)}$$

used in Ref. [10]. Using the same parameters with Eq. (2.38) yields the classical normalized torsional buckling parameter $\bar{\tau}_c$ as a function of Batdorf's Z-parameter displayed in the upper part of Fig. 15.

Once for a given Z-value the lowest torsional buckling parameter $\bar{\tau}_c$ and the corresponding wave numbers m and n and the skewedness parameter τ_K have been found, one can use Eq. (4.72) to evaluate the b-factors displayed in the lower part of Fig. 15.

A direct comparison of the results of Figs. 14 and 15 shows that the buckling loads obtained by DISDECO agree reasonably well with those of Ref. [10]. For a more detailed comparison based on the results listed in Tables 3 and 4 one sees that the agreement between the results of Ref. [10] and the ones obtained by DISDECO is better, as expected, for the higher Z-values (the longer shells).

6.3 Effect of axisymmetric imperfections

As pointed out by Koiter [1] and by others [2,9] whenever the lowest buckling load is axisymmetric, the postbuckling coefficient b is identically equal to zero. In this case, once the classical buckling load Λ_c is attained the perfect shell deflects in its axisymmetric buckling mode under constant load Λ_c until an asymmetric bifurcation from the axisymmetric state of deformation occurs. Following bifurcation the load falls below Λ_c with increasing axisymmetric and nonaxisymmetric deflections.

Axially compressed cylinders with outside axial stiffeners

Hutchinson and Amazigo [9] have carried out calculations to determine the effect of an imperfection in the shape of the axisymmetric buckling mode

$$\bar{W} = -\delta \sin \pi \frac{x}{L} \tag{6.2}$$

on the bifurcation load λ_s. The results of this original paper (Ref. [9], Fig 4) are here reproduced as Fig. 16. Notice that in Ref. [9] W is positive outward, thus Eq. (6.2) represents an inward imperfection. From the nondimensional parameters given in Fig. 16 and the geometric properties listed in Table 1 one can calculate the quantities A_s, I_s and e_s needed in Eq. (3.20) to calculate the bifurcation load λ_s of the imperfect shell.

Results of calculations using Eq. (3.20) as the characteristic equation did not agree at all with those obtained in Ref. [9]. This is due to the fact that in the derivation of Eq. (3.20) the following half-wave cosine axisymmetric imperfection

$$\bar{W} = t\bar{\xi}_1 \cos i\pi \frac{x}{L} \tag{3.6}$$

is used. One can argue that when i is large enough (for long shells thus) it must not matter whether one uses half-wave sine of half-wave cosine representation. For i=1, however, the two representations are not equivalent.

In order to investigate this disagreement further, in Appendix E in [12] the bifurcation buckling problem of a cylindrical shell with axisymmetric imperfection was solved again using

$$\bar{W} = t\bar{\xi}_1 \sin i\pi \frac{x}{L} \tag{6.3}$$

also a half-wave sine axisymmetric imperfection. The results of the calculations using Eq. (E.18) as the characteristic equation are shown in Fig. 17 for Z=60, Z=80 and Z=100. In all cases the asymmetric buckling mode has many waves in the circumferential direction (between 40 and 65). Although at the first sight these results appear to agree quite well with those of Ref. [9] shown in Fig. 16, a closer examination reveals that in some cases the present results are up to about 35% lower. The reason for the discrepancy is unknown at the present time.

7. NUMERICAL RESULTS FOR LAYERED COMPOSITE SHELLS

To demonstrate the accuracy and reliability of the Level-1 computational modules of DISDECO when calculating the classical buckling load and the imperfection sensitivity of layered composite shells, initially an attempt will be made to reproduce the results published by Khot and Venkayya [3] in 1970. They used shells of either glass-epoxy or boron-epoxy composites. The geometric properties of the shells are given in Tables 5 and 6, respectively. To illustrate the effect of fiber orientation on the buckling load and on the initial postbuckling behavior two different lay-ups will be used. The fiber orientations in outer, middle and inner layers are for

case 1: $-\theta, \theta, 0$ and case 2: $-\theta, \theta, 90$.

The value of θ is varied from 0 to 90 at ten degree increments. The convention used in defining the angle θ is shown in Fig. 1. Notice that when θ is 0 all fibers are oriented axially and when θ is 90 all fibers are oriented circumferentially.

The results of the buckling load and imperfection sensitivity calculations are listed in Tables 7 through 10 and Tables 11 through 14, respectively. To obtain the buckling loads Eq. (2.14) is used, whereby one must remember that the eigenvalue $\lambda_{mn\tau}$ must not only be minimized with respect to m and n (the number of axial half-waves and the number of circumferential full waves, respectively), but also with respect to the real number τ_K. As can be seen from Eq. (2.11) τ_K is a measure of the skewedness of the buckling pattern. It was first introduced by Khot and Venkayya in Ref. [3].

Once for a given configuration the minimum buckling load $\lambda_c = \lambda_{min}$ with the corresponding wave numbers m and n skewedness parameter τ_K is found, one can proceed and investigate the expected postbuckling behavior. This is done by evaluating the b-factor using Eq. (4.36) and by calculating the angle of the initial slope just after buckling θ^* from Eqs. (5.21) and (5.22).

When one compares the results obtained with DISDECO (present work) with the results published in Ref. [3], it appears that the agreement is better when in the present analysis one is using a τ_K which is close to zero. This apparent discrepancy will be discussed in more detail later.

The results of the original paper (Ref. [3], Fig. 3) showing the effect of fiber orientation on the b-factor is here reproduced as Fig. 18. Notice that the curves for glass-epoxy and boron-epoxy are shown on the same figure to illustrate the effect of elastic properties of the material. As mentioned earlier a negative b-factor can be used as an indicator for the degree of imperfection sensitivity of a structure. From Fig. 18 one can see that the magnitude of the b-factor is greater for glass-epoxy than for boron-epoxy shells. This indicates that the glass-epoxy shells are more imperfection sensitive than the boron-epoxy shells.

Figures 19 through 26 show the results obtained by DISDECO for the same cases, whereby on each figure the effect of fiber orientation on both the buckling load λ_c and the postbuckling coefficient b is displayed. Notice that for each case there are two figures presented. In the first figure τ_K is varied such that for each configuration the minimum buckling load λ_c is found. For this case an increase of the classical buckling load with fiber orientation is not accompanied by an increase in imperfection sensitivity. In the second figure τ_K is set equal to zero. Notice that for this case an increase of the classical buckling load with fiber orientation is well accompanied by an increase in imperfection sensitivity.

To elucidate this question further, in Fig. 27 the variation of the buckling load $\lambda_{mn\tau}$ with Khot's skewedness parameter τ_K for a specific configuration (glass-epoxy, case 1: –40 , 40 , 0) is plotted. As one can see there are 3 distinct local minima at $\tau_K = -0.76$ (with $\lambda_c = 0.540706$ and $n = 12$), at $\tau_K = -0.12$ (with $\lambda_{6,12,\tau} = 0.561107$) and at $\tau_K = +0.76$ (with $\lambda_{1,11,\tau} = 0.593021$). A look at the contour plots of the corresponding buckling modes shown in Figures 28, 31 and 34, respectively, confirms the existence of these nearly simultaneous, but distinct eigenvalues. By the way, it appears that in most cases Khot and Venkayya found only the local minimum closest to $\tau_K = 0$ in Reference [3], which explains why there is a better agreement with the present results in Tables 7 through 14 if a τ_K close to zero is used.

If one wants to compare the results obtained with the Level-1 computational modules presented in this report with similar results obtained with the more accurate Level-2 computational module ANILISA [8], it is helpful to have information available about the shape of the buckling and postbuckling modes belonging to the different buckling loads calculated. Figures 29 through 30, 32 through 33 and 35 through 36 display this information for the 3 nearly simultaneous buckling loads discussed earlier.

8. CONCLUSIONS

It is by-now widely accepted that Koiter's general theory of elastic stability [1] has greatly contributed to our understanding of the sometimes perplexing stability behavior of thin-walled structures. However, due to its mathematical complexity it is not always easy for the practicing structural engineer to find the information he wants for the particular structure at hand from the many publications that are available. What he needs is a computational module that enables him to obtain the desired information readily.

It has been shown that within the context of Koiter's initial postbuckling theory the Level-1 computational modules described in this report can be used successfully to investigate the imperfection sensitivity of the buckling loads of isotropic, orthotropic and of fully anisotropic cylindrical shells under combined axial compression, external of internal pressure and torsion taking into account the effect of different initial imperfection shapes.

As a "building block" of the hierarchical design and analysis system "DISDECO" these computational modules make the first step towards acquiring of detailed understanding of the expected instability behavior of different cylindrical shell configurations feasible. This knowledge is a prerequisite for the development of discrete nonlinear computational models which can reliably predict the load carrying capability of the structure.

It must be stressed that the predictions of the Level-1 computational modules provide only a first indication of the expected nonlinear behavior and all its findings must be evaluated within the context of the fundamental assumptions involved in the theory.

Thus, as mentioned earlier, it is known that also boundary conditions [5] and nonlinear modal interactions [6] can have a profound effect on the imperfection sensitivity prediction. Further, it has been shown by Tennyson et al [7] that with short anisotropic shells for reliable prediction of the postbuckling behavior one must use a rigorous prebuckling analysis.

To take also these effects in account, the user must switch to other more advanced computational modules available within "DISDECO" such as ANILISA [8], which can handle the above mentioned effects.

REFERENCES

1. Koiter, W.T., "On the Stability of Elastic Equilibrium", Ph.D. Thesis 1945, in Dutch, TH-Delft, The Netherlands, H.J. Paris, Amsterdam. (English translation issued as NASA TT F-10, pp. 833, 1967.)
2. Budiansky, B. and Hutchinson, J.W. (1964). "Dynamic Buckling of Imperfection Sensitive Structures". Proceedings XI Intern. Congr. Appl. Mech. (H. Görtler, Ed.), Springer Verlag, Berlin, pp. 636-651.
3. Khot, N.S. and Venkayya, V.B. (1970). "Effect of Fiber Orientation on Initial Postbuckling Behavior and Imperfection Sensitivity of Composite Cylindrical Shells". Report AFFDL-TR-70-125, Air Force Flight Dynamics Laboratory, Wright-Patterson Air Force Base, Ohio.
4. Cohen, G.A. (1968). "Effect of a Nonlinear Prebuckling State on the Postbuckling Behavior and Imperfection Sensitivity of Elastic Structures". AIAA Journal, Vol. 6, No. 8, pp. 1616-1619.
5. Hutchinson, J.W. and Frauenthal, J.C. (1969). "Elastic postbuckling behavior of stiffened and barreled cylindrical shells". ASME J. Appl. Mech., Vol. 36, pp. 784-790.
6. Byskov, E. and Hutchinson, J.W. (1977). "Mode Interaction in Axially Stiffened Cylindrical Shells". AIAA J. Vol. 15, No. 7, pp. 941-948.
7. Tennyson, R.C., Booton, M. and Chan, K.H. (1978). "Buckling of Short Cylinders under Combined Loading". ASME J. Appl. Mech., Vol. 45, pp. 574-578.
8. Arbocz, J. and Hol, J.M.A.M. (1990). "Koiter's Stability Theory in a Computer-Aided Engineering (CAE) Environment". Int. Solids & Structures, Vol. 26, No 9/10, pp. 945-973.
9. Hutchinson, J.W. and Amazigo, J.C. (1967). "Imperfection sensitivity of eccentrically stiffened cylindrical shells". AIAA Journal, Vol. 5, No. 3, pp. 392-401.
10. Budiansky, B. (1969). "Post-Buckling Behavior of Cylinders in Torsion". Proceedings 2nd IUTAM Symposium Theory of Thin Shells (F.I. Niordson, Ed.), Springer Verlag, Berlin, pp. 212-233.
11. Ashton, J.E., Halpin, J.C. and Petit, P.H. (1969). "Primer on Composite Materials: Analysis". Technomic Publishing Co., Stamford, Connecticut.
12. Arbocz, J. (1992). "The Effect of Initial Imperfection on Shell Stability - an Updated Review". Report LR-695, Delft University of Technology, Faculty of Aerospace Engineering.

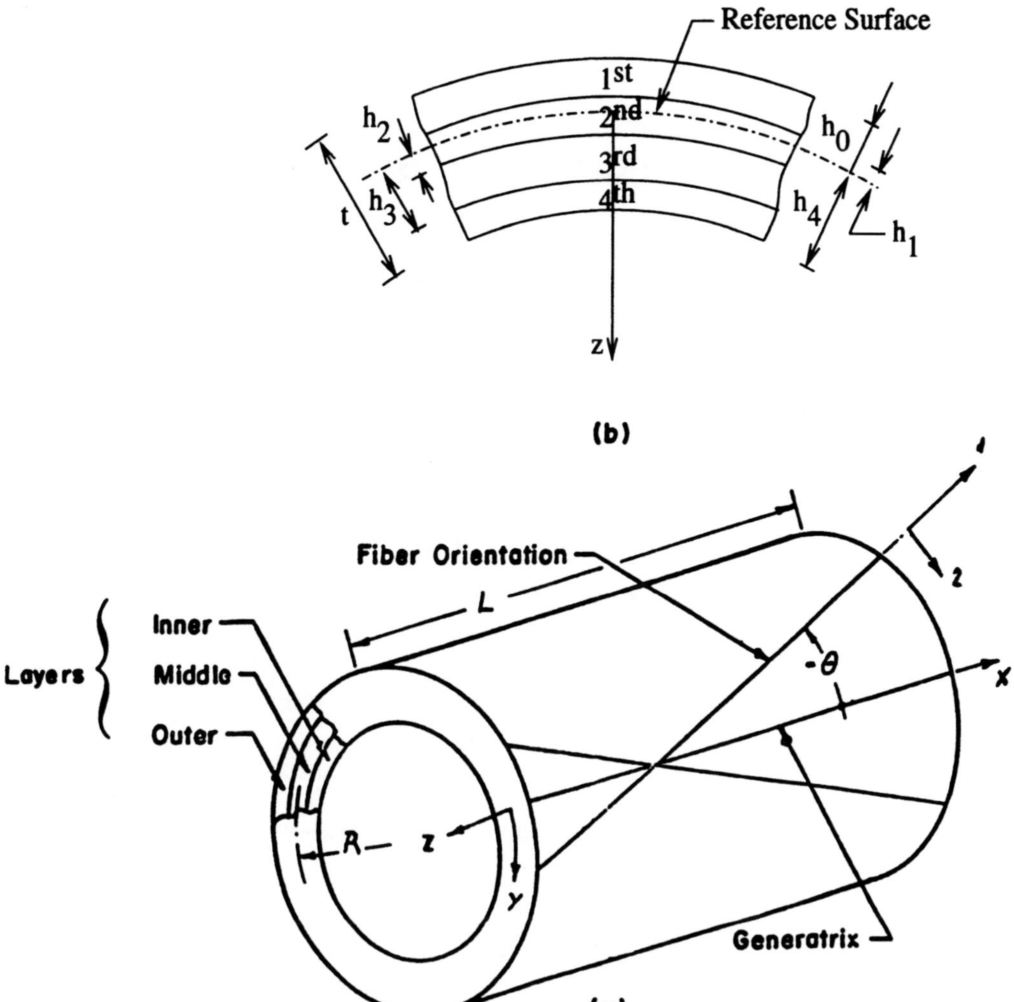

Fig. 1. Notation and sign convention

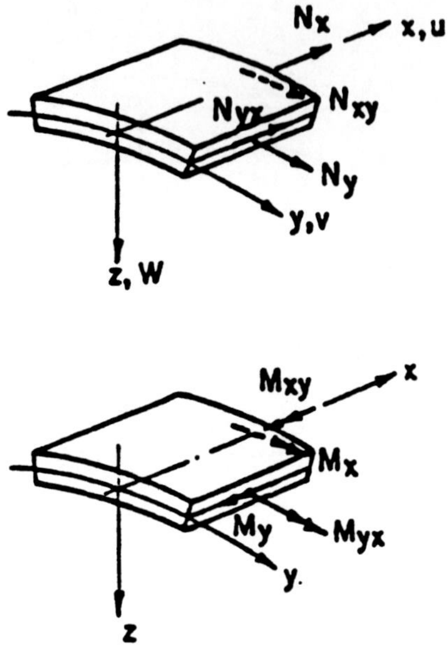

Fig. 2. Definition of stress- and moment resultants

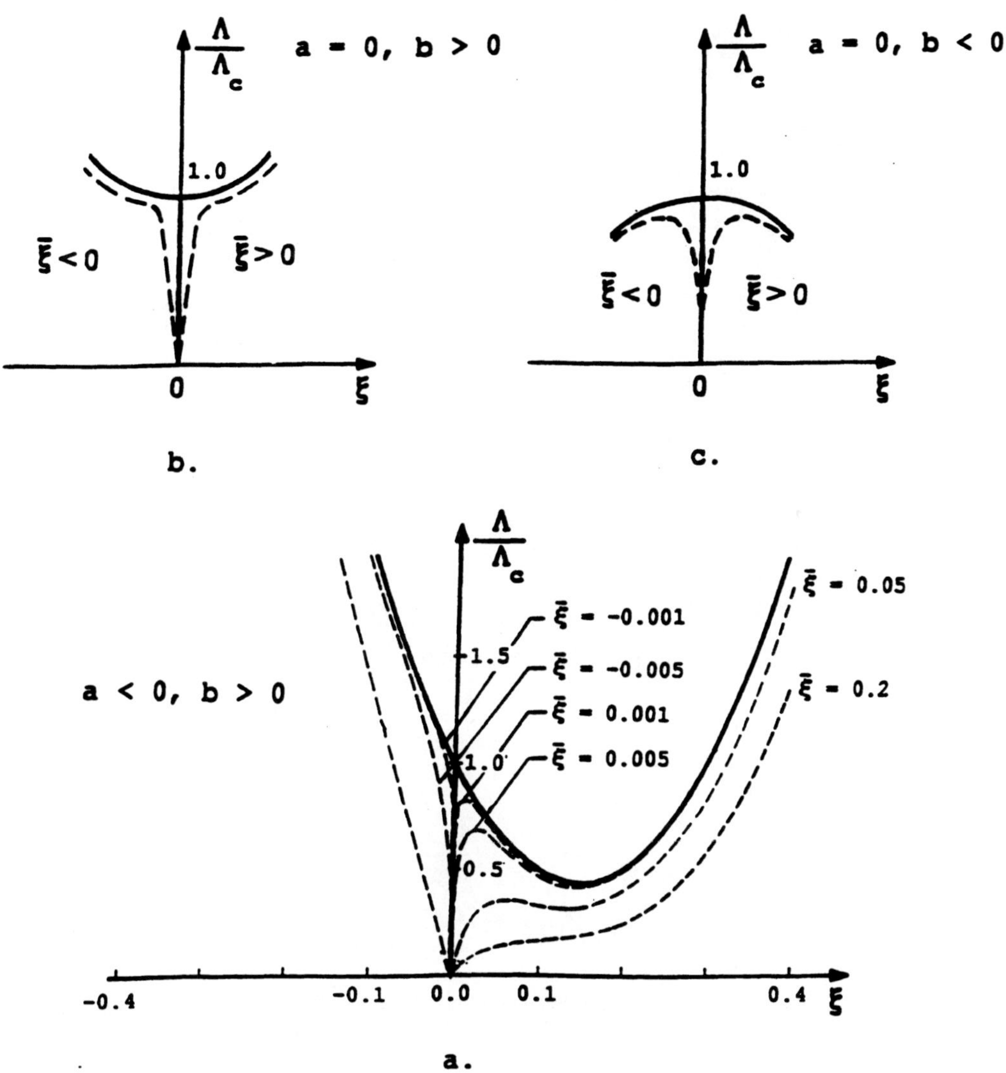

Fig. 3. Nondimensional load versus additional displacement relationships

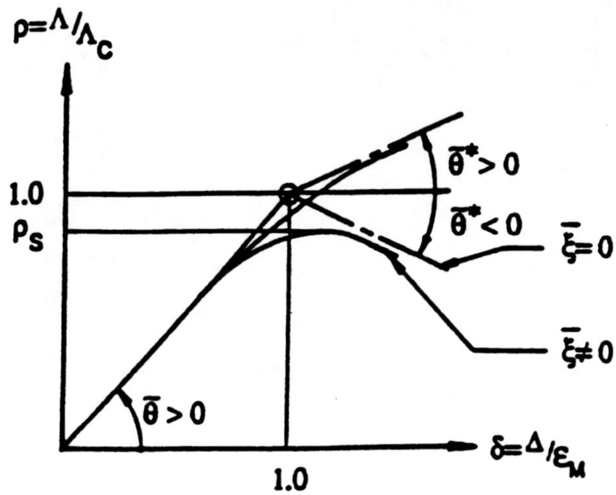

Fig. 4. Generalized "Load-Shortening" curve

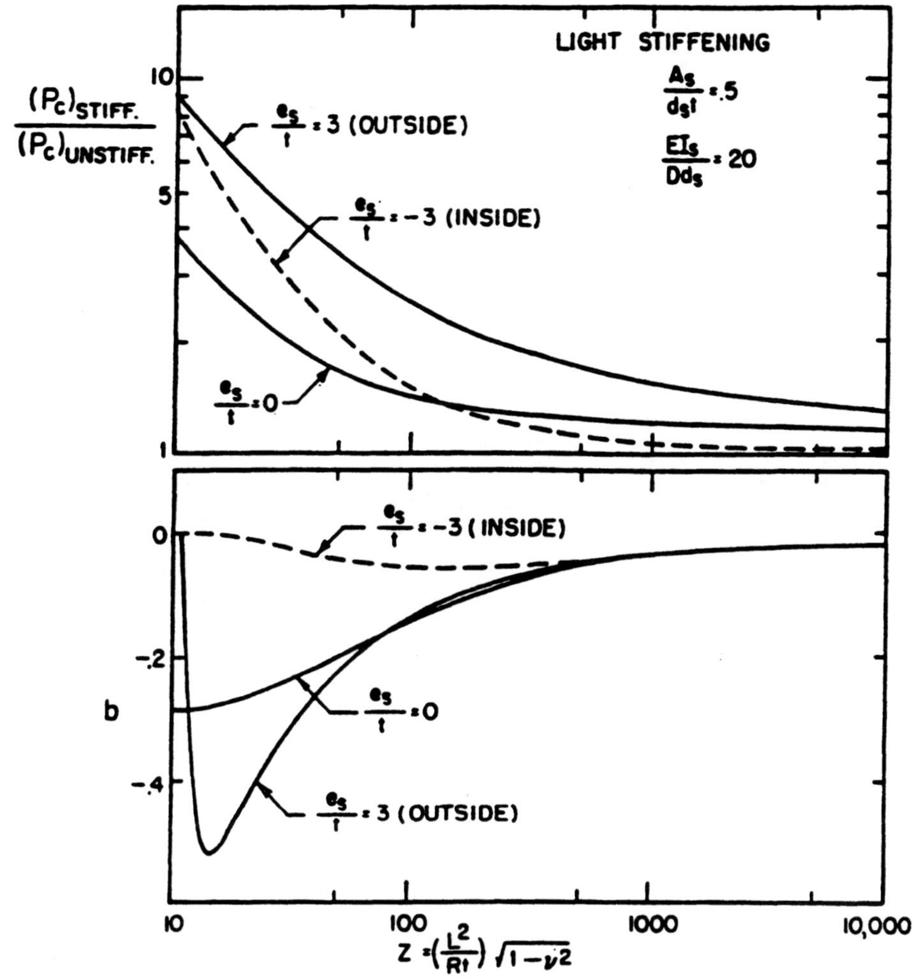

Fig. 5. Buckling and imperfection-sensitivity of simply supported, axially stiffened cylinders under axial compression [9] (light stiffening)

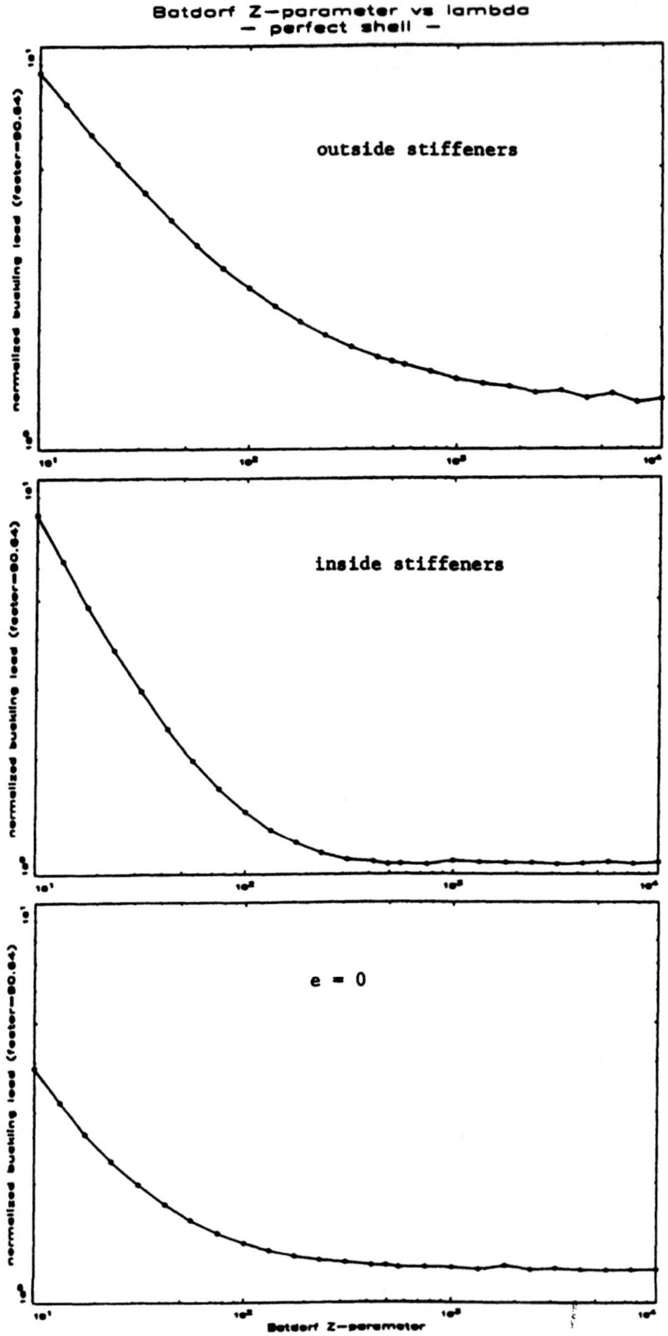

Fig. 6. Buckling of simply supported, axially stiffened cylinders under axial compression (light stiffening)

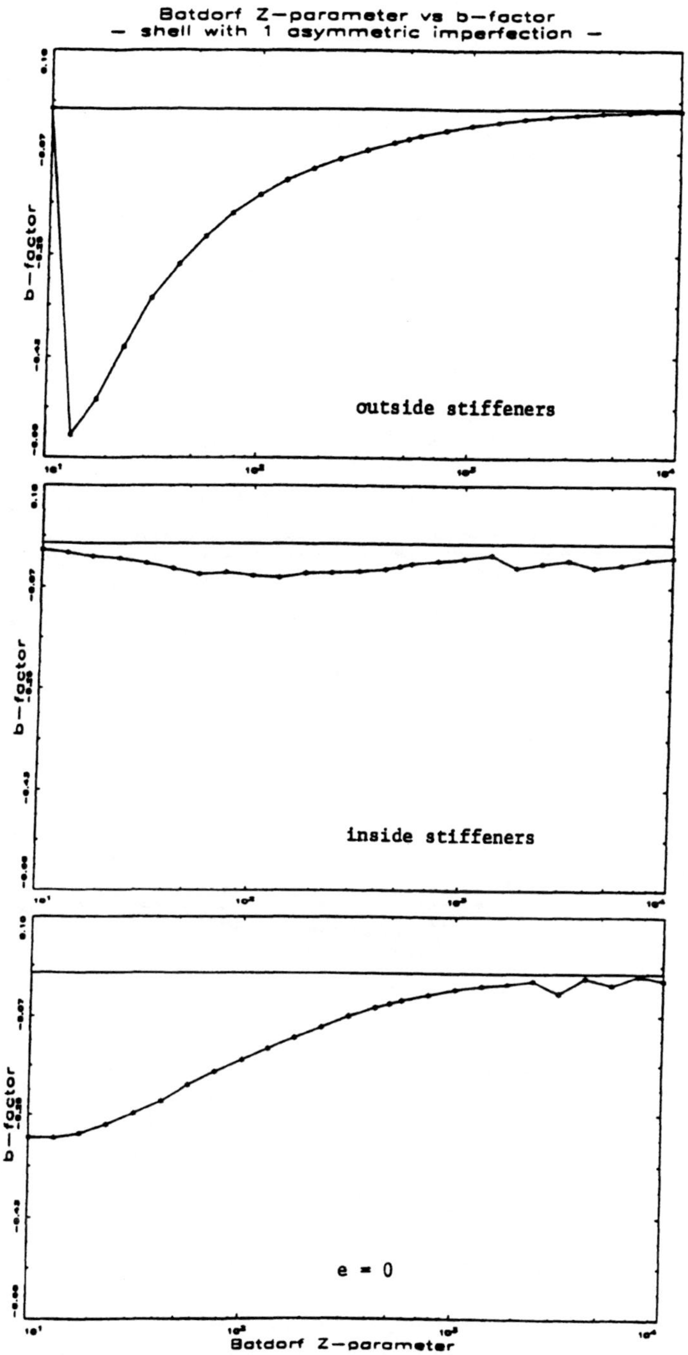

Fig. 7. Imperfection sensitivity of simply supported, axially stiffened cylinders under axial compression (light stiffening)

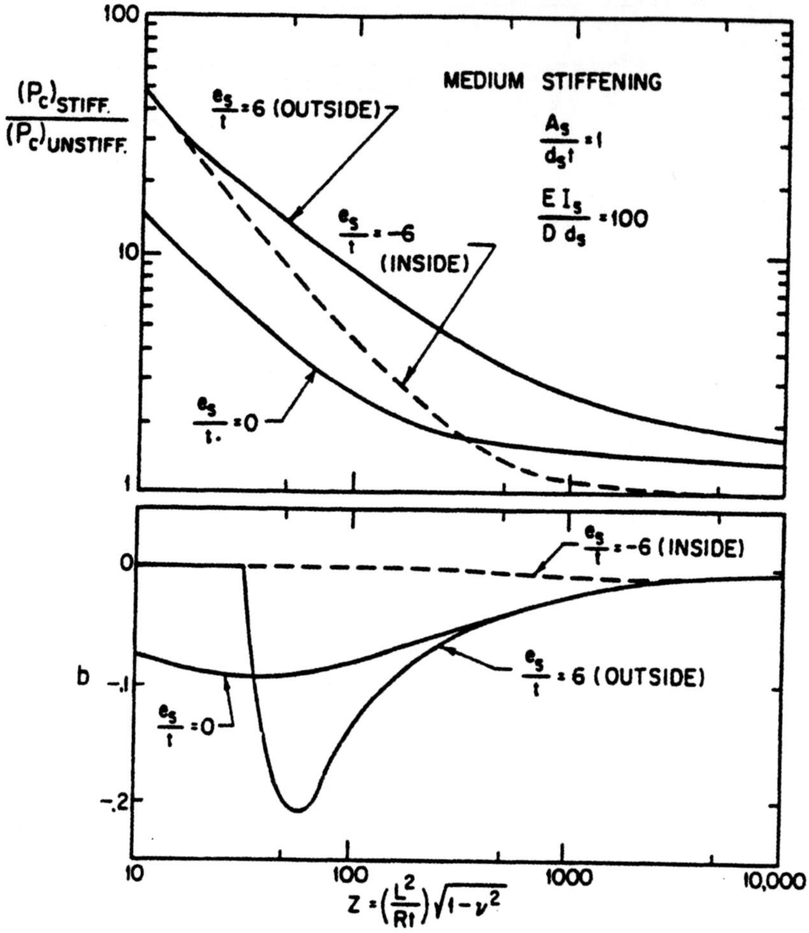

Fig. 8. Buckling and imperfection-sensitivity of simply supported, axially stiffened cylinders under axial compression [9] (medium stiffening)

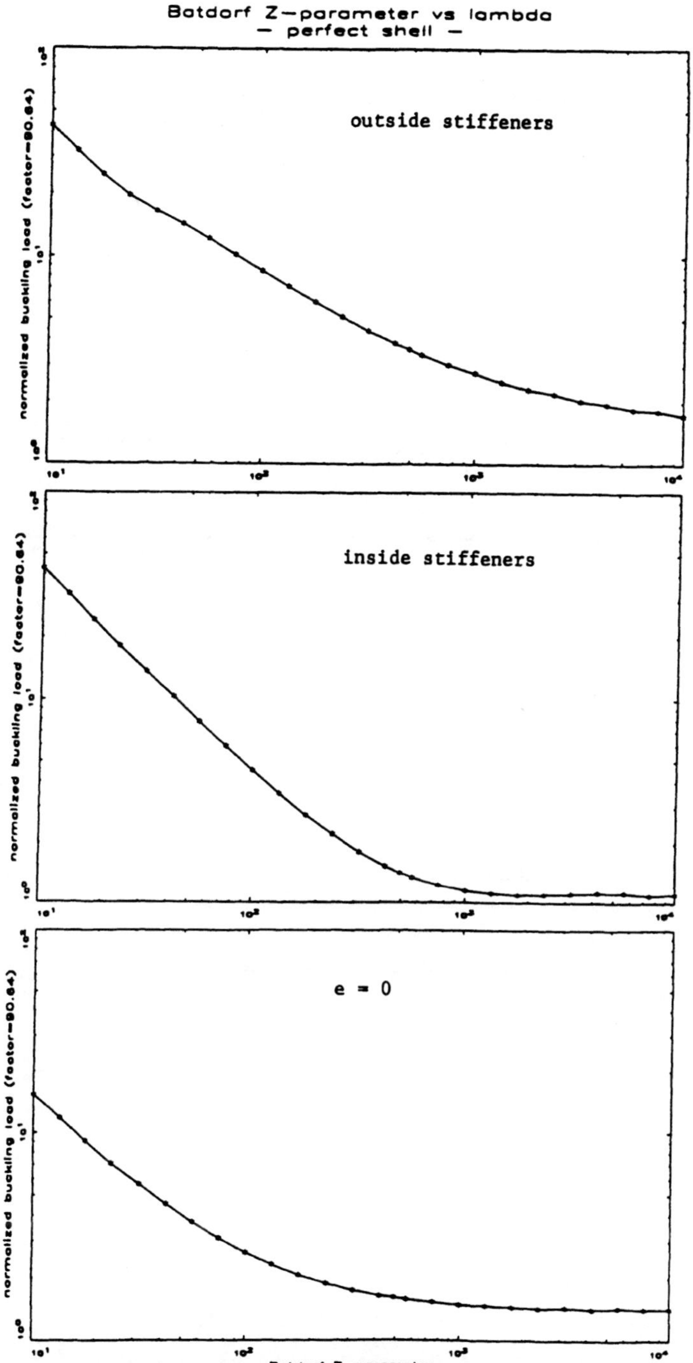

Fig. 9. Buckling of simply supported, axially stiffened cylinders under axial compression (medium stiffening)

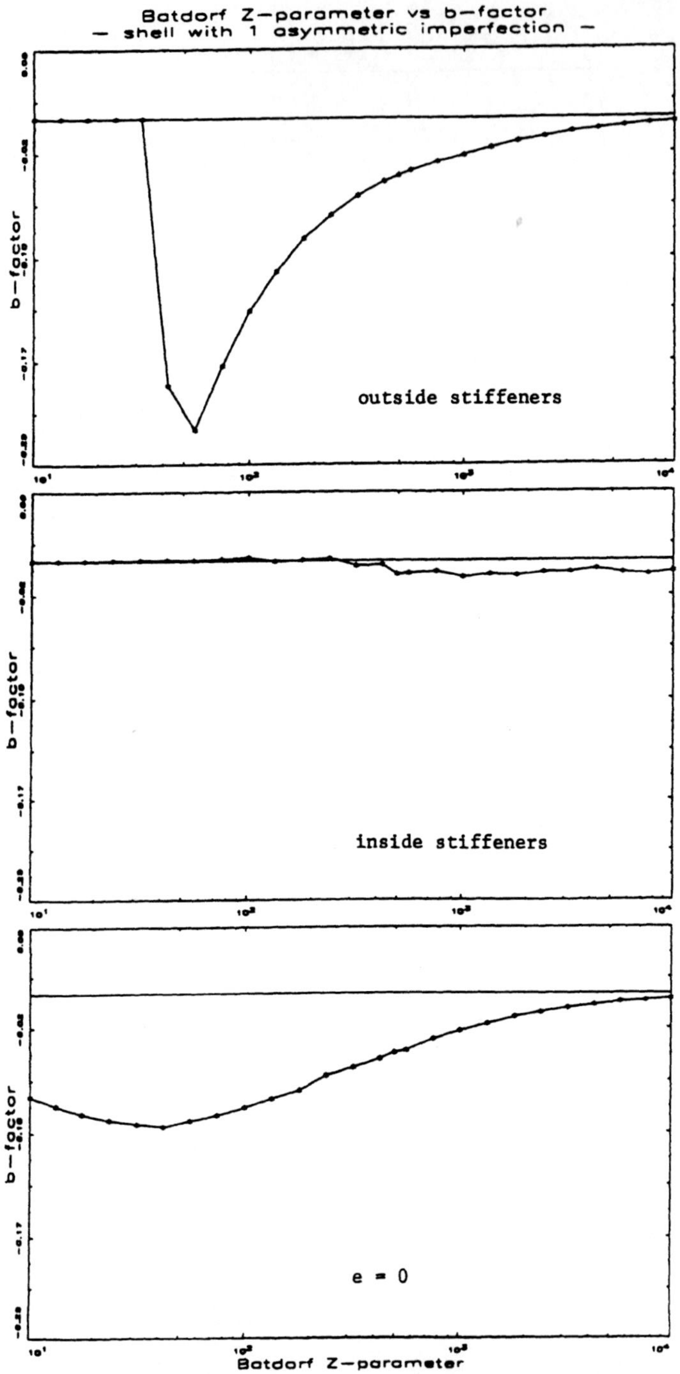

Fig. 10. Imperfection sensitivity of simply supported, axially stiffened cylinders under axial compression (medium stiffening)

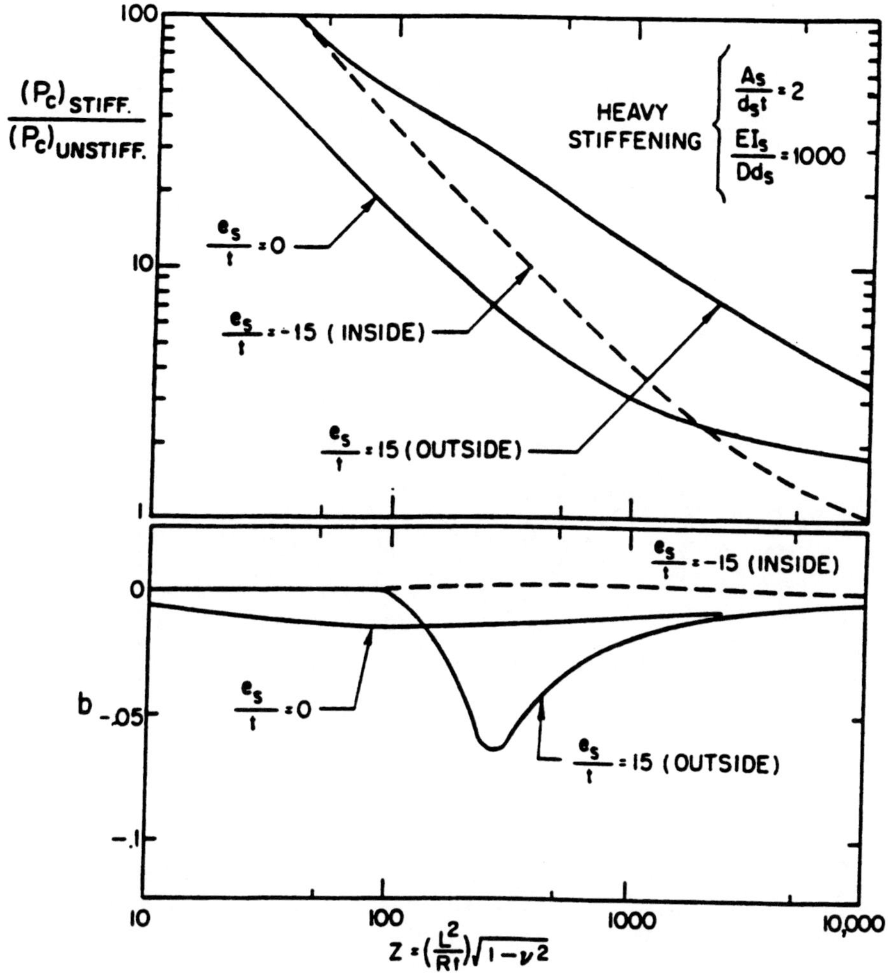

Fig. 11. Buckling and imperfection sensitivity of simply supported, axially stiffened cylinders under axial compression [9] (heavy stiffening)

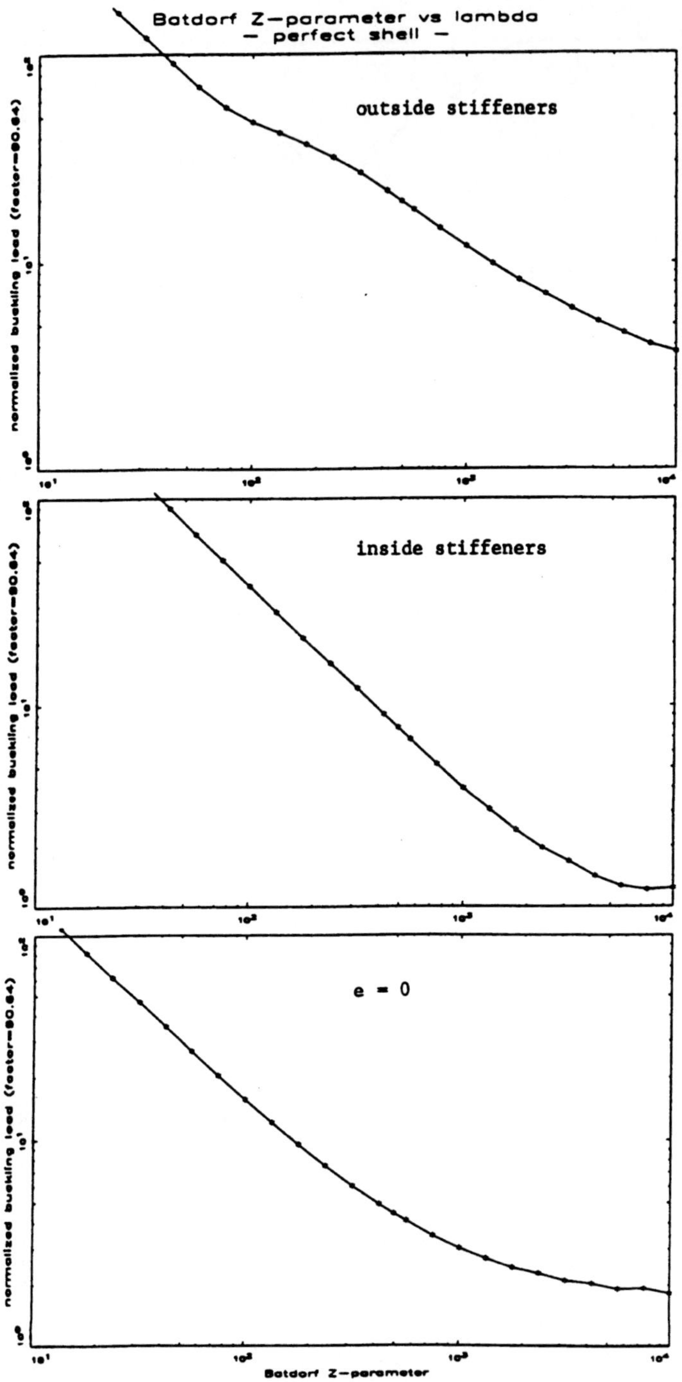

Fig. 12. Buckling of simply supported, axially stiffened cylinders under axial compression (heavy stiffening)

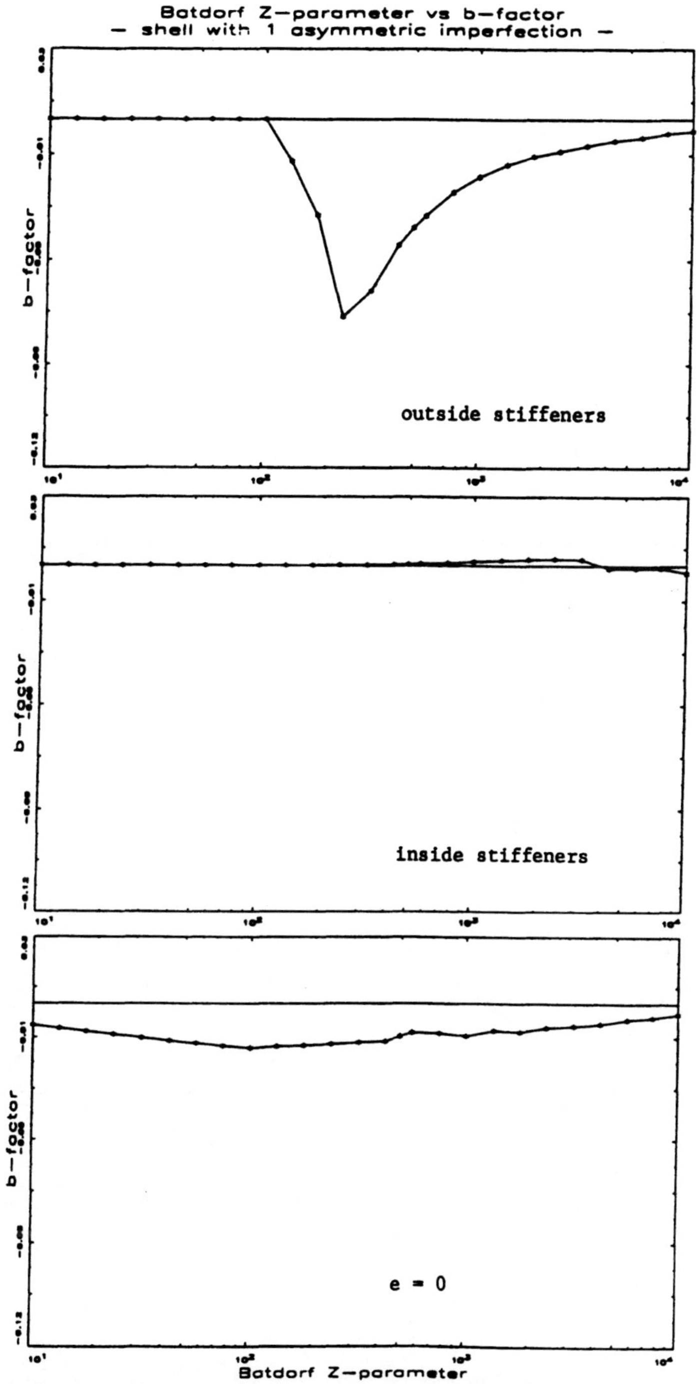

Fig. 13. Imperfection sensitivity of simply supported, axially stiffened cylinders under axial compression (heavy stiffening)

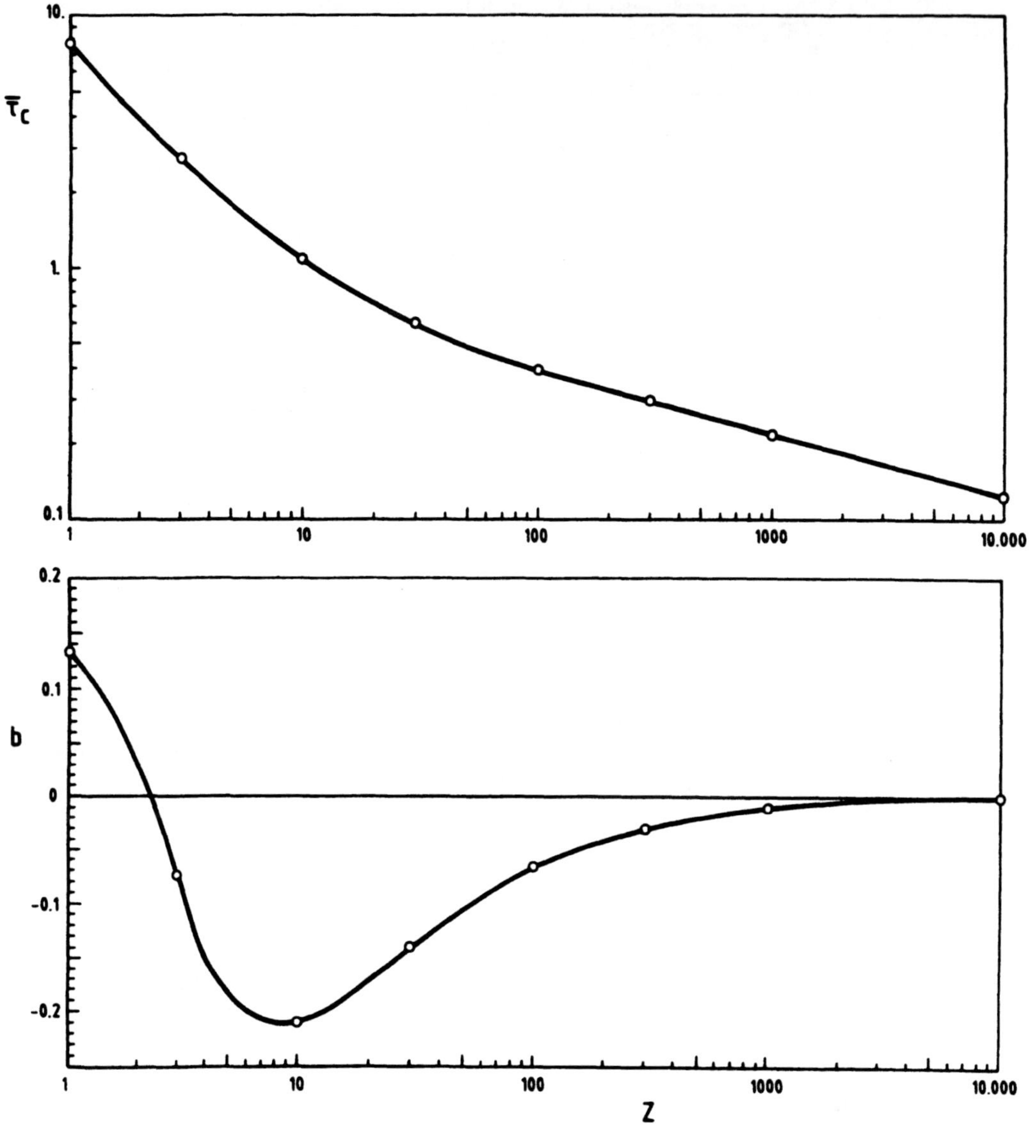

Fig. 14. Buckling and imperfection sensitivity of simply supported, isotropic cylinders, under torsion [10]

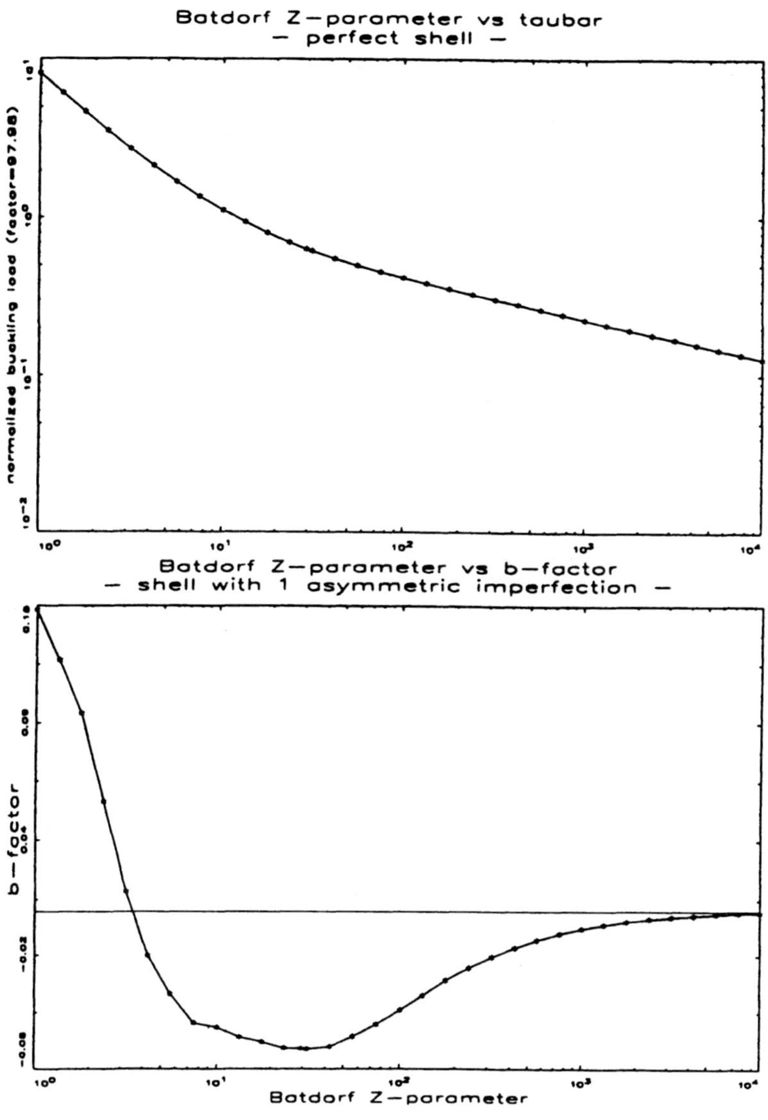

Fig. 15. Buckling and imperfection sensitivity of simply supported, isotropic cylinders, under torsion

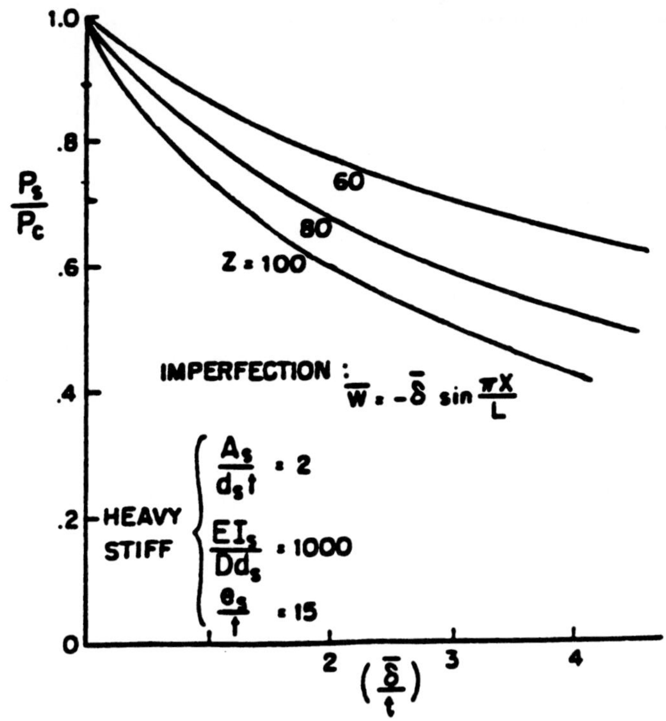

Fig. 16. Effect of axisymmetric imperfections on axially compressed cylinders with outside axial stiffeners [9]

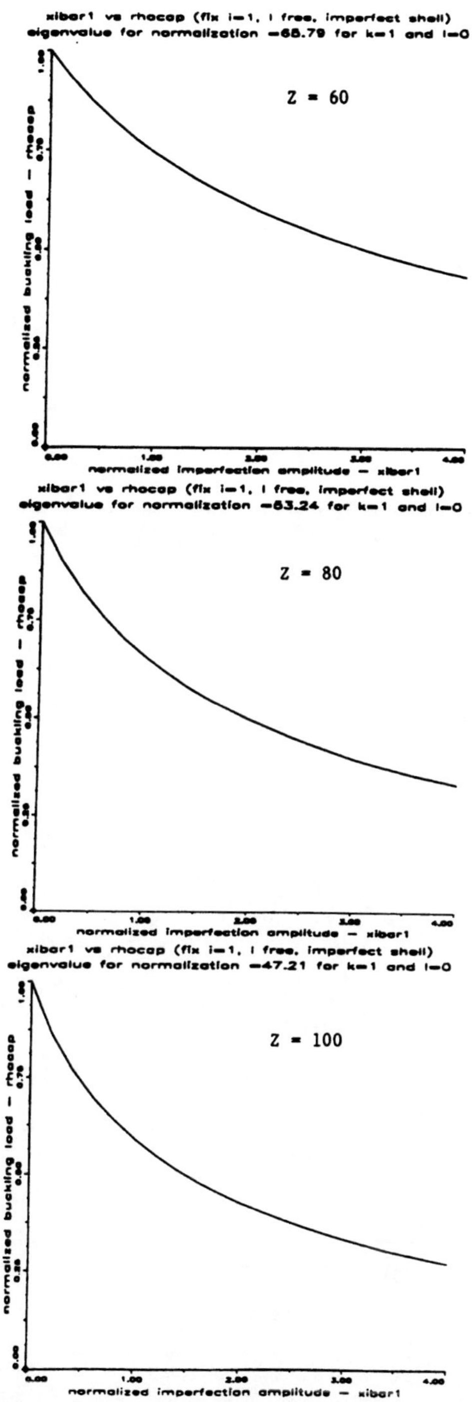

Fig. 17. Effect of axisymmetric imperfections on axially compressed cylinders with outside axial stiffeners

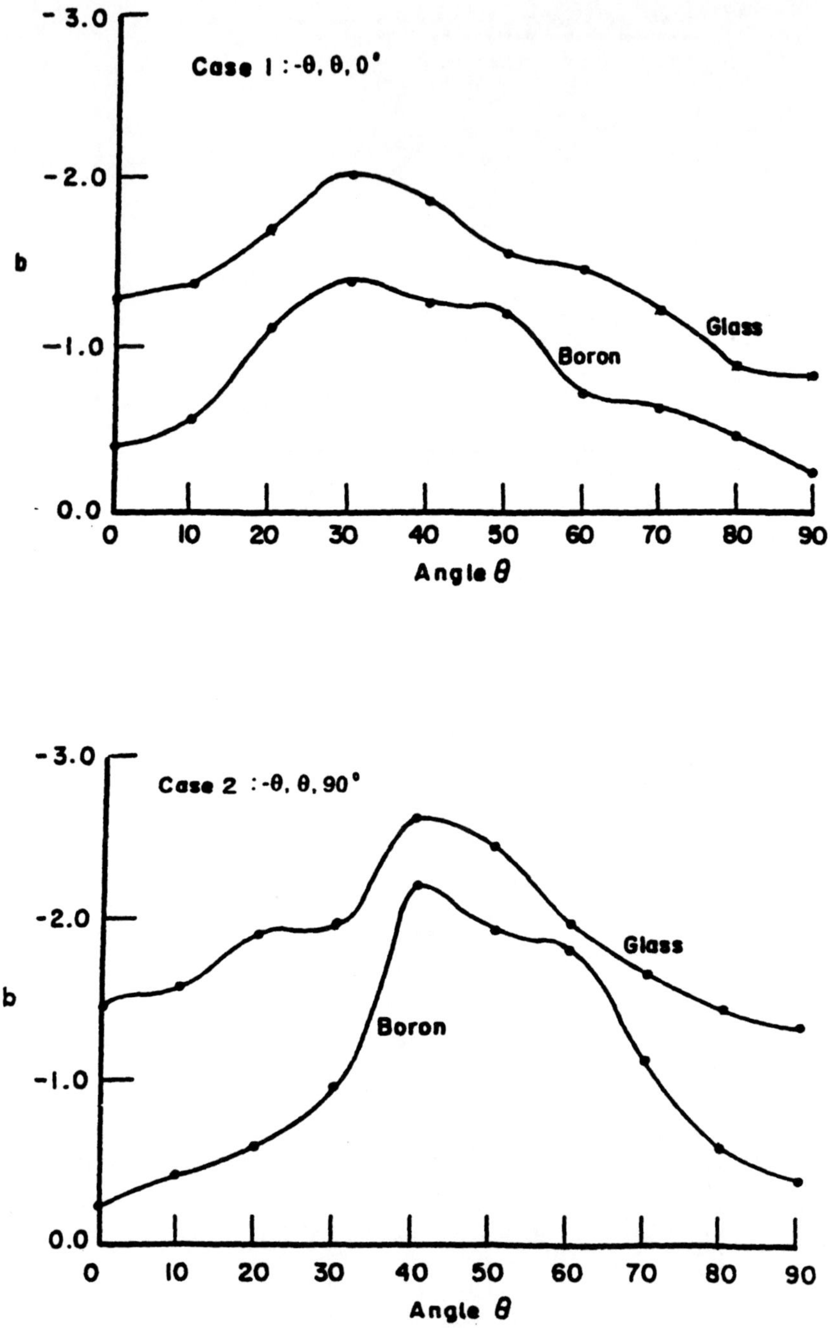

Fig. 18. Effect of fiber orientation on the postbuckling coefficient b [3]

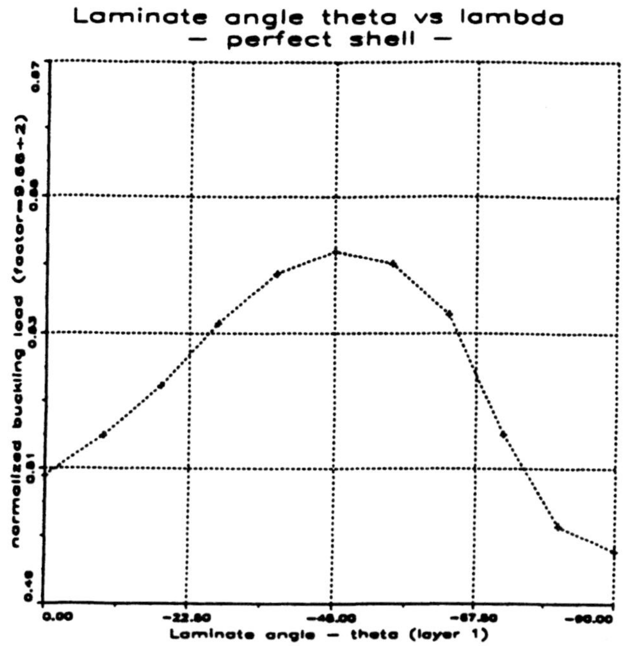

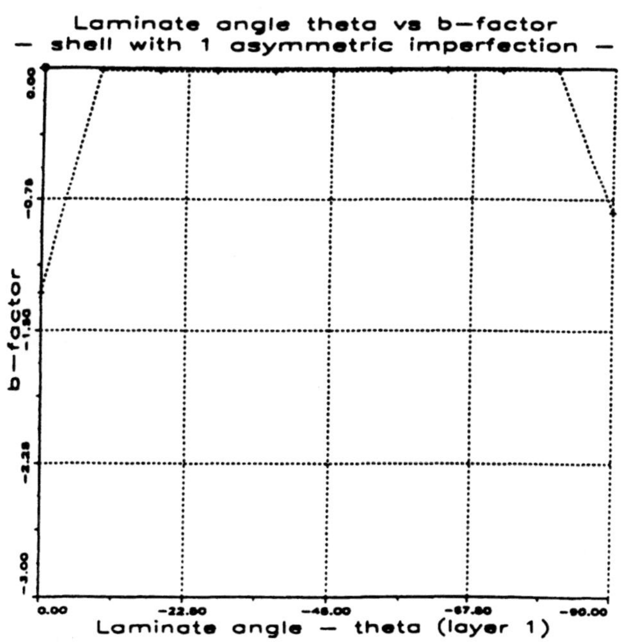

Fig. 19. Effect of fiber orientation on the buckling load λ_c and on the postbuckling coefficient b.

(Glass epoxy; case 1: $-\theta, \theta, 0^\circ$; τ_K is variable)

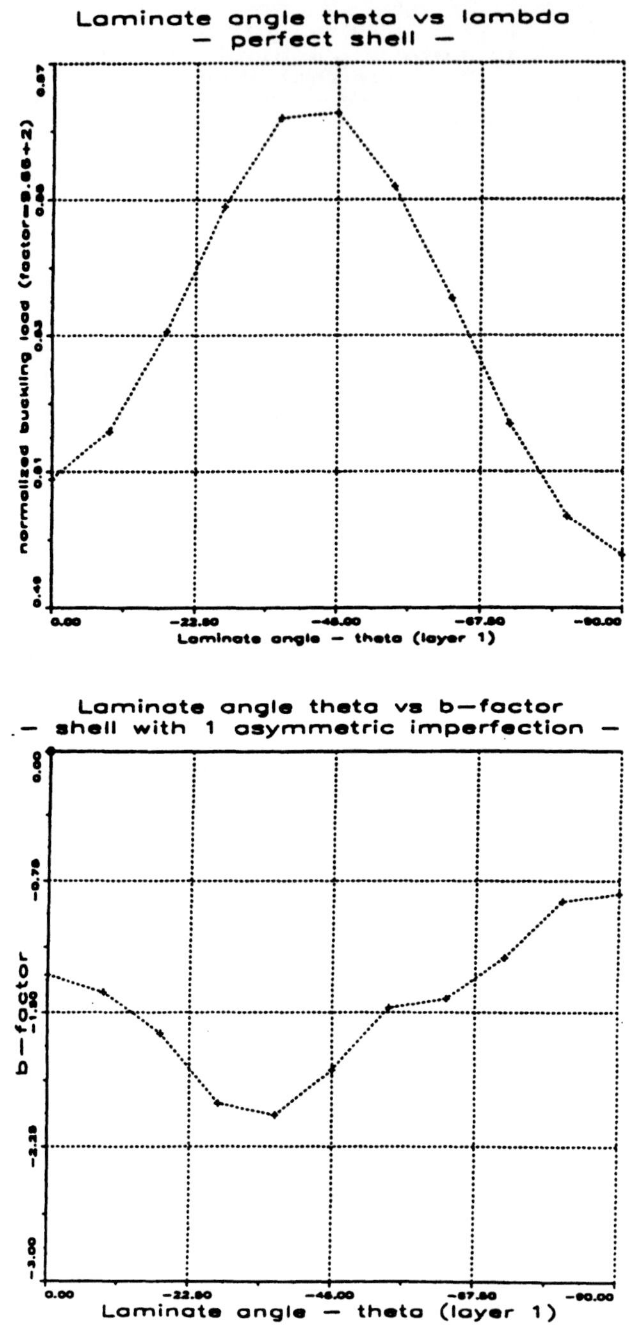

Fig. 20. Effect of fiber orientation on the buckling load λ_c and on the postbuckling coefficient b.

(Glass epoxy; case 1: $-\theta, \theta, 0°$; $\tau_K = 0$)

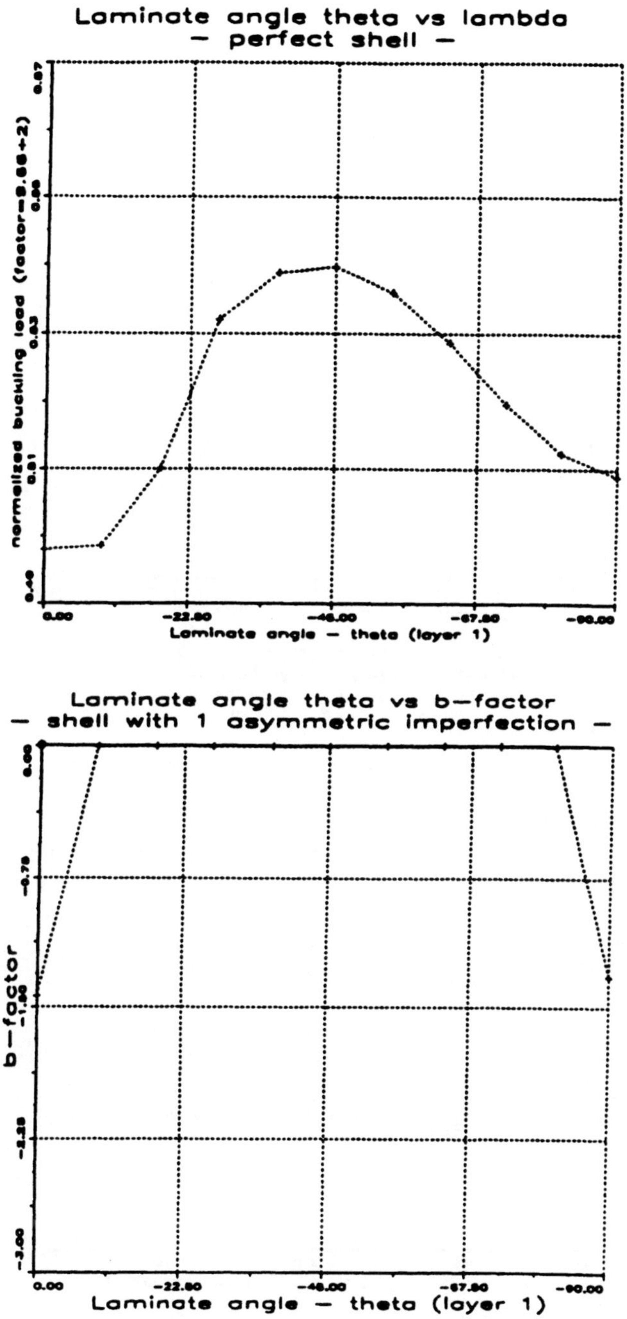

Fig. 21. Effect of fiber orientation on the buckling load λ_c and on the postbuckling coefficient b. (Glass epoxy; case 2: $-\theta, \theta, 90^\circ$; τ_K is variable)

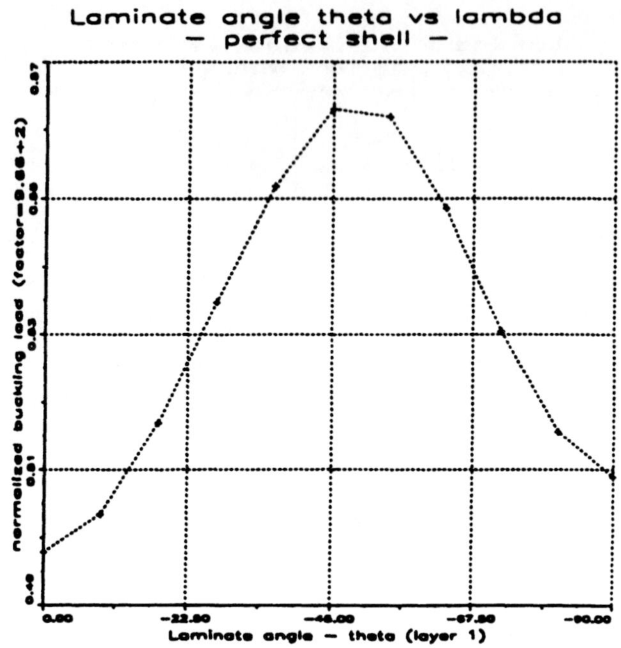

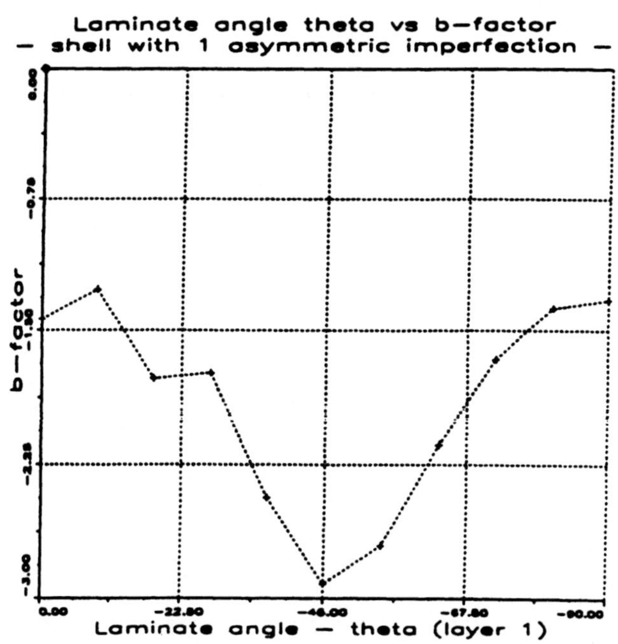

Fig. 22. Effect of fiber orientation on the buckling load λ_c and on the postbuckling coefficient b.

(Glass epoxy; case 2: $-\theta, \theta, 90^\circ$; $\tau_K = 0$)

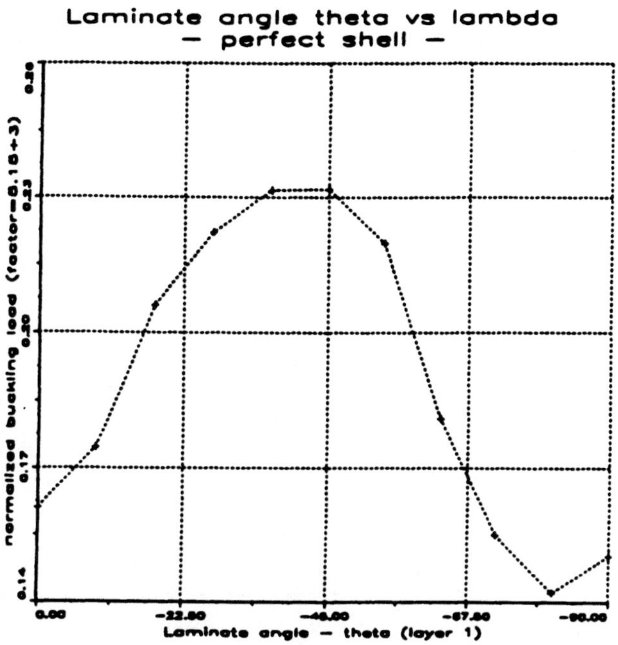

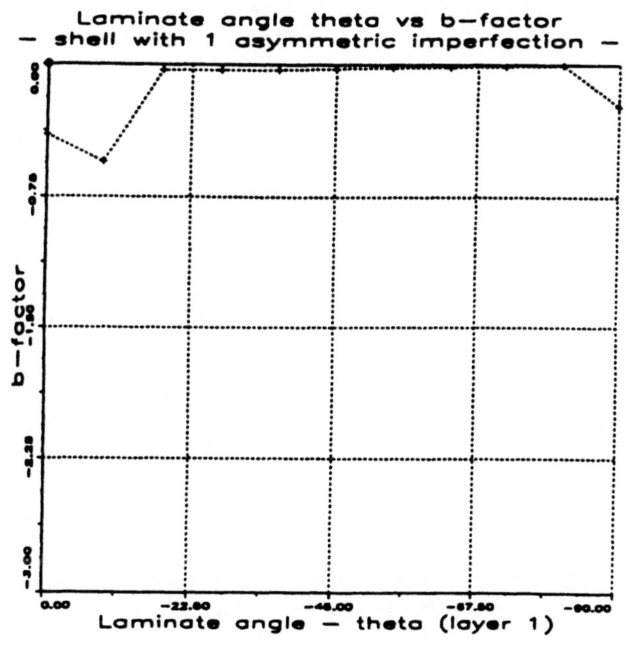

Fig. 23. Effect of fiber orientation on the buckling load λ_c and on the postbuckling coefficient b.

(Boron epoxy; case 1: $-\theta$, θ, 0°; τ_K is variable)

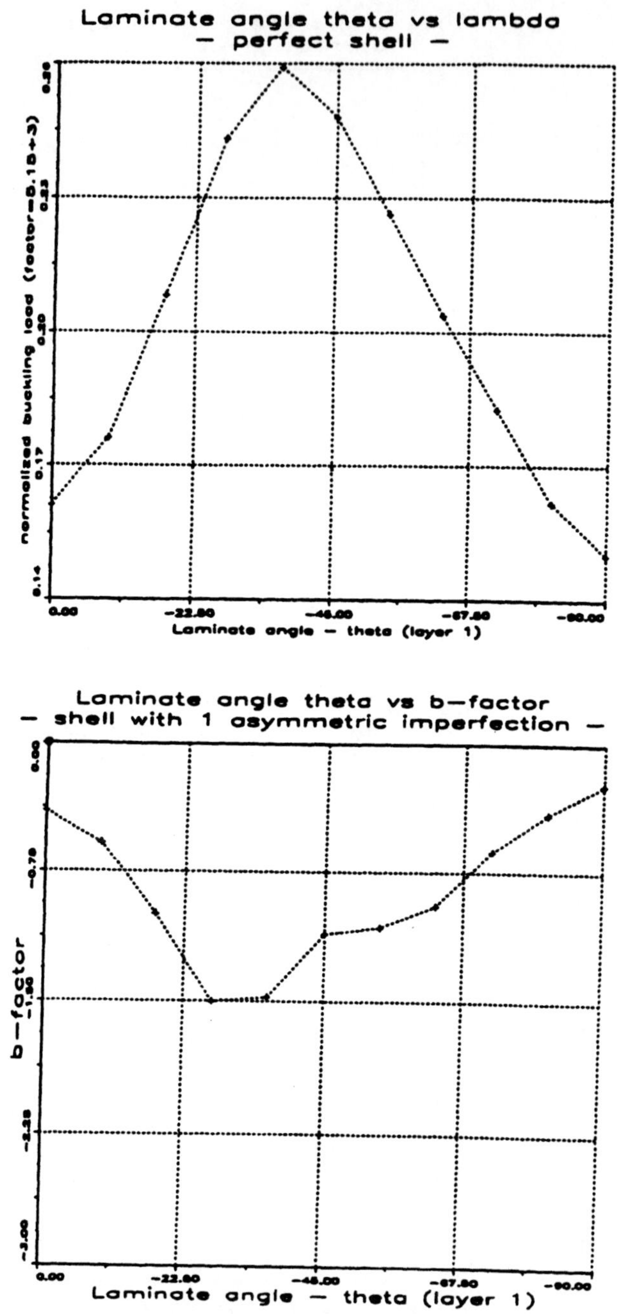

Fig. 24. Effect of fiber orientation on the buckling load λ_c and on the postbuckling coefficient b.

(Boron epoxy; case 1: $-\theta, \theta, 0^\circ$; $\tau_K = 0$)

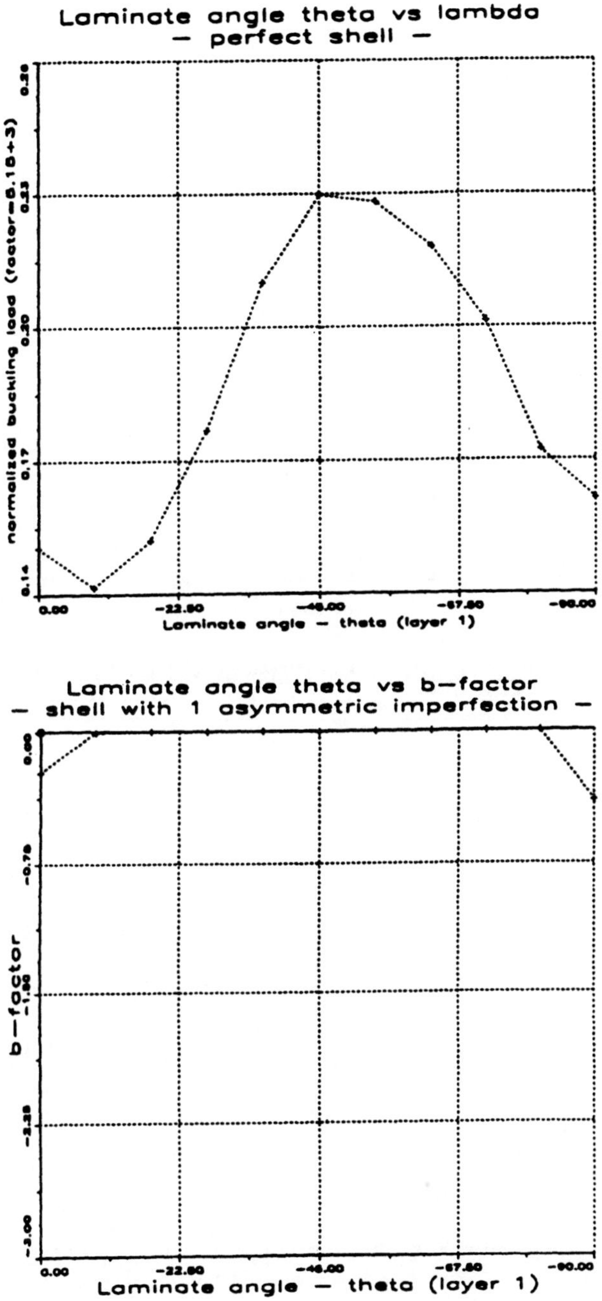

Fig. 25. Effect of fiber orientation on the buckling load λ_c and on the postbuckling coefficient b.

(Boron epoxy; case 2: $-\theta, \theta, 90^\circ$; τ_K is variable)

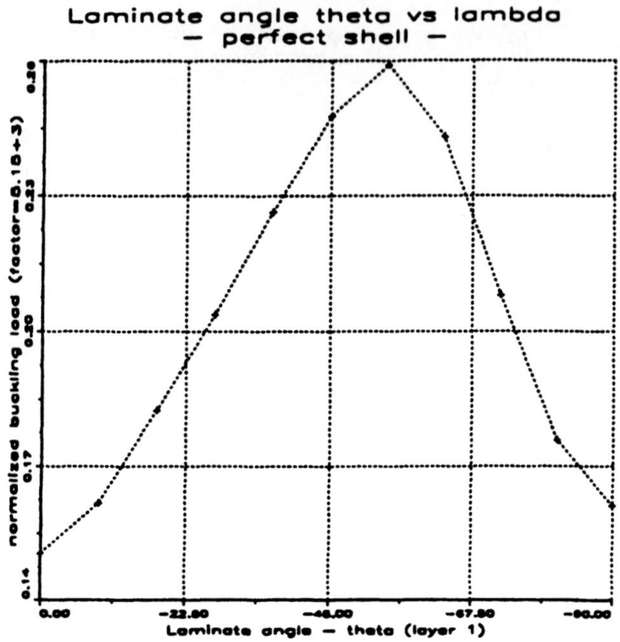

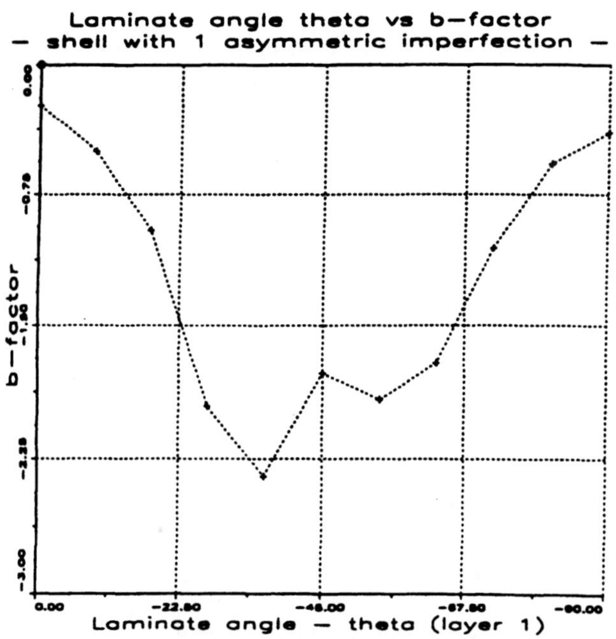

Fig. 26. Effect of fiber orientation on the buckling load λ_c and on the postbuckling coefficient b.

(Boron epoxy; case 2: $-\theta, \theta, 90^{\circ}$; $\tau_K = 0$)

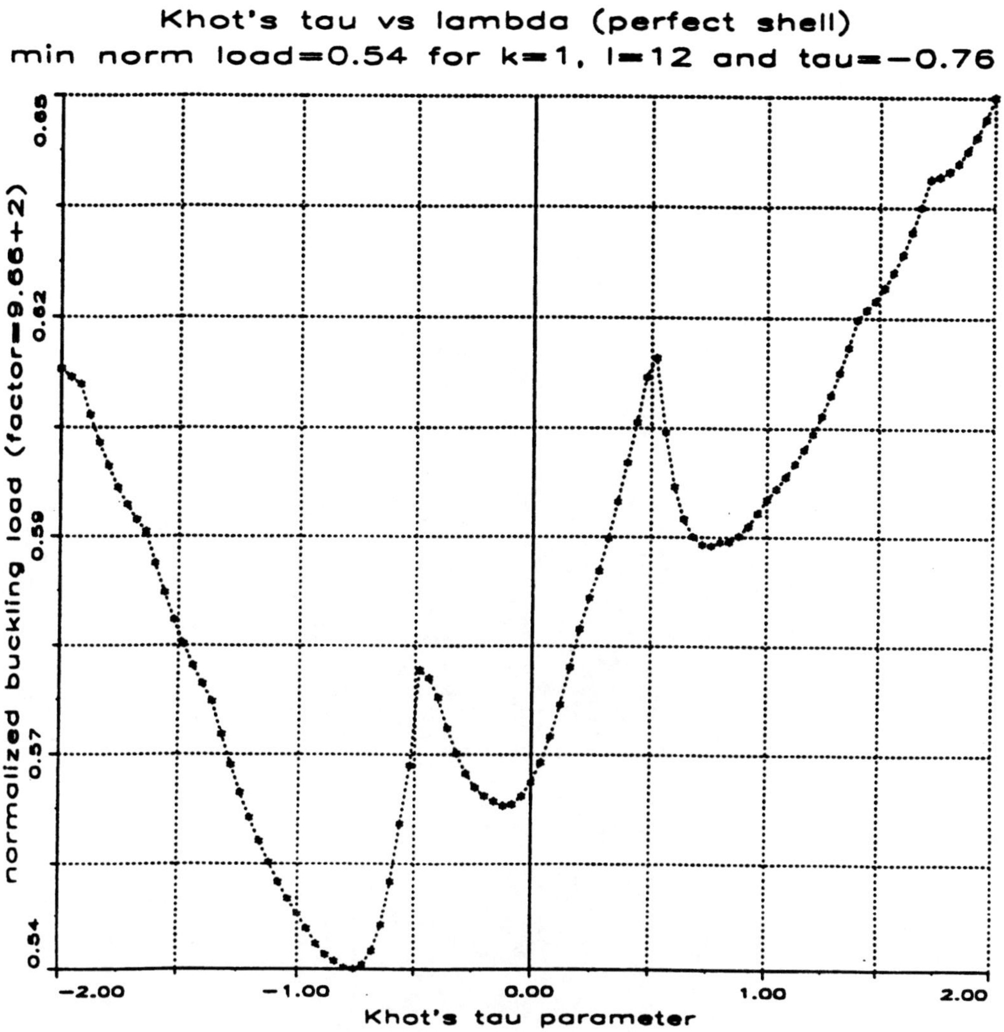

Fig. 27. Variation of the buckling load $\lambda_{mn\tau}$ with Khot's skewedness parameter τ_K for $\theta = 40°$.

(Glass epoxy; case 1: $-40°, 40°, 0°$)

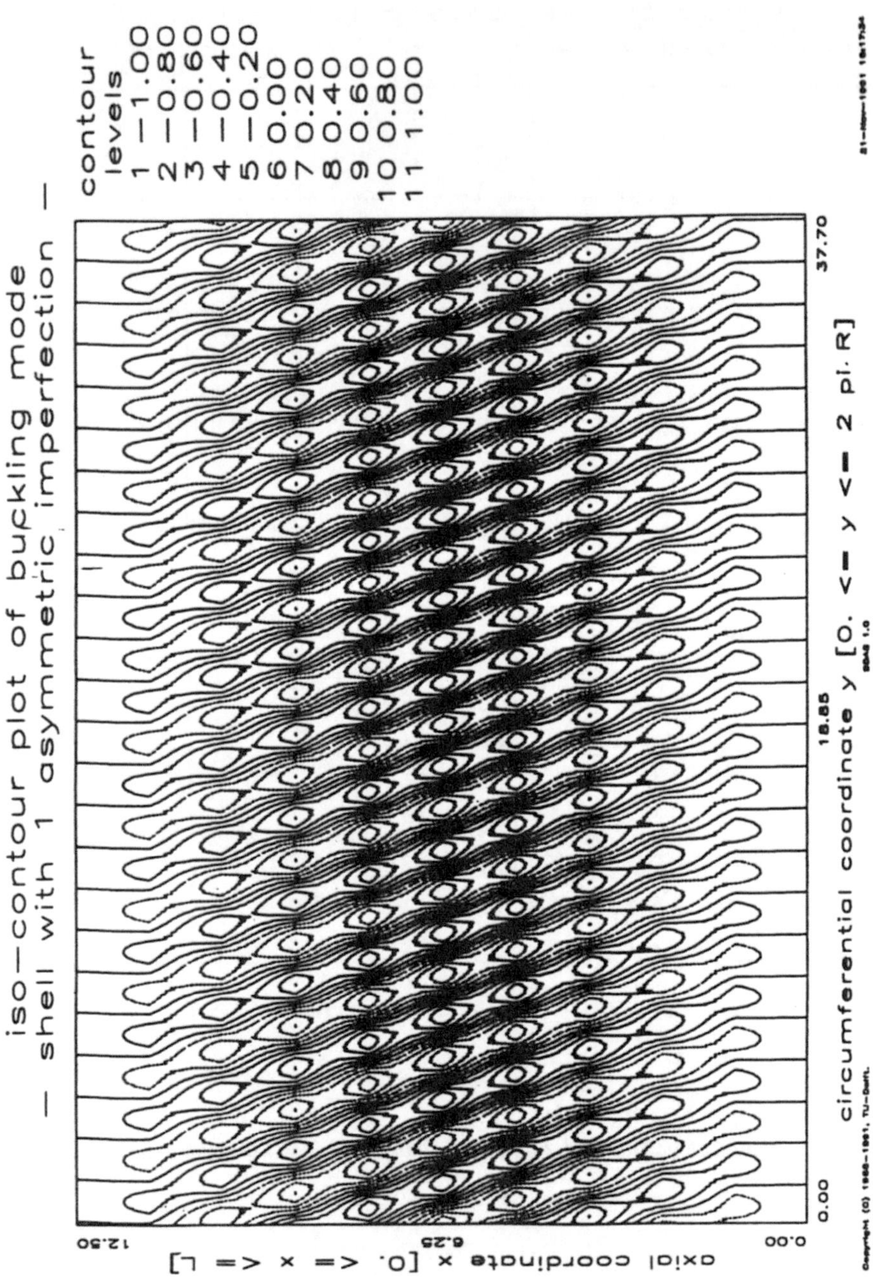

Fig. 28. Contour plot of the lowest buckling mode ($\lambda_c = 0.540706$ at $\tau_K = -0.76$, n = 12)

(Glass epoxy; case 1: −40°, 40°, 0°)

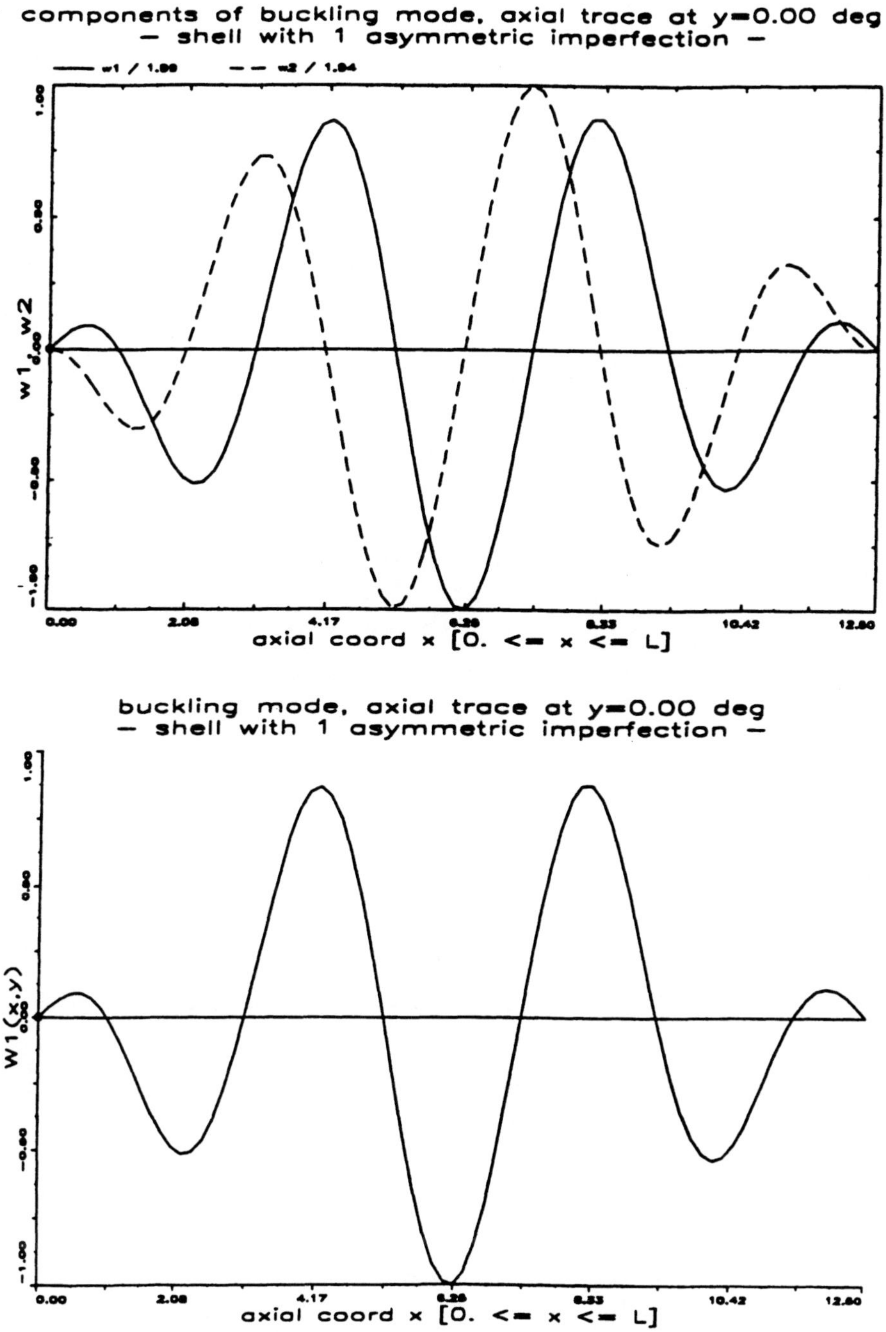

Fig. 29. Axial traces of the lowest buckling mode ($\lambda_c = 0.540706$ at $\tau_K = -0.76$, n = 12)

(Glass epoxy; case 1: –40°, 40°, 0°)

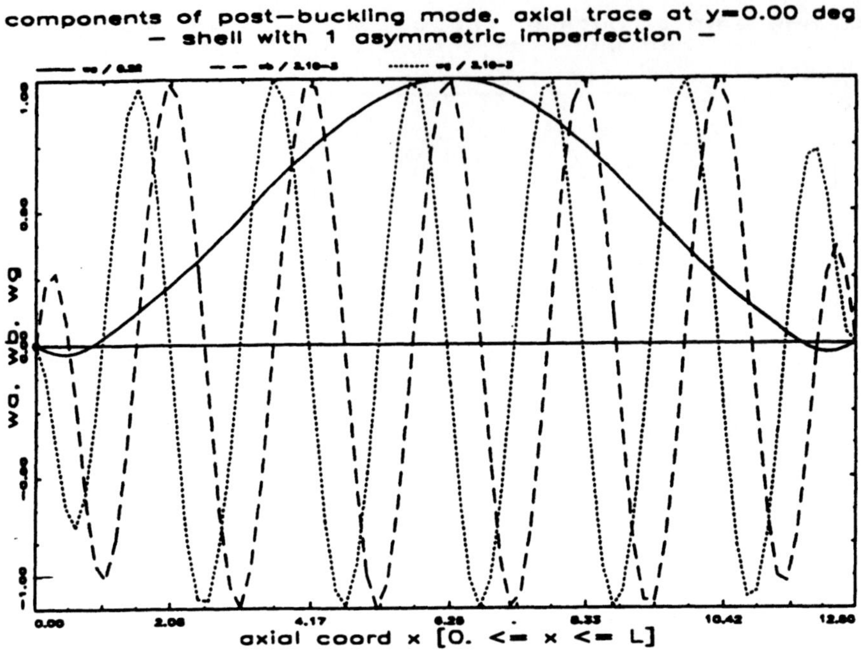

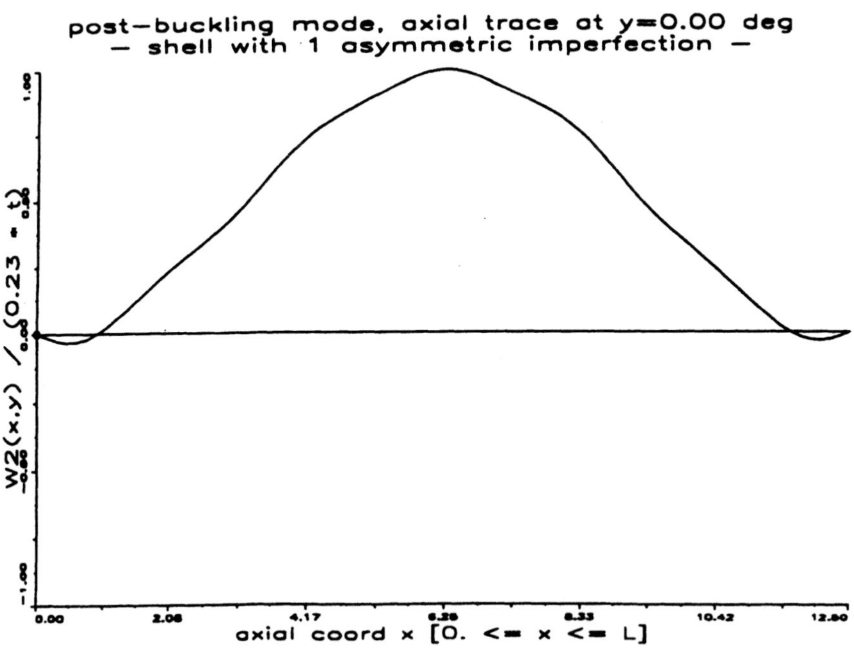

Fig. 30. Axial traces of the postbuckling modes ($\lambda_c = 0.540706$ at $\tau_K = -0.76$, n = 12)

(Glass epoxy; case 1: $-40°$, $40°$, $0°$)

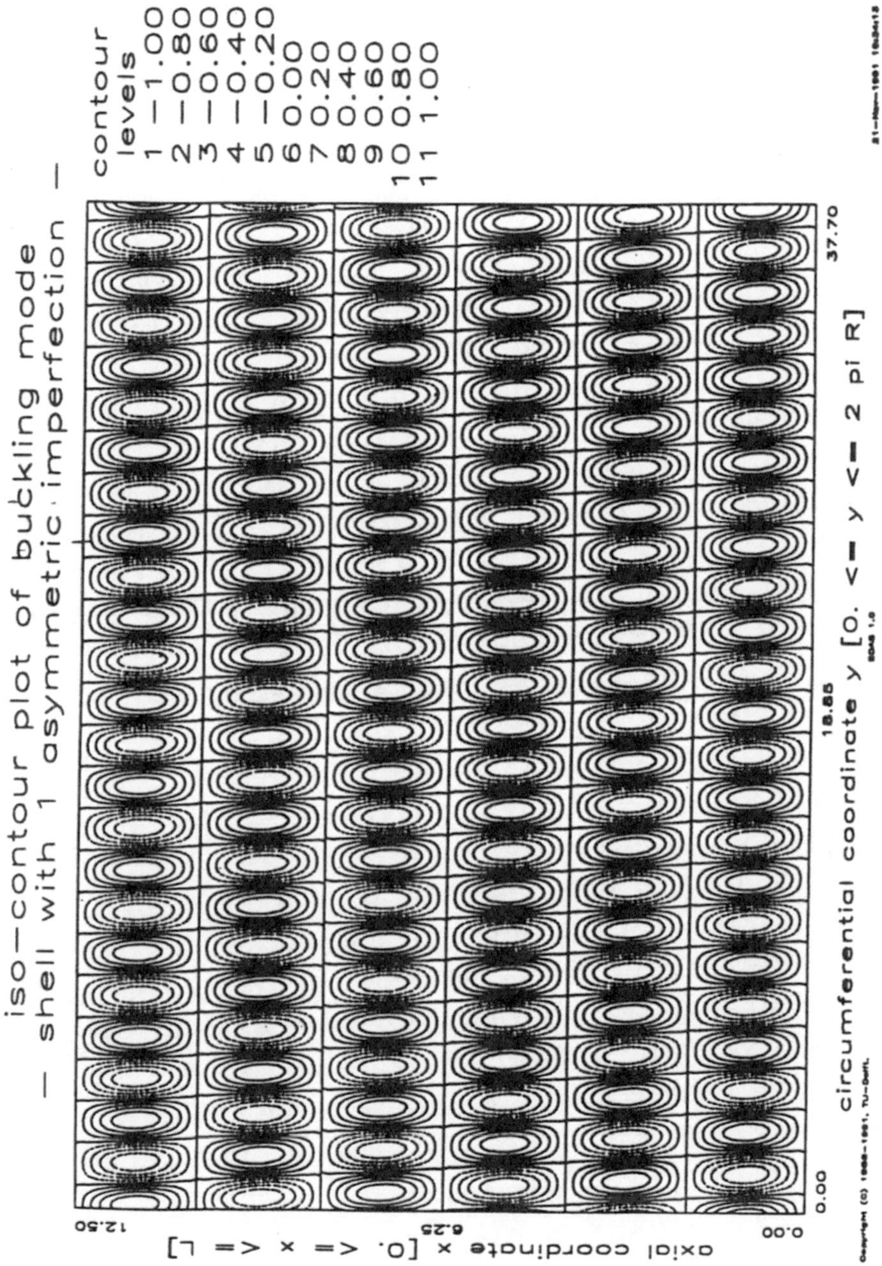

Fig. 31. Contour plot of the buckling mode at $\tau_K = -0.12$ ($\lambda_{6,12,\tau} = 0.561107$)

(Glass epoxy; case 1: $-40^\circ, 40^\circ, 0^\circ$)

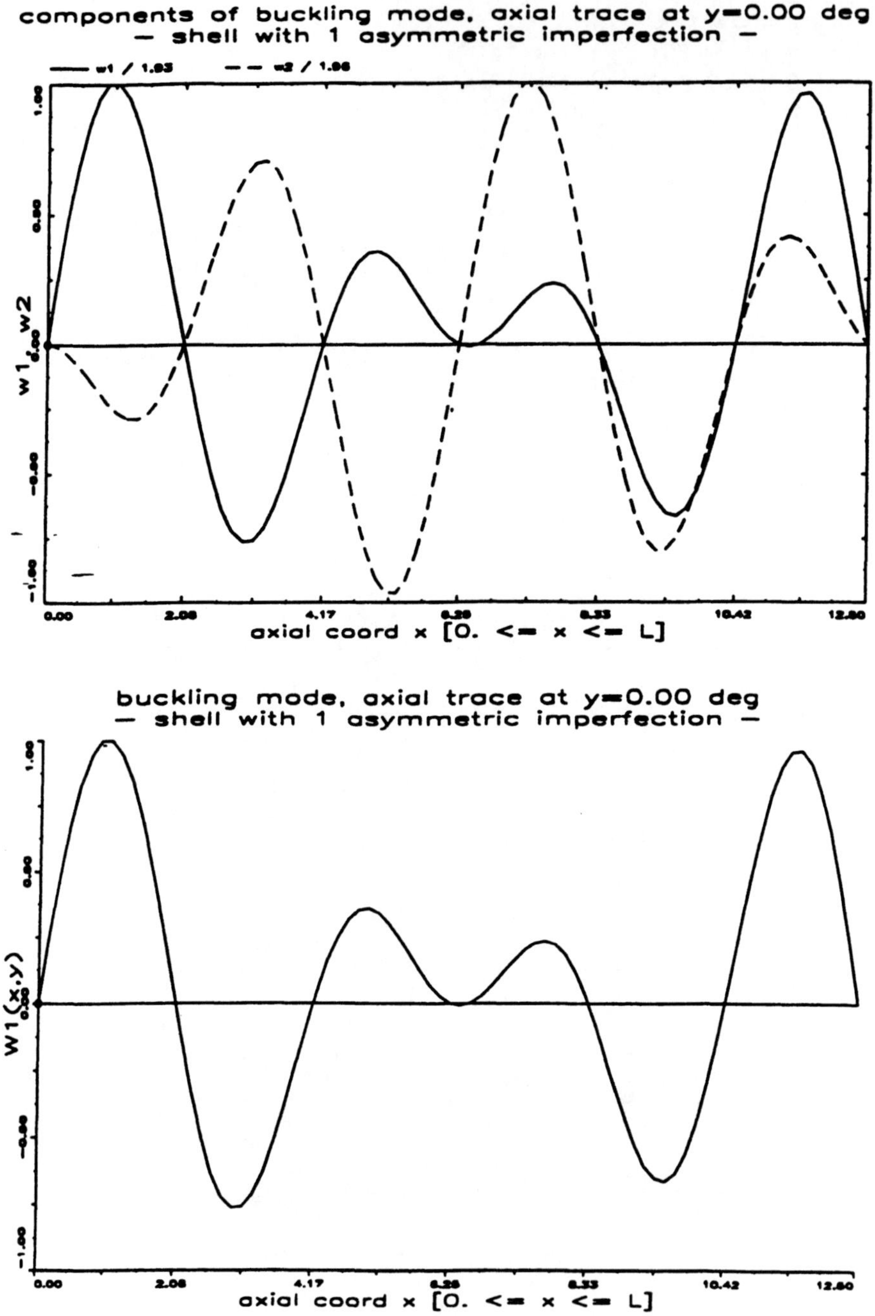

Fig. 32. Axial traces of the buckling mode at $\tau_K = -0.12$ ($\lambda_{6,12,\tau} = 0.561107$)

(Glass epoxy; case 1: $-40^\circ, 40^\circ, 0^\circ$)

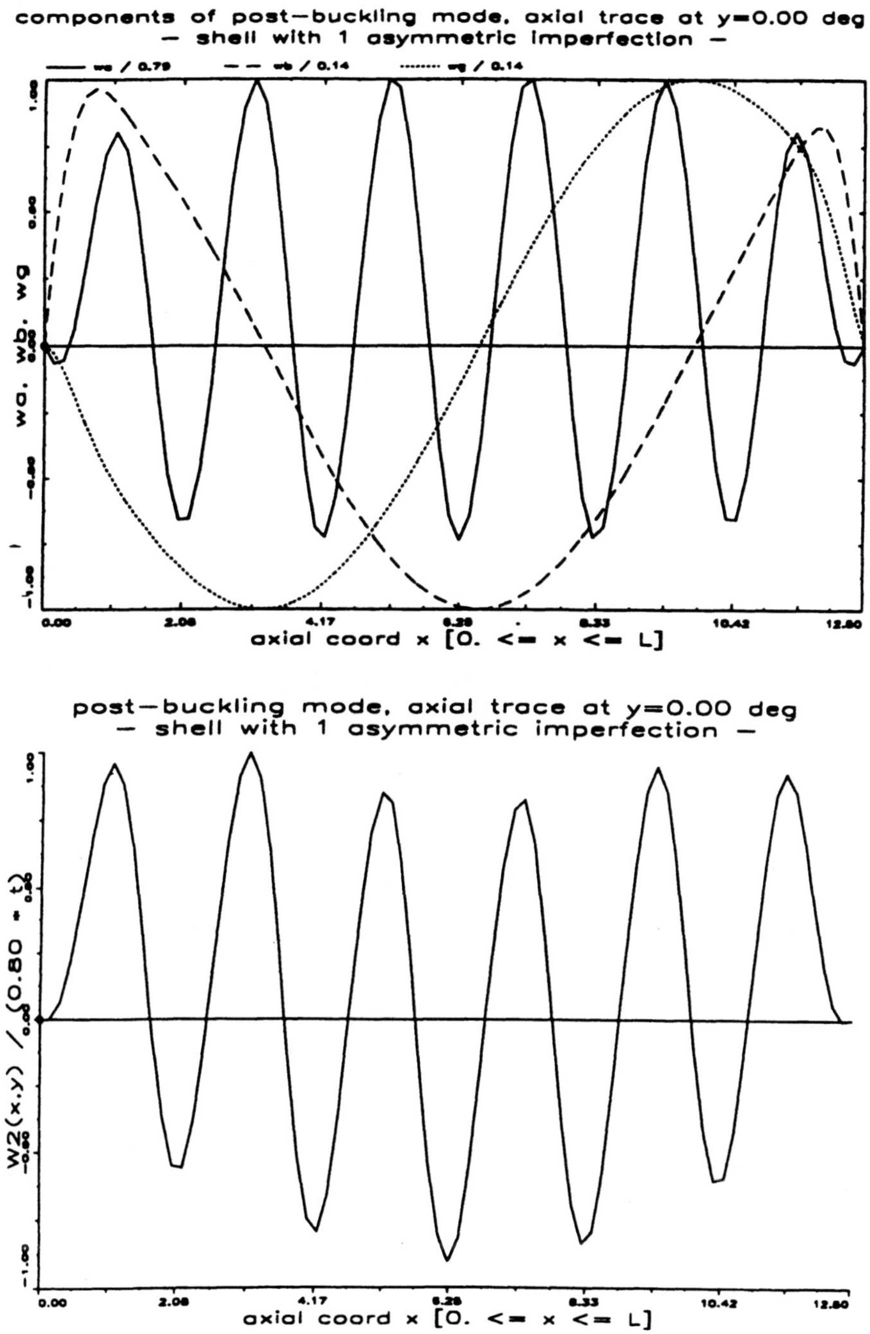

Fig. 33. Axial traces of the postbuckling mode at $\tau_K = -0.12$ ($\lambda_{6,12,\tau} = 0.561107$)

(Glass epoxy; case 1: $-40°, 40°, 0°$)

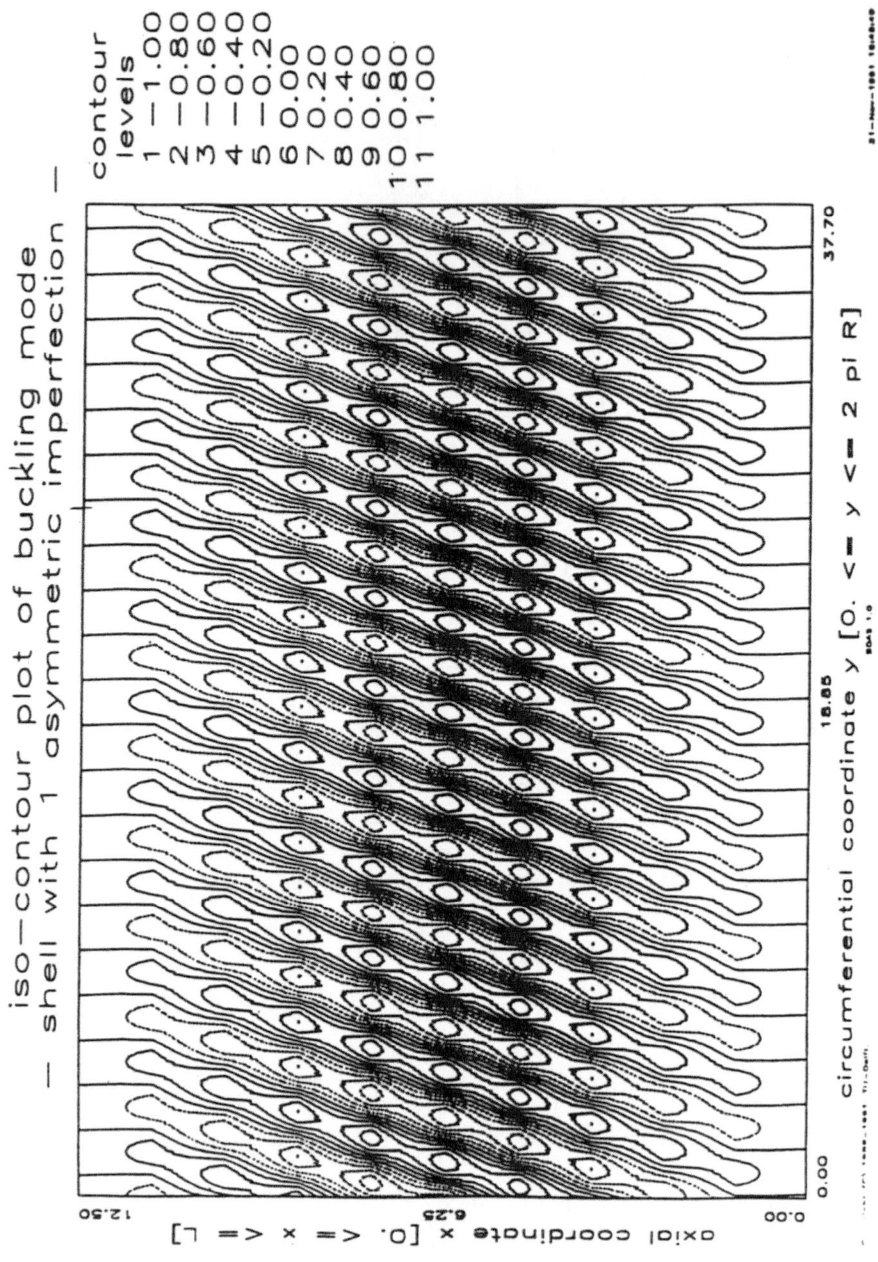

Fig. 34. Contour plot of the buckling mode at $\tau_K = +0.76$ ($\lambda_{1,11,\tau} = 0.593021$)

(Glass epoxy; case 1: $-40^\circ, 40^\circ, 0^\circ$)

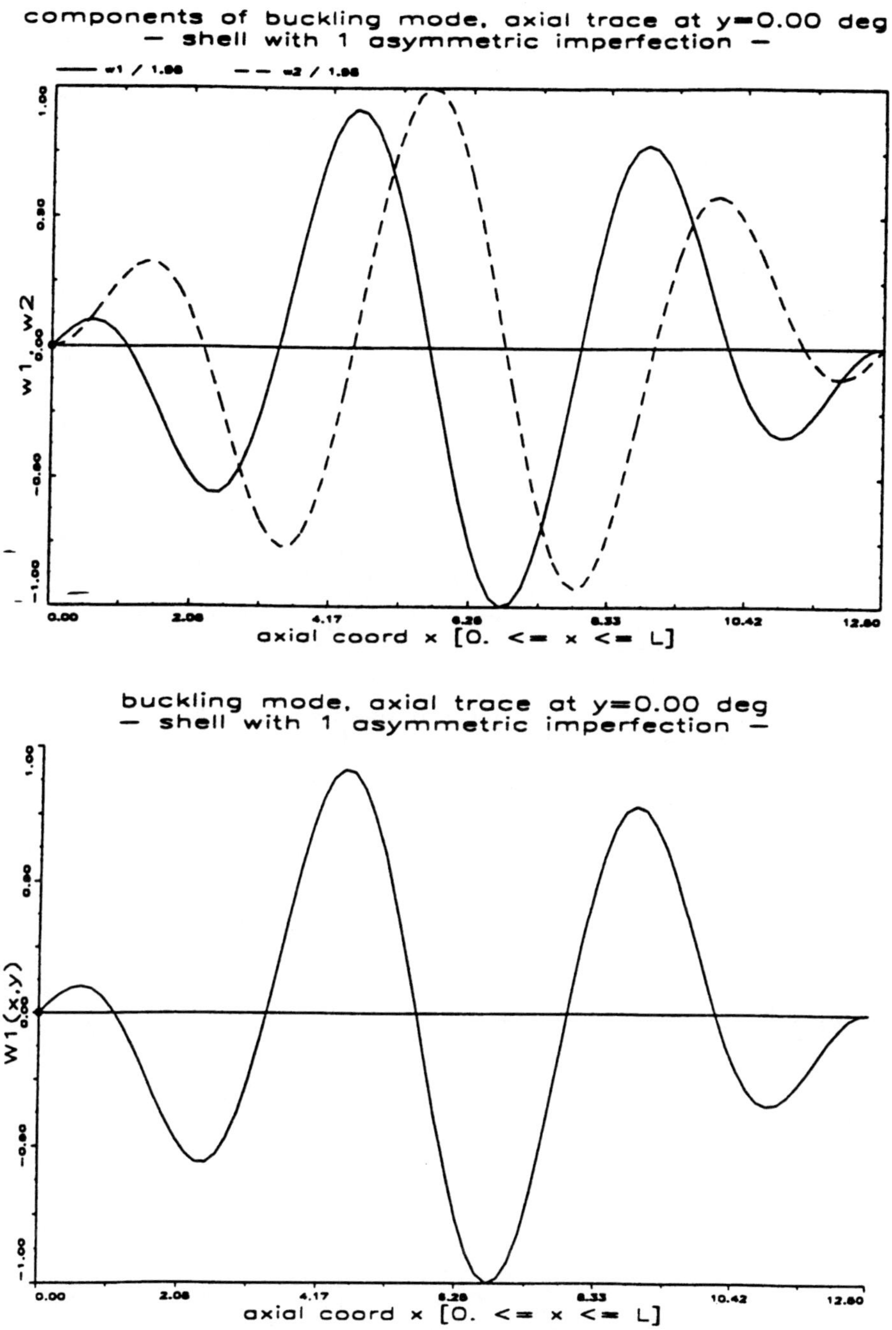

Fig. 35. Axial traces of the buckling mode at $\tau_K = +0.76$ ($\lambda_{1,11,\tau} = 0.593021$)

(Glass epoxy; case 1: $-40°, 40°, 0°$)

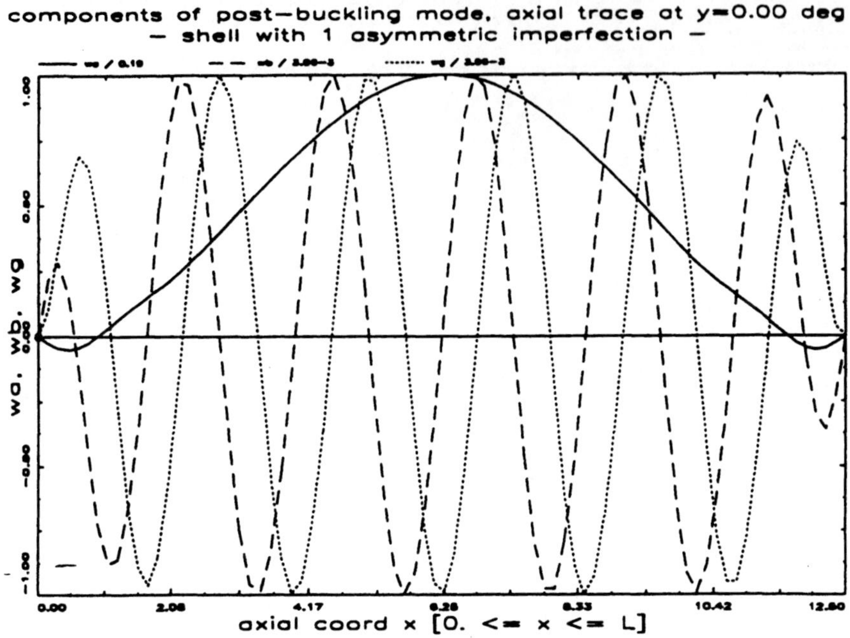

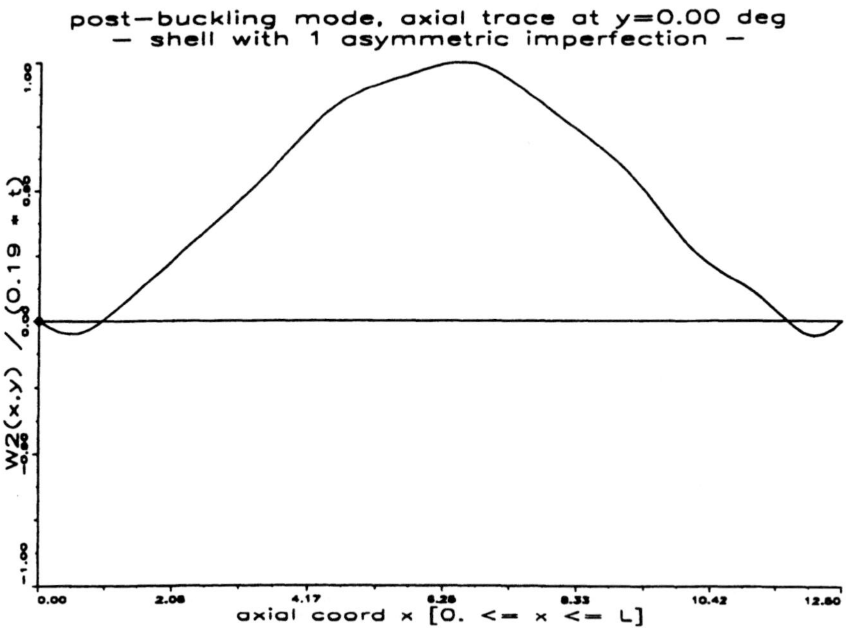

Fig. 36. Axial traces of the postbuckling mode at $\tau_K = +0.76$ ($\lambda_{1,11,\tau} = 0.593021$)

(Glass epoxy; case 1: $-40°, 40°, 0°$)

TABLE OF CONTENTS

LIST OF SYMBOLS

a	first postbuckling coefficient (see Eq. 4.1)
A_{ij}	extensional stiffness matrix (see Eq. A.7)
A^*_{ij}	semi-inverted extensional stiffness matrix (see Eq. A.9)
$\bar{A}^*_{ij}$	nondimensional $A^*_{ij}(\bar{A}^*_{ij} = Et\, A^*_{ij})$
b	second postbuckling coefficient (see Eq. 4.1)
B_{ij}	bending-stretching coupling matrix
B^*_{ij}	semi-inverted bending-stretching coupling matrix
$\bar{B}^*_{ij}$	nondimensional $B^*_{ij}(\bar{B}^*_{ij} = (2c/t)B^*_{ij})$
c	$=\sqrt{3(1-\nu^2)}$
D_{ij}	flexural stiffness matrix
D^*_{ij}	semi-inverted flexural stiffness matrix
$\bar{D}^*_{ij}$	nondimensional $D^*_{ij}(\bar{D}^*_{ij} = (4c^2/Et^3)D^*_{ij})$
E	arbitrarily chosen reference Young's modulus
F	Airy stress function
$F^{(0)}, F^{(1)}, F^{(2)}$	zeroth order, first order, second order fields, respectively
h_k	thickness of the kth layer
K_c	slope of the variable load vs generalized displacement curve just before buckling
K^*_c	slope of the variable load vs generalized displacement curve just after buckling
$L_{A^*}(\ )$	linear operator defined by Eq. (1.34)
$L_{B^*}(\ )$	linear operator defined by Eq. (1.35)

$L_{D^*}(\,)$	linear operator defined by Eq. (1.36)
$L_{NL}(\,)$	nonlinear operator defined by Eq. (1.37)
M_x, M_y, M_{xy}	moment resultants
n	number of full waves in the circumferential direction
N_x, N_y, N_{xy}	stress resultants
$\bar{N}_{\alpha\beta}$	variable applied stress resultant (see Eq. (5.1)
p	pressure
$\bar{p}$	nondimensional external pressure $(\bar{p} = (cR^2/Et^2)p)$
P_c	buckling load of the "perfect" structure
q	axial load eccentricity measured from the midsurface of the shell wall - positive inward
$\bar{q}$	nondimensional load eccentricity $(\bar{q} = 4cRq/t^2)$
Q_{ij}	specially orthotropic laminar stiffness matrix
$\bar{Q}_{ij}$	generally orthotropic laminar stiffness matrix
R	shell radius
t	shell wall-thickness
u, v	displacement components in the x and y directions, respectively
W	radial displacement (positive inward)
$W^{(0)}, W^{(1)}, W^{(2)}$	zeroth order, first order, second order fields, respectively
$\bar{W}$	initial radial imperfection $(\bar{W} = \xi \hat{W})$
$\hat{W}$	shape of the initial radial imperfection
W_v	axial Poisson's effect $(W_v = \bar{A}^*_{12}\lambda/c)$
W_p	radial Poisson's effect $(W_p = \bar{A}^*_{22}\bar{p}/c)$
W_t	circumferential Poisson's effect $(W_t = -\bar{A}^*_{26}\bar{\tau}/c)$

x, y	axial and circumferential coordinates on the middle surface of the shell, respectively
$\overline{x}, \overline{y}$	nondimensional coordinates $(\overline{x} = x/R,\ \overline{y} = y/R)$
z	coordinate normal to the middle surface of the shell (positive inward)
$\overline{Z}$	modified Batdorf parameter $(\overline{Z} = L^2/Rt)$
α	first imperfection form factor (see Eq. 4.11)
β	second imperfection form factor (see Eq. 4.11)
γ_{xy}	shearing strain
Δ	generalized displacement
ε_c	membrane strain which corresponds to the applied variable load (see Eq. 5.10)
$\varepsilon_{\alpha\beta}$	linear strain which corresponds to the variable applied stress resultant (see Eq. 5.2)
$\varepsilon_x, \varepsilon_y$	normal strains
$E_{\alpha\beta}$	nonlinear strain (see Eq. 5.2)
Δ	generalized displacement (see Eq. 5.1)
θ	circumferential coordinate $(\theta = y/R)$
θ	angle of the fundamental path $(\theta = \tan^{-1}(K_c))$
θ^*	angle of the initial slope just after buckling $(\theta^* = \tan^{-1}(K_c^*))$
$\kappa_x, \kappa_y, \kappa_{xy}$	curvature components
λ	nondimensional axial load parameter $(\lambda = (cR/Et^2)N_o)$
Λ	nondimensional variable load factor
Λ_c	nondimensional variable load factor evaluated at the bifurcation point
Λ_s	nondimensional variable load factor evaluated at the limit point
ν	arbitrarily chosen reference Poisson's ratio

ξ	perturbation parameter (see Eq. 4.1)
$\bar{\xi}$	amplitude of the initial imperfection (see Eq. 6.1)
ρ_s	normalized variable load factor $(\rho_s = \Lambda_s / \Lambda_c)$
σ_x, σ_y	normal stresses
$\bar{\tau}$	nondimensional torque parameter $\bar{\tau} = (cR / Et^2)N_{xy})$ - positive counter-clockwise
τ_K	Khot's skewedness parameter (see Eq. (2.11)
τ_{xy}	shearing stress

Part V. Bifurcations of Eigenvalues and Stability Problems in Mechanics

Alexander P. Seyranian

Institute of Mechanics, Moscow State Lomonosov University,
Michurynski pr. 1, Moscow 117192, Russia
e-mail: seyran@imec.msu.ru

Abstract. In this Part of the Lecture Notes we study bifurcations of eigenvalues of nonsymmetrical matrix operators depending on parameters with applications to the stability study in different mechanical problems. Section 1 gives general analysis of bifurcations for eigenvalues with geometric interpretation in two- and three-dimensional spaces. Main attention is focused on simple and double eigenvalues, and strong and weak interactions of eigenvalues are distinguished. In Section 2 stability and catastrophes in one-parameter circulatory systems with simple mechanical examples are studied. It is proven that in general they are subjected to catastrophes of three types: flutter, divergence, and transition from flutter to divergence and vice versa. In Section 3 properties of two-parameter circulatory systems are studied, and the explicit formulas describing metamorphoses of frequency curves are derived. These formulas use information on the system only at a merging point of the frequencies, and allow qualitative as well as quantitative analysis of behavior of frequency curves near that point with a change of parameters. In particular, development of "a bubble of instability" is analyzed. In Section 4 the Keldysh problem of aeroelastic stability of a wing with bracing struts is discussed, and the effect of disappearance of flutter instability revealed by Keldysh (1938) is explained. It is shown that this effect is connected with the convexity of the flutter domain in the parameter space. In Section 5 we study a problem of maximizing the critical buckling load of an elastic column of given length and volume assuming elastic supports at both ends of the column. This problem was first formulated by J.-L. Lagrange in 1773 for simply supported columns and has been considered by many authors for various boundary conditions. We obtain the bimodal optimal solutions and investigate their post-buckling behavior. Section 6 concerns instability domains for Hill's equation with damping under assumption of small excitation amplitude and damping coefficient. It is shown that these domains are halves of cones in the three-dimensional parameter space. One of the important applications of Hill's equation is connected with the stability study of periodic motions for nonlinear dynamical systems. It is shown how to find stable and unstable regimes for harmonically excited Duffing's equation.

1 Bifurcations of Eigenvalues Depending on Parameters

1.1 Introduction

This Section mostly based on the papers by Seyranian (1991, 1993) is devoted to bifurcation analysis of eigenvalues of nonsymmetrical matrix operators depending on parameters. Main attention is focused on simple and double eigenvalues. Strong and weak interactions of eigenvalues are distinguished and geometric interpretation in two- and three-dimensional space is given. It is shown that the strong interaction is a typical catastrophe. The bifurcation analysis of eigenvalues is based on perturbation theory of non-selfadjoint operators developed in Vishik and Lyusternik (1960).

1.2 Formulation of the problem

Consider an eigenvalue problem

$$\mathbf{A}\mathbf{u} = \lambda \mathbf{u}\,, \tag{1}$$

where $\mathbf{A}$ is a real nonsymmetrical $m \times m$ matrix smoothly depending on a vector of real parameters $\mathbf{p}^{\mathrm{T}} = (p_1, p_2, ..., p_n)$, λ is an eigenvalue, and $\mathbf{u}$ is a corresponding eigenvector of dimension m. It is assumed that at fixed $\mathbf{p} = \mathbf{p}_0$, λ_0 is an eigenvalue of $\mathbf{A}(\mathbf{p}_0)$, and we look for a change of the corresponding eigenvalue λ depending upon change of the vector of parameters $\mathbf{p}$. For this purpose we take to the vector $\mathbf{p}_0$ an increment $\mathbf{p} = \mathbf{p}_0 + \varepsilon \mathbf{e}$, where $\varepsilon > 0$ is a small parameter, and $\mathbf{e}$ is a real vector of variation of dimension n. As a result, the matrix $\mathbf{A}$ takes the increment

$$\mathbf{A}(\mathbf{p}_0 + \varepsilon \mathbf{e}) = \mathbf{A}_0 + \varepsilon \mathbf{A}_1 + \varepsilon^2 \mathbf{A}_2 + ... \tag{2}$$

where the matrices $\mathbf{A}_0$, $\mathbf{A}_1$, and $\mathbf{A}_2$ are given by the relations

$$\mathbf{A}_0 = \mathbf{A}(\mathbf{p}_0)\,, \quad \mathbf{A}_1 = \sum_{s=1}^{n} \frac{\partial \mathbf{A}}{\partial p_s} e_s\,, \quad \mathbf{A}_2 = \frac{1}{2} \sum_{s,k=1}^{n} \frac{\partial^2 \mathbf{A}}{\partial p_s \partial p_k} e_s e_k\,. \tag{3}$$

Due to variation of the vector $\mathbf{p}$ the eigenvalues λ and the eigenvectors $\mathbf{u}$ also take increments. According to the perturbation theory of non-selfadjoint operators, developed by Vishik and Lyusternik (1960), these increments can be expressed as series of integer or fractional powers of ε, depending on a Jordan structure of the matrix $\mathbf{A}_0$.

1.3 Simple Eigenvalue

We assume that λ_0 is a simple eigenvalue of the matrix $\mathbf{A}_0$, and $\mathbf{u}_0$ is a corresponding eigenvector. In this case λ and $\mathbf{u}$ are smooth functions of ε and can be expressed as Taylor series, see Vishik and Lyusternik (1960)

$$\lambda = \lambda_0 + \varepsilon \lambda_1 + \varepsilon^2 \lambda_2 + \dots$$

$$\mathbf{u} = \mathbf{u}_0 + \varepsilon \mathbf{w}_1 + \varepsilon^2 \mathbf{w}_2 + \dots \tag{4}$$

Substitution of expansions (2), (4) into equation (1) yields

$$\begin{aligned}(\mathbf{A}_0 - \lambda_0 \mathbf{I})\mathbf{u}_0 &= \mathbf{0}\\ (\mathbf{A}_0 - \lambda_0 \mathbf{I})\mathbf{w}_1 &= \lambda_1 \mathbf{u}_0 - \mathbf{A}_1 \mathbf{u}_0\\ (\mathbf{A}_0 - \lambda_0 \mathbf{I})\mathbf{w}_2 &= \lambda_1 \mathbf{w}_1 - \mathbf{A}_1 \mathbf{w}_1 + \lambda_2 \mathbf{u}_0 - \mathbf{A}_2 \mathbf{u}_0 .\end{aligned} \tag{5}$$

where $\mathbf{I}$ is the identity matrix. We introduce the inner product $(\mathbf{a}, \mathbf{b}) = \sum_{i=1}^{m} a_i \bar{b}_i$ for vectors $\mathbf{a}, \mathbf{b} \in C^m$ with the bar over a symbol denoting complex conjugate quantity. Then the operator $\mathbf{L}^* = \mathbf{A}_0^{\mathrm{T}} - \bar{\lambda}_0 \mathbf{I}$ is the adjoint to $\mathbf{L} = \mathbf{A}_0 - \lambda_0 \mathbf{I}$ as it satisfies the relation $(\mathbf{Lu}, \mathbf{v}) = (\mathbf{u}, \mathbf{L}^* \mathbf{v})$ holding for arbitrary vectors $\mathbf{u}$ and $\mathbf{v}$.

For the following presentation we introduce the *adjoint* eigenvalue problem

$$(\mathbf{A}_0^{\mathrm{T}} - \bar{\lambda}_0 \mathbf{I})\mathbf{v}_0 = 0, \tag{6}$$

and assume that the eigenvector $\mathbf{v}_0$ satisfies the normality condition

$$(\mathbf{u}_0, \mathbf{v}_0) = 1. \tag{7}$$

For given $\mathbf{u}_0$ this equality defines $\mathbf{v}_0$ uniquely. For the perturbed vector $\mathbf{u}$ we will use the normality condition $(\mathbf{u}, \mathbf{v}_0) = 1$ which is necessary to uniquely determine all the terms λ_i and $\mathbf{w}_i$. Substituting (4) to this condition and using (7) we obtain

$$(\mathbf{w}_i, \mathbf{v}_0) = 0, \quad i = 1,2,... \tag{8}$$

The solvability condition for the second and third of equations (5) requires that the right hand side be orthogonal to a solution of adjoint problem (6). Thus, with the use of normality conditions (7), (8) the equalities $(\lambda_1 \mathbf{u}_0 - \mathbf{A}_1 \mathbf{u}_0, \mathbf{v}_0) = 0$ and $(\lambda_2 \mathbf{u}_0 - \mathbf{A}_2 \mathbf{u}_0 + \lambda_1 \mathbf{w}_1 - \mathbf{A}_1 \mathbf{w}_1, \mathbf{v}_0) = 0$ give

$$\lambda_1 = (\mathbf{A}_1 \mathbf{u}_0, \mathbf{v}_0), \tag{9}$$

$$\lambda_2 = (\mathbf{A}_2 \mathbf{u}_0, \mathbf{v}_0) + (\mathbf{A}_1 \mathbf{w}_1, \mathbf{v}_0). \tag{10}$$

The vectors $\mathbf{w}_1$, $\mathbf{w}_2$ are found from the second and third of equations (5)

$$\mathbf{w}_1 = \mathbf{G}_0(\lambda_1 \mathbf{u}_0 - \mathbf{A}_1 \mathbf{u}_0), \tag{11}$$

$$\mathbf{w}_2 = \mathbf{G}_0(\lambda_2 \mathbf{u}_0 - \mathbf{A}_2 \mathbf{u}_0 + \lambda_1 \mathbf{w}_1 - \mathbf{A}_1 \mathbf{w}_1) \tag{12}$$

where $\mathbf{G}_0$ is an inverse operator to $\mathbf{A}_0 - \lambda_0 \mathbf{I}$. This operator can be given in the form, see Yakubovich and Starzhinskii (1987)

$$\mathbf{G}_0 = [\mathbf{A}_0 - \lambda_0 \mathbf{I} - \bar{\mathbf{v}}_0 \mathbf{v}_0^{\mathrm{T}}]^{-1} \tag{13}$$

Note that adding of the term $-\bar{\mathbf{v}}_0 \mathbf{v}_0^{\mathrm{T}}$ makes the matrix $\mathbf{A}_0 - \lambda_0 \mathbf{I}$ nonsingular.

Using formulas (3), (7-11), and (13) we can derive first and second derivatives of simple eigenvalues with respect to parameters in the form

$$\frac{\partial \lambda}{\partial p_j} = \left(\frac{\partial \mathbf{A}}{\partial p_j} \mathbf{u}_0, \mathbf{v}_0 \right), \tag{14}$$

$$\frac{\partial^2 \lambda}{\partial p_s \partial p_k} = \left(\frac{\partial^2 \mathbf{A}}{\partial p_s \partial p_k} \mathbf{u}_0, \mathbf{v}_0 \right) + \left(\frac{\partial \mathbf{A}}{\partial p_s} \mathbf{u}_0, \mathbf{v}_0 \right) \left(\frac{\partial \mathbf{A}}{\partial p_k} \mathbf{G}_0 \mathbf{u}_0, \mathbf{v}_0 \right) +$$

$$\left(\frac{\partial \mathbf{A}}{\partial p_k} \mathbf{u}_0, \mathbf{v}_0 \right) \left(\frac{\partial \mathbf{A}}{\partial p_s} \mathbf{G}_0 \mathbf{u}_0, \mathbf{v}_0 \right) - \left(\frac{\partial \mathbf{A}}{\partial p_s} \mathbf{G}_0 \frac{\partial \mathbf{A}}{\partial p_k} \mathbf{u}_0, \mathbf{v}_0 \right) - \left(\frac{\partial \mathbf{A}}{\partial p_k} \mathbf{G}_0 \frac{\partial \mathbf{A}}{\partial p_s} \mathbf{u}_0, \mathbf{v}_0 \right)$$

Note that to calculate these derivatives we only need to know the derivatives of the matrix $\mathbf{A}$ with respect to parameters, the matrix $\mathbf{G}_0$, and the eigenvectors $\mathbf{u}_0$ and $\mathbf{v}_0$ taken at the point $\mathbf{p}_0$. In particular case of the symmetric matrix $\mathbf{A}$ in all the preceding formulas we should put $\mathbf{A}^{\mathrm{T}} = \mathbf{A}$, $\bar{\lambda} = \lambda$, $\mathbf{v} = \mathbf{u}$.

1.4 Double λ_0: Weak Interaction

Let us consider the case of a double root λ_0 of the characteristic equation $\mathbf{det}(\mathbf{A}_0 - \lambda_0 \mathbf{I}) = 0$ with two linearly independent eigenvectors $\mathbf{u}_1$ and $\mathbf{u}_2$, so that

$$\mathbf{A}_0 \mathbf{u}_1 = \lambda_0 \mathbf{u}_1 \quad , \quad \mathbf{A}_0 \mathbf{u}_2 = \lambda_0 \mathbf{u}_2 \tag{15}$$

Note that this is always the case for the symmetric matrices $\mathbf{A}_0$.

The adjoint eigenvalue problem (6) is also characterized by two linearly independent eigenvectors $\mathbf{v}_1$ and $\mathbf{v}_2$. Any linear combination of the vectors $\mathbf{u}_1, \mathbf{u}_2$ and $\mathbf{v}_1, \mathbf{v}_2$ is an eigenvector in the relevant problem. Assuming that $\mathbf{u}_1$ and $\mathbf{u}_2$ are fixed, we can choose the vectors $\mathbf{v}_1$ and $\mathbf{v}_2$ such that the normality condition is satisfied

$$(\mathbf{u}_i, \mathbf{v}_j) = \delta_{ij}, \quad i, j = 1,2 \tag{16}$$

where δ_{ij} is the Kronecker symbol.

In this case, expansions of λ and $\mathbf{u}$ are sought in integer power series of ε, see Vishik and Lyusternik (1960)

$$\begin{gathered} \lambda = \lambda_0 + \varepsilon \lambda_1 + \varepsilon^2 \lambda_2 + \ldots \\ \mathbf{u} = \mathbf{w}_0 + \varepsilon \mathbf{w}_1 + \varepsilon^2 \mathbf{w}_2 + \ldots \end{gathered} \tag{17}$$

Substituting expansions (17) and (2) in (1) we obtain

$$\begin{gathered} (\mathbf{A}_0 - \lambda_0 \mathbf{I}) \mathbf{w}_0 = \mathbf{0}, \\ (\mathbf{A}_0 - \lambda_0 \mathbf{I}) \mathbf{w}_1 = \lambda_1 \mathbf{w}_0 - \mathbf{A}_1 \mathbf{w}_0 \end{gathered} \tag{18}$$

First equation (18) means that $\mathbf{w}_0$ is an eigenvector corresponding to λ_0. Hence

$$\mathbf{w}_0 = \gamma_1 \mathbf{u}_1 + \gamma_2 \mathbf{u}_2 \,, \tag{19}$$

where γ_1 and γ_2 are constants.

Taking the inner product of the second equation (18) with $\mathbf{v}_1$ and $\mathbf{v}_2$, respectively, and using (16), (19) we get

$$\sum_{i=1}^{2} [(\mathbf{A}_1 \mathbf{u}_i, \mathbf{v}_j) - \lambda_1 \delta_{ij}]\gamma_i = 0 \,, \quad j = 1,2 \tag{20}$$

This is a system of linear homogeneous equations to find the constants γ_1 and γ_2. A nontrivial solution of these equations exists if the following condition is fulfilled

$$\det [(\mathbf{A}_1 \mathbf{u}_i, \mathbf{v}_j) - \lambda_1 \delta_{ij}] = 0 \tag{21}$$

This is a quadratic equation on λ_1. Two roots λ_1 correspond to two pairs γ_1 and γ_2 according to (20). Having obtained λ_1 and $\mathbf{w}_0$ from (19), we can find the next terms in expansions (17).

We will call the case in which a double λ_0 corresponds to two linearly independent eigenvectors a *weak interaction* of two eigenvalues, see Seyranian (1991,1993).

1.5 Double λ_0: Strong Interaction

Consider now the case of a double eigenvalue λ_0 with the length of Jordan chain equal to 2. This means that at $\mathbf{p} = \mathbf{p}_0$ there exist an eigenvector $\mathbf{u}_0$ and an associated vector $\mathbf{u}_1$ corresponding to λ_0 and governed by the equations

$$(\mathbf{A}_0 - \lambda_0 \mathbf{I})\mathbf{u}_0 = \mathbf{0} \tag{22}$$

$$(\mathbf{A}_0 - \lambda_0 \mathbf{I})\mathbf{u}_1 = \mathbf{u}_0 \,.$$

For the adjoint eigenvalue problem we have

$$(\mathbf{A}_0^{\mathrm{T}} - \bar{\lambda}_0 \mathbf{I})\mathbf{v}_0 = \mathbf{0}$$

$$(\mathbf{A}_0^{\mathrm{T}} - \bar{\lambda}_0 \mathbf{I})\mathbf{v}_1 = \mathbf{v}_0 \,. \tag{23}$$

The vectors $\mathbf{u}_0, \mathbf{u}_1, \mathbf{v}_0, \mathbf{v}_1$ are related by the following conditions

$$(\mathbf{u}_0, \mathbf{v}_0) = 0 \,, \; (\mathbf{u}_1, \mathbf{v}_0) \equiv (\mathbf{u}_0, \mathbf{v}_1) \neq 0 \,, \tag{24}$$

that can be proved by means of equations (22) and (23).

Taking a variation of the vector of parameters $\mathbf{p} = \mathbf{p}_0 + \varepsilon \mathbf{e}$, we obtain a variation of the matrix $\mathbf{A}$ in the form (2). In the case of a Jordan block, expansions for eigenvalues and eigenvectors contain terms with fractional powers $\varepsilon^{j/l}$, $j = 0, 1, 2, \ldots$, where l is the length of the Jordan chain, see Vishik and Lyusternik (1960). The case of a double eigenvalue λ_0 with $l = 2$ yields

$$\lambda = \lambda_0 + \varepsilon^{1/2}\lambda_1 + \varepsilon\lambda_2 + \varepsilon^{3/2}\lambda_3 + \ldots$$

$$\mathbf{u} = \mathbf{u}_0 + \varepsilon^{1/2}\mathbf{w}_1 + \varepsilon\,\mathbf{w}_2 + \varepsilon^{3/2}\mathbf{w}_3 + \ldots \tag{25}$$

Substituting (25) and (2) into (1), we get the equations for determining the coefficients of expansions for the eigenvalue and the eigenvector

$$(\mathbf{A}_0 - \lambda_0\mathbf{I})\mathbf{w}_1 = \lambda_1\mathbf{u}_0$$

$$(\mathbf{A}_0 - \lambda_0\mathbf{I})\mathbf{w}_2 = -\mathbf{A}_1\mathbf{u}_0 + \lambda_2\mathbf{u}_0 + \lambda_1\mathbf{w}_1\,. \tag{26}$$

Assuming that the vector $\mathbf{u}_0$ is fixed we take normality conditions for $\mathbf{v}_0, \mathbf{v}_1$ and $\mathbf{u}$ in the form

$$(\mathbf{u}_0, \mathbf{v}_1) = 1,\ (\mathbf{u}, \mathbf{v}_1) = 1\,. \tag{27}$$

From the first of equations (26) using (22) we obtain that $\mathbf{w}_1 = \lambda_1\mathbf{u}_1 + \gamma\mathbf{u}_0$ with an unknown constant γ. The second of normality conditions (27) yields $\gamma = -\lambda_1(\mathbf{u}_1, \mathbf{v}_1)$. Then, the solvability condition for the second of equations (26) with (24), (27) gives

$$\lambda_1^2 = (\mathbf{A}_1\mathbf{u}_0, \mathbf{v}_0)\,. \tag{28}$$

If the right hand side of (28) is not zero, it yields two different nonzero roots $\lambda_1 = \pm\sqrt{(\mathbf{A}_1\mathbf{u}_0, \mathbf{v}_0)}$ with the corresponding vectors $\mathbf{w}_1 = \lambda_1(\mathbf{u}_1 - (\mathbf{u}_1, \mathbf{v}_1)\mathbf{u}_0)$.

We will call the case in which a double eigenvalue λ_0 corresponds to the Jordan chain with the length $l = 2$ a *strong interaction* of two eigenvalues, see Seyranian (1991,1993).

Degenerate case. Consider now a *degenerate* case when $(\mathbf{A}_1\mathbf{u}_0, \mathbf{v}_0) = 0$. Then it is easy to verify that the terms with fractional powers disappear in (25), and we have

$$\lambda = \lambda_0 + \varepsilon\lambda_2 + \varepsilon^2\lambda_4 + \ldots$$

$$\mathbf{u} = \mathbf{u}_0 + \varepsilon\mathbf{w}_2 + \varepsilon^2\mathbf{w}_4 + \ldots \tag{29}$$

Substituting (29) and (2) into (1), we obtain equations for determining the coefficients of the expansions

$$(\mathbf{A}_0 - \lambda_0\mathbf{I})\mathbf{w}_2 = \lambda_2\mathbf{u}_0 - \mathbf{A}_1\mathbf{u}_0$$

$$(\mathbf{A}_0 - \lambda_0\mathbf{I})\mathbf{w}_4 = \lambda_2\mathbf{w}_2 - \mathbf{A}_1\mathbf{w}_2 + \lambda_4\mathbf{u}_0 - \mathbf{A}_2\mathbf{u}_0\,. \tag{30}$$

The solvability condition for the second of equations (30) with (24) yields

$$\lambda_2(\mathbf{w}_2, \mathbf{v}_0) - (\mathbf{A}_1\mathbf{w}_2, \mathbf{v}_0) - (\mathbf{A}_2\mathbf{u}_0, \mathbf{v}_0) = 0 \tag{31}$$

The term $(\mathbf{w}_2, \mathbf{v}_0)$ can be found from the first of equations (30) with the use of (23) and normality condition (27)

$$(\mathbf{w}_2, \mathbf{v}_0) = \lambda_2 - (\mathbf{A}_1\mathbf{u}_0, \mathbf{v}_1). \tag{32}$$

Substituting (32) and $\mathbf{w}_2 = \mathbf{G}_0(\lambda_2\mathbf{u}_0 - \mathbf{A}_1\mathbf{u}_0)$ into (31) we obtain a quadratic equation on λ_2

$$\lambda_2^2 + \lambda_2 a_1 + a_2 = 0 \tag{33}$$

with the coefficients

$$a_1 = -(\mathbf{A}_1\mathbf{u}_0, \mathbf{v}_1) - (\mathbf{A}_1\mathbf{u}_1, \mathbf{v}_0),$$
$$a_2 = -(\mathbf{A}_2\mathbf{u}_0, \mathbf{v}_0) + (\mathbf{G}_0(\mathbf{A}_1\mathbf{u}_0), \mathbf{A}_1^T\mathbf{v}_0). \tag{34}$$

Equation (33) defines in the first approximation bifurcation of a double eigenvalue λ_0 in the degenerate case.

1.6 Geometric Interpretation of Weak and Strong Interactions

For geometric interpretation we consider the case of only one parameter. Denoting this by p we consider the cases of weak and strong interactions that occur at $p = p_0$. The matrix $\mathbf{A_1}$ takes the form

$$\mathbf{A_1} = e\, d\mathbf{A} / dp \tag{35}$$

In this expression $e = \pm 1$, depending on the sign of $\Delta p = p - p_0$. If $\Delta p > 0$, then $e = 1$, and if $\Delta p < 0$, then $e = -1$. As a small parameter we take the quantity $\varepsilon = |\Delta p|$.

Weak interaction. It can be seen from (17) and (21) that in spite of the sign of Δp the expansion for λ yields

$$\lambda = \lambda_0 + \lambda_1 \Delta p + o(|\Delta p|) \tag{36}$$

where the two roots λ_1 are determined from the equation

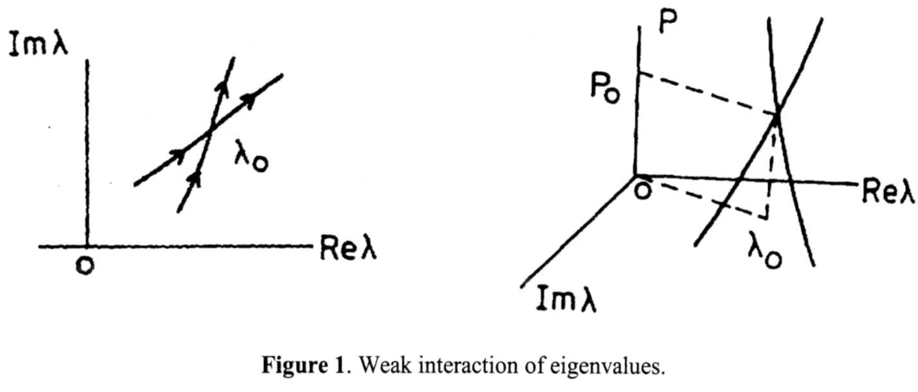

Figure 1. Weak interaction of eigenvalues.

$$\mathbf{det}\left[\left(\frac{d\mathbf{A}}{dp}\mathbf{u}_i, \mathbf{v}_j\right) - \lambda_1 \delta_{ij}\right] = 0, \quad i, j = 1,2 \tag{37}$$

Interaction of eigenvalues λ on the complex plane and in the three-dimensional space is shown in Fig 1. Arrows indicate paths of λ when the parameter p is increased. The eigenvalues approach, merge, and then diverge without change of the direction of motion. Note that the weak interaction in the three-dimensional space occurs on the same plane.

Strong interaction. First we consider $\Delta p = p - p_0 > 0$. Then we insert $e = 1$ in (29) and use the notation

$$\left(\frac{d\mathbf{A}}{dp}\mathbf{u}_0, \mathbf{v}_0\right) = R\,\mathbf{exp}(i\varphi) \tag{38}$$

where R is the modulus and φ is the argument of a complex number, and i is the imaginary unit. Using (38) and (35) in (28) we find

$$\lambda_1 = \pm\sqrt{R}\,\mathbf{exp}(i\varphi / 2) \tag{39}$$

Taking $\Delta p < 0$ and inserting $e = -1$ in (35) we get

$$\lambda_1 = \pm i\sqrt{R}\,\mathbf{exp}(i\varphi / 2) \tag{40}$$

According to (39), (40) the four roots λ_1 are placed on the circumference of radius $\sqrt{R}$, the

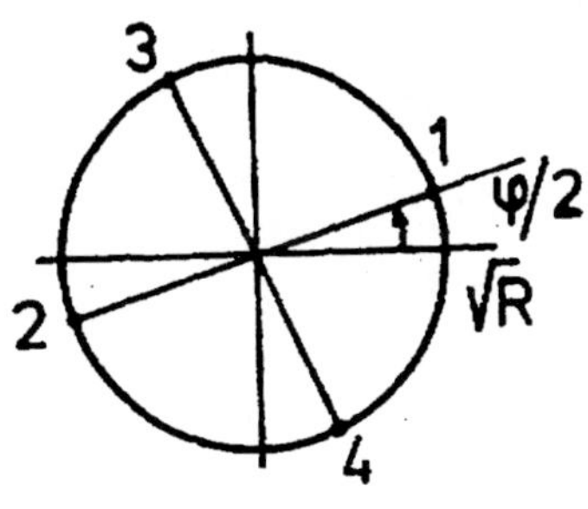

Figure 2. The four roots λ_1.

difference between arguments of the adjacent roots being equal to $\pi / 2$, see Fig 2.

Substituting in (25) expressions (39) and (40) and taking $\varepsilon = |p - p_0|$ we obtain

$$\begin{aligned} \lambda &= \lambda_0 \pm \sqrt{R}\,\mathbf{exp}(i\varphi / 2)\sqrt{p - p_0} + O(|\Delta p|), \quad & p > p_0 \\ \lambda &= \lambda_0 \pm i\sqrt{R}\,\mathbf{exp}(i\varphi / 2)\sqrt{p_0 - p} + O(|\Delta p|), \quad & p < p_0 \end{aligned} \tag{41}$$

Thus, when the parameter p is increased $(p < p_0)$ the eigenvalues λ approach on the straight line, merge at $p = p_0$ and then, at $p > p_0$, diverge along the line, perpendicular to the line of approach, see Fig 3a. The arrows show motion of λ when p is increased. The eigenvalues collide, and then diverge at the right angles in opposite direction.

Strong interaction in the three-dimensional space is shown in Fig 3b. Note that the plane of the strong interaction of λ at $p < p_0$ is changed through the interaction to the orthogonal plane at $p > p_0$. This is the mechanism of strong interaction, which is described by expansions (25). Remember that strong interaction is characterized by coincidence both of the eigenvalues and eigenvectors, while weak interaction is characterized by coincidence of the eigenvalues only.

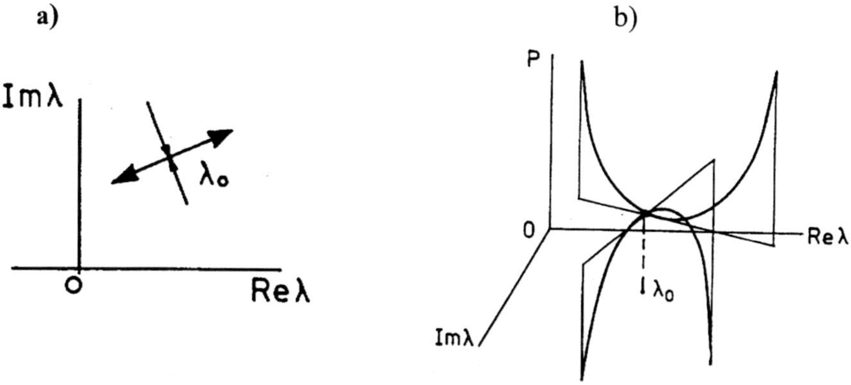

Figure 3. Strong interaction of eigenvalues.

Expressions (41) can be given in a unified form as

$$\lambda = \lambda_0 \pm \sqrt{R}\,\mathbf{exp}\,(i\varphi / 2)\sqrt{p - p_0} + O(|\Delta p|) \tag{42}$$

If λ_0 is real, then in relations (38)-(42) one should put $\varphi = 0$. This means that the roots λ_1 in this case are real or purely imaginary.

Strong interaction of eigenvalues is a typical catastrophe, see Arnold (1984, 1989) and Thompson (1982). Velocity of growth of catastrophe tends to infinity when the catastrophe is approaching $(d\lambda / d\varepsilon \to \infty$ when $\varepsilon \to 0)$. Fig 3b is an expressive symbol of the catastrophe theory, see Seyranian (1993).

Degenerate case. It is easy to see that in the degenerate case according to (29) we have

$$\lambda = \lambda_0 + \lambda_2 \Delta p + o(|\Delta p|) \tag{43}$$

where the two roots λ_2 are found from the quadratic equation (33) with the coefficients

$$\begin{aligned} a_1 &= -\left(\frac{d\mathbf{A}}{dp}\mathbf{u}_0, \mathbf{v}_1\right) - \left(\frac{d\mathbf{A}}{dp}\mathbf{u}_1, \mathbf{v}_0\right) \\ a_2 &= -\frac{1}{2}\left(\frac{d^2\mathbf{A}}{dp^2}\mathbf{u}_0, \mathbf{v}_0\right) + \left(\mathbf{G}_0 \frac{d\mathbf{A}}{dp}\mathbf{u}_0, \frac{d\mathbf{A}^{\mathrm{T}}}{dp}\mathbf{v}_0\right) \end{aligned} \tag{44}$$

We note that in this case behavior of eigenvalues on the complex plane is similar to the weak interaction given by formulas (36), (37) and shown in Fig 1.

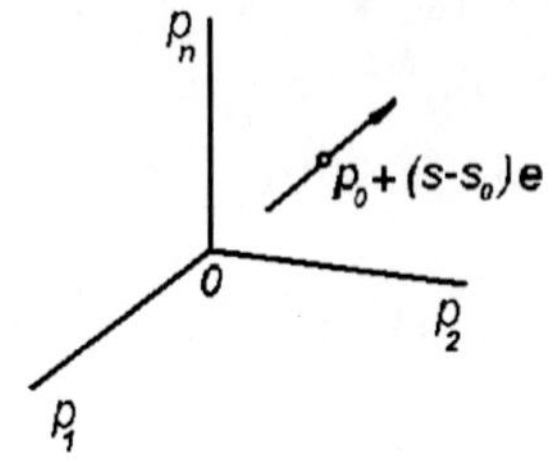

Figure 4. A ray in the parameter space.

In conclusion we note that Fig 1 and Fig 3 characterize the interaction of λ along any ray $\mathbf{p} = \mathbf{p}_0 + (s - s_0)\mathbf{e}$ in the parameter space for small $\varepsilon = |s - s_0|$, see Fig 4.

2 Stability and Catastrophes in One-Parameter Circulatory Systems

2.1 Introduction

Stability problems for circulatory systems have been considered in several books, see for example Bolotin (1963), Panovko and Gubanova (1965), Ziegler (1968), Huseyin (1978), Leipholz (1980), Thompson (1982), Merkin (1997), as well as in many papers. Most of these problems consist in studying the stability of circulatory systems on a load parameter – the magnitude of a follower force, the velocity or density of flow, and so on. In this sense these problems are one-parametric.

This Section, based on the paper by Seyranian (1994), is devoted to generalization of the results on the stability of one-parameter circulatory systems. The analysis is based on certain results of catastrophe theory, see Arnold (1983, 1990), Galin (1972), and the perturbation technique of the spectrum of non-selfadjoint operators developed by Vishik and Lyusternik (1960).

2.2 One-Parameter Circulatory Systems

Let us consider a linear autonomous system with nonconservative positional forces

$$\mathbf{M}\ddot{\mathbf{q}} + \mathbf{C}\mathbf{q} = \mathbf{0} \tag{1}$$

Here, $\mathbf{M}$ is a symmetric positive definite mass matrix of dimension $n \times n$, $\mathbf{C}$ is a nonsymmetrical stiffness matrix of the same order, and $\mathbf{q}$ is a vector of generalized coordinates. System (1) is usually termed *circulatory* system, see Ziegler (1968).

Finding a solution of (1) in the form $\mathbf{q} = \mathbf{u}\exp(\lambda t)$ we obtain the eigenvalue problem

$$[\mathbf{C} + \lambda^2\mathbf{M}]\mathbf{u} = \mathbf{0} \tag{2}$$

The eigenvalues λ^2 are determined from the characteristic equation $\det[\mathbf{C} + \lambda^2\mathbf{M}] = 0$. Since the matrices $\mathbf{M}$ and $\mathbf{C}$ are real it is easy to see that if λ is a solution to the characteristic equation, then the quantities $-\lambda,\ \bar{\lambda}, -\bar{\lambda}$ are also solutions to this equation.

This means that system (1) can be stable only when all the characteristic exponents λ belong to the imaginary axis.

We consider the case in which the matrices $\mathbf{M}$ and $\mathbf{C}$ depend on only one parameter p. Using the notation

$$\mu = -\lambda^2, \quad \mathbf{A} = \mathbf{M}^{-1}\mathbf{C} \tag{3}$$

instead of (2), we consider the eigenvalue problem for the matrix $\mathbf{A}(p)$

$$\mathbf{A}\mathbf{u} = \mu\mathbf{u} \tag{4}$$

According to Arnold (1983) and Galin (1972), the generic (typical) case of one-parametric family of real matrices is characterized by the simple (distinct) eigenvalues μ and, at some isolated points of the parameter $p = p_0$, by the double real μ_0 with a Jordan chain of the second order. More complicated Jordan structures may be destroyed by infinitely small shift in the family of matrices $\mathbf{A}(p)$.

Now we study bifurcation of eigenvalues using the perturbation theory of nonsymmetric operators, see Vishik and Lyusternik (1960). Let a double real eigenvalue μ_0 correspond to the Jordan chain of the second order. This means that μ_0 corresponds to an eigenvector $\mathbf{u}_0$ and an associated vector $\mathbf{u}_1$, satisfying the equations

$$\begin{aligned} \mathbf{A}\mathbf{u}_0 &= \mu_0\mathbf{u}_0, \\ \mathbf{A}\mathbf{u}_1 &= \mu_0\mathbf{u}_1 + \mathbf{u}_0 \end{aligned} \tag{5}$$

Along with (5), we consider the adjoint eigenvalue problem, which is necessary for further analysis

$$\begin{aligned} \mathbf{A}^{\mathbf{T}}\mathbf{v}_0 &= \mu_0\mathbf{v}_0, \\ \mathbf{A}^{\mathbf{T}}\mathbf{v}_1 &= \mu_0\mathbf{v}_1 + \mathbf{v}_0 \end{aligned} \tag{6}$$

The vectors $\mathbf{u}_0, \mathbf{u}_1, \mathbf{v}_0, \mathbf{v}_1$ in (5) and (6) are related by the conditions

$$(\mathbf{u}_0, \mathbf{v}_0) = 0, \quad (\mathbf{u}_0, \mathbf{v}_1) = (\mathbf{u}_1, \mathbf{v}_0) \neq 0 \tag{7}$$

We assume that the matrices $\mathbf{M}, \mathbf{C}$, and hence also $\mathbf{A}$, depend smoothly on the parameter p. If we take an increment $p = p_0 + \Delta p$, then the matrix $\mathbf{A}$ can be expressed in the form

$$\mathbf{A} = \mathbf{A}_0 + \Delta p\mathbf{A}_1 + \ldots \tag{8}$$

where $\mathbf{A}_0 = \mathbf{A}(p_0), \mathbf{A}_1 = (d\mathbf{A}/dp)_{p=p_0}$. The matrices $\mathbf{A}_0$ and $\mathbf{A}_1$ are related to $\mathbf{C}$ and $\mathbf{M}$ as

$$\begin{gathered} \mathbf{A}_0 = \mathbf{M}_0^{-1}\mathbf{C}_0, \quad \mathbf{A}_1 = \mathbf{M}_0^{-1}\mathbf{C}_1 - \mathbf{M}_0^{-1}\mathbf{M}_1\mathbf{M}_0^{-1}\mathbf{C}_0 \\ \mathbf{C}_0 = \mathbf{C}(p_0), \mathbf{M}_0 = \mathbf{M}(p_0), \mathbf{C}_1 = (d\mathbf{C}/dp)_{p=p_0}, \mathbf{M}_1 = (d\mathbf{M}/dp)_{p=p_0} \end{gathered} \tag{9}$$

Due to the variation Δp the eigenvalues and eigenvectors of (4) also take increments. In the case of the Jordan chain of the second order, the disturbed eigenvalues are expressed in fractional powers of the small parameter $\varepsilon > 0$, see Vishik and Lyusternik (1960)

$$\mu = \mu_0 + \varepsilon^{1/2}\mu_1 + \varepsilon\mu_2 + ... \tag{10}$$

In our case, Δp plays the role of the small parameter $\varepsilon = \Delta p$ if $\Delta p > 0$, and $\varepsilon = -\Delta p$ if $\Delta p < 0$. Using (4)-(10) we obtain, see Section 1

$$\mu_1^2 = -F,\ \Delta p > 0 \quad and \quad \mu_1^2 = F,\ \Delta p < 0, \tag{11}$$

$$F = -(\mathbf{A}_1\mathbf{u}_0, \mathbf{v}_0)/(\mathbf{u}_1, \mathbf{v}_0) \tag{12}$$

The vectors $\mathbf{u}_0, \mathbf{u}_1, \mathbf{v}_0$ are determined by (5) and (6). Due to the assumption that μ_0 is real, the vectors $\mathbf{u}_0, \mathbf{u}_1, \mathbf{v}_0, \mathbf{v}_1$ are also real. Hence, F is a real constant, positive or negative. According to (10), (11), bifurcation of μ up to terms $o(|\Delta p|)$ is described by

$$\mu = \mu_0 \pm i\sqrt{F\Delta p}\ ,\ \Delta p > 0 \quad and \quad \mu = \mu_0 \pm \sqrt{F\Delta p}\ ,\ \Delta p < 0$$

These two relations can be combined into

$$\mu = \mu_0 \pm i\sqrt{F\Delta p} \tag{13}$$

Note that this formula is valid only for the constant $F \neq 0$, see Vishik and Lyusternik (1960).

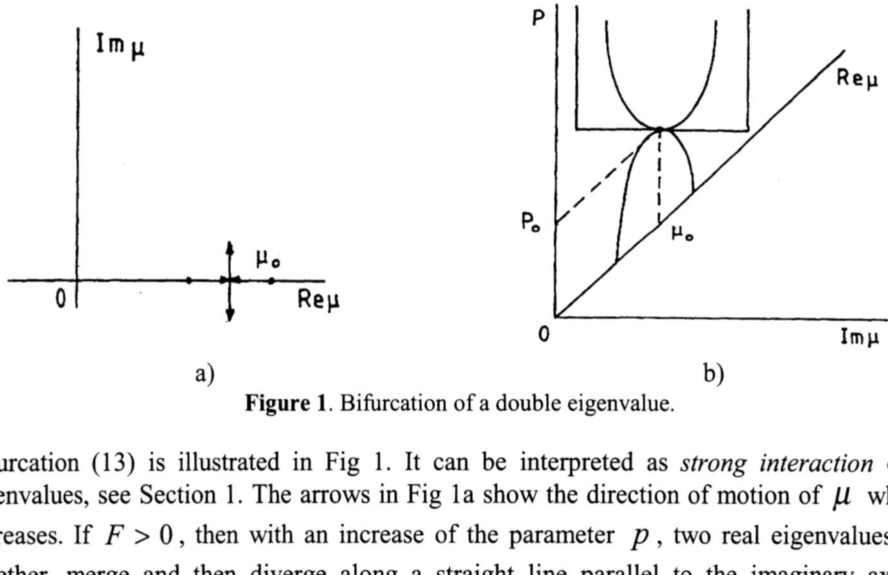

Figure 1. Bifurcation of a double eigenvalue.

Bifurcation (13) is illustrated in Fig 1. It can be interpreted as *strong interaction* of two eigenvalues, see Section 1. The arrows in Fig 1a show the direction of motion of μ when p increases. If $F > 0$, then with an increase of the parameter p, two real eigenvalues come together, merge and then diverge along a straight line parallel to the imaginary axis. If $F < 0$, the direction of motion of μ changes to the opposite. This means that with an increase of p, two complex conjugate μ approach each other, merge to a real μ_0, and then diverge along the real axis in opposite directions.

Bifurcation (13) in the three-dimensional space (for $F > 0$) is shown in Fig 1b. At $p \approx p_0$ the intersecting curves are quadratic parabolas of the same curvature lying in the orthogonal planes $\mathrm{Im}\,\mu = 0$ and $\mathbf{Re}\,\mu = \mu_0$.

Due to the fact that the matrix $\mathbf{A}$ is real, the interaction pictures in Fig 1 are symmetric with respect to the axis and plane $\mathbf{Im}\,\mu = 0$, respectively. It is easy to see that if μ is a simple real eigenvalue, it remains real with a change of p. Indeed, in the other case, the complex conjugate eigenvalue $\overline{\mu}$ also appears, which means the increase of the total number of roots of the characteristic equation. Therefore, with a change of p, eigenvalues are able to leave the real axis (or the plane $\mathrm{Im}\,\mu = 0$) only when they meet and interact strongly according to (13), see Fig 1. Despite the simplicity, this fact has been recognized only in Seyranian (1994).

Let us return to problem (1), (2). Because $\mu = -\lambda^2$, positive eigenvalues μ correspond to pure imaginary $\lambda = \pm i\sqrt{\mu}$, and negative μ correspond to real $\lambda = \pm\sqrt{|\mu|}$. Thus, a positive $\mu_0 = -\lambda_0^2$ at the bifurcation point (13) means transition of stability of system (1) to kinetic instability (flutter), or vice versa (depending on the sign of F), and a negative μ_0 corresponds to transition of kinetic instability (flutter) to aperiodic instability (divergence), or vice versa.

Bifurcation (13) can be expressed through λ. Substituting $p - p_0$ instead of Δp and using $\mu_0 = -\lambda_0^2$ we obtain

$$\lambda = \pm\lambda_0\left(1 \pm \frac{i\sqrt{F(p-p_0)}}{2\lambda_0^2}\right) \tag{14}$$

Interaction (14) is illustrated in Fig 2a,b for $F > 0$. Direction of the arrows corresponds to the increase of p. If $F < 0$, then the arrows must be reversed.

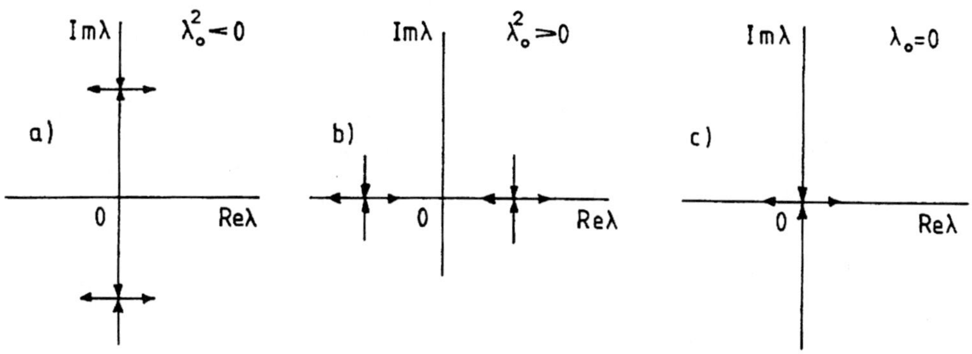

Figure 2. Catastrophes of the characteristic exponents λ.

Besides catastrophes (13), (14), system (1) can lose stability statically, which means transition of positive eigenvalues μ to negative values (divergence). Expansion of simple eigenvalues in the vicinity of $\mu_0 = 0$ up to higher order terms leads to the relation

$$\mu = -D\Delta p, \quad D = -(\mathbf{A}_1\mathbf{u}_0, \mathbf{v}_0)/(\mathbf{u}_0, \mathbf{v}_0) \tag{15}$$

Here $\mathbf{u}_0$ and $\mathbf{v}_0$ corresponding to $\mu_0 = 0$ are defined by (5) and (6), respectively. In terms of λ we obtain

$$\lambda = \pm\sqrt{D(p - p_0)} \tag{16}$$

For $D > 0$, the behavior of λ with an increase of p is shown in Fig 2c. If $D < 0$, the direction of motion of λ changes to the opposite.

The main result of this subsection can be formulated as follows.

Theorem. *One-parameter circulatory systems are subjected to catastrophes of three types: flutter, transition of flutter to divergence (or vice versa), and divergence, described by relations (14), (16) and shown in Fig 2 a,b,c, respectively.*

The three-dimensional plots $\lambda - p$ are similar to Fig 1.

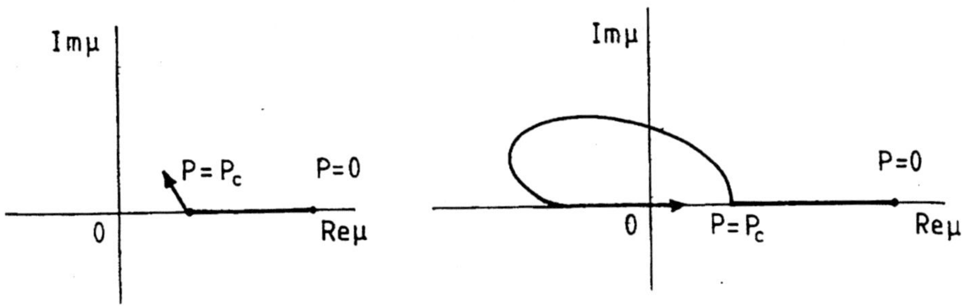

Figure 3. Motion of an eigenvalue μ on the complex plane depending on the parameter p.

One-parameter circulatory systems have been considered by many authors. In the book by Leipholz (1987) the behavior of eigenvalues of matrices depending on a single parameter has been studied. Fig 3 reproduces Fig 66c,d from that book. In this figure, a qualitative pattern of motion of the eigenvalue μ of a nonsymmetrical matrix, depending on the parameter p, is plotted in the complex plane with p_c denoting the critical value. Note that such behavior of μ contradicts the results presented above, and is therefore impossible.

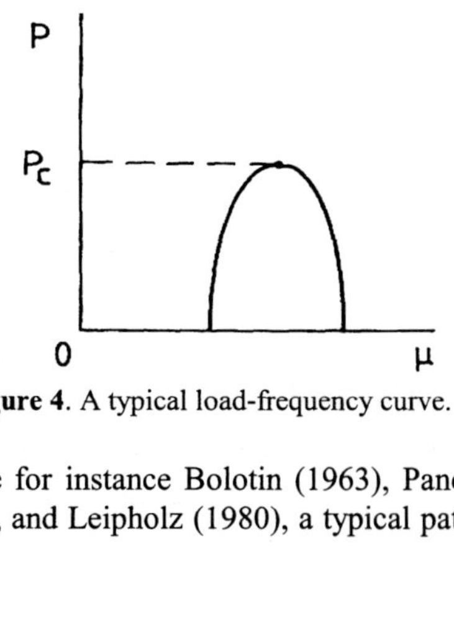

Figure 4. A typical load-frequency curve.

In many books and papers, see for instance Bolotin (1963), Panovko and Gubanova (1965), Ziegler (1968), Huseyin (1978), and Leipholz (1980), a typical pattern of dependence of p on

μ, reproduced in Fig 4, is plotted. Note that, at $p > p_c$ the eigenvalues do not disappear. In Plaut (1972) and Huseyin (1978, p. 62-63) the equality $dp/d\mu = 0$ is used as the flutter condition. We emphasize that at the point of multiplicity the eigenvalues are not differentiable in common sense. That is why the last equality can not be used as the condition of the onset of flutter.

2.3 Linear Dependence on the Parameter

In this section we consider the case in which the matrix $\mathbf{M}$ is constant and the matrix $\mathbf{C}$ linearly depends on the parameter p

$$\mathbf{C} = \mathbf{B} - p\mathbf{H} \tag{17}$$

Here, $\mathbf{B}$ is a symmetric positive definite matrix, $\mathbf{H}$ is a nonsymmetrical matrix, and $p \geq 0$. Substituting (17) in (2) we obtain the eigenvalue problem

$$[p\mathbf{H} - \mathbf{B}]\mathbf{u} = \lambda^2 \mathbf{M}\mathbf{u} \tag{18}$$

We are interested in studying the behavior of λ on p. For this purpose we take the scalar product of (18) with $\mathbf{u}$ and get

$$\lambda^2 = (ph - b)/m$$

Here, $h = (\mathbf{Hu},\mathbf{u})$, $b = (\mathbf{Bu},\mathbf{u})$, $m = (\mathbf{Mu},\mathbf{u})$. Due to the positive definiteness of $\mathbf{B}$ and $\mathbf{M}$ we have $b > 0,\ m > 0$. At $p = 0$, $\lambda^2 = -b/m$. Hence, all the characteristic exponents λ are purely imaginary $\lambda = \pm i\omega$, and the corresponding eigenvectors $\mathbf{u}$ are real (up to arbitrary multipliers).

Now we calculate the first derivative of the simple λ with respect to p. For this purpose we consider the eigenvalue problem adjoint to (18)

$$[p\mathbf{H}^T - \mathbf{B}]\mathbf{v} = \bar{\lambda}^2 \mathbf{M}\mathbf{v} \tag{19}$$

The expression for the derivative takes the form, see Section 1

$$\frac{d\lambda}{dp} = \frac{(\mathbf{Hu},\mathbf{v})}{2\lambda(\mathbf{Mu},\mathbf{v})} \tag{20}$$

At $p = 0$, we have $\lambda = i\omega$, $\bar{\lambda} = -i\omega$, and $\mathbf{v} = \mathbf{u}$. Thus, at $p = 0$, the derivative $d\lambda/dp$ is a purely imaginary quantity. This is in good agreement with the result of subsection 2.2 that simple λ do not leave the imaginary axis with an increase of p. They are able to leave the imaginary axis only when they meet and interact strongly, see Fig 2a,b.

First we consider the flutter case, see Fig 2a. This case, taking place at $p = p_0$, is characterized by the double eigenvalue $-\lambda_0^2$ (and the double characteristic exponent $\lambda_0 = i\omega_0$) with the corresponding eigenvectors $\mathbf{u}_0, \mathbf{v}_0$ of problems (18), (19). The orthogonality condition takes the form

$$(\mathbf{Mu}_0, \mathbf{v}_0) = 0 \tag{21}$$

Sometimes, (21) is referred as *the flutter condition*. When p tends to p_f then, according to (20) $d\lambda / dp \to \infty$. The same is also true for the divergence case, because $\lambda \to 0$. This means infinite velocity of growth of the catastrophe, which is typical in the catastrophe theory, see Arnold (1984).

According to (14), near the point $p = p_f$ we have

$$\lambda = \pm i\omega_0 \left[1 \pm \frac{\sqrt{G(p-p_f)}}{2\omega_0^2} \right] \quad , \quad G = \frac{(\mathbf{Hu}_0, \mathbf{v}_0)}{(\mathbf{Mu}_1, \mathbf{v}_0)} \tag{22}$$

Here, the eigenvectors $\mathbf{u}_0$ and $\mathbf{v}_0$ are determined from (18), (19), where one should put $\lambda = i\omega_0$, $p = p_f$. The associated vector $\mathbf{u}_1$ is found from the equation

$$[p_f \mathbf{H} - \mathbf{B}]\mathbf{u}_1 + \omega_0^2 \mathbf{Mu}_1 = \mathbf{Mu}_0 \tag{23}$$

Let us consider the case of divergence, see Fig 2c. According to (16) we obtain

$$\lambda = \pm\sqrt{K(p-p_d)} \quad , \quad K = \frac{(\mathbf{Hu}_0, \mathbf{v}_0)}{(\mathbf{Mu}_0, \mathbf{v}_0)} \tag{24}$$

The eigenvectors $\mathbf{u}_0, \mathbf{v}_0$ in this relation satisfy the equations

$$\mathbf{Bu}_0 = p_d \mathbf{Hu}_0 \ , \ \mathbf{Bv}_0 = p_d \mathbf{H}^{\mathbf{T}} \mathbf{v}_0 \tag{25}$$

The minimum positive eigenvalue in (25) is the critical divergence value p_d.

We consider now a particular case in which the matrix $\mathbf{H}$ is skew-symmetric $\mathbf{H}^{\mathbf{T}} = -\mathbf{H}$. We shall show that divergence instability is impossible in this case. Indeed, in the other case from (25) we can obtain $p_d(\mathbf{Hu}_0, \mathbf{u}_0) - (\mathbf{Bu}_0, \mathbf{u}_0) = 0$. Because $\mathbf{u}_0$ is a real vector, we have $(\mathbf{Hu}_0, \mathbf{u}_0) = 0$, and hence $(\mathbf{Bu}_0, \mathbf{u}_0) = 0$. The last equality is impossible due to the assumption $\mathbf{B} > 0$. Thus, divergence instability is possible only in the case when $\mathbf{B}$, which is the symmetric part of the matrix $\mathbf{C}$, has zero eigenvalues. A similar result was obtained by Zevin (1988).

2.4 Illustrative Example

As an illustrative example we consider a flutter problem of airfoil in a simplified formulation, see Ziegler (1968). A rigid rectangular panel replaces the airfoil, see Fig 5, with two degrees of freedom: a vertical displacement z of the line passing through the point L and normal to the

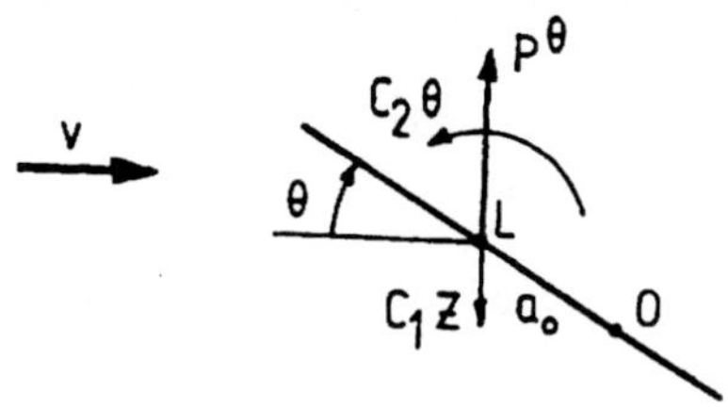

Figure 5. Rigid airfoil vibrating in a flow.

plane of the figure, and a rotation θ about this line. It is assumed that the aerodynamic lift force $p\theta$, applied to the point L, is proportional to the angle of attack θ and to the square of the velocity v of flight. Point O denotes the centroid of the section. The differential equations of motion are, see Ziegler (1968)

$$\begin{aligned} m\ddot{z} - ma_0\ddot{\theta} + c_1 z - p\theta &= 0 \\ -ma_0\ddot{z} + m(i_0^2 + a_0^2)\ddot{\theta} + c_2\theta &= 0 \end{aligned} \tag{26}$$

In these equations m is the mass of the panel of unit length, c_1 and c_2 are the stiffness coefficients, p is a load parameter, proportional to the square of the velocity v, and i_0 is the radius of gyration about a normal to the plane of the figure through the point O. The radius of gyration about the point L is $i_L^2 = i_0^2 + a_0^2$.

Let us analyze the stability of system (26) on the parameter $p \geq 0$ regarding all other parameters as fixed quantities. It is easy to see that system (26) is circulatory. The characteristic equation for it takes the form

$$b_0\lambda^4 + b_2\lambda^2 + b_4 = 0 \tag{27}$$

where $b_0 = m^2 i_0^2$, $b_2 = c_2 m + c_1 m i_L^2 - pma_0$, $b_4 = c_1 c_2$. The discriminant of equation (27) $D_s = b_2^2 - 4b_0 b_4$ is a quadratic parabola with respect to the parameter p. It is easy to see that it has two positive roots p_1 and p_2, and $D_s < 0$ for $p_1 < p < p_2$. The coefficient $b_2(p)$ is a linear function of p, $b_2(p) < 0$ at $p > p_3$, where $p_1 < p_3 < p_2$. Hence, the roots λ^2 of equation (27) are negative and different when $0 \leq p < p_1$, complex conjugate when $p_1 < p < p_2$, and positive when $p > p_2$. The values p_1 and p_2 are equal to

$$p_1 = c_1 a_0 + \left(\sqrt{c_2} - i_0\sqrt{c_1}\right)^2 / a_0, \quad p_2 = c_1 a_0 + \left(\sqrt{c_2} + i_0\sqrt{c_1}\right)^2 / a_0 \tag{28}$$

The behavior of the characteristic exponents λ depending on p is shown in Fig 6. The panel

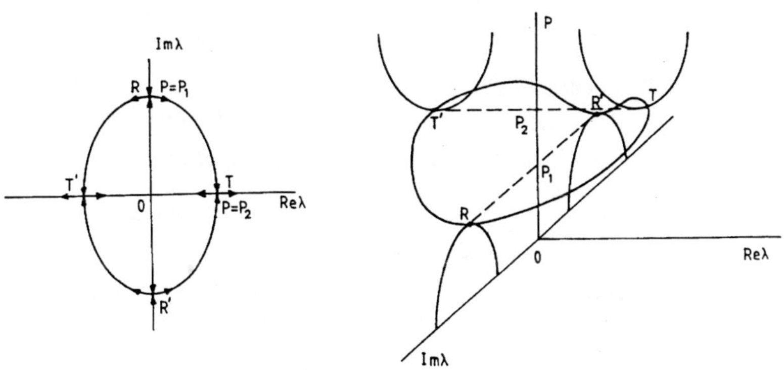

Figure 6. Behavior of the characteristic exponents λ.

is stable at $0 \le p < p_1$. When p lies between the limits $p_1 \le p < p_2$ it is the flutter domain, and when p becomes greater than p_2 the flutter is changed by the divergence instability.

3 Metamorphoses of Frequency Curves in Nonconservative Systems

3.1 Introduction

In this Section we study metamorphoses of frequency curves in nonconservative systems. The frequency curve is dependence of a frequency of the system on a chosen parameter, say, on a parameter of nonconservative load. In nonconservative systems depending on parameters a phenomenon of overlapping of frequency curves has been observed by Claudon (1975), Hanaoka and Washizu (1980), Kounadis and Katsikadelis (1980), Seyranian and Sharanyuk (1983), Langthjem and Sugiyama (1999). This phenomenon is that with a change of other parameters of the system the frequency curves come together, merge at some point, and then overlap forming a "bubble of instability", Seyranian (1991, 1993). Besides, the overlapping is accompanied by a discontinuity of the critical load.

In the present Section, based on a recent paper by Kirillov and Seyranian (2001), the explicit formulas describing metamorphoses of frequency curves in two-parameter nonconservative systems are derived. These formulas use information on a system only at the merging point of the frequencies, and allow qualitative, as well as quantitative analysis of behavior of frequency curves near that point. It is shown that the metamorphoses of frequency curves are related to the convexity properties of the flutter instability boundary. A quadratic approximation of the flutter boundary in the parameter space is obtained.

3.2 Two-Parameter Circulatory Systems

Consider a linear autonomous nonconservative mechanical system with no velocity-dependent forces governed by the equation

$$\mathbf{M}\ddot{\mathbf{q}} + \mathbf{C}\mathbf{q} = \mathbf{0}, \tag{1}$$

where $\mathbf{M} = \mathbf{M}^{\mathrm{T}} > 0$ and $\mathbf{C} \neq \mathbf{C}^{\mathrm{T}}$ are the real $m \times m$ mass and stiffness matrices, and $\mathbf{q}$ is the vector of generalized coordinates of dimension m. Dots mean differentiation with respect to time t. System (1) is usually called a circulatory system, see Bolotin (1963) and Ziegler (1968). Let ω be a frequency of oscillations. Substituting $\mathbf{q} = \mathbf{u}\exp(i\omega t)$ into (1) and introducing the notation $\mathbf{A} = \mathbf{M}^{-1}\mathbf{C}$, $\lambda = \omega^2$ we obtain the eigenvalue problem

$$\mathbf{A}\mathbf{u} = \lambda\mathbf{u}. \tag{2}$$

System (1) is stable if all the eigenvalues of problem (2) are positive and semisimple. If all λ are real and some of them negative, then system (1) is statically unstable (divergence). If at least one eigenvalue λ is complex it means flutter instability.

Suppose that the matrix $\mathbf{A}$ smoothly depends on a vector of two parameters $\mathbf{p} = (p_1, p_2)$. It is known, see Kirillov and Seyranian (1999), that in a generic case the stability boundary of two-parameter circulatory system consists of smooth curves, at which the matrix $\mathbf{A}$ possesses either a simple zero eigenvalue or a positive double eigenvalue with the Jordan chain of length 2. At some isolated points of the stability boundary singularities of two types (cusps and nodes) are possible. These singularities correspond to matrices with more complicated Jordan structure, see Arnold (1983).

Consider a point $\mathbf{p} = \mathbf{p}_0$ on the stability boundary, where the matrix $\mathbf{A}_0 = \mathbf{A}(\mathbf{p}_0)$ possesses a positive double eigenvalue λ_0 with the Jordan chain of length 2. At this point the eigenvalue λ_0 has an eigenvector $\mathbf{u}_0$ and associated vector $\mathbf{u}_1$ governed by the equations

$$(\mathbf{A}_0 - \lambda_0 \mathbf{I})\mathbf{u}_0 = \mathbf{0}, \quad (\mathbf{A}_0 - \lambda_0 \mathbf{I})\mathbf{u}_1 = \mathbf{u}_0, \tag{3}$$

and an eigenvector $\mathbf{v}_0$ and associated vector $\mathbf{v}_1$ of the adjoint problem

$$(\mathbf{A}_0^{\mathrm{T}} - \lambda_0 \mathbf{I})\mathbf{v}_0 = \mathbf{0}, \quad (\mathbf{A}_0^{\mathrm{T}} - \lambda_0 \mathbf{I})\mathbf{v}_1 = \mathbf{v}_0, \tag{4}$$

where $\mathbf{I}$ is the identity matrix. Vectors $\mathbf{u}_0, \mathbf{u}_1, \mathbf{v}_0, \mathbf{v}_1$ are related by means of the orthogonality and normality conditions

$$(\mathbf{u}_0, \mathbf{v}_0) = 0, \ (\mathbf{u}_1, \mathbf{v}_0) \equiv (\mathbf{u}_0, \mathbf{v}_1) = 1. \tag{5}$$

Parentheses in (5) denote the Hermitian scalar product $(\mathbf{a}, \mathbf{b}) = \sum_{i=1}^{m} a_i \bar{b}_i$ of the vectors $\mathbf{a}$, $\mathbf{b} \in C^m$.

We take smooth variations $\mathbf{p}(\varepsilon) = \mathbf{p}_0 + \varepsilon \mathbf{e} + \varepsilon^2 \mathbf{d}$ of the vector of parameters near the point $\mathbf{p}_0$. Here $\mathbf{e}$, $|\mathbf{e}| = 1$ and $\mathbf{d} \in R^2$ are the vectors of variation and $\varepsilon > 0$ is the small parameter. As a result of the perturbation of parameters the matrix $\mathbf{A}_0$ takes an increment

$$\mathbf{A}(\mathbf{p}_0 + \varepsilon \mathbf{e} + \varepsilon^2 \mathbf{d}) = \mathbf{A}_0 + \varepsilon \mathbf{A}_1 + \varepsilon^2 \mathbf{A}_2 + \ldots, \tag{6}$$

where the matrices $\mathbf{A}_1$ and $\mathbf{A}_2$ are

$$\mathbf{A}_1 = \sum_{s=1}^{2} \frac{\partial \mathbf{A}}{\partial p_s} e_s, \quad \mathbf{A}_2 = \sum_{i=1}^{2} \frac{\partial \mathbf{A}}{\partial p_i} d_i + \frac{1}{2} \sum_{s,t=1}^{2} \frac{\partial^2 \mathbf{A}}{\partial p_s \partial p_t} e_s e_t . \tag{7}$$

Derivatives in (7) are taken at the initial point $\mathbf{p} = \mathbf{p}_0$.

Due to variation of the vector of parameters an eigenvalue λ_0 and a corresponding eigenvector $\mathbf{u}_0$ also take increments. According to the perturbation theory of nonselfadjoint operators by Vishik and Liusternik (1960), in the case of a double eigenvalue with the Jordan chain of length 2 expansions for the eigenvalue and the eigenvector contain terms with fractional powers of the small parameter ε

$$\lambda = \lambda_0 + \varepsilon^{1/2}\lambda_1 + \varepsilon \lambda_2 + \varepsilon^{3/2}\lambda_3 + \ldots; \ \mathbf{u} = \mathbf{u}_0 + \varepsilon^{1/2}\mathbf{w}_1 + \varepsilon \mathbf{w}_2 + \varepsilon^{3/2}\mathbf{w}_3 + \ldots \tag{8}$$

Substituting expansions (6) and (8) into eigenvalue problem (2), we get equations for determining perturbations of λ_0 and $\mathbf{u}_0$. For the first order coefficient λ_1 we have

$$\lambda_1 = \pm\sqrt{(\mathbf{A}_1\mathbf{u}_0, \mathbf{v}_0)}. \tag{9}$$

Using expression (7) for the matrix $\mathbf{A}_1$ we introduce a vector $\mathbf{f}$ with the components

$$f^s = \left(\frac{\partial \mathbf{A}}{\partial p_s}\mathbf{u}_0, \mathbf{v}_0\right), \quad s = 1, 2. \tag{10}$$

As λ_0 is a real eigenvalue, the vector $\mathbf{f}$ is also real. With the use of (10) we obtain a formula describing splitting of the double eigenvalue , see Seyranian (1991, 1993)

$$\lambda = \lambda_0 \pm \sqrt{\varepsilon\langle \mathbf{f}, \mathbf{e}\rangle} + O(\varepsilon), \tag{11}$$

where $\langle \mathbf{a}, \mathbf{b}\rangle = \sum_{s=1}^{2} a_s b_s$ is the scalar product of the vectors $\mathbf{a}$, $\mathbf{b} \in R^2$ in the parameter space. Formula (11) is valid until the radicand is not zero. If $\langle \mathbf{f}, \mathbf{e}\rangle < 0$, then the double λ_0 splits into a complex conjugate pair (flutter instability). And if $\langle \mathbf{f}, \mathbf{e}\rangle > 0$ two positive eigenvalues appear (stability). Therefore, the vector $\mathbf{f}$ is a normal vector to the stability boundary lying in the stability domain, Fig 1a.

Let us consider *a degenerate* case

$$\langle \mathbf{f}, \mathbf{e}_*\rangle = 0, \tag{12}$$

This condition means that the curves $\mathbf{p}(\varepsilon) = \mathbf{p}_0 + \varepsilon\mathbf{e}_* + \varepsilon^2\mathbf{d}$ are tangent to the stability boundary as $\varepsilon \to 0$. Then in expansions (8) the terms with fractional powers disappear, see Vishik and Liusternik (1960)

$$\lambda = \lambda_0 + \varepsilon\lambda_2 + \varepsilon^2\lambda_4 + \ldots; \; \mathbf{u} = \mathbf{u}_0 + \varepsilon\mathbf{w}_2 + \varepsilon^2\mathbf{w}_4 + \ldots \tag{13}$$

The coefficient λ_2 is a solution of the following quadratic equation, see Section 1

$$\lambda_2^2 + \lambda_2 a_1 + a_2 = 0, \tag{14}$$

with the coefficients

$$a_1 = -(\mathbf{A}_1\mathbf{u}_0, \mathbf{v}_1) - (\mathbf{A}_1\mathbf{u}_1, \mathbf{v}_0), \; a_2 = -(\mathbf{A}_2\mathbf{u}_0, \mathbf{v}_0) + \left(\mathbf{G}_0(\mathbf{A}_1\mathbf{u}_0), \mathbf{A}_1^{\mathrm{T}}\mathbf{v}_0\right). \tag{15}$$

Here the operator $\mathbf{G}_0$, inverse to $\mathbf{A}_0 - \lambda_0\mathbf{I}$, is defined by the expressions

$$(\mathbf{A}_0 - \lambda_0\mathbf{I})\psi = \varphi, \quad \psi = \mathbf{G}_0\varphi,$$

where φ, $\psi \in C^m$, and it is supposed that the solvability condition $(\varphi, \mathbf{v}_0) = 0$ is satisfied. In our case the solvability condition has the form $(\mathbf{A}_1\mathbf{u}_0, \mathbf{v}_0) = 0$, and it is valid for degenerate directions (12) of the vector $\mathbf{e}$. Note that the matrix $\mathbf{G}_0$ can be given in the form, see Section 1

$$\mathbf{G}_0 = \left(\mathbf{A}_0 - \lambda_0\mathbf{I} - \overline{\mathbf{v}_0}\mathbf{v}_0^{\mathrm{T}}\right)^{-1}$$

Using expressions (7) for the matrices $\mathbf{A}_1$ and $\mathbf{A}_2$ the coefficients a_1 and a_2 can be written as

$$a_1 = \langle \mathbf{h}, \mathbf{e}_* \rangle, \quad a_2 = \langle \mathbf{H}\mathbf{e}_*, \mathbf{e}_* \rangle - \langle \mathbf{f}, \mathbf{d} \rangle, \tag{16}$$

where the real vector $\mathbf{h}$ and the matrix $\mathbf{H}$ are defined by (15).

3.3 Metamorphosis Equation

We substitute expressions (16) into equation (14) and then multiply the result by ε^2. Introducing the notation $\Delta\lambda = \lambda_2 \varepsilon$ we get

$$\Delta\lambda^2 + \varepsilon \langle \mathbf{h}, \mathbf{e}_* \rangle \Delta\lambda + \varepsilon^2 \langle \mathbf{H}\mathbf{e}_*, \mathbf{e}_* \rangle = \varepsilon^2 \langle \mathbf{f}, \mathbf{d} \rangle. \tag{17}$$

Note that no restrictions were imposed on the vector $\mathbf{d}$. It is convenient to consider this vector as a collinear to the normal vector $\mathbf{f}$, i.e. put $\mathbf{d} = \gamma \mathbf{f}$. Denoting $\rho = \gamma |\mathbf{f}| \varepsilon^2$ we see that ε and ρ are the coordinates of $\Delta\mathbf{p} = \mathbf{p} - \mathbf{p}_0$ in the orthonormalized basis of the vectors $\mathbf{e}_*$ and $\mathbf{f}/|\mathbf{f}|$, related by means of orthogonality condition (12). Indeed, $\Delta\mathbf{p} \equiv \varepsilon \mathbf{e}_* + \varepsilon^2 \mathbf{d} = \varepsilon \mathbf{e}_* + \rho \mathbf{f}/|\mathbf{f}|$, Fig 1a.

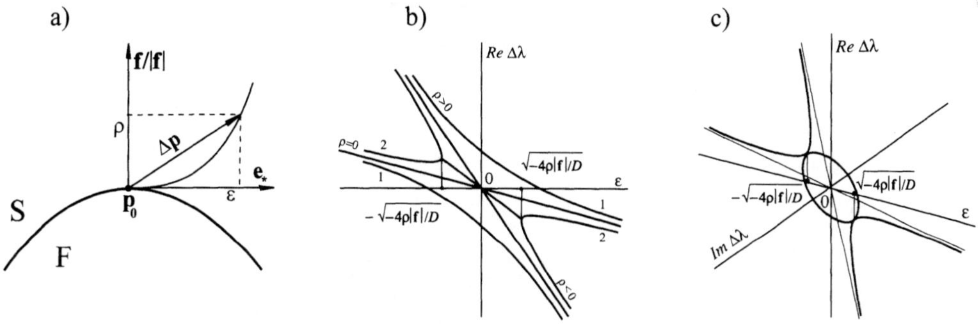

Figure 1. Metamorphosis of frequency curves near flutter boundary.

Now we transform the right hand side of equation (17). We have

$$\varepsilon^2 \langle \mathbf{f}, \mathbf{d} \rangle = \langle \mathbf{f}, \Delta\mathbf{p} \rangle = \rho |\mathbf{f}|. \tag{18}$$

Substituting (18) into (17) finally yields

$$\Delta\lambda^2 + \varepsilon \langle \mathbf{h}, \mathbf{e}_* \rangle \Delta\lambda + \varepsilon^2 \langle \mathbf{H}\mathbf{e}_*, \mathbf{e}_* \rangle = \rho |\mathbf{f}|. \tag{19}$$

This is the metamorphosis equation for splitting of the double eigenvalue λ_0 due to variation of the parameters ε and ρ near the stability boundary. In particular, the approximate equation for the stability boundary in the vicinity of the point $\mathbf{p} = \mathbf{p}_0$ follows from (19). Really, the boundary between stability and flutter domains consists of the points corresponding to matrices containing positive double eigenvalues with the Jordan chain of length 2. Therefore, the discriminant of quadratic equation (19) has to be zero at the boundary points. This condition gives a quadratic approximation of the boundary in the local coordinates ε and ρ

$$\rho = -\varepsilon^2 \frac{D}{4|\mathbf{f}|}, \tag{20}$$

$$D = \langle \mathbf{h}, \mathbf{e}_* \rangle^2 - 4\langle \mathbf{H}\mathbf{e}_*, \mathbf{e}_* \rangle. \tag{21}$$

Parabola (20) has convexity downward $(D < 0)$, or upward $(D > 0)$ depending on the sign of D, Fig 1a. It immediately follows from equation (19) that the flutter instability domain is defined by the inequality

$$\rho < -\varepsilon^2 \frac{D}{4|\mathbf{f}|}. \tag{22}$$

If the double λ_0 at the point $\mathbf{p_0}$ is less than zero, then equation (20) describes the boundary between flutter and divergence domains, and inequality (22) as earlier defines the flutter domain.

At any fixed value of ρ equation (19) describes behavior of two eigenvalues λ along a line, parallel to the tangent direction to the boundary at the point $\mathbf{p_0}$. The change of the sign of the parameter ρ means passing through the flutter boundary, Fig 1a. It is natural to expect that such a change will be accompanied by qualitative changes in behavior of the frequency curves $\lambda(\varepsilon)$. Equation (19) allows to investigate metamorphoses of the frequency curves both qualitatively and quantitatively using information only at the point $\mathbf{p} = \mathbf{p_0}$.

Now we take perfect square of the left-hand side of (19) and study the cases of positive and negative values of the discriminant D defined in (21)

$$\left(\Delta\lambda + \frac{1}{2}\varepsilon\langle \mathbf{h}, \mathbf{e}_* \rangle\right)^2 - \left(\frac{\varepsilon\sqrt{D}}{2}\right)^2 = \rho|\mathbf{f}|, \quad D > 0; \tag{23}$$

$$\left(\Delta\lambda + \frac{1}{2}\varepsilon\langle \mathbf{h}, \mathbf{e}_* \rangle\right)^2 + \left(\frac{\varepsilon\sqrt{-D}}{2}\right)^2 = \rho|\mathbf{f}|, \; D < 0. \tag{24}$$

Consider equation (23) when $D > 0$ and the flutter domain is convex, Fig 1a. If $\rho > 0$, then for any ε the eigenvalues λ are real and are situated on the hyperbola 1, Fig 1b. When the parameter ρ tends to zero the two branches of hyperbola 1 approach each other, so that at $\rho = 0$ the eigenvalues are located on two real asymptotes $\operatorname{Re}\Delta\lambda(\varepsilon) = -\frac{1}{2}\varepsilon\left(\langle \mathbf{h}, \mathbf{e}_* \rangle \pm \sqrt{D}\right)$.

For $\rho < 0$ the set of real solutions of equation (23) consists of two branches of the adjacent hyperbola 2, Fig 1b, while for ε satisfying the inequality $\varepsilon^2 < -4\rho|\mathbf{f}|/D$ the eigenvalues λ are complex and belong to an ellipse forming the so called *bubble of instability*, see

Seyranian (1991, 1993). The projections of the bubble on the real and imaginary planes are given by the equations

$$\operatorname{Re}\Delta\lambda = -\frac{\varepsilon}{2}\langle \mathbf{h}, \mathbf{e}_* \rangle, \tag{25}$$

$$(\operatorname{Im}\Delta\lambda)^2 + \left(\frac{\varepsilon\sqrt{D}}{2}\right)^2 = -\rho|\mathbf{f}|. \tag{26}$$

The ellipse and the two branches of hyperbola 2 have common points at $\varepsilon_{1,2} = \pm\sqrt{-4\rho|\mathbf{f}|/D}$, Fig 1c, at which the matrix $\mathbf{A}$ acquires double real eigenvalues

$$\lambda_{1,2} = \lambda_0 \mp \frac{1}{2}\langle \mathbf{h}, \mathbf{e}_* \rangle\sqrt{-4\rho|\mathbf{f}|/D}.$$

Hyperbola (23) and ellipse (25), (26) lie in the orthogonal planes since the projection of the ellipse (25) on the real plane is a straight line, Fig 1b. Therefore, with a change of the parameter ε the eigenvalues λ bifurcate at the points ε_1 and ε_2: two eigenvalues being in the same plane merge, and then come out of the plane in the orthogonal direction to it. Such behavior of eigenvalues is called *strong interaction*, see Seyranian (1991,1993), and is typical for passing through the flutter boundary. Note that the inequality giving the *bubble of instability* exactly coincides with the approximation of flutter domain (22). Thereby, the overlapping phenomenon is closely related to convexity properties of the flutter domain.

Consider now the case $D < 0$. Then families of ellipses (24) describe metamorphoses of frequency curves. There are no real solutions $\Delta\lambda(\varepsilon)$ if $\rho < 0$ (flutter). At $\rho = 0$ the set of real solutions consists of an isolated point. If $\rho > 0$, then the set of real solutions $\Delta\lambda(\varepsilon)$ is ellipse (24), whose boundaries are defined by the inequality $\varepsilon^2 < 4\rho|\mathbf{f}|/(-D)$. It is easy to see that such metamorphosis corresponds to a concave flutter domain (22).

Analogously, using variations along the curves $\mathbf{p}(\varepsilon) = \mathbf{p}_0 + \varepsilon\mathbf{e} + \varepsilon^2\mathbf{d}$ it is possible to investigate metamorphoses of frequency curves near the boundary between the stability and divergence domains.

It should be also noted that the basic equations, describing bifurcations of eigenvalues (11), (17) and metamorphoses of frequency curves (23), (24), are valid also for linear differential operators $\mathbf{A}$ with homogeneous linear boundary conditions depending on parameters. The difference is that the vectors $\mathbf{f}$, $\mathbf{h}$ and the matrix $\mathbf{H}$ are defined through eigenfunctions and associated functions of the operator and through derivatives of the differential expression and boundary forms with respect to parameters.

3.4 Mechanical Example

As an example, we consider a simple problem of stability of a rigid panel of infinite span oscillating in airflow, see Panovko and Gubanova (1965). The panel is maintained on two elastic supports with stiffness coefficients c_1 and c_2 per unit span having two degrees of freedom: a vertical displacement y and a rotation angle φ, Fig 2a. Small oscillations of the panel are described by the following equations given in the nondimensional form, see Kirillov and Seyranian (1999)

$$\begin{pmatrix} \ddot{y} \\ \ddot{\varphi} \end{pmatrix} + \mathbf{A}\begin{pmatrix} y \\ \varphi \end{pmatrix} = 0, \; \mathbf{A} = \begin{bmatrix} 1 & c-q \\ 12c & 3-3q \end{bmatrix}, \tag{27}$$

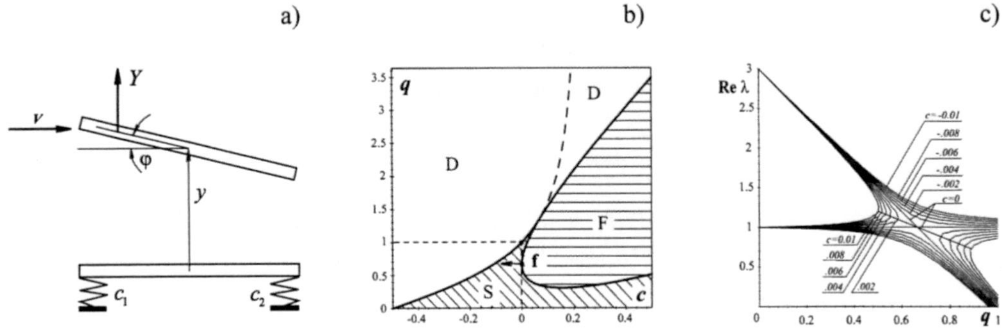

Figure 2. Stability of vibrations of a rigid plate in a flow.

where $q = 1/2\, c_y^a \rho v^2 (c_1 + c_2)^{-1}$ is the load parameter proportional to the dynamic pressure of the flow, and $c = 1/2(c_1 - c_2)(c_1 + c_2)^{-1}$ is the parameter characterizing relation between the support stiffnesses. Thus, circulatory system (27) depends on a vector of two parameters $\mathbf{p} = (c, q)$. From physical point of view it follows that $q \geq 0$, and $-0.5 \leq c \leq 0.5$.

Finding a solution to equation (27) in the form $\begin{pmatrix} y \\ \varphi \end{pmatrix} = \mathbf{u}e^{i\omega\tau}$ we obtain eigenvalue problem (2). The corresponding characteristic equation is

$$\lambda^2 + (3q-4)\lambda + 12cq - 3q - 12c^2 + 3 = 0, \; \lambda = \omega^2. \tag{28}$$

Equations for the curves subdividing the plane of parameters c and q into stability, flutter, and divergence domains follow directly from (28), Fig 2b, with the boundaries given by

$$q_d(c) = \frac{1-4c^2}{1-4c}, \; q_f(c) = \frac{2}{3}\left(1 + 4c \pm 2\sqrt{c(c+2)}\right). \tag{29}$$

Consider a point on the boundary between stability and flutter domains $(c = 0,\; q = 2/3)$, corresponding to the double eigenvalue $\lambda = 1$, Fig 2b. Characteristic equation (28) can be easily transformed into

$$\left(\lambda - 1 + \frac{3}{2}q - 1\right)^2 - \left(-4c + \frac{3}{2}q - 1\right)^2 = -8c - 4c^2. \tag{30}$$

For $c = 0$ equation (30) has two solutions

$$\lambda = 1, \;\; \lambda = 3 - 3q. \tag{31}$$

Two lines (31) intersect at the point $(q = 2/3, \; \lambda = 1)$. If $c \neq 0$, then equation (30) describes a family of hyperbolae with asymptotes (31), Fig 2c. For small $c < 0$ and $0 \leq q \leq 1$ solutions $\lambda(q)$ of equation (30) belong to the real plane. One of the two eigenvalues remains positive for any q, $0 \leq q \leq 1$, while another eigenvalue changes its sign at some $q_d < 1$, Fig 2b,c. Thus, for $c < 0$ and rather large values of q system (27) loses stability statically (divergence). Change of the sign of the parameter c results in metamorphosis of frequency curves accompanied by the origination of a complex eigenvalues region. After metamorphosis the system loses stability at the values q_f, when two positive λ merge with origination of a double eigenvalue with the Jordan chain of length 2 (flutter).

Now we show that metamorphosis equation (30) can be approximated by formula (23) with the coefficients calculated only by using information on the system at the point $\mathbf{p}_0 = (0, 2/3)$. The eigenvectors and associated vectors of the double eigenvalue $\lambda = 1$ are

$$\mathbf{u}_0 = \begin{pmatrix} 1 \\ 0 \end{pmatrix}, \; \mathbf{v}_0 = \begin{pmatrix} 0 \\ -2/3 \end{pmatrix}, \; \mathbf{u}_1 = \begin{pmatrix} 0 \\ -3/2 \end{pmatrix}, \; \mathbf{v}_1 = \begin{pmatrix} 1 \\ 0 \end{pmatrix} \tag{32}$$

Formulae (10) and (12) give the normal vector $\mathbf{f}$ to the flutter boundary at the point $\mathbf{p}_0 = (0, 2/3)$, and therefore, a tangent vector $\mathbf{e}_*$

$$\mathbf{f} = \begin{pmatrix} -8 \\ 0 \end{pmatrix}, \; \mathbf{e}_* = \begin{pmatrix} 0 \\ 1 \end{pmatrix}. \tag{33}$$

Finding the vector $\mathbf{h}$ and matrix $\mathbf{H}$ by formulae (15), (16) and using expression (21) we get $\langle \mathbf{h}, \mathbf{e}_* \rangle = 3$, $D = 9$. The vector of variation of parameters is $\Delta\mathbf{p} = (c, q - 2/3)$. Thus, the parameter ρ in equations (23), (24) is $\langle \mathbf{f}, \Delta\mathbf{p} \rangle / |\mathbf{f}| = -c$. Since $D > 0$, the metamorphosis of the frequency curves is described by equation (23), which in this case takes the form

$$\left(\lambda - 1 + \frac{3}{2}q - 1 \right)^2 - \left(\frac{3}{2}q - 1 \right)^2 = -8c. \tag{34}$$

Comparison of precise equation (30) with its approximation (34) shows that the asymptotes $\lambda = 1$, $\lambda = 3 - 3q$ coincide completely in both equations. Approximations of the frequency curves are rather good for small values of the parameter c. The quadratic approximation of the flutter domain near the point $\mathbf{p}_0 = (0, 2/3)$ is given by formula (22)

$$c > \frac{9}{32} \left(q - \frac{2}{3} \right)^2. \tag{35}$$

Approximation (35) means convex flutter domain, Fig 2b, which is in good agreement with the precise expression $q_f(c)$ of flutter boundary (29).

4 The Keldysh Problem on Aeroelastic Stability of a Wing with Bracing Struts

4.1 Introduction

The aeroelastic stability of the unswept high-aspect-ratio wing braced by struts has been considered by Keldysh (1938). The strut is supposed to be a perfectly rigid rod connecting the point P of the wing with the fuselage, see Fig 1. The presence of the strut implies that P is a fixed point, which imposes extra boundary conditions on the functions describing the vibration modes.

For the particular case of a rectangular wing with a single strut attached to the stiffness axis of the wing (an A-type strut), Keldysh (1938) made calculations and came to the conclusion that "at about $h = 0.47l$, the critical speed becomes imaginary; consequently, vibrations of the wing with the strut turn out to be impossible at $h > 0.47l$". Evidently, it means that the wing becomes stable. Keldysh (1938) made a similar conclusion also for the B-type strut (two struts bracing the cross section h of the wing and holding it fixed during vibrations): "at $h/l > 0.8$, the critical speed for the wing with a B-type strut does not exist."

These conclusions, which Keldysh made on the basis of the Bubnov-Galerkin one-term approximation, appear to be incorrect. In this Section, based on the papers by Mailybaev and Seyranian (1996, 1998), the problem of aeroelastic stability of the wing is reduced to the study of behavior of eigenvalues λ for linearized equations of motion of the wing on the complex plane as functions of the flow speed V and the distance h between the point of the wing, the strut is attached to, and the wing base. In this way, the critical speeds for the vibrational (flutter) and static (divergence) types of loss of stability are determined, and the domains of stability, flutter, and divergence are plotted on the h - V plane. It is shown that in the case of the A-type strut, the flutter instability is replaced by the static form, the divergence, and at $h = 0.471l$ the critical speed has a discontinuity: its value jumps from $V_f = 55.7$ m/s to $V_d = 61.3$ m/s, see Fig 3. In the case of the B-strut, the critical speed turns out to be finite and continuous and, at $h = 0.76\ l$, reaches the maximum $V_{cr} = 119$ m/s that is almost four times greater than the critical speed of the unbraced wing $V_{cr} = 30.3$ m/s, see Fig 4. The performed calculations show that the use of the struts can effectively improve the aeroelastic stability characteristics.

4.2 Aeroelastic Stability Problem

Let us consider vibrations of a thin high-aspect-ratio wing braced by an A-type strut, see Fig 1, in the airflow. The wing is represented by an elastic beam subjected to torsion and bending with a straight elastic axis Oy (stiffness axis) perpendicular to the fuselage. Deformation of the wing is described by the deflection function $w(y,t)$ and the angle of rotation $\theta(y,t)$ about the elastic axis, where t is the time. The linearized equations of motion of the wing in the flow have the form, see Grossman (1937), and Fung (1955):

$$\frac{\partial^2}{\partial y^2}\left(EI\frac{\partial^2 w}{\partial y^2}\right)+m\frac{\partial^2 w}{\partial t^2}-m\sigma\frac{\partial^2\theta}{\partial t^2}=L_a$$
$$-\frac{\partial}{\partial y}\left(GJ\frac{\partial\theta}{\partial y}\right)-m\sigma\frac{\partial^2 w}{\partial t^2}+I_m\frac{\partial^2\theta}{\partial t^2}=M_a \tag{1}$$

In these equations EI and GJ are the bending and torsion stiffnesses of the wing, m and I_m are the specific (per unit length of the span) mass and moment of inertia with respect to the elastic axis, and σ is the distance between the center of stiffness and the center of gravity of the cross

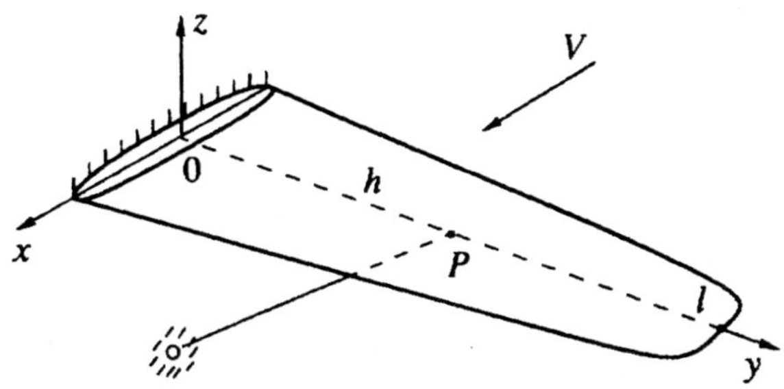

Figure 1. Wing braced by a strut.

section of the wing. The aerodynamic force L_a and moment M_a per unit span are determined on the basis of the steady-state hypothesis, see Grossman (1937), and Fung (1955). The expressions for L_a and M_a have the form

$$L_a=C_y^\alpha\rho V^2 b\left[\theta+\frac{b}{V}\left(\frac{3}{4}-\frac{x_0}{b}\right)\frac{\partial\theta}{\partial t}-\frac{1}{V}\frac{\partial w}{\partial t}\right]$$
$$M_a=C_m^\alpha\rho V^2 b^2\left[\theta+\frac{b}{V}\left(\frac{3}{4}-\frac{x_0}{b}-\frac{\pi}{16C_m^\alpha}\right)\frac{\partial\theta}{\partial t}-\frac{1}{V}\frac{\partial w}{\partial t}\right] \tag{2}$$

where b is the chord of the wing, x_0 is the distance between the front edge and the elastic axis, V is the flow speed, ρ is the air density, and C_y^α and C_m^α are the aerodynamic coefficients.

Assuming that the wing is a cantilever, clamped at the fuselage, we write the boundary conditions for the functions w and θ at the clamped ($y = 0$) and free ($y = l$) ends of the wing in the form:

$$y=0:\quad w=\frac{\partial w}{\partial y}=\theta=0,$$
$$y=l:\quad EI\frac{\partial^2 w}{\partial y^2}=\frac{\partial}{\partial y}\left(EI\frac{\partial^2 w}{\partial y^2}\right)=GJ\frac{\partial\theta}{\partial y}=0 \tag{3}$$

From the assumption that the wing is braced by the strut at the point P lying on the elastic axis, and from the continuity conditions for the angle of torsion, the slope of the deflection function, and the torsional and bending moments in the h-section, we have, see Keldysh (1938)

$$y=h:\ w_-=w_+=0\ ,\theta_-=\theta_+\ ,\left(\frac{\partial w}{\partial y}\right)_-=\left(\frac{\partial w}{\partial y}\right)_+ ,$$
$$\left(GJ\frac{\partial\theta}{\partial y}\right)_-=\left(GJ\frac{\partial\theta}{\partial y}\right)_+ ,\left(EI\frac{\partial^2 w}{\partial y^2}\right)_-=\left(EI\frac{\partial^2 w}{\partial y^2}\right)_+ \tag{4}$$

Note that + and – correspond to the right and left limits when y tends to h.

The system of equations (1)-(4) is a linear homogeneous boundary value problem. We seek a solution of equations (1)-(4) in the form

$$w(y,t)=f(y)\exp(\lambda t)\,,\ \ \theta(y,t)=\varphi(y)\exp(\lambda t) \tag{5}$$

where λ is an eigenvalue, and $f(y)$ and $\varphi(y)$ are eigenfunctions. Substituting (5) in (1)-(4), we obtain a system of ordinary differential equations on the functions $f(y)$ and $\varphi(y)$ in the form

$$\begin{pmatrix} L_{11} & L_{12} \\ L_{21} & L_{22}\end{pmatrix}\begin{pmatrix} f \\ \varphi\end{pmatrix}=0 \tag{6}$$

where L_{ij} are the linear differential operators depending on λ and given by

$$\begin{aligned}
L_{11}&=\frac{d^2}{dy^2}\left(EI\frac{d^2}{dy^2}\right)+m\lambda^2+\lambda VC_y^\alpha\rho b\,,\\
L_{12}&=-m\sigma\lambda^2-C_y^\alpha\rho V^2 b-C_y^\alpha\rho\lambda Vb^2\left(\frac{3}{4}-\frac{x_0}{b}\right),\\
L_{21}&=-m\sigma\lambda^2+C_m^\alpha\lambda\rho Vb^2\,,\\
L_{22}&=-\frac{d}{dy}\left(GJ\frac{d}{dy}\right)+I_m\lambda^2-C_m^\alpha\rho V^2b^2-C_m^\alpha\rho\lambda Vb^3\left(\frac{3}{4}-\frac{x_0}{b}-\frac{\pi}{16C_m^\alpha}\right)
\end{aligned} \tag{7}$$

Boundary conditions for the functions f and φ are the same as (3) and (4) for the functions w and θ, respectively.

Due to the fact that the problem is nonconservative (non-selfadjoint) the eigenvalues are in general complex quantities $\lambda=\alpha+i\omega$. Depending on the flow speed V, the amplitudes of solutions (5), as functions of the time, can decrease ($\operatorname{Re}\lambda<0$, stability), be constant ($\operatorname{Re}\lambda=0$, stability boundary), or increase ($\operatorname{Re}\lambda>0$, instability). The flutter critical speed V_f is defined by the relations $\operatorname{Re}\lambda=0$ and $\operatorname{Im}\lambda=\omega\neq 0$, where ω is the flutter frequency, and the divergence critical speed V_d is defined by the equality $\lambda=0$. The critical speed of the loss of stability of the system V_{cr} is equal to the lower of the speeds V_f and V_d.

For the bracing strut of type A the divergence critical speed can be found directly from equations (6), (7) with the boundary conditions (3) and (4) setting there $\lambda = 0$. As a result, we arrive at the following eigenvalue problem:

$$\frac{d}{dy}\left(GJ\frac{d\varphi}{dy}\right) + C_m^\alpha \rho V_d{}^2 b^2 \varphi = 0\,, \tag{8}$$

$$\begin{gathered}\varphi(0) = 0\,,\quad \left(GJ\frac{d\varphi}{dy}\right)_{y=l} = 0\,,\\ y = h:\quad \varphi_- = \varphi_+\,,\quad \left(GJ\frac{d\varphi}{dy}\right)_- = \left(GJ\frac{d\varphi}{dy}\right)_+\end{gathered} \tag{9}$$

This is a selfadjoint eigenvalue problem with V_d^2 as an eigenvalue. Hence, the critical divergence speed is the minimum eigenvalue of problem (8), (9). Using variational formulation, we find

$$V_d^2 = \min_{\varphi} \frac{\int_0^l GJ(d\varphi/dy)^2 dy}{\int_0^l C_m^\alpha \rho b^2 \varphi^2 dy} \tag{10}$$

where the trial function φ is continuously differentiable and has to satisfy only the first of the boundary conditions (9). The remaining boundary conditions are natural for functional (10). From (10) it follows that V_d^2 is independent on the location of the strut *h*. This is natural to expect due to the above assumption that the bracing point on the wing lies on the elastic axis and, consequently, does not affect the wing torsion.

To solve eigenvalue problem (6), we use the Bubnov-Galerkin method. With this aim, we choose two systems of linearly independent coordinate functions $f_1,..., f_n$ and $\varphi_1,..., \varphi_n$ that are differentiable on the intervals $(0, h)$ and (h, l) four and two times, respectively, and satisfy boundary conditions (3) and (4) of the problem. In accordance with the Bubnov-Galerkin method, the eigenfunctions f and φ of system (6) may be represented in the form of linear combinations of the coordinate functions with unknown coefficients a_j and b_j, respectively:

$$f(y) = \sum_{j=1}^{n} a_j f_j(y), \qquad \varphi(y) = \sum_{j=1}^{n} b_j \varphi_j(y) \tag{11}$$

Upon substituting these expansions in equations (6), multiplying the left-hand sides of these equations by f_k and φ_k, $k = 1,..., n$, respectively, and integrating with respect to y from 0 to l, we obtain $2n$ linear homogeneous equations on the coefficients a_j and b_j which constitute an algebraic eigenvalue problem of the form

$$[\lambda^2 \mathbf{M} + \lambda V \mathbf{C} + \mathbf{K} + V^2 \mathbf{D}]\mathbf{c} = \mathbf{0} \tag{12}$$

where $\mathbf{c}$ is the column vector, $\mathbf{c}^{\mathrm{T}} = (a_1,..., a_n, b_1,..., b_n)$, consisting of unknown coefficients, and **M, C, K, D** are the square matrices of order $2n$. A nontrivial solution to problem (12) exists only if

$$\det[\lambda^2\mathbf{M} + \lambda V\mathbf{C} + \mathbf{K} + V^2\mathbf{D}] = 0 \tag{13}$$

This equation serves to determine the eigenvalues λ as functions of the flow speed V.

4.3 Behavior of Eigenvalue Branches in the Complex Plane

Using the method described above, we studied the stability of the rectangular wing braced with the A-type strut as a function of the parameters V and h. We took the same input data as Keldysh (1938). The expansions (11) were realized using from five to seven functions that were chosen among the eigenfunctions of pure bending-torsional vibrations of the wing with bracing strut in the vacuum (at $V = 0$).

Figure 2 shows some of trajectories of motion of the eigenvalues on the complex plane for typical values of h when the speed V varies in the range from 0 to 155 m/s. The arrows in Fig 2 indicate the direction in which the speed increases. As the matrices in equation (9) are real, the eigenvalues lie symmetrically relative to the real axis in the complex plane. The modes are numbered according to the numbers of eigenfrequencies of the conservative system at $V = 0$. As calculations showed, the branches corresponding to higher modes, which are not shown, are located on the left half of the complex plane for all h, i.e. they are stable. The numbers at the imaginary axis are the values of the critical speed for the corresponding mode.

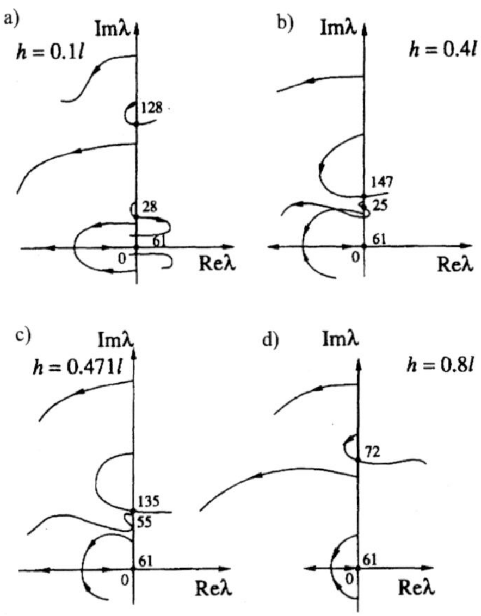

Figure 2. Behavior of eigenvalue branches $\lambda(V)$ depending on h.

For small h, a loss of stability occurs in the first, second, and fourth modes, Fig 2a. The first mode is divergent: two complex-conjugate eigenvalues approach each other, collide, and then diverge in opposite directions along the real axis (strong interaction). At the speed $V_d = 61.3$ m/s, one of them crosses the imaginary axis. The second and fourth modes correspond to flutter: the appropriate eigenvalues cross the imaginary axis at the points with $\operatorname{Im}\lambda \neq 0$ at V_f

equal to 28 and 128 m/s, respectively. Thus, the second mode is a critical one. Fig 2a shows that the second mode branch crosses the imaginary axis twice. It means that starting with a certain value of the speed, this mode becomes stable again.

With an increase of *h,* the following changes are observed, see Fig 2b: the fourth mode becomes stable, and the third unstable (V_f =147 m/s). The second mode branch shifts to the left. As in the first case, the critical speed is a flutter speed of the second mode. With further increase of *h*, the second mode branch continues to come into the left half-plane, so that at $h = 0.471l$ (Fig 2c), it only touches (but does not intersect) the imaginary axis, and then, it becomes straighter (Fig 2d). On the contrary, the branch, corresponding to the third mode, shifts to the right with the value of V_f decreasing from 135 m/s at $h = 0.471l$ to 72 m/s at $h = 0.8l$. Starting with the value $h = 0.471l$, that corresponds to the transition of the second mode branch into the region $\mathrm{Re}\,\lambda < 0$ (stability), the divergence speed of the first mode V_d = 61.3 m/s becomes the critical one.

The phenomenon of transference of instability between branches, which is connected with a double eigenvalue with only one eigenvector, was revealed at $h = 0.145l$ and $h = 0.935l$. This phenomenon happens in linear vibrational systems depending on multiple parameters, see Seyranian (1994), and Seyranian and Pedersen (1995): two eigenvalues approach each other in the complex plane, collide at the point of a double λ, and then diverge at right angles to the line of their approach. In this case, the fourth and fifth modes interact at $h = 0.145l$, and the second and third modes interact at $h = 0.935l$.

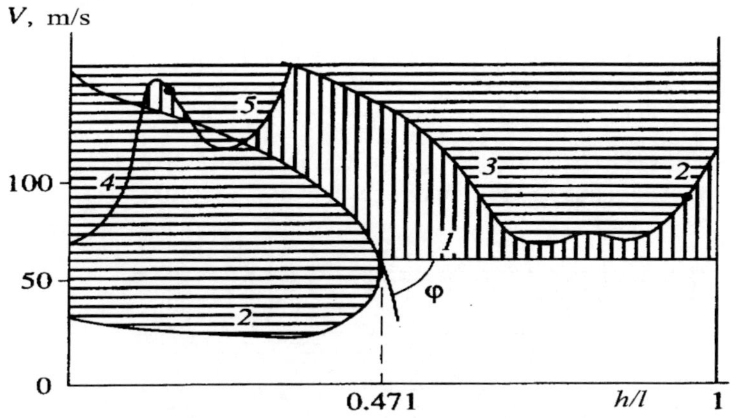

Figure 3. Stability, flutter and divergence domains for bracing strut *A*.

The results of the calculations allowed us to plot the domains of stability, flutter, and divergence on the plane of parameters *V* and *h,* see Fig 3. Note that divergence instability corresponds only to real unstable eigenvalues, while flutter means existence of at least one pair unstable complex eigenvalue. The flutter and divergence domains are hatched with horizontal and vertical lines, respectively. The numbers indicate the number of the mode that becomes unstable on transition through the corresponding boundary.

The flutter domain is convex at $h \leq 0.471l$ and has a vertical tangent $h = 0.471l$. At the intersection, the flutter and divergence domains form a sharp angle φ in the stability domain.

This implies a discontinuity of the critical speed at $h = 0.471l$ and the existence of a small stability domain at $V > V_{cr}$ within the small range $0.470l < h < 0.471l$, Fig 3. The discontinuity takes place at the transition of the second (critical) mode branch into the stable half-plane $\mathrm{Re}\,\lambda < 0$. The discontinuity point corresponds to the critical mode branch touching the imaginary axis at $h = 0.471l$ with the critical flutter speed $V_f = 55.7$ m/s , Fig 2c.

Comparing our results with the results of Keldysh (1938, Fig 4), we conclude that they are in a good agreement at $h < 0.471l$. Whereas at $h > 0.471l$, the critical speed has a discontinuity and becomes equal to the critical divergence speed rather than disappearing, as it was stated in Keldysh (1938).

Keldysh (1938) did not study the divergence instability. If the strut is attached to the elastic axis of a wing, the divergence critical speed does not depend on h and, in the case of a rectangular wing, can be evaluated analytically

$$V_d = \frac{\pi}{2bl}\sqrt{\frac{GJ}{C_m^\alpha \rho}} = 61.3 m/s \tag{14}$$

We note that a flutter domain also exists at $h > 0.471l$, it is hatched with horizontal lines in Fig 3.

Let us investigate now the stability of the wing braced with the B-type strut that fixes the cross section of the wing at $y = h$. In this case, the boundary conditions take the form, see Keldysh (1938):

$$y = h: \; w_- = w_+ = 0 \;, \theta_- = \theta_+ = 0 \,,$$
$$\left(\frac{\partial w}{\partial y}\right)_- = \left(\frac{\partial w}{\partial y}\right)_+ , \left(EI\frac{\partial^2 w}{\partial y^2}\right)_- = \left(EI\frac{\partial^2 w}{\partial y^2}\right)_+ \tag{15}$$

The critical speed was calculated for the rectangular wing, see Keldysh (1938), and the results are presented in Fig 4. It turns out that, for any position of the strut $0 \le h \le l$, the wing loses

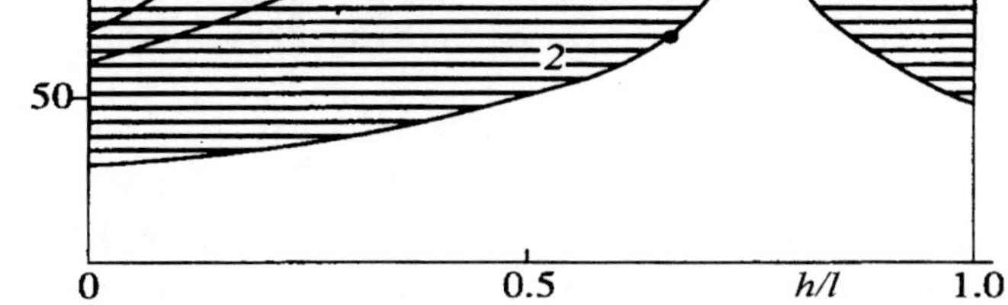

Figure 4. Stability, flutter and divergence domains for bracing strut *B*.

stability by flutter: at $0 \le h \le 0.65l$ and $0.74l \le h \le l$ the second mode is critical, and at

$0.65l \le h \le 0.74l$ the stability is lost owing to the third mode. The transference of instability between branches (modes) takes place at the points $h = 0.65l$ and $h = 0.74l$. With an increase of h from zero, the flutter critical speed increases reaching a maximum at $h = 0.76l$, and then monotonically decreases (Fig 4). The maximum of the critical speed is equal to V_f =119 m/s, which is about four times greater than the critical speed of the unbraced wing.

Comparison with the results of Keldysh (1938, Fig 8) shows that the results at $0 \le h \le 0.7l$ are in satisfactory agreement, but at $h > 0.7l$, the results disagree. So, the conclusion that "the wing becomes nonvibratory at $h > 0.8l$", see Keldysh (1938), is not confirmed. The results obtained evidence that use of the struts, especially of the B-type, can effectively improve the aeroelastic stability characteristics of the wings.

4.4 Generalization

In this section we give a general description and explanation of the effect of "flutter disappearance" with a jump in the critical flutter speed presented above. For this purpose we consider an eigenvalue problem for a real nonsymmetrical matrix $\mathbf{A}$ of dimension $n \times n$, smoothly depending on two real parameters h and V

$$\mathbf{Au} = \lambda \mathbf{u} \tag{16}$$

Consider behavior of an eigenvalue branch $\lambda(V)$ on the complex plane with a change of the second parameter h, arrows in Fig 5 show motion of the eigenvalue λ with increase of V. We assume that at the point $h = h_0, V = V_0$ in the parameter space we have a simple eigenvalue λ_0, and the following conditions are satisfied

$$h = h_0\,,\, V = V_0: \qquad \mathbf{Re}\lambda_0 = 0\,, \quad \frac{\partial\, \mathbf{Re}\lambda}{\partial V} = 0\,, \quad \frac{\partial^2\, \mathbf{Re}\,\lambda}{\partial V^2} < 0, \quad \frac{\partial\, \mathbf{Re}\lambda}{\partial h} < 0 \tag{17}$$

The first three conditions mean that this is a transition (bifurcation) point on the complex plane, and the last condition implies that the flutter instability disappears at $h > h_0$, see Fig 5. Since λ_0 is the simple eigenvalue, it is a differentiable function. Hence, we can write a Taylor expansion

$$\lambda = \lambda_0 + \frac{\partial \lambda}{\partial h}\Delta h + \frac{\partial \lambda}{\partial V}\Delta V + \frac{1}{2}\frac{\partial^2 \lambda}{\partial h^2}(\Delta h)^2 + \frac{\partial^2 \lambda}{\partial h \partial V}\Delta h \Delta V + \frac{1}{2}\frac{\partial^2 \lambda}{\partial V^2}(\Delta V)^2 \ldots \tag{18}$$

If we take the real part of both sides of (18) then at the stability boundary $\mathbf{Re}\,\lambda = 0$ we obtain

$$\mathbf{Re}\left[\frac{\partial \lambda}{\partial h}\Delta h + \frac{1}{2}\frac{\partial^2 \lambda}{\partial h^2}(\Delta h)^2 + \frac{\partial^2 \lambda}{\partial h \partial V}\Delta h \Delta V + \frac{1}{2}\frac{\partial^2 \lambda}{\partial V^2}(\Delta V)^2 \ldots\right] = 0 \tag{19}$$

The zero and linear terms with respect to ΔV vanish due to the first two conditions of (17).

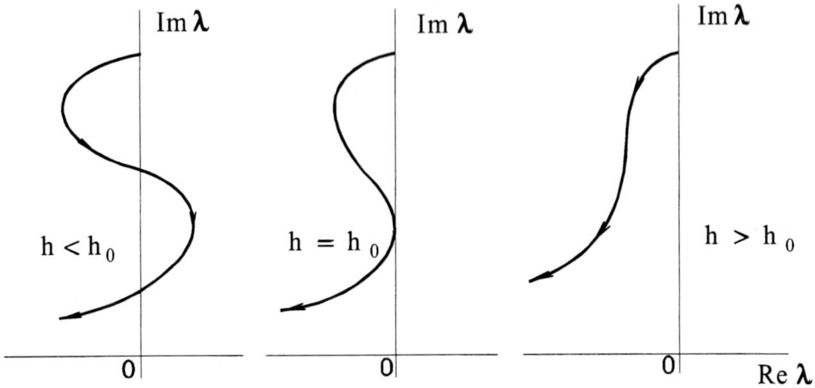

Figure 5. Behavior of the eigenvalue branch $\lambda(V)$ near the critical point.

Formula (19) is an approximate equation for the stability boundary in the vicinity of the point $h = h_0, V = V_0$. Assuming that $\Delta h = a(\Delta V)^k$, a and k being unknown constants, we immediately find from (19) up to the terms of second order

$$\Delta h = a(\Delta V)^2, \tag{20}$$

$$a = -\frac{1}{2}\left[\mathbf{Re}\left(\frac{\partial^2 \lambda}{\partial V^2}\right) \Big/ \mathbf{Re}\left(\frac{\partial \lambda}{\partial h}\right)\right] \tag{21}$$

Relation (20) is a quadratic parabola. Due to assumptions (17) the coefficient a is negative, $a < 0$. It is easy to see that in the vicinity of the point h_0, V_0, a corresponding flutter domain $\mathbf{Re}\,\lambda > 0$ is given by the inequality $h < h_0 + a(V - V_0)^2$. This means convexity of the flutter domain with a vertical tangent at the boundary point h_0, V_0, see Fig 6a. It also shows that at this point a jump in the critical flutter speed can happen due to presence of other eigenvalue branches.

Note that if we assume that at the point h_0, V_0 the derivative $\partial\, \mathbf{Re}\,\lambda / \partial h > 0$, then $a < 0$, and the flutter domain is given by $h > h_0 + a(V - V_0)^2$, Fig 6b.

The only concern now is how to calculate the constant a. According to (21) we have to know the first and second derivatives of the eigenvalue λ with respect to the parameters taken at the point h_0, V_0. These derivatives have been derived in Seyranian et al. (1999, 2000), see

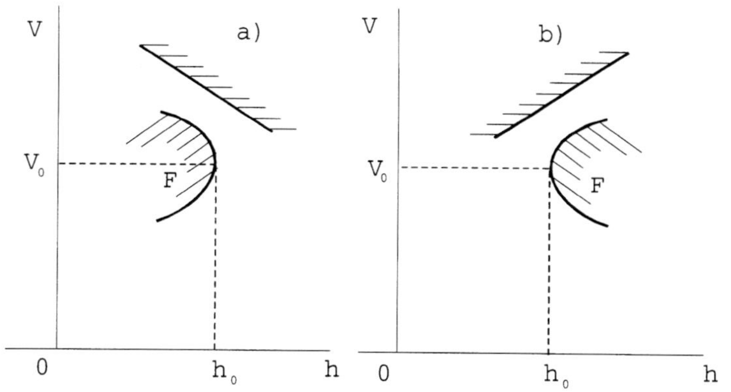

Figure 6. Convexity of the flutter domain with a vertical tangent at the boundary point h_0, V_0.

also Section 1. To calculate them we need to know the eigenvectors of the adjoint eigenvalue problems $\mathbf{u}_0$ and $\mathbf{v}_0$, as well as the first and second derivatives of the matrix $\mathbf{A}$ with respect to the parameters.

5 Post-buckling Behavior of Bimodal Optimal Columns

5.1 Introduction

We consider a problem of maximizing the critical buckling load of an elastic column of given length and volume. This problem was first formulated by J.-L. Lagrange in 1773 and was considered by many authors for various boundary conditions. The long history of this problem and some of the solutions can be found in Tadjbakhsh and Keller (1962), Olhoff and Rasmussen (1977), Cox and Overton (1992), Seyranian et al. (1994), and Seyranian (2000). It turns out that in certain cases, for instance, for clamped–clamped supports, the optimal columns are bimodal, i.e. possess two linearly independent eigenmodes. In such cases the post-buckling behavior of optimal columns is of particular interest. Some authors, see for instance Thompson (1972), and Thompson and Hunt (1973), stated that the existence of several buckling modes, which is typical for optimal structures, inevitably leads to dangerous nonlinear coupling of modes and high sensitivity to imperfections.

In the present Section, based on the recent publications by Seyranian (1995, 2000), Seyranian and Privalova (1999), Privalova and Seyranian (1999), we consider the Lagrange problem for the case of elastic supports at both ends of the column. This is a natural generalization of the boundary conditions considered earlier. It turns out that this type of the boundary conditions always admits the bimodal solutions. Then we study the initial post-buckling behavior of the bimodal optimal columns. It is shown that this behavior, being always supercritical, is determined by four solutions branching off from the trivial equilibrium state. The stability of the new equilibrium states is analyzed near the bifurcation point λ_0. It is shown that only two new equilibrium states are stable, while two others are unstable, this conclusion being valid for all considered support stiffnesses.

5.2 Optimization problem

Let us consider a thin elastic column of circular length-dependent cross section subjected to a compressive axial force P. The column is assumed to be elastically supported at both ends with stiffness coefficients C_1 and C_2, respectively. In a linear approximation buckling of the column in dimensionless variables is described by the following eigenvalue problem:

$$(a^2 y'')'' + \lambda y'' = 0, \quad 0 < x < 1 \tag{1}$$

$$y(0) = (a^2 y'' - c_1 y')_{x=0} = 0, \quad y(1) = (a^2 y'' + c_2 y')_{x=1} = 0 \tag{2}$$

Here, $a(x)$, $y(x)$, and λ are the cross-sectional area of the column, the deflection function, and the axial force, respectively, the positive constants c_1 and c_2 are the stiffness coefficients, and the prime indicates differentiation with respect to x. The dimensionless variables are expressed in terms of dimensional ones as follows:

$$x = S/L, \qquad y = Y(x)/L, \quad a(x) = A(x)L/V,$$

$$\lambda = 4\pi PL^4/(EV^2), \quad c_i = 4\pi C_i L^3/(EV^2) \geq 0, \quad i = 1,2$$

where L, V, and A (x) are the length, volume, cross-sectional area of the column, respectively, and E is Young's modulus.

Eigenvalue problem (1), (2) is selfadjoint, and all eigenvalues λ are real and positive for smooth and positive functions $a(x)$.

The condition of constant volume is

$$\int_0^1 a(x)\,dx = 1 \tag{3}$$

The optimization problem involves the determination of the cross-sectional area $a_0(x) \geq 0$ such that the critical buckling load (i.e. the minimum eigenvalue λ_0) is maximal provided that the condition of constant volume (3) is satisfied. The optimization problem was first formulated by Lagrange (1868) for simply supported boundary conditions $c_1 = c_2 = 0$. Subsequently, several authors considered boundary conditions of different types; a historical review of their progress is given in Cox and Overton (1992), and Seyranian (2000).

We transform eigenvalue problem (1), (2) to an eigenvalue problem with second order equation with respect to the bending moment $M(x)$. Using the substitution $M = a^2 y''$ and the notation $M(0) = M_-$, $M'(0) = M'_-$, $M(1) = M_+$, $M'(1) = M'_+$ we obtain

$$M'' + \lambda a^{-2} M = 0, \quad 0 < x < 1 \tag{4}$$

$$M'_- = M_+ - M_-(1 + \lambda/c_1), \quad M'_+ = M_+(1 + \lambda/c_2) - M_- \tag{5}$$

Boundary conditions (5) are derived by integration of the equation $M'' + \lambda y'' = 0$ and using boundary conditions (2). The deflection function $y(x)$ is expressed through $M(x)$ as

$$y(x) = M(x) + [M(0) - M(1)]x - M(0) \tag{6}$$

Since eigenvalue problems (1), (2) and (4), (5) are linear and homogeneous, the eigenfunctions $y(x)$ and $M(x)$ are determined up to arbitrary multipliers. Eigenvalue problem (4), (5) is equivalent to (1), (2) for $\lambda \neq 0$. However, unlike (1), (2), the eigenvalue λ in problem (4), (5) is involved in the boundary conditions, and $\lambda = 0$ is the eigenvalue.

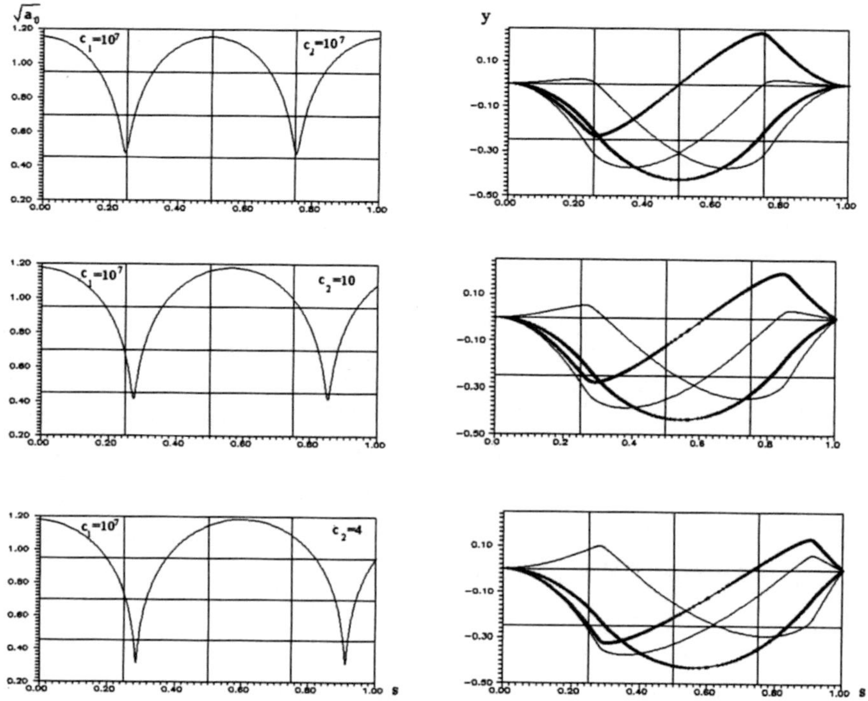

Figure 1. The optimum distribution $\sqrt{a_0(s)}$ and the supercritical equilibrium states $y(s)$.

Problem (4), (5) satisfies the conditions of Kamke's oscillatory theorem, see Kamke (1967, p. 266-267), for continuous and positive functions $a(x)$, $0 \leq x \leq 1$ and positive constants c_1 and c_2. According to that theorem all eigenvalues of problem (4), (5) are real and can be placed in two infinitely growing sequences

$$\lambda_1 < \lambda_2 < \lambda_3 < \cdots, \quad \tilde{\lambda}_1 < \tilde{\lambda}_2 < \tilde{\lambda}_3 < \cdots$$

with the inequality

$$\lambda_n \leq \tilde{\lambda}_n < \lambda_{n+1}, \qquad n = 1, 2, \ldots$$

If $\lambda_n = \tilde{\lambda}_n$ this is the double eigenvalue with two linearly independent eigenfunctions $M_n(x)$, $\tilde{M}_n(x)$, and if $\lambda_n < \tilde{\lambda}_n$ these are the simple eigenvalues. If $N(n)$ denotes the number of zeros on the semi-interval $0 < x \leq 1$ of the functions $M_n(x)$, $\tilde{M}_n(x)$, corresponding to the eigenvalues λ_n, $\tilde{\lambda}_n$, then $N(n)$ is equal to $2n-2$ or $2n-1$. Note, that in this case we have $\lambda_1 = \tilde{\lambda}_1 = 0$, $M_1 = 1$, $\tilde{M}_1 = x - 1/2$. Thus, for elastically supported columns it is possible to have solutions with one or two buckling modes with the eigenfunctions $M(x)$ having on the semi-interval $0 < x \leq 1$ two or three zeros.

It has been shown in Seyranian (2000) that for any $c_1 > 0$, $c_2 > 0$ a unimodal solution leads to vanishing cross-sectional area of the column $a(x)$ at some interior points of the interval $0 < x < 1$. These points can be treated as hinges with possible discontinuity of the slope in a buckling mode. It turns out that the unimodal column buckles with a mode having discontinuities of the slope and at the lower load than λ_0, corresponding to the mode with no discontinuities assumed as an optimal. Hence, the unimodal solution is not optimal.

The bimodal optimal solution $a_0(x)$ was found in the form of an elliptic integral, see Seyranian (1984, 1995)

$$x - x_* = \pm\frac{3}{2}\int_{a_*}^{a_0} \frac{t^2 dt}{\sqrt{D_1 t^3 - 3\lambda_0 t^4 - D_2^2}} \tag{7}$$

where D_1, D_2, λ_0 and x_*, a_* are the constants to be determined from some transcendental equations and condition (3). The bimodal optimal solutions for different values of the stiffness coefficients c_1, c_2 are presented in Fig 1. When $c_1 = c_2$ the solution is symmetric with respect to the center of the column. The right-hand hump of the optimal shape diminishes with a decrease of c_2, and disappears when it tends to zero.

5.3 Postbuckling behavior

Consider now finite deflections of an inextensible elastic column loaded by an axial force, see Fig 2. The deflection curve of the column is described by the dimensionless functions $x(s)$, $y(s)$, $0 \leq s \leq 1$. The boundary conditions for the elastic supports are

$$\begin{aligned} y(0) = 0, \quad (a^2\theta' - c_1\theta)_{s=0} = 0, \\ y(1) = 0, \quad (a^2\theta' + c_2\theta)_{s=1} = 0, \end{aligned} \tag{8}$$

where $\theta(s)$ is the slope of the deflection curve. The prime denotes differentiation with respect to s.

Since $x' = \cos\theta$ and $y' = \sin\theta$, we have

$$x(s) = \int_0^s \cos\theta \, ds, \qquad y(s) = \int_0^s \sin\theta \, ds \tag{9}$$

Therefore, the boundary condition $y(1) = 0$ can be represented in the form

$$\int_0^1 \sin\theta \, ds = 0 \tag{10}$$

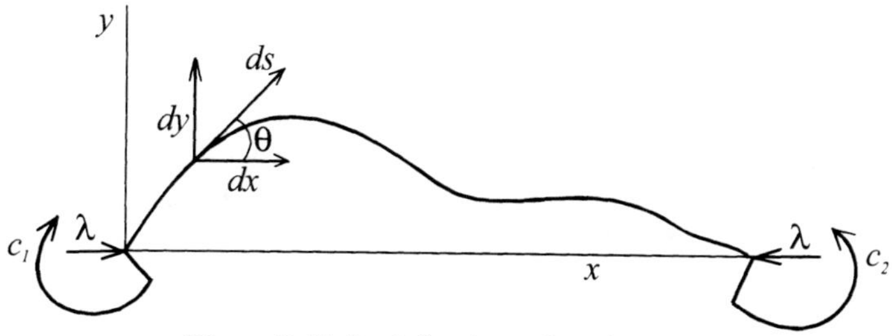

Figure 2. Finite deflections of a column.

The total potential energy of the column is given by the integral, see Thompson and Hunt (1973)

$$U = \frac{1}{2}\int_0^1 a^2\theta'^2 ds - \lambda\int_0^1 (1-\cos\theta)\, ds + \frac{c_1}{2}\theta^2(0) + \frac{c_2}{2}\theta^2(1) \tag{11}$$

Equilibrium of the column is characterized by the minimum of the potential energy provided that the isoperimetric condition (10) is satisfied. We add the integral of (10) multiplied by an unknown constant N to U and equate the first variation of the resulting functional to zero:

$$\delta\left(U + N\int_0^1 \sin\theta\, ds \right) = 0$$

From this equality we find the nonlinear differential equation for the column

$$(a^2\theta')' + \lambda\sin\theta - N\cos\theta = 0 \tag{12}$$

with the boundary conditions

$$(a^2\theta' - c_1\theta)_{s=0} = 0, \qquad (a^2\theta' + c_2\theta)_{s=1} = 0 \tag{13}$$

Equation (12) can also be obtained from the condition of static equilibrium. In this case, N is the vertical force at the left support. By integrating eqn. (12) from 0 to 1 and using isoperimetric condition (10) and boundary condition (13), we express the constant N in the form

$$N = -[c_1\theta(0) + c_2\theta(1)] \Big/ \int_0^1 \cos\theta\, ds \tag{14}$$

The nonlinear integro–differential equations (12) – (14) with condition (10) permit us to determine the function $\theta(s)$ for given force λ and stiffnesses $a(s)$, c_1 and c_2. Then, from (9), we can calculate the deflection curve $x(s)$, $y(s)$ of the column.

We use the perturbation technique to analyze the postbuckling behavior of the bimodal columns with the cross-sectional area $a_0(s)$. At the bifurcation point λ_0, the trivial state of equilibrium $\theta(s) \equiv 0$ splits into several new states of equilibrium. We seek the solution in the form of series in ε

$$\theta(s) = \varepsilon[\varphi(s) + \varepsilon^2\psi(s) + \cdots], \qquad \lambda = \lambda_0 + \varepsilon^2\mu + \cdots, \tag{15}$$

where the small parameter is equal to the norm

$$\varepsilon = \left(\int_0^1 \theta^2(s)\, ds \right)^{1/2} \tag{16}$$

Expansions (15) are due to the specific nonlinearities of equations (12) – (14) and (10).

Substituting expansions (15) into (12) – (14) and (10) and using (16), we obtain in the first approximation in ε the following eigenvalue problem

$$\begin{gathered} (a_0^2\varphi')' + \lambda_0\varphi + c_1\varphi(0) + c_2\varphi(1) = 0 \\ (a_0^2\varphi' - c_1\varphi)_{s=0} = 0, \qquad (a_0^2\varphi' + c_2\varphi)_{s=1} = 0 \end{gathered} \tag{17}$$

$$\int_0^1 \varphi\, ds = 0, \qquad \int_0^1 \varphi^2 ds = 1 \tag{18}$$

If we differentiate the first of equations (17) once and use the notation $y' = \varphi,\ y(0) = 0$ we arrive at conventional linear problem (1), (2) describing the buckling of the column. Since λ_0 is the eigenvalue of this problem corresponding to two linearly independent buckling modes $y_1(s),\ y_2(s)$, a general solution to problem (17) can be represented by the following linear combination

$$\varphi(s) = \gamma_1\varphi_1(s) + \gamma_2\varphi_2(s), \tag{19}$$

where γ_1, γ_2 are unknown constants, and $\varphi_1 = y_1',\ \varphi_2 = y_2'$ are the eigenfunctions. We impose the normalization condition

$$\int_0^1 \varphi_i\varphi_j ds = \delta_{ij}, \qquad i, j = 1, 2 \tag{20}$$

Taking into account (18) - (20) we obtain the relation

$$\gamma_1^2 + \gamma_2^2 = 1 \tag{21}$$

From the next approximation with respect to ε in (12) – (14), (10) we obtain a linear boundary value problem for the function $\psi(s)$ with the right-hand side depending on $\varphi(s)$. The solvability conditions for that boundary value problem imply

$$\begin{gathered} \frac{1}{6}\lambda_0 \int_0^1 \varphi^3\varphi_i ds + \frac{1}{2}[c_1\varphi(0) + c_2\varphi(1)] \int_0^1 \varphi^2\varphi_i ds + \\ + \frac{1}{6}[c_1\varphi_i(0) + c_2\varphi_i(1)] \int_0^1 \varphi^3 ds = \mu \int_0^1 \varphi\varphi_i ds, \quad i = 1, 2 \end{gathered} \tag{22}$$

Substituting (19) to (22) and using (20) we obtain the following system of nonlinear equations on γ_1, γ_2 and μ

$$\begin{gathered} \gamma_1^3 A_1 + \gamma_1^2\gamma_2 A_2 + \gamma_1\gamma_2^2 A_3 + \gamma_2^3 A_4 = \mu\gamma_1 \\ \gamma_1^3 B_1 + \gamma_1^2\gamma_2 B_2 + \gamma_1\gamma_2^2 B_3 + \gamma_2^3 B_4 = \mu\gamma_2 \end{gathered} \tag{23}$$

Here, the coefficients A_i, B_i, $(i = 1,\ldots 4)$ are integrals of the eigenfunctions $\varphi_1(s), \varphi_2(s)$ according to (22). Condition (21) completes system of equations (23) on γ_1, γ_2 and μ.

Symmetric case $\mathbf{c}_1 = \mathbf{c}_2$. If the boundary conditions are symmetric $c_1 = c_2 = c$, the optimal solutions are symmetric $a_0(s) = a_0(1-s)$. In this case, the symmetric and antisymmetric eigenfunctions $\varphi_1(s) = -\varphi_1(1-s)$ and $\varphi_2(s) = \varphi_2(1-s)$ can be used as the basis functions. This leads to the equalities $A_2 = A_4 = 0$ and $B_1 = B_3 = 0$, which simplify equations (23). As a result, we obtain the nonlinear equations for the unknowns γ_1, γ_2 and μ in the form

$$\begin{aligned} \gamma_1^3 A_1 + \gamma_1 \gamma_2^2 A_3 &= \mu \gamma_1 \\ \gamma_1^2 \gamma_2 B_2 + \gamma_2^3 B_4 &= \mu \gamma_2, \end{aligned} \tag{24}$$

where

$$\begin{gathered} A_1 = \frac{1}{6}\lambda_0 \int_0^1 \varphi_1^4 ds, \quad A_3 = B_2 = \frac{1}{2}\lambda_0 \int_0^1 \varphi_1^2 \varphi_2 ds + 2c\varphi_2(0) \int_0^1 \varphi_1^2 \varphi_2 ds, \\ B_4 = \frac{1}{6}\lambda_0 \int_0^1 \varphi_2^4 ds + \frac{4}{3} c\varphi_2(0) \int_0^1 \varphi_2^3 ds \end{gathered} \tag{25}$$

System (24) plus relation (21) has four groups of solutions branching off the trivial solution $\varphi \equiv 0$ at the bifurcation point λ_0. These solutions have the form

$$\text{I.} \quad \gamma_1 = \pm 1, \quad \gamma_2 = 0, \quad \mu_1 = A_1 > 0, \quad \theta_1(s) = \pm \varepsilon \varphi_1(s) \tag{26}$$

$$\text{II.} \quad \gamma_1 = 0, \quad \gamma_2 = \pm 1, \quad \mu_2 = B_4, \quad \theta_2(s) = \pm \varepsilon \varphi_2(s) \tag{27}$$

$$\text{III–IV.} \quad \begin{gathered} \gamma_1 = \sqrt{\frac{B_4 - B_2}{A_1 + B_4 - 2B_2}}, \qquad \gamma_2 = \sqrt{\frac{A_1 - B_2}{A_1 + B_4 - 2B_2}}, \\ \mu_3 = \mu_4 = \frac{A_1 B_4 - B_2^{\,2}}{A_1 + B_4 - 2B_2}, \end{gathered} \tag{28}$$

$$\theta_3(s) = \pm\varepsilon[\gamma_1\varphi_1(s) + \gamma_2\varphi_2(s)], \quad \theta_4(s) = \pm\varepsilon[\gamma_1\varphi_1(s) - \gamma_2\varphi_2(s)]$$

In (28), the radical sign stands for nonnegative square root.

The first two groups of solutions correspond to the pure antisymmetric and symmetric buckling modes, while the third and fourth groups correspond to the mixed buckling modes. The solutions $\theta_3(s)$ and $\theta_4(s)$ are related by the expression

$$\theta_4(1-s) = \pm\varepsilon[\gamma_1\varphi_1(1-s) - \gamma_2\varphi_2(1-s)] = \pm\varepsilon\,[-\gamma_1\varphi_1(s) - \gamma_2\varphi_2(s)] = \theta_3(s)$$

implying that they are the mirror images of each other with respect to the center of the column. Note that these functions exist only if the radicands in (28) are nonnegative. It turns out that for any $c > 0$, the values of μ are positive (supercritical bifurcations arising at $\lambda \geq \lambda_0$). A typical dependence of λ on ε is shown in Fig 3.

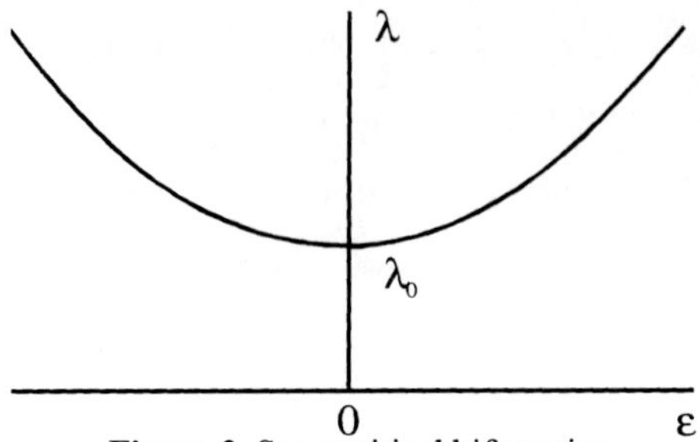

Figure 3. Supercritical bifurcation

Nonsymmetrical case $\mathbf{c}_1 \neq \mathbf{c}_2$. In the general case, one can eliminate μ from (23) and obtain the following equation quadric in the quantity $\gamma = \gamma_1/\gamma_2$

$$A_4\gamma^4 + (A_3 - B_4)\gamma^3 + (A_2 - B_3)\gamma^2 + (A_1 - B_2)\gamma - B_1 = 0 \qquad (29)$$

Finding a real root γ from this equation, we determine (up to the sign) the constants γ_1, γ_2 from condition (21), and then find uniquely the value of μ from (23). According to (15), (19), the constants γ_1, γ_2 determine, in a first approximation, new equilibrium states branching off the trivial state $\theta \equiv 0$ at the bifurcation point λ_0.

The results for calculations of the quantities γ_1, γ_2 and μ for various stiffness coefficients c_1 and c_2 are listed in Tab 1, and the corresponding deflection functions $y_n(s)$, $n = 1,\ldots,4$ are shown in Fig 1. It is seen from Tab 1 that the number of equilibrium states corresponding to the considered values of c_1 and c_2 equals four: all the roots of equation (29) are real, and all the bifurcations are supercritical $\mu_n > 0$, $n = 1,\ldots,4$, Fig 3. Consequently, the supercritical behavior of the bimodal optimal columns shows that the effect of a high sensitivity to imperfections is not observed. However, a question arises: which of the new equilibrium states are stable and which are not?

5.4 Stability of the equilibrium states

A state of equilibrium is stable if the total potential energy (11) reaches the local minimum provided that the isoperimetric condition (10) is satisfied. Hence, in a state of stable equilibrium $\theta(s)$, the following necessary conditions must be satisfied: the first variation of the functional

$$W = U + N\int_0^1 \sin\theta\, ds$$

is equal to zero, $\delta W = 0$, and the second variation of the functional is nonnegative, $\delta^2 W \geq 0$, for arbitrary smooth variations $\delta\theta(s)$ satisfying the constraints

$$\int_0^1 \cos\theta\,\delta\theta\,ds = 0, \tag{30}$$

$$\int_0^1 (\delta\theta)^2\,ds = 1 \tag{31}$$

Equation (30) corresponds to the isoperimetric condition (10), while eqn. (31) is the normalization condition. The quantity N is an unknown Lagrange's multiplier.

The second variation of the functional W has the form

$$\begin{aligned}\delta^2 W = \frac{1}{2}\int_0^1 [a_0^2(\delta\theta')^2 - \lambda\cos\theta(\delta\theta)^2 - N\sin\theta(\delta\theta)^2]ds + \\ +\frac{1}{2}[c_1\delta\theta^2(0) + c_2\delta\theta^2(1)]\end{aligned} \tag{32}$$

The Euler–Lagrange equation and the transversality conditions for the extremal of the functional $\delta^2 W$ with constraints (30) and (31) lead to the eigenvalue problem

$$\begin{aligned}&(a_0^2\delta\theta')' + [\lambda\cos\theta + N\sin\theta]\delta\theta + \nu\delta\theta + \beta\cos\theta = 0, \\ &(a_0^2\delta\theta' - c_1\delta\theta)_{s=0} = 0, \qquad (a_0^2\delta\theta' + c_2\delta\theta)_{s=1} = 0,\end{aligned} \tag{33}$$

where ν and β are unknown Lagrange's multipliers, ν being the eigenvalue, and $\delta\theta$ the eingenfunction. This can be shown by integration of (33) from 0 to 1 and using the boundary conditions and condition (30). The quantity β is a linear functional on $\delta\theta$

$$\beta = -\left\{\int_0^1 [N\sin\theta + \nu]\delta\theta\,ds + [c_1\delta\theta(0) + c_2\delta\theta(1)]\right\}\left(\int_0^1 \cos\theta\,ds\right)^{-1} \tag{34}$$

The minimum of the functional $\delta^2 W$ under constraints (30) and (31) is equal to the minimum eigenvalue $\nu_{\min}$ of problem (33). Hence, the necessary condition $\delta^2 W \geq 0$ under constraints (30) and (31) is reduced to the nonnegativity of the eigenvalue $\nu_{\min}$.

Table 1. The coefficients for postbuckling behavior.

c_1	c_2	λ_0	γ_1	γ_2	μ_1	μ_2	μ_3	μ_4
			1	0	15,4			
10^7	10^7	52,356	0	1		17,2		
			0,721	0,693			27,43	27,43
			0,999	-0,043	12,8			
10^7	10	43,27	0,088	-0,996		17,4		
			0,66	0,75			27,4	
			0,67	-0,74				22,8
			0,998	-0,067	10,937			
10^7	4	37,5	0,233	-0,972		22,4		
			0,58	0,81			32,4	
			0,55	-0.83				23,3

We will analyze the stability of the states of equilibrium represented in the form of expansions

$$\theta(s)=\varepsilon\,[\varphi(s)+\varepsilon^2\psi(s)+\cdots], \quad \lambda=\lambda_0+\varepsilon^2\mu+\cdots, \tag{35}$$

The first order approximation functions $\varphi_{(k)}(s)=\gamma_1^{(k)}\varphi_1(s)+\gamma_2^{(k)}\varphi_2(s)$ and parameters μ_k ($k=1,\ldots,4$) were determined by solving equations (26) – (29) with the results presented in Tab 1.

We substitute expansions (35) into eigenvalue problem (33) and condition (30) and, in the second order approximation with respect to ε, obtain

$$(a_0^2\delta\theta')'+[\lambda_0+\varepsilon^2 g_{(k)}(s)]\delta\theta+\nu_{\min}\delta\theta+\beta\left(1-\varepsilon^2\varphi_{(k)}^2/2\right)=0, \tag{36}$$

$$g_{(k)}(s)=\mu_k-\lambda_0\varphi_{(k)}^2(s)/2-[c_1\varphi_{(k)}(0)+c_2\varphi_{(k)}(1)]\,\varphi_{(k)}(s), \tag{37}$$

$$\int_0^1(1-\varepsilon^2\varphi_{(k)}^2/2)\,\delta\theta\,ds=0, \qquad \int_0^1\delta\theta^2 ds=1 \tag{38}$$

The boundary conditions of (33) for the function $\delta\theta$ remain the same.

We will solve problem (36) – (38) using the perturbation technique:

$$\delta\theta(s)=\delta\theta_0(s)+\varepsilon^2\chi(s)+\ldots, \quad \nu_{\min}=\varepsilon^2\nu_1+\ldots, \quad \beta=\beta_0+\varepsilon^2\beta_1+\ldots \tag{39}$$

For the first approximation coefficient of the eigenvalue ν_1 we obtain the quadratic equation, see Seyranian and Privalova (1999)

$$\det[A_{ij}-\nu_1\delta_{ij}]=0, \quad i,j=1,2 \tag{40}$$

with the symmetric matrix

$$A_{ij}=\frac{1}{2}[c_1\varphi_i(0)+c_2\varphi_i(1)]\int_0^1\varphi_{(k)}^2\varphi_j ds+\frac{1}{2}[c_1\varphi_j(0)+c_2\varphi_j(1)]\int_0^1\varphi_{(k)}^2\varphi_i ds-$$
$$-\int_0^1 g_{(k)}\varphi_i\varphi_j ds \tag{41}$$

The coefficient ν_1 is equal to the minimum of the second variation δ^2W for the second approximation in ε. Hence, the necessary condition of nonnegativity of δ^2W under constraints (30) and (31) is reduced to the nonnegativity of both roots ν_1 of the quadratic equation (40). Then, the matrix A_{ij} must be positive semi-definite, i.e., the Sylvester conditions

$$A_{11}\geq 0, \quad A_{22}\geq 0, \quad A_{11}A_{22}-A_{12}^2\geq 0 \tag{42}$$

ought to be satisfied for a stable solution. Conditions (42) were checked for the solutions $\theta_k(s)$, μ_k, $k=1,\ldots,4$ found in Section 5.3. If the rigorous inequalities (42) for a solution $\theta_k(s)$, μ_k are satisfied, the solution is stable. And if conditions (42) are not fulfilled, this state of equilibrium is unstable.

First, we analyze the stability of the trivial state of equilibrium $\theta(s) \equiv 0$ for $\lambda = \lambda_0 + \varepsilon^2 \mu$. We substitute $\varphi_{(k)}(s) = 0$ and $g_{(k)} = \mu$ into (41) and use condition (20) to obtain $A_{ij} = -\mu\delta_{ij}$. Then, from (40), we find the double root $v_1 = -\mu$. Thus, stability conditions (42) are reduced to $\mu \le 0$. This implies that the trivial solution $\theta(s) \equiv 0$ is stable for $\lambda < \lambda_0$ and unstable for $\lambda > \lambda_0$, as expected.

In the symmetric case $c_1 = c_2 = c$ application of conditions (42) to bifurcations (26) – (28) shows that only first two solutions $\theta_1(s) = \pm\varepsilon\varphi_1(s)$ and $\theta_2(s) = \pm\varepsilon\varphi_2(s)$, emanating from the pure antisymmetric and symmetric buckling modes, respectively, are stable. Two remaining solutions $\theta_3(s)$ and $\theta_4(s)$, corresponding to the mixed buckling modes, are unstable, this conclusion being valid for any $c > 0$.

In the nonsymmetrical case verification of conditions (42) shows that only two solutions $y(s)$ of four are stable. The stable solutions are shown in Fig 1 by thick lines, while the unstable solutions – by thin lines.

Thus, we have investigated the post-buckling behavior of the bimodal optimal columns with cross-sectional area $a_0(s)$ elastically supported at both ends taking into account the geometrical nonlinearity. This study reveals that all bifurcations $\theta_k(s)$, $k = 1,\dots,4$ branching off the trivial state of equilibrium $\theta(s) \equiv 0$ are supercritical, i.e., $\mu_k > 0$, $k = 1,\dots,4$, and dangerous coupling of modes does not occur. However, only two solutions of the four are stable, and the other two are unstable. These results turn out to be valid for all considered support stiffnesses.

6 Instability Domains for Hill's Equation with Damping and Stability of Periodic Solutions

6.1 Introduction

In this Section based on the publications by Seyranian et al. (1999, 2000) and Seyranian (2001) with the use of derivatives of a monodromy matrix with respect to parameters the instability domains for Hill's equation with damping are determined under the assumption of small excitation amplitude and damping coefficient. A description of these domains (halves of cones) in three-dimensional parameter space is given. As an example the domains of parametric resonance of a pendulum with vertically vibrating hanging point are obtained. Another important application of Hill's equation is connected with the stability study of periodic motions in nonlinear dynamical systems. It is shown how to find stable and unstable regimes for harmonically excited Duffing's equation.

6.2 Derivatives of the Floquet Matrix with Respect to Parameters

We consider a system of linear differential equations

$$\dot{\mathbf{x}} = \mathbf{G}\mathbf{x}, \tag{1}$$

where $\mathbf{G} = \mathbf{G}(t, \mathbf{p})$ is a real square matrix of dimension m, which is smoothly depending on a vector of real parameters $\mathbf{p} = (p_1, p_2, \ldots p_n)$ and is a continuous periodic function of the time $\mathbf{G}(t, \mathbf{p}) = \mathbf{G}(t+T, \mathbf{p})$, T being a period. We denote linear independent solutions of system (1) as $\mathbf{x}_1(t), \mathbf{x}_2(t), \ldots, \mathbf{x}_m(t)$ and form out of them a fundamental matrix $\mathbf{X}(t) = [\mathbf{x}_1(t), \mathbf{x}_2(t), \ldots, \mathbf{x}_m(t)]$. The $\mathbf{X}(t)$ matrix satisfying the equations

$$\dot{\mathbf{X}} = \mathbf{G}\mathbf{X}, \quad \mathbf{X}(0) = \mathbf{I}, \tag{2}$$

where $\mathbf{I}$ is the identity matrix of dimension m, is called a matriciant, and the matrix $\mathbf{F} = \mathbf{X}(T)$ is called a monodromy matrix.

According to the Floquet theory, see Nayfeh and Mook (1979) and Yakubovich and Starzhinskii (1987), stability of system (1) is determined by multipliers (eigenvalues of the monodromy matrix): if all the multipliers for their absolute values are less than one, the system is asymptotically stable, and if at least one of them is greater than one, the system becomes unstable.

Along with (2) we consider an adjoint system of equations for the matrix $\mathbf{Y}$

$$\dot{\mathbf{Y}} = -\mathbf{G}^{\mathbf{T}}\mathbf{Y}, \quad \mathbf{Y}(0) = \mathbf{I}. \tag{3}$$

The solutions $\mathbf{X}(t)$ and $\mathbf{Y}(t)$ of equations (2) and (3) are related by the equality, see Yakubovich and Starzhinskii (1987)

$$\mathbf{X}^{\mathbf{T}}(t)\,\mathbf{Y}(t) = \mathbf{I}, \tag{4}$$

allowing to express $\mathbf{Y}(t) = [\mathbf{X}^{\mathbf{T}}(t)]^{-1}$.

Now we assume that the vector of parameters takes a variation $\mathbf{p} = \mathbf{p_0} + \Delta\mathbf{p}$. Hence, the $\mathbf{G}$ matrix, and therefore the matriciant $\mathbf{X}(t)$ obtain variations. This accordingly leads to a change of the monodromy matrix $\mathbf{F}$. The formulas for the first and second derivatives of a monodromy matrix with respect to parameters were derived in Seyranian et al. (1999, 2000) in the form of integrals over the period

$$\frac{\partial \mathbf{F}}{\partial p_k} = \mathbf{F_0}\int_0^T \mathbf{Y_0^T}\frac{\partial \mathbf{G}}{\partial p_k}\mathbf{X_0}\,dt, \quad k = 1,\ldots,n, \tag{5}$$

$$\frac{\partial^2 \mathbf{F}}{\partial p_i \partial p_j} = \mathbf{F_0}\{\int_0^T \mathbf{Y_0^T}\frac{\partial^2 \mathbf{G}}{\partial p_i \partial p_j}\mathbf{X_0}\,dt + \int_0^T \mathbf{Y_0^T}\frac{\partial \mathbf{G}}{\partial p_i}\mathbf{X_0}\left(\int_0^\tau \mathbf{Y_0^T}\frac{\partial \mathbf{G}}{\partial p_j}\mathbf{X_0}\,d\zeta\right)d\tau +$$

$$+\int_0^T \mathbf{Y_0^T}\frac{\partial \mathbf{G}}{\partial p_j}\mathbf{X_0}\left(\int_0^\tau \mathbf{Y_0^T}\frac{\partial \mathbf{G}}{\partial p_i}\mathbf{X_0}\,d\zeta\right)d\tau\}, \quad i, j = 1,\ldots,n, \tag{6}$$

where the zero subscript means that the corresponding value is taken at $\mathbf{p} = \mathbf{p_0}$. Note that to find derivatives (5) and (6) it is necessary to know only the matriciants $\mathbf{X_0}(t)$, $\mathbf{Y_0^T}(t) = \mathbf{X_0^{-1}}(t)$

and the derivatives of the **G** matrix with respect to the parameters taken at $\mathbf{p} = \mathbf{p}_0$. Using derivatives (5) and (6) a variation of the monodromy matrix can be given in the form

$$\mathbf{F}(\mathbf{p}_0 + \Delta\mathbf{p}) = \mathbf{F}_0 + \sum_{k=1}^{n} \frac{\partial \mathbf{F}}{\partial p_k} \Delta p_k + \frac{1}{2} \sum_{i,j=1}^{n} \frac{\partial^2 \mathbf{F}}{\partial p_i \partial p_j} \Delta p_i \Delta p_j + \cdots \tag{7}$$

Knowing the derivatives of the monodromy matrix we can calculate this matrix in the vicinity of the initial point $\mathbf{p}_0$, and therefore estimate behavior of the multipliers responsible for the stability of system (1) when the problem parameters are changed.

6.3 Resonance Domains for Hill's Equation with Damping

In this section we consider Hill's equation with damping

$$\ddot{y} + \beta \dot{y} + [\omega^2 + \varepsilon\varphi(t)]y = 0, \tag{8}$$

where β is the damping coefficient, ε is the excitation amplitude, ω is the natural frequency, and $\varphi(t)$ is a continuous periodic function of the time with the period 2π. Equation (8) has been considered by many authors, see Nayfeh and Mook (1979). A thorough analysis of damped Hill's equation was carried out by Pedersen (1980, 1985).

Our aim is to find analytically the domains of instability of the trivial solution $y = 0$ (the domains of parametric resonance) in the case of small excitation amplitude ε, damping coefficient β, and arbitrary natural frequency $\omega \neq 0$. For this purpose we represent equation (8) in the form (1) with

$$\mathbf{x} = \begin{pmatrix} y \\ \dot{y} \end{pmatrix}, \quad \mathbf{G}(t,\mathbf{p}) = \begin{bmatrix} 0 & 1 \\ -\omega^2 - \varepsilon\varphi(t) & -\beta \end{bmatrix}. \tag{9}$$

This system contains three parameters $\mathbf{p} = (\varepsilon, \beta, \omega)$. If we substitute $\varepsilon = 0,\ \beta = 0$ in (8), (9), then it is easy to find the matriciants from (2) and (3)

$$\mathbf{X}_0(t) = \begin{bmatrix} \cos\omega t & \omega^{-1}\sin\omega t \\ -\omega\sin\omega t & \cos\omega t \end{bmatrix}, \quad \mathbf{Y}_0(t) = \begin{bmatrix} \cos\omega t & \omega\sin\omega t \\ -\omega^{-1}\sin\omega t & \cos\omega t \end{bmatrix}. \tag{10}$$

Note that $\mathbf{X}_0^{\mathbf{T}}(t)\,\mathbf{Y}_0(t) = \mathbf{I}$ in agreement with (4). So, when $\varepsilon = 0, \beta = 0$ we have the monodromy matrix in the form

$$\mathbf{F}_0 = \mathbf{X}_0(2\pi) = \begin{bmatrix} \cos 2\pi\omega & \omega^{-1}\sin 2\pi\omega \\ -\omega\sin 2\pi\omega & \cos 2\pi\omega \end{bmatrix}. \tag{11}$$

The eigenvalues of this matrix (the multipliers) are

$$\rho_{1,2} = \cos 2\pi\omega \pm i\sin 2\pi\omega. \tag{12}$$

When $\omega \neq k/2,\ k = 1,2,\ldots$ the multipliers are complex conjugate quantities lying on the unit circle (stability). When $\beta = 0$ system (8) is Hamiltonian. This means that simple multipliers can not leave the unit circle when the parameters ω and ε are changed, see Yakubovich and Starzhinskii (1987). Let's show that adding of small damping $(\beta > 0)$ leads to the asymptotic stability of the system. For this purpose we find the approximate value of the monodromy matrix $\mathbf{F}$ using relations (5), (7), and (9)-(11). Up to the terms of order β we obtain

$$\mathbf{F}=\mathbf{F}_0(\mathbf{I}+\mathbf{A}\beta), \tag{13}$$

where $\mathbf{F}_0$ is defined in (11), $\mathbf{I}$ is the identity matrix, and

$$\mathbf{A}=\begin{bmatrix} \dfrac{\sin 4\pi\omega}{4\omega}-\pi & \dfrac{\sin^2 2\pi\omega}{2\omega^2} \\ \dfrac{\sin^2 2\pi\omega}{2} & -\dfrac{\sin 4\pi\omega}{4\omega}-\pi \end{bmatrix}. \tag{14}$$

From (13) and (14) we find that

$$\det\mathbf{F}=\det\mathbf{F}_0\det(\mathbf{I}+\mathbf{A}\beta)=1+\beta\,\mathbf{Tr}\mathbf{A}+o(\beta)=1-2\pi\beta+o(\beta). \tag{15}$$

Since the determinant of the monodromy matrix is equal to the product of the pair of complex conjugate multipliers, then for rather small $\beta>0$ it follows from (15) that $|\rho_{1,2}|<1$. This means asymptotic stability.

Therefore, the instability (resonance) can take place only in the vicinity of the points

$$\mathbf{p}_0:\ \varepsilon=0,\ \beta=0,\ \omega=k/2,\ \ k=1,2,\ldots, \tag{16}$$

at which the multipliers are double.

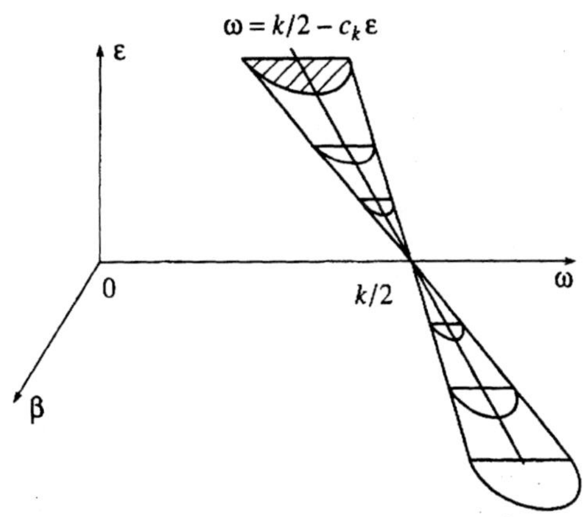

Figure 1. The domains of parametric resonance.

For finding domains of parametric resonance we expand the monodromy matrix $\mathbf{F}$ in Taylor's series in the vicinity of the points $\mathbf{p}_0$ with respect to the parameters ε, β, and $\Delta\omega=\omega-k/2$

$$\mathbf{F}(\mathbf{p})=\mathbf{F}(\mathbf{p}_0)+\frac{\partial\mathbf{F}}{\partial\varepsilon}\varepsilon+\frac{\partial\mathbf{F}}{\partial\beta}\beta+\frac{\partial\mathbf{F}}{\partial\omega}\Delta\omega+\cdots. \tag{17}$$

According to formula (5) and using (9)-(11) we calculate the derivatives $\partial\mathbf{F}/\partial\varepsilon$, $\partial\mathbf{F}/\partial\beta$, and $\partial\mathbf{F}/\partial\omega$ at $\mathbf{p}=\mathbf{p}_0$. Then, up to the accuracy of first order (17) we get

$$\mathbf{F(p)} = \cos \pi k \begin{bmatrix} 1 + 2\pi a_k \varepsilon - \pi\beta & 4\pi(\Delta\omega + (c_k - b_k)\varepsilon)/k \\ -k\pi(\Delta\omega + (c_k + b_k)\varepsilon) & 1 - 2\pi a_k \varepsilon - \pi\beta \end{bmatrix}. \quad (18)$$

Here we have introduced the notation for the coefficients

$$a_k = \frac{1}{2\pi k}\int_0^{2\pi} \varphi(t)\sin kt dt, \quad b_k = \frac{1}{2\pi k}\int_0^{2\pi} \varphi(t)\cos kt dt, \quad c_k = \frac{1}{2\pi k}\int_0^{2\pi} \varphi(t) dt, \quad k = 1,2,\ldots \quad (19)$$

which are directly related to the coefficients of Fourier series of the periodic function $\varphi(t)$.

The eigenvalues of the matrix (18) (the multipliers) can be found approximately as

$$\rho_{1,2} = (-1)^k (1 - \pi\beta) \pm 2\pi\sqrt{D}, \quad (20)$$

$$D = r_k^2 \varepsilon^2 - (\Delta\omega + c_k\varepsilon)^2, \quad r_k = \sqrt{a_k^2 + b_k^2}\ . \quad (21)$$

The system is unstable if the absolute value of at least one multiplier is greater than one, see Nayfeh and Mook (1979) and Merkin (1997). This condition is fulfilled for $\beta < 0$, and the system becomes unstable. But if $\beta \geq 0$ this condition is satisfied only when $2\pi\sqrt{D} > \pi\beta$. Hence, using (21) we obtain that the domain of parametric resonance is the half of the cone

$$(\omega - k/2 + c_k\varepsilon)^2 + \beta^2/4 < r_k^2\varepsilon^2, \quad \beta \geq 0, \quad (22)$$

joining the halfspace $\beta < 0$, Fig 1. Putting in (22) $\beta = 0$ we obtain the domains of parametric resonance with no damping

$$-c_k - r_k < \frac{\omega - k/2}{\varepsilon} < -c_k + r_k. \quad (23)$$

The cross section of the domain (22) by the plane $\varepsilon = const$ is the half of the ellipse with the semiaxes $|\omega - \omega_0| = r_k\varepsilon$, $\omega_0 = k/2 - c_k\varepsilon$ and $\beta = 2r_k\varepsilon$, the centers of the ellipses lying on the axis $\omega = k/2 - c_k\varepsilon$, see Fig 1, 2. Note that with an increase of the damping coefficient β the width of the resonance domain by the frequency ω decreases, and for $\beta > 2r_k\varepsilon$ disappears at all.

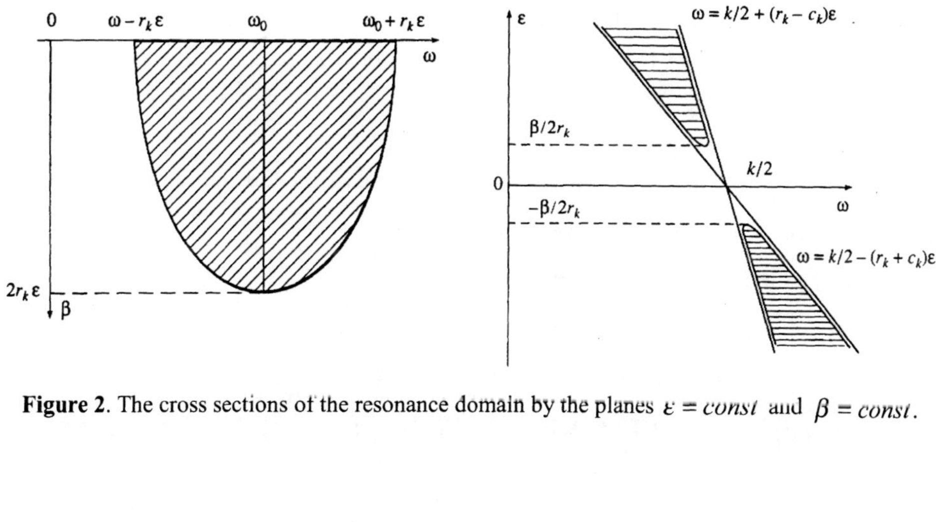

Figure 2. The cross sections of the resonance domain by the planes $\varepsilon = const$ and $\beta = const$.

The cross section of the halfcone (22) by the plane $\beta = const, \beta \geq 0$ gives the domains of parametric resonance limited by two hyperbolas, see Fig 2. The asymptotes of these hyperbolas are found from inequalities (23). When damping is included $(\beta > 0)$ the minimal by absolute value excitation amplitude of the resonance according to (22) is $\varepsilon = \pm\beta/(2r_k)$, and the appropriate resonance frequency is shifted from $k/2$ by the value $\mp c_k\beta/(2r_k)$, Fig 2. With an increase of the resonance number k the coefficients r_k and c_k tend to zero. This means that when k is increasing the axis of the cone (22) is approaching the vertical line, and the cone itself is narrowing. Note that when the average value of the periodic function is equal to zero ($c_k = 0$) the axes of the cones are vertical lines for all k, and the domains of resonance are symmetric with respect to the planes $\omega = k/2$.

Formula (22) agrees with those obtained earlier for some specific cases. For instance, if we put in (8) $\beta = 0, \varphi(t) = \cos t$ we get the Mathieu equation. In this case for the domain of the first resonance $k = 1$ we get from (19) $a_1 = 0,\ b_1 = r_1 = 1/2,\ c_1 = 0$, and from (23) we obtain the well known relation $1-\varepsilon < 2\omega < 1+\varepsilon$, see Nayfeh and Mook (1979) and Merkin (1997).

The presented method of analysis of the parametric resonance domains using derivatives of the monodromy matrix with respect to parameters seems to be more simple and evident in comparison with the methods based on the search of periodic solutions on the boundaries of the stability domains, see Nayfeh and Mook (1979) and Merkin (1997). Besides, it readily gives the resonance domains (not only the boundaries) and the multipliers describing the character of unstable motion.

6.4 Stability of a Pendulum with Vibrating Support

As an example we consider a pendulum with vertically vibrating hanging point as

$$z = a\phi(\Omega t), \tag{24}$$

where a is the amplitude, $\phi(\tau)$ is a periodic function with the period 2π, and Ω is the excitation frequency. We investigate stability and instability of the lower equilibrium position of the pendulum depending on the excitation frequency Ω and small quantities: the excitation amplitude a and the damping coefficient γ.

The linearized equation of motion of the pendulum with damping has the form, see Nayfeh and Mook (1979) and Merkin (1997)

$$ml\ddot{\theta} + \gamma l\dot{\theta} + m(g - \ddot{z})\theta = 0, \tag{25}$$

where θ is the deviation angle of the pendulum measured from the equilibrium position, m is the mass, l is the length, and g is the gravitational acceleration. According to (24) we have $\ddot{z} = a\Omega^2\ddot{\phi}(\Omega t)$. Putting this equality into (25) and introducing the notation $\tau = \Omega t,\ \varphi(\tau) = -\ddot{\phi}(\tau)$, we transform equation (25) to the form (8)

$$\frac{d^2\theta}{d\tau^2} + \frac{\gamma}{\Omega m}\frac{d\theta}{d\tau} + \left[\frac{g}{\Omega^2 l} + \frac{a}{l}\varphi(\tau)\right]\theta = 0. \tag{26}$$

It follows from the results of subsection 6.3 that for small excitation amplitude a and damping coefficients γ the lower equilibrium position of the pendulum becomes unstable at excitation frequencies Ω close to the critical values

$$\Omega_k = \frac{2}{k}\sqrt{\frac{g}{l}}\ , \quad k = 1,2.... \tag{27}$$

According to (22) the instability (parametric resonance) domains are determined by the half-cone

$$\left[\frac{k}{2}\left(1 - \frac{\Omega}{\Omega_k}\right) + \frac{c_k a}{l}\right]^2 + \frac{\gamma^2}{4\Omega_k^2 m^2} < \frac{r_k^2 a^2}{l^2}, \quad \gamma \geq 0, \tag{28}$$

where Ω_k is given by (27), and the coefficients c_k and r_k of the periodic function $\varphi(\tau)$ are defined in (19), (21). Formula (28) can be transformed to the form

$$\left(1 - \frac{\Omega}{\Omega_k} + \frac{2c_k a}{kl}\right)^2 + \frac{\gamma^2 l}{4gm^2} < \frac{4r_k^2 a^2}{k^2 l^2}. \tag{29}$$

From this expression it is easy to see how fast with the growth of k the instability cone becomes narrow and vertical. Inequality (29) gets even simpler for periodic functions with zero mean value $(c_k = 0)$.

6.5 Stability of Periodic Solutions for Nonlinear Systems

In this subsection using the results on the instability domains for Hill's equation with damping we show how to study stability of periodic solutions for nonlinear dynamical systems. For this purpose we consider harmonically excited Duffing's equation, see Nayfeh and Mook (1979) and Thomsen (1997)

$$\ddot{u} + 2\mu\dot{u} + \omega_0^2 u + \alpha u^3 = f\cos\Omega t\ . \tag{30}$$

Here we assume that the excitation frequency is close to the natural frequency $\Omega = \omega_0 + \sigma$ (primary resonance), and the constants μ, α, f and σ are small quantities of the same order $o(1)$. Equation (30) admits an approximate periodic solution, see Thomsen (1997)

$$u_0(t) = a\cos(\Omega t - \gamma) + \frac{\alpha a^3}{32\omega_0^2}\cos(3\Omega t - 3\gamma)\,, \tag{31}$$

where the amplitude a and phase γ are constants of the order $O(1)$ satisfying the equations

$$\left[\left(\Omega - \omega_0 - \frac{3\alpha a^2}{8\omega_0}\right)^2 + \mu^2\right]a^2 = \frac{f^2}{4\omega_0^2}\,, \tag{32}$$

$$\gamma = \arcsin\left(\frac{2\mu a\omega_0}{f}\right). \tag{33}$$

For given constants $\Omega, \omega_0, \alpha, \mu, f$ (32) is a cubic equation on the unknown squared amplitude a^2. The plot of a as a function of Ω for given other quantities is called a *frequency-response curve.* To draw this curve it is convenient to express Ω from (32) as a function of a

$$\Omega = \omega_0 + \frac{3\alpha a^2}{8\omega_0} \pm \sqrt{\frac{f^2}{4\omega_0^2 a^2} - \mu^2}\,. \tag{34}$$

This formula gives two branches Ω_+ and Ω_- lying respectively on the right and left sides of the parabola $\Omega = \omega_0 + 3\alpha a^2 / 8\omega_0$, see Fig 3. A typical frequency-response curve for $\alpha > 0$ is presented in the same figure. Note that the cases $\alpha > 0$ and $\alpha < 0$ are symmetric with respect to the axis $\Omega = \omega_0$. Such multivaluable behavior $a(\Omega)$ is typical for nonlinear systems. For some values Ω it is possible to have three different regimes a and γ of the periodic solution (31). Some of them might be stable, and some unstable.

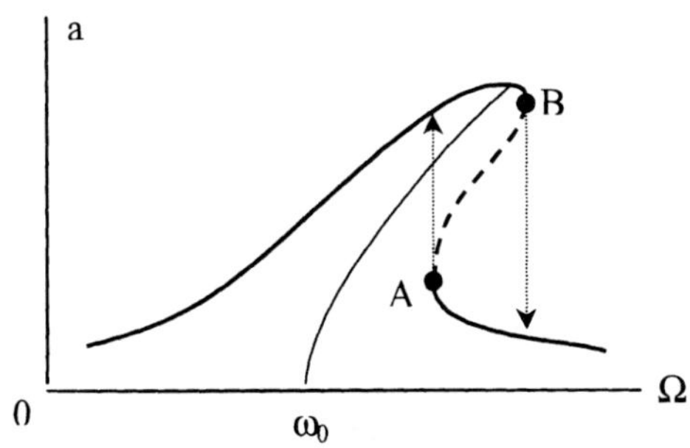

Figure 3. The frequency-response curve for Duffing's equation.

To study the stability of the periodic solution (31) we take

$$u(t) = u_0(t) + v(t)\,, \tag{35}$$

substitute this expression into (30), and then linearize with respect to $v(t)$. Using (32) and (33) we get, see Nayfeh and Mook (1979)

$$\ddot{v} + 2\mu\dot{v} + [\omega_0^2 + 3\alpha a^2 \cos^2(\Omega t - \gamma)]v = 0\,. \tag{36}$$

The stability of the periodic solution (31) is governed by equation (36). If the solution $v(t)$ to this equation is stable, the periodic motion (31) is stable, and vice versa. Introducing the notation $\tau = \Omega t - \gamma$ we obtain Hill's equation in the form (8)

$$\frac{d^2v}{d\tau^2} + \frac{2\mu}{\omega_0}\frac{dv}{d\tau} + \left[1 - \frac{2\sigma}{\omega_0} + \frac{3\alpha a^2}{\omega_0^2}\cos^2\tau\right]v = 0\,. \tag{37}$$

Comparison with equation (8) implies

$$\beta = \frac{2\mu}{\omega_0}, \quad \omega^2 = 1 - \frac{2\sigma}{\omega_0}, \quad \varepsilon = \frac{3\alpha a^2}{\omega_0^2}, \quad \varphi(\tau) = \cos^2\tau, \quad k = 2. \tag{38}$$

Calculating the coefficients according to (19) and (21) we find

$$a_2 = 0, \quad b_2 = \frac{1}{8}, \quad c_2 = \frac{1}{4}, \quad r_2 = \frac{1}{8}. \tag{39}$$

Then the instability condition (22) yields the cone

$$\left(\Omega - \omega_0 - \frac{3\alpha a^2}{4\omega_0}\right)^2 + \mu^2 < \left(\frac{3\alpha a^2}{8\omega_0}\right)^2. \tag{40}$$

However, unlike the linear case the excitation frequency Ω and the amplitude a are not independent quantities due to relation (34). Condition (40) can be transformed to

$$\left(\Omega - \omega_0 - \frac{3\alpha a^2}{8\omega_0}\right)\left(\Omega - \omega_0 - \frac{9\alpha a^2}{8\omega_0}\right) + \mu^2 < 0. \tag{41}$$

In this form the instability condition was derived in Nayfeh and Mook (1979) and Thomsen (1997) using the method of multiple scales. Verifying condition (41) for the branch Ω_- from (34) for $\alpha > 0$ we obtain the inequality of the opposite sign implying the stability of the periodic solution (31). Substituting in (41) the branch Ω_+ from (34) we obtain that for $\alpha > 0$ it is satisfied only when

$$a_* < a < a_{**}, \tag{42}$$

where a_* and a_{**} are the two roots of the equation

$$\frac{f^4}{9\alpha^2\omega_0^2 a^8} = \frac{f^2}{4\omega_0^2 a^2} - \mu^2. \tag{43}$$

We note that equation (43) is equivalent to the condition $d\Omega_+ / da = 0$ indicating that the unstable regime (42) lies between the points A and B shown in Fig 3 by dashed line. Other parts of the branch Ω_+, shown in Fig 3 by solid line, correspond to the stable periodic solutions (31). The points A and B are the catastrophe points due to a jump of the amplitude a. We emphasize that for $\omega_0 > 0, \mu > 0, \alpha > 0, f \neq 0$ these two points are always existing since equation (43) possesses the two roots $a_* < a_{**}$. The case $\alpha < 0$ can be treated similarly.

7 Acknowledgement

The author expresses his gratitude to Pauli Pedersen for long-term cooperation, as well as to Oleg N. Kirillov, Alexei A. Mailybaev, Olga G. Privalova and Anton O. Belyakov for their help and fruitful discussions.

References

Arnold, V.I. (1983). *Geometrical Methods of the Theory of Ordinary Differential Equations*. New York: Springer-Verlag.

Arnold, V.I. (1984). *Catastrophe Theory*. Berlin: Springer-Verlag.

Arnold, V.I. (1989). Bifurcations and singularities in mathematics and mechanics. In *Theoretical and Applied Mechanics,* Paris: Elsevier.1-25.

Bolotin, V.V. (1963). *Nonconservative Problems of the Theory of Elastic Stability*. New York: Pergamon.

Claudon, J.L. (1975). Characteristic curves and optimum design of two structures subjected to circulatory loads. *Journal de Mechanique* 14: 531-543.

Cox, S.J., and Overton, M.L. (1992). On the optimal design of columns against buckling. *SIAM Journal of Mathematical Analysis* 23: 287-325.

Fung, Y.C. (1955). *An Introduction to the Theory of Aeroelasticity*. New York: Wiley.

Grossman, E.P. (1937). Flutter. *Trudy TsAGI*, no.284.

Galin, A.M. (1972). On real matrices depending on parameters. *Uspekhi matematicheskikh nauk* 27(1): 241 – 242.

Hanaoka, M., and Washizu, K. (1980). Optimum design of Beck's column. *Computers and Structures* 11: 473-480.

Huseyin, K. (1978). *Vibrations and Stability of Multiple Parameter Systems.* Alphen aan den Rijn: Sijthoff & Nordhoff.

Kamke, E. (1967). *Differentialgleichungen.* Leipzig: Verlagsgesellschaft.

Keldysh, M.V. (1938). Vibrations of a wing with bracing struts in airflow. *Trudy TsAGI*, no. 357. See also in Keldysh, M.V. (1985). *Mechanics. Selected Works.* Moscow: Nauka, 304-341.

Kirillov, O.N., and Seyranian, A.P. (1999). *On the Stability Boundaries of Circulatory Systems*. Moscow State Lomonosov University. Preprint 51- 99, Institute of Mechanics.

Kirillov, O.N., and Seyranian, A.P. (2001). Overlapping of frequency curves in nonconservative systems. *Physics-Doklady* 377(1): 44-49.

Kounadis, A.N., and Katsikadelis, J.T. (1980). On the discontinuity of the flutter load for various types of cantilevers. *International Journal of Solids and Structures*, 16: 375-383.

Lagrange, J.-L. (1868). Sur la figure des colonnes. In *Ouvres de Lagrange*, v.2. Paris: Gauthier -Villars, 125-170.

Langthjem, M.A., and Sugiyama, Y. (1999). Optimum shape design against flutter of a cantilevered column with an end-mass of finite size subjected to a non-conservative load. *Journal of Sound and Vibration* 226: 1-23.

Leipholz, H. (1980). *Stability of Elastic Systems*. Alphen aan den Rijn: Sijthoff & Nordhoff.

Leipholz, H. (1987). *Stability Theory.* Stuttgart: John Wiley& Sons and B.G.Teubner.

Mailybaev, A.A., and Seyranian, A.P. (1996). The Keldysh problem on aeroelastic stability of a strut braced wing. *Physics-Doklady* 41(10): 484-487.

Mailybaev, A.A., and Seyranian, A.P. (1998). Aeroelastic stability of a wing with bracing struts (Keldysh problem). *Fluid Dynamics* 33(1): 124-134.

Merkin, D. (1997). *Introduction to the Theory of Stability*. New York: Springer-Verlag.

Nayfeh, A.H., and Mook, D.T. (1979). *Nonlinear Oscillations*. New York: John Wiley and Sons.

Olhoff, N., and Rasmussen, S.H. (1977). On single and bimodal optimum buckling loads of clamped columns. *International Journal of Solids and Structures* 13: 605-614.

Panovko, Ya.G., and Gubanova, I.I. (1965). *Stability and Oscillations of Elastic Systems*. New York: Consultants Bureau.

Pedersen, P. (1980). Stability of the solutions to Mathieu-Hill equations with damping. *Ingenieur-Archiv* 49: 15-29.

Pedersen, P. (1985). On stability diagrams for damped Hill equations. *Quarterly of Applied Mathematics*

42: 477-495.

Plaut, R.H. (1972). Determining the nature of instability in nonconservative problems. *AIAA Journal* 10: 967- 968.

Privalova, O.G., and Seyranian, A.P. (1999). Supercritical behavior of bimodal optimal rods. *Mechanics of Solids* 342 (2): 142-150.

Seyranian, A.P. (1991). Sensitivity analysis of eigenvalues and development of instability. *Stroinicki Casopis* 42: 193-208.

Seyranian, A.P. (1993). Sensitivity analysis of multiple eigenvalues. *Mechanics of Structures and Machines* 21: 261-284.

Seyranian, A.P. (1994). Bifurcations in single-parameter circulatory systems. *Izv. RAN. Mekhanika Tverdogo Tela* 29:142-148.

Seyranian, A.P. (1994). Collision of eigenvalues in linear oscillatory systems. *Journal of Applied Mathematics and Mechanics* 58: 805-813.

Seyranian, A.P. (1995). New solutions to Lagrange's problem. *Physics-Doklady* 40(5): 251-253.

Seyranian, A.P. (2000). *The Lagrange Problem on Optimal Column.* Moscow State Lomonosov University, Institute of Mechanics. Preprint No.60

Seyranian, A.P. (2001). Resonance domains for the Hill equation with allowance for damping. *Physics-Doklady* 46 (1): 41-44.

Seyranian, A.P., Lund, E., and Olhoff, N. (1994). Multiple eigenvalues in structural optimization problems. *Structural Optimization* 8: 207-227.

Seyranian, A.P., and Pedersen, P. (1995). On two effects in fluid/structure interaction theory. In Bearman, P.W., ed., *Flow-Induced Vibration*. Rotterdam: A.A.Balkema. 565-576.

Seyranian, A.P., and Privalova, O.G. (1999). Supercritical behavior of optimal bars with two modes of buckling. *Doklady Physics* 44 (6): 368-372.

Seyranian, A.P., and Sharanyuk, A.V. (1983). Sensitivity and optimization of critical parameters in dynamic stability problems. *Izv. AN SSSR. Mekhanika Tverdogo Tela* 18(5): 173-182.

Seyranian, A.P., Solem, F., and Pedersen, P. (1999). Stability analysis for multi-parameter linear periodic systems. *Archive of Applied Mechanics* 69: 160-180.

Seyranian, A.P., Solem, F., and Pedersen, P. (2000). Multi-parameter linear periodic systems: sensitivity analysis and applications. *Journal of Sound and Vibration* 229: 89-111.

Seyranian, A.P., Solem, F., and Pedersen, P. (2000). Sensitivity analysis for the Floquet multipliers. In Chernousko, F.L., and Fradkov, A.L., eds., *Proceedings of 2nd International Conference on Control of Oscillations and Chaos.* St.Petersburg. 404-407.

Tadjbakhsh, I. and Keller, J.B. (1962). Strongest columns and isoperimetric inequalities for eigenvalue. *Transactions of ASME, Ser.E. Journal of Applied Mechanics* 29: 159–164.

Thompson, J.M.T. (1972). Optimization as a generator of structural instability. Letter to the Editor. *International Journal of Mechanical Sciences* 14: 627-629.

Thompson, J.M.T. (1982). *Instabilities and Catastrophes in Science and Engineering.* London: John Wiley & Sons.

Thompson, J.M.T., and Hunt, G.W. (1973). *A General Theory of Elastic Stability.* London: John Wiley & Sons.

Thomsen, J.J. (1997). *Vibrations and Stability. Order and Chaos*. London: McGraw-Hill.

Vishik, M.I., and Lyusternik, L.A. (1960). Solution of some perturbation problems in the case of matrices and selfadjoint and non-selfadjoint equations. *Russian Mathematical Surveys* 15(3): 1-73

Yakubovich, V.A., and Starzhinskii, V.M. (1987). *Parametric Resonance in Linear Systems.* Moscow: Nauka (in Russian).

Zevin, A.A. (1988). On the theory of linear nonconservative systems. Journal of Applied Mathematics and Mechanics 52(3): 386-391.

Ziegler, H. (1968). *Principles of Structural Stability*. Waltham, Massachusetts: Blaisdell.

Part VI. Sensitivity Analysis for Dynamic Stability Problems

Pauli Pedersen

Department of Mechanical Engineering, Solid Mechanics, Technical University of Denmark
Nils Koppels Allè, Building 404, DK-2800 Kgs.Lyngby, Denmark, email: pauli@mek.dtu.dk

0.1 Goal of the notes

These notes are neither chapters in a textbook, nor short research papers. They are something in between and they include results known for some years as well as just published results. The six notes are written to be fairly independent without extensive mutual reference and the page limit for each note is set to 10. This page limit means that extensive examples are not included.

The main goal is to communicate to the course participants:

• Knowledge about sensitivity analysis, i.e. gradient information obtained without re-analysis. Especially for eigenvalue problems. The simplicity and the practical use of sensitivity analysis is today not widely known.

• Classification of systems and solutions for choosing the right tools to solve specific problems.

• The versatility in the application of sensitivity analysis, which however, has a tendency to make these notes somewhat disconnected.

0.2 Comments to the individual notes/lectures

1) Note number 1 is rather general and abstract, hopefully not too theoretical. Restricted to linear systems we classify the systems as well as the possible instability behaviour. Then general variational analysis for non-selfadjoint systems is presented with focus on energy interpretation. Specific treatment is given to finite element discretization and to global Galerkin discretization. The necessary procedure for sensitivity analysis with multiple eigenvalues is derived for the selfadjoint case.

2) In note number 2 we are more specific, dealing with two rather simple two-degrees-of-freedom problems to exemplify the case of double eigenvalues with only one eigenvector. Analytical analysis of isolated eigenvalues explains in a more complete way the earlier reported results. The goal of this note is also to point out the importance of the Routh-Hurwitz stability criteria.

3) In note number 3 we concentrate on linear periodic systems and describe two possible methods of stability analysis for the solutions. For the case of a single linear differential equation with periodic coefficients, it is suggested to set up general stability diagram from which a more complete overview can be obtained. Actual examples of relevant physical problems are listed and classifications are described. For the case of multi-degrees-of-freedom, the Floquet method is suggested and therefore this methods is described in more details. This description also gives the background for note number 4.

4) In note number 4 we primarily establish the sensitivity analysis which is necessary to obtain the derivatives of the Floquet matrix, the reason being that the eigenvalues of the Floquet matrix give the information about the stability of solutions. The Floquet matrix is real but non-symmetric. Thus the eigenvalues of the Floquet matrix may be complex and the sensitivity analysis of these eigenvalues must include the possibility for both multi-modal solutions with full eigenvector space as well as not complete eigenvector space.

5) Applications of sensitivity analysis are described in note number 5 and we have concentrated on three specific examples.

• The results of numerical studies of the influence of external damping are explained analytically by sensitivity analysis.

• Sensitivity analysis is applied to laminate inverse problems and a possible iteration procedure is presented, although this example is not related to an eigenvalue problem.

• In material identification problems, eigenfrequencies are often used as response quantity, here sensitivities to material parameters are involved.

6) Although sensitivity analysis has many domains of application, the domain of optimal design still shows the major interest. In note number 6 we as an introduction shortly give the background for the two major optimal design solution procedures in terms of sensitivities. Then four specific examples are presented:

- Beam design with several eigenvalue constraints.
- Material orientation of composite plates for optimization of supersonic flutter.
- Column design with follower forces.
- Beam design with periodic loads.

0.3 Conclusion

All in all, the available time for writing these notes has been limited and much can be improved. However, it is hoped that the course participants will find a red line through the notes and hopefully be inspired to go deeper into the interesting area of sensitivity analysis.

0.4 Acknowledgments

Thanks to Niels L. Pedersen, to Alexander P. Seyranian and to Jon Juel Thomsen for helpful comments.

1 Sensitivity Analysis for Non-Conservative Problems

1.1 Introduction

The dynamic behaviour of systems subjected to non-conservative loads is often against our physical intuition and can only be understood with a deep insight. As examples: vibration frequency may increase with increasing load and damping may act destabilizing. The non-conservative forces make it necessary to deal with complex functionals in the description, and the intuitive understanding of the physical results is not easy because the indirect effects may be the determining ones. However, we shall in these notes give high priority to the physical interpretations.

The usual (familiar) analysis for the displacements etc., could be based either on the equations of equilibrium, set up directly, or on the stationarity of an energy-functional. For the non-selfadjoint problems such functionals are now also available in the literature and we term them here the mutual energy functionals - mutual in the sense that displacements of the physical as well as of the adjoint problem are involved. The sensitivity analysis may also be based directly on the equations of equilibrium, but the simplicity is striking when we base it on the mutual energy, because then the variations of displacements do not have to be considered.

The response analysis and sensitivity analysis can be formulated in either a continuous or in a discrete form. As almost all problems are solved numerically with only finite degrees of freedom, the sensitivity analysis is presented in a matrix formulation, which has the advantage that boundary conditions are an integrated part. Furthermore, this formulation relates directly to the practical methods of analysis, such as the finite element method and the Galerkin method.

1.2 Classification of dynamic systems and their behaviour

Restricting ourselves to linear, elastic system, we can in matrix notation write the dynamic equilibrium by

$$[M]\{\ddot{D}\} + [C]\{\dot{D}\} + [S]\{D\} = \{A\} \tag{1}$$

where $[M]$ is termed the mass matrix, being the coefficient matrix to the vector $\{\ddot{D}\}$ of accelerations. Thus derivative with respect to time t is denoted by the $\dot{}$ (dot) superscript. The matrix $[C]$ is termed the damping matrix, being the coefficient matrix to the vector $\{\dot{D}\}$ of velocities. The stiffness matrix $[S]$ is the coefficient matrix to the displacements $\{D\}$, and the external forces $\{A\}$ then make the system inhomogeneous.

System classification. The system classification will follow the classic book by Ziegler, 1968, and we shall here immediately identify the class termed in-stationary systems. If only one element in one of the matrices $[M]$, $[C]$, $[S]$, $\{A\}$ is explicitly a function of time t, then the system is termed **in-stationary**. These systems will be covered in other lectures. With stationary external forces $\{A\}$, we can concentrate on the **homogeneous system** that describes the dynamics relative to the static equilibrium $[S]\{D\} = \{A\}$. Thus we continue our classification based on system description by

$$[M]\{\ddot{D}\} + [C]\{\dot{D}\} + [S]\{D\} = \{0\} \tag{2}$$

and furthermore we shall assume this to be normalized in such a way that $[M]$ is symmetric and positive definite. Now, the remaining classification relates directly to the contents of the matrices $[C]$ and $[S]$. Writing these matrices as a sum of their symmetric and skew-symmetric parts

$$[C] = \frac{1}{2}([C] + [C]^T) + \frac{1}{2}([C] - [C]^T), \quad [S] = \frac{1}{2}([S] + [S]^T) + \frac{1}{2}([S] - [S]^T) \tag{3}$$

we can directly define our remaining four classes of systems.

A **conservative, non-gyroscopic system** has no velocity depending terms and a symmetric stiffness matrix, which in physics may describe simple, free, undamped vibrations. The characteristics of the possible solutions are well known, and these solutions often serve as expansions for more complicated systems. A **conservative, gyroscopic system** may in addition include gyroscopic forces which are reflected in the skew-symmetric part of the damping matrix which in physics often will be a possible description for a rotor system. The non-conservative systems are systems with dissipative forces and/or with circulatory forces. The dissipative forces are described by the symmetric part of the damping matrix and the circulatory forces are described by the skew-symmetric part of the stiffness matrix. Thus for a **dissipative system** we have $([C] + [C]^T) \neq [0]$ without further condition on our general system. For a **circulatory system** we have $([S] - [S]^T) \neq [0]$ which means that non-potential forces are included in the system.

Classification of the behaviour. We are concerned about the stability of the solution $\{D\} = \{0\}$ of (2) to small disturbances. Now to classify as far as possible the behaviour of the above classes of systems, we assume the time separation by

$$\{D\} = \{\Phi\} \exp((\alpha + \imath\omega)t) = \{\Phi\} \exp(\alpha t)(\cos\omega t + \imath \sin\omega t) \tag{4}$$

and thus in this separation from the initial formulation uses a complex exponential factor $\lambda = \alpha + \imath\omega$ as time coefficient. With this formulation we can avoid the physical nonsense of complex frequency. Two distinct classes of instability follow from this separation. The first one is **flutter instability** (also termed dynamic instability), it corresponds to solutions $\alpha > 0, \omega \neq 0$. The second one is **divergent instability** (also termed static instability), it corresponds to solutions $\alpha > 0, \omega = 0$. Stability is in these notes defined by solutions having only negative α (real part), and a critical solution is the terminology for the case of a solution with at least one $\alpha = 0$ and negative α values for the remaining solutions. For $\omega \neq 0$ this corresponds to a harmonic vibration part of the solution. To be more specific about stability limits we should include gradient information. With p as a load parameter we define a **flutter load** p_F by the conditions that at least for one solution we have

$$\omega_F \neq 0, \quad \alpha_F = 0, \quad (\partial\alpha/\partial p)_F > 0 \tag{5}$$

and non-positive values for the remaining α solutions. In parallel a divergence load p_D is obtained for

$$\omega_D = 0, \quad \alpha_D = 0, \quad (\partial\alpha/\partial p)_D > 0, \quad (\partial\omega/\partial p)_D = 0 \tag{6}$$

Now, one of the most important theorems in stability tells that a conservative, non-gyroscopic system can only loose its stability by a divergent instability. This means that the definiteness of the stiffness matrix gives complete information about stability, but naturally not about the dynamic behaviour in general. Thus, such a system is stable if and only if $[S]$ is positive definite.

1.3 Sensitivity analysis

The notion of sensitivity analysis is a very wide one, even when we restrict it to response sensitivity for dynamic systems. It has been argued that sensitivity is just a new word for partial derivative, and to some extent this is correct. When our response is measured by a scalar like an eigenvalue and we ask for the sensitivity with respect to another scalar then it is a partial derivative. However, we may also ask for the sensitivity with respect to a function, and then the notion of gradient-function is often used. Furthermore, the response may be a function like an eigenfunction or another displacement description, and then we need a notion for the change of a function with respect to another function. So sensitivity analysis is a notion that covers the work of determining all this important information. The description in this section follows closely to Pedersen, 1984.

General variational analysis. We are interested in studying the dynamic behaviour, and with the separation (4) the spatial problem is described by the homogeneous matrix equation

$$[L]\{\Phi\} = \{0\} \quad with \quad [L] = [L(\alpha + \imath\omega, p, h)] \tag{7}$$

where the system matrix $[L]$ for the present analysis depends on the complex eigenvalue λ with α as a stability measure and ω as frequency according to the separation by $\{D\} = \{\Phi\}\exp(\alpha + \imath\omega t)$. Furthermore $[L]$ depends on the load level described by the real parameter p and on design, load distribution, damping etc., all of which we symbolize by the real quantity h, which - in individual cases - may be a scalar parameter or a spatial parameter function. In addition to the physical system we also analyse the adjoint problem

$$[L]^T\{\Psi\} = \{0\} \quad or \quad \{\Psi\}^T[L] = \{0\}^T \tag{8}$$

i.e. we need both right $\{\Phi\}$ and left $\{\Psi\}$ eigenvectors, for the same eigenvalue. We also remember that with a symmetric system matrix (self-adjoint problem) we have $\{\Phi\} = \{\Psi\}$ The results we shall obtain in this section are expressed as ratios between complex functionals, and in order to make it less abstract, an interpretation in terms of specific, mutual energies is presented. The term specific means that the energy may be per area or per volume and a (for the sensitivity analysis) non-important factor may be omitted. The term mutual relates to the fact that two different eigenvectors determine the energy. The functional, termed the total, specific, mutual energy Π is defined by

$$\Pi := \{\Psi\}^T[L]\{\Phi\} = 0 \tag{9}$$

where the zero follows from (7) as well as from (8). Taking general variations of Π, which is also zero with the variations, we get from $\delta\Pi = 0$

$$\{\delta\Psi\}^T[L]\{\Phi\} + \{\Psi\}^T[\delta L]\{\Phi\} + \{\Psi\}^T[L]\{\delta\Phi\} = 0 \tag{10}$$

We find that the variations of the eigenvectors $\{\Phi\}, \{\Psi\}$ disappear, which is naturally the reason for introducing the adjoint eigenvector and for taking variations of the mutual energy, $\delta\Pi = 0$, instead of variations relative to the equilibrium (7).

In relation to the sensitivity analysis where $[\delta L] \neq [0]$ we write (10) more specifically in the variations of the involved independent parameters δp and δh. To have a shorter notation we define the important functionals (normally complex)

$$\mathcal{A} = \{\Psi\}^T(\partial[L]/\partial\lambda)\{\Phi\}, \quad \mathcal{B} = \{\Psi\}^T(\partial[L]/\partial p)\{\Phi\}, \quad \mathcal{C} = \{\Psi\}^T(\partial[L]/\partial h)\{\Phi\} \tag{11}$$

and then get

$$\mathcal{A}(\delta\alpha + \imath\delta\omega) + \mathcal{B}\delta p + \mathcal{C}\delta h = 0 \tag{12}$$

Direct interpretations of this equation (12) give the results needed.

Sensitivity with respect to load level. With a fixed "design" $\delta h = 0$ the sensitivity from (12) makes possible a more rigorous definition of terms like critical load and flutter load. Assuming $\mathcal{A} \neq 0$ we read from (12) with $\delta h = 0$

$$\partial\alpha/\partial p = -Re(\mathcal{B}/\mathcal{A}), \quad \partial\omega/\partial p = -Im(\mathcal{B}/\mathcal{A}) \tag{13}$$

where the notation for real part is $Re()$ and the notation for imaginary part is $Im()$. To clarify the term instability initiation, we must in addition to $\alpha = 0$ require $\partial\alpha/\partial p > 0$. By (13) this instability condition will be

$$\alpha = 0 \quad and \quad Re(\mathcal{B}/\mathcal{A}) < 0 \tag{14}$$

In the case of $\omega \neq 0$ the condition (14) corresponds to initiation of flutter and in the case of $\omega = 0$ together with $Im(\mathcal{B}/\mathcal{A}) = 0$ the condition (14) corresponds to initiation of divergence.

Sensitivity with respect to design. With a fixed load level $\delta p = 0$ we read from (12), again assuming $\mathcal{A} \neq 0$

$$\partial\alpha/\partial p = -Re(\mathcal{C}/\mathcal{A}), \quad \partial\omega/\partial p = -Im(\mathcal{C}/\mathcal{A}) \tag{15}$$

We are often interested in knowing whether a certain parameter h acts in a stabilizing or destabilizing manner, which, from (15), we find to

$$Re(\mathcal{C}/\mathcal{A}) > 0 \Rightarrow h \ stabilizing, \quad Re(\mathcal{C}/\mathcal{A}) < 0 \Rightarrow h \ destabilizing \tag{16}$$

Flutter load sensitivity to design. A main question is related to the change in instability level as a function of design, i.e. all variations in (12) are involved. Since we focus on a load level of instability initialization, the variations are restricted by

$$\alpha = 0 \quad and \quad \delta\alpha = 0 \tag{17}$$

Again for $\mathcal{A} \neq 0$ we from (12) divided by $\mathcal{A}$ with (17) and using that $\imath\delta\omega$ is purely imaginary, get

$$\partial p_F/\partial h = -Re(\mathcal{C}/\mathcal{A})/Re(\mathcal{B}/\mathcal{A}) \tag{18}$$

Note that by (18) the sign of $\partial p_F/\partial h$ is equal to the sign of $Re(\mathcal{C}/\mathcal{A})$ This simply states the natural fact from (12) that if h is stabilizing, then p_F increases with h . If $\omega = \delta\omega = 0$ the load p_F , here termed flutter load, will as a special case be a divergence load. No specific analysis for this is therefore needed. The change in flutter frequency is in a similar way obtained with $\delta\alpha = 0$ and we get

$$\partial\omega_F/\partial h = -Im(\mathcal{C}/\mathcal{B})/Re(\mathcal{A}/\mathcal{B}) \tag{19}$$

because δp_F is a real quantity. In (19) we have used that $\mathcal{B} \neq 0$ which follows from (14).

Results with finite element formulation. We now restrict the description to system matrices $[L]$ given by

$$[L] = \lambda^2[M] + \lambda[C] + [S] + p[K] \tag{20}$$

with λ as the only complex quantity. In order to get convenient short notation we define specific, mutual energies: kinetic $\mathcal{T}$, dissipative $\mathcal{D}$, elastic $\mathcal{U}$ and external $\mathcal{W}$ by

$$\mathcal{T} := \{\Psi\}^T[M]\{\Phi\},\ \ \mathcal{D} := \{\Psi\}^T[C]\{\Phi\},\ \ \mathcal{U} := \{\Psi\}^T[S]\{\Phi\},\ \ \mathcal{W} := -\{\Psi\}^T[K]\{\Phi\} \tag{21}$$

and with (20) we may express $\mathcal{W}$ by

$$\mathcal{W} = (\lambda^2\mathcal{T} + \lambda\mathcal{D} + \mathcal{U})/p \tag{22}$$

The important functionals $\mathcal{A}$ and $\mathcal{B}$ can then directly be expressed in the mutual energies

$$\mathcal{A} = 2\lambda\mathcal{T} + \mathcal{D},\quad \mathcal{B} = -\mathcal{W} = -(\lambda^2\mathcal{T} + \lambda\mathcal{D} + \mathcal{U})/p \tag{23}$$

and thus the results for pure load change (13) are directly available

$$\partial\lambda/\partial p = \partial\alpha/\partial p + \imath\partial\omega/\partial p = (\lambda^2\mathcal{T} + \lambda\mathcal{D} + \mathcal{U})/(p(2\lambda\mathcal{T} + \mathcal{D})) \tag{24}$$

To obtain the functional $\mathcal{C}$ we need more specific information, and even for the case of $[K]$ independent of h , the evaluation of

$$\mathcal{C} = \{\Psi\}^T(\lambda^2\partial[M]/\partial h + \lambda\partial[C]/\partial h + \partial[S]/\partial h)\{\Phi\} \tag{25}$$

seems rather complicated. For a finite element model we have

$$[M] = \sum[M_e],\quad [C] = \sum[C_e],\quad [S] = \sum[S_e] \tag{26}$$

This has to be read symbolically because $[M], [C]$ and $[S]$ are matrices of order say 10^3 while $[M_e], [C_e]$ and $[S_e]$ are element matrices of order 10 . Often a parameter h_e only influences a specific element and then (25) is brought down to

$$\mathcal{C} = \{\Psi_e\}^T(\lambda^2\partial[M_e]/\partial h_e + \lambda\partial[C_e]/\partial h_e + \partial[S_e]/\partial h_e)\{\Phi_e\} \tag{27}$$

where the components of the lower order vectors $\{\Psi_e\}, \{\Phi_e\}$ are contained in the higher order vectors $\{\Psi\}, \{\Phi\}$. Thus the sensitivity analysis is brought down to the element level. Furthermore, the h_e dependence is often homogeneous, such that

$$\partial[M_e]/\partial h_e = m[M_e]/h_e,\quad \partial[C_e]/\partial h_e = l[C_e]/h_e,\quad \partial[S_e]/\partial h_e = k[S_e]/h_e \tag{28}$$

where m, l, k are given factors. Then (27) can be written directly in element mutual energies defined in parallel to (21)

$$\mathcal{C} = (m\lambda^2\mathcal{T}_e + l\lambda\mathcal{D}_e + k\mathcal{U}_e)/h_e \tag{29}$$

With the assumption behind (29), localized sensitivities are available , and the general results (15), (18), and (19) simplify to

$$\frac{\partial p_F}{\partial h_e} = \frac{Re((m\lambda^2\mathcal{T}_e + l\lambda\mathcal{D}_e + k\mathcal{U}_e)/(2\lambda\mathcal{T} + \mathcal{D}))}{h_e Re((\lambda^2\mathcal{T} + \lambda\mathcal{D} + \mathcal{U})/(2\lambda\mathcal{T} + \mathcal{D}))} \tag{30}$$

and

$$\frac{\partial\omega_F}{\partial h_e} = \frac{-Im((m\lambda^2\mathcal{T}_e + l\lambda\mathcal{D}_e + k\mathcal{U}_e)/(\lambda^2\mathcal{T} + \lambda\mathcal{D} + \mathcal{U}))}{h_e Re((2\lambda\mathcal{T} + \mathcal{D})/(\lambda^2\mathcal{T} + \lambda\mathcal{D} + \mathcal{U}))} \tag{31}$$

These formulas are directly suited for evaluation or programming.

Results with Galerkin modelling (global expansion). An alternative to the finite element method, which we can interpret as a Galerkin method with local expansion functions, we have methods with global expansion function often named Galerkin, Ritz and weighted residual. The advantages of these analysis methods is the moderate number of degrees of freedom, and the disadvantage is that expansion function valid over the whole structural domain V must be obtainable. With x as the space parameter in V , we assume given expansion functions $u_j = u_j(x)$ for the physical eigenfunction and $v_j = v_j(x)$ for the adjoint eigenfunction. Let us rewrite (27) without matrix notation

$$\mathcal{C} = \sum\sum(\lambda^2 \partial m_{ij}/\partial h + \lambda \partial c_{ij}/\partial h + \partial s_{ij}/\partial h)\Psi_i \Phi_j \tag{32}$$

where m_{ij}, c_{ij}, s_{ij} are matrix elements of $[M], [C], [S]$, respectively. These elements are for the global Galerkin discretization determined by

$$m_{ij} = \int mH(u_i, v_j)dx, \quad c_{ij} = \int mG(u_i, v_j)dx, \quad s_{ij} = \int mF(u_i, v_j)dx, \tag{33}$$

with F, G, H as given expressions in the functions u, v and their spatial derivatives. The factors $m = m(x), c = c(x), s = s(x)$, depend on the design function $h = h(x)$. By the partial derivative with respect to $h(x)$ we mean change of h at position x in space. Assuming only local influence we get

$$\frac{\partial m_{ij}}{\partial h(x)} = \frac{\partial m(x)}{\partial h(x)} H(u_i, v_j), \quad \frac{\partial c_{ij}}{\partial h(x)} = \frac{\partial c(x)}{\partial h(x)} G(u_i, v_j), \quad \frac{\partial s_{ij}}{\partial h(x)} = \frac{\partial s(x)}{\partial h(x)} F(u_i, v_j) \tag{34}$$

Inserting this in (32) we have

$$\mathcal{C}(x) = \sum\sum(\lambda^2 \frac{\partial m}{\partial h(x)} H(u_i, v_j) + \lambda \frac{\partial c}{\partial h(x)} G(u_i, v_j) + \frac{\partial s}{\partial h(x)} F(u_i, v_j))\Psi_i \Phi_j \tag{35}$$

The sensitivities are then a function of space and we term them gradient functions. For example in relation to the most important sensitivity (18) we have

$$g(x) := \partial p_F / \partial h(x) = -Re(\mathcal{C}(x)/\mathcal{A})/Re(\mathcal{B}/\mathcal{A}) \tag{36}$$

and the resulting flutter load variation is obtained by integration

$$\delta p_F = \int g(x)\delta h(x)dx \tag{37}$$

The following example has results evaluated by these formulas.

Example of a cantilever beam. In figure 1 (left) we have shown a cantilever beam of non-uniform mass $m(x)$, bending stiffness $s(x)$, Kelvin-Voigt internal damping with coefficient γ, external damping with coefficient β, and loaded according to the distribution $pq(x)$ as follower forces.

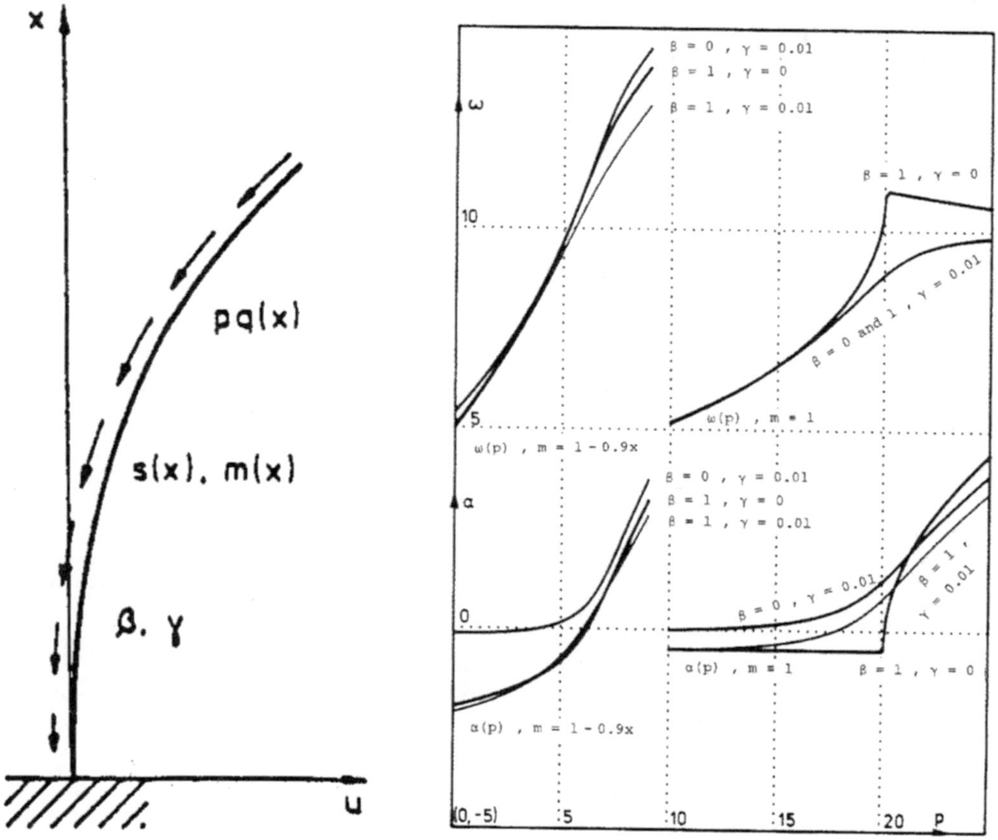

Figure 1. Left: extended Leipholz and Hauger columns including external and internal damping. Right: characteristic curves for the Becks column for a uniform well as for a linearly tapered column.

With eigenvalue λ and eigenfunction $u(x)$, the continuous eigenvalue problem is

$$((su'')'' + pQu'') + \lambda(\beta u + \gamma(su'')'') + \lambda^2 mu \equiv 0, \quad Q := \int pq(\xi)d\xi, \tag{38}$$

with the boundary conditions

$$u(0) = u'(0) = 0, \quad (1 + \lambda\gamma)(su'')_{x=1} = (1 + \lambda\gamma)(su'')'_{x=1} = 0 \tag{39}$$

Special cases are known as Leipholz's column and Haugers column, and with an end load Beck's column. Discretizing this problem by Galerkin expansions as given in details in Pedersen and Seyranian (1983) we obtain the analysis results shown in figure 1 (right) for the Becks column and with very similar results for the Leipholz and Hauger columns. The sensitivity analysis follows the theory described and the specific functional will for this problem be

$$\mathcal{C}(x) = \sum\sum((1 + \lambda\gamma)2mv_i''u_j'' + \lambda^2 v_i u_j)\Psi_i\Phi_j \tag{40}$$

and figure 2 shows results from sensitivity analysis. Note, that at positions (horizontal axes) where the curves are below the zero level we have negative flutter load gradient, and at these beam positions it is advantageous to decrease the beam area.

Multiple eigenvalues. The case of multiple eigenvalues is often excluded in sensitivity formulations, but in reality it is a physical important example that must be properly dealt with. The most simple case of a conservative, non-gyroscopic system for which

$$[L] = -\omega^2[M] + [S] \tag{41}$$

With $[M]$ and $[S]$ symmetric we will always have a complete set of eigenvectors, even when some of them have the same eigenvalue. All eigenvalues and eigenvectors will be real with left and right eigenvectors being identical (self-adjoint). A straight forward sensitivity analysis as derived from (25) gives for the specific solution $(\omega^2, \{\Phi\})$

$$\frac{\partial \omega^2}{\partial h} = \{\Phi\}^T(-\omega^2 \frac{\partial [M]}{\partial h} + \frac{\partial [S]}{\partial h})\{\Phi\} \tag{42}$$

when the normalization

$$\{\Phi\}^T[M]\{\Phi\} = \mathcal{T} = 1 \tag{43}$$

is assumed. This result can be found in Wittrick, 1962 and in many later references. The problem is that when the eigenvector $\{\Phi\}$ is not unique, the sensitivity is only defined as a directional derivative (h positive scalar coefficient to a specified design direction), and furthermore $\{\Phi\}$ can be any linear combination of the set of eigenvectors all with the same eigenvalue ω^2 . Only for some specific $\{\hat{\Phi}\}$ will the sensitivity (42) have a physical meaning. To find the specific $\{\hat{\Phi}\}$ vectors let us deal only with the bimodal case, where with $\{\Phi_1\}, \{\Phi_2\}$ being two arbitrary, but mutual orthogonal eigenvectors. We have

$$\{\Phi\} = c_1\{\Phi_1\} + c_2\{\Phi_2\}, \ \ \{\Phi_1\}[M]\{\Phi_2\} = 0, \ \ c_1^2 + c_2^2 = 1 \ \Rightarrow \ \{\Phi\}^T[M]\{\Phi\} = 1 \tag{44}$$

Inserting this expansion for $\{\Phi\}$ in (42) we get

$$\partial\omega^2/\partial h = c_1^2 g_{11} + c_2^2 g_{22} + 2c_1c_2g_{12} \ \ with \ \ g_{mn} := \{\Phi_m\}^T(-\omega^2\partial[M]/\partial h + \partial[S]/\partial h)\{\Phi_n\} \tag{45}$$

and thus depending on c_1, c_2. Extreme values of $\partial\omega^2/\partial h$ can be found by solving the eigenvalue problem constituted by these sensitivities

$$\begin{vmatrix} g_{11} - g & g_{12} \\ g_{12} & g_{22} - g \end{vmatrix} = 0 \tag{46}$$

The eigenvalues of this matrix are the true sensitivities, with corresponding eigenvectors $\{\hat{\Phi}\}$. When the eigenvalues g_a, g_b to (46) are found, we also have the corresponding eigenvectors $\{\Phi_a\}, \{\Phi_b\}$ that will return $g_{ab} = g_{ba} = 0$, which means that the mutual sensitivities are zero.

Note, that this case learns us the primary importance of eigenmodes as compared to eigenvalues. A recent, detailed discussion of the multiple modal problem is given by Seyranian et al., 1994. This reference includes several specific examples.

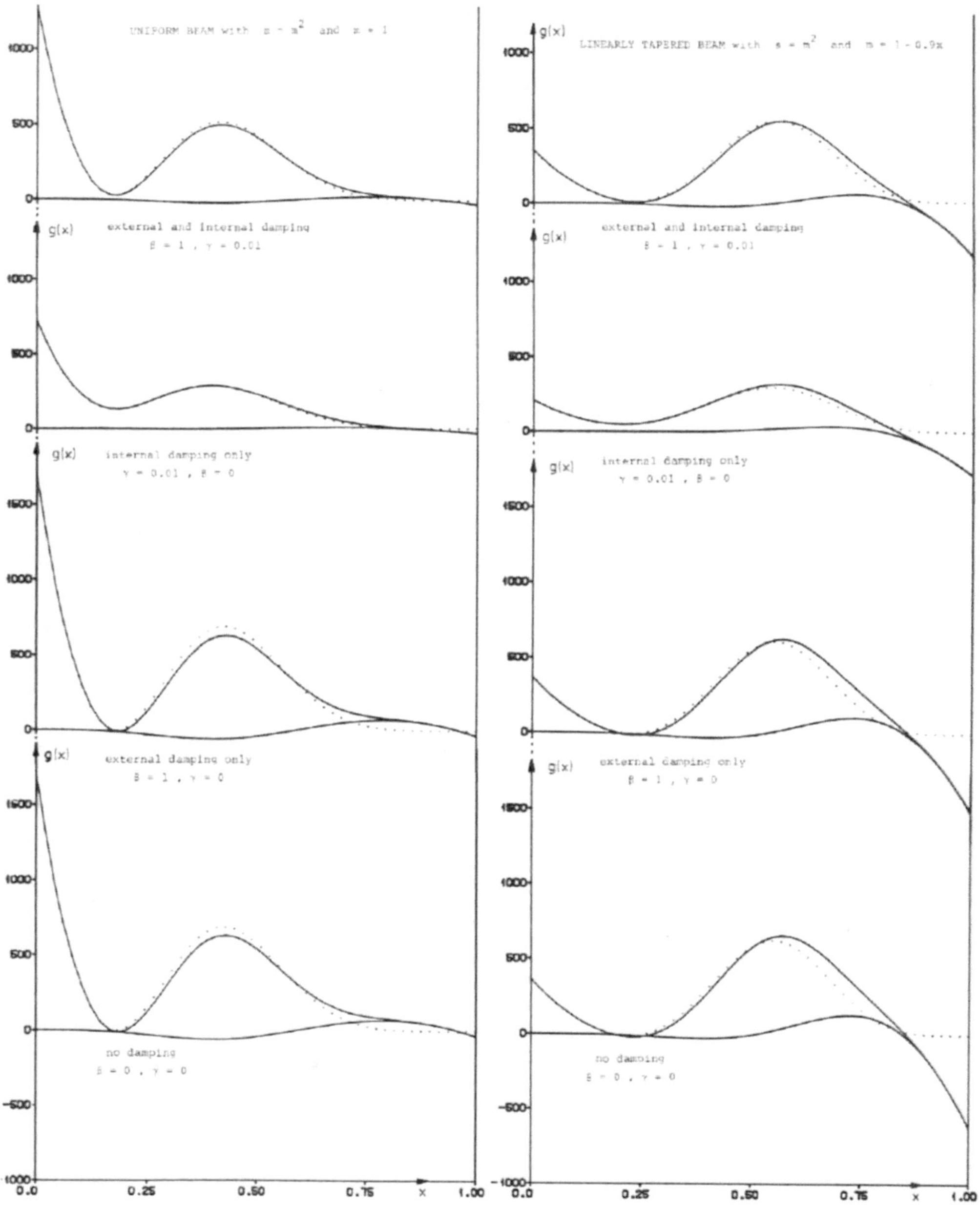

Figure 2. Gradient functions $g(x)$ for Hauger columns. Left: Uniform beam. Right: Linearly tapered beam. 1st row: both external and internal damping. 2nd row: only internal damping. 3rd row: only external damping. 4th row; no damping. For more details see Pedersen and Seyranian 1983.

Double eigenvalue with only a single eigenvector. For non-selfadjoint problems the sensitivity analysis may be further complicated by the fact that we do not have a complete set of eigenvectors. Thus the case of a double eigenvalue with the same eigenvector has to be dealt with. Often this is necessary to describe important physical problems like flutter initialization and interaction of eigenvalue branches. The paper Seyranian and Pedersen, 1995 treats this in details, and it will be described in another note.

2 Explanations for two Fluid/Structure Interaction Effects

2.1 Introduction

In this note two interesting mechanical phenomena are considered. The intention is to illustrate specific stability analysis, illustrate the use of Routh-Hurwitz stability criteria, and most of all it shows important analytical information available from sensitivity analysis. The first one is the phenomenon of transference of instability between branches. The second phenomenon is the destabilization of a non-conservative system due to infinitely small damping. These two phenomena are considered and explained from the point of view of the behaviour of eigenvalue branches in the vicinity of a double eigenvalue with a single corresponding eigenvector. It is found that the interaction of the eigenvalues near a double root is described by hyperbolas. To calculate the coefficients of these hyperbolas it is necessary to find the eigenvectors of main and adjoint systems as well as the associated vector corresponding to the double root. This supplies not only qualitative but also quantitative information.

The phenomenon of transference of instability between branches has often been discussed in the literature, as by Bunkov, 1969, Bishop and Fawzy, 1976 and Sugiyama and Noda, 1981. It turns out that a stable eigenvalue branch of a system subjected to non-conservative loading suddenly becomes unstable and vice verse with a change of problem parameters. Qualitatively, this effect is illustrated in figure 3, p being a load parameter.

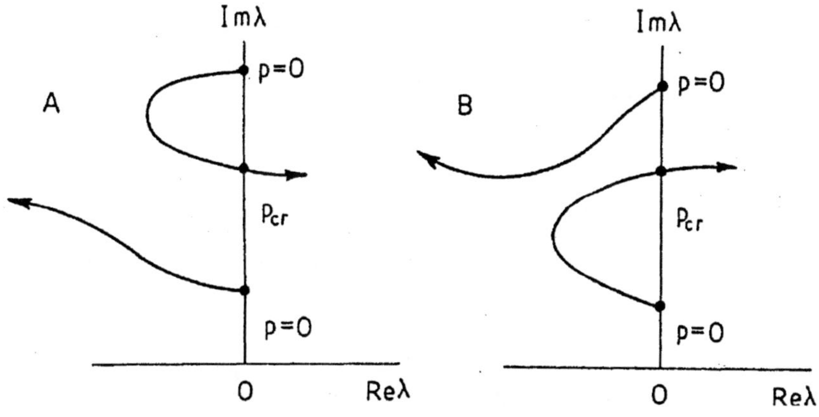

Figure 3. Transference of instability between branches. Behaviour A may change to behaviour B with change of problem parameters.

The second phenomenon considered is the destabilization effect due to infinitely small damping. This has been discussed in several papers: Ziegler, 1952, Herrmann and Jong, 1965, Gregory and Paidoussis, 1966, Herrmann, 1967, Bolotin, 1963, Bolotin and Zhinzher, 1969, Seyranian, 1990, Lottati and Kornecki, 1986, Huseyin, 1978 and Pedersen and Seyranian, 1983. It turns out that the critical load parameter of a non-conservative system with small damping in the general case is smaller than the critical load of the system with no damping. The additional references on these two effects can be found in a review by Paidoussis and Li, 1993.

These questions are close connected with sensitivity analysis of non-conservative vibrational systems subjected to instability phenomena. Sensitivity analysis of critical loads with respect to changes of problem parameters and simple eigenvalues were studied in earlier work by Pedersen and Seyranian, 1983. A continuation of this study to the behaviour of non-conservative systems exhibiting multiple roots is presented in Seyranian and Pedersen, 1995, which constitutes most of the present note.

2.2 Pipes conveying fluid

The dynamic behaviour of pipes conveying fluid has been studied in a number of papers, of which our main references are Benjamin, 1961, Bishop and Fawzy, 1976, Sugiyama and Noda, 1981. Further references are given in the reviews by Paidoussis and Li, 1993 and by Langthjem and Sugiyama, 2000a.

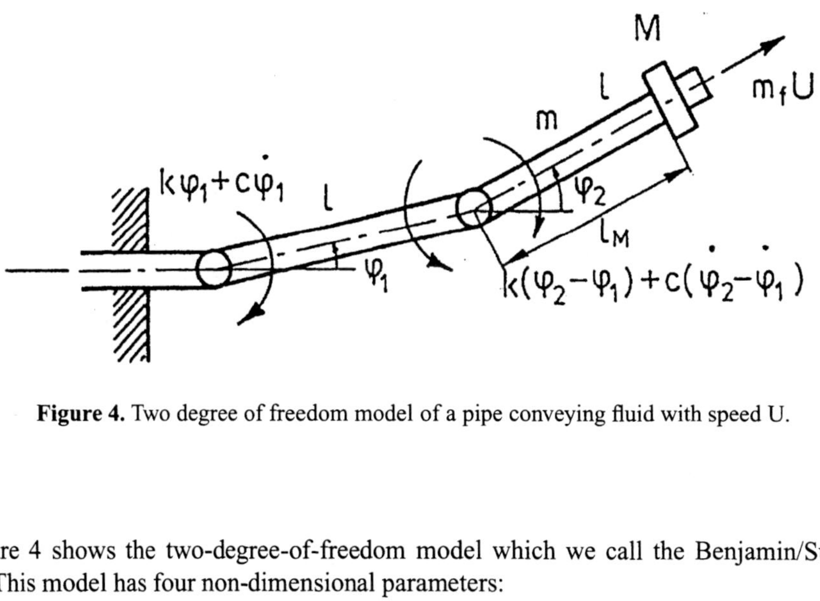

Figure 4. Two degree of freedom model of a pipe conveying fluid with speed U.

Figure 4 shows the two-degree-of-freedom model which we call the Benjamin/Sugiyama model. This model has four non-dimensional parameters:

$$\begin{aligned} non-dimensional\ fluid\ mass: \eta &:= \sqrt{m_f/(m+m_f)}\ for\ 0<\eta<1 \\ relative\ concentrated\ mass: \mu &:= M/((m+m_f)l)\ for\ 0<\mu<1 \\ position\ for\ concentrated\ mass: \xi &:= l_M/l\ for\ 0\le\xi\le 1 \\ relative\ damping: \gamma &:= c/\sqrt{k(m+m_f)l^3}\ for\ 0\le\gamma<\infty \end{aligned} \tag{47}$$

The quantities involved are: mass per length m of pipe and m_f of fluid, length of half the pipe l, length to concentrated mass l_M, concentrated mass M, stiffness coefficient k, viscous damping coefficient c, all shown in figure 4. The fluid speed is U, from which the non-dimensional speed u is defined

$$u := Ul\sqrt{m_f/(kl)}\ for\ 0 < u < \infty \tag{48}$$

and the non-dimensional time τ is defined from the absolute time t

$$\tau := t\sqrt{k/((m + m_f)l^3)} \tag{49}$$

The equations of free vibrations as stated in Sugiyama and Noda, 1981 are

$$\frac{1}{6}\begin{bmatrix} 8+6\mu & 3+6\mu\xi \\ 3+6\mu\xi & 2+6\mu\xi^2 \end{bmatrix}\begin{Bmatrix} \ddot{\phi}_1 \\ \ddot{\phi}_2 \end{Bmatrix} + \begin{bmatrix} \eta u + 2\gamma & 2\eta u - \gamma \\ -\gamma & \eta u + \gamma \end{bmatrix}\begin{Bmatrix} \dot{\phi}_1 \\ \dot{\phi}_2 \end{Bmatrix} +$$
$$\begin{bmatrix} 2-u^2 & -1+u^2 \\ -1 & 1 \end{bmatrix}\begin{Bmatrix} \phi_1 \\ \phi_2 \end{Bmatrix} = \begin{Bmatrix} 0 \\ 0 \end{Bmatrix} \tag{50}$$

The three matrices are: a symmetric, positive definite mass matrix, a damping matrix with gyroscopic and dissipative forces, and a stiffness matrix with circulatory forces, as classified by Ziegler, 1968. Assuming an exponential time dependence $\exp{(\alpha + \imath\omega)\tau}$ we get the characteristic equation

$$a_0(\alpha + \imath\omega)^4 + a_1(\alpha + \imath\omega)^3 + a_2(\alpha + \imath\omega)^2 + a_3(\alpha + \imath\omega) + a_4 = 0 \tag{51}$$

Onset of flutter - critical fluid speed. With the coefficients $a_0 - a_4$ in (51) being real quantities, we then have the two conditions for $\alpha = 0$

$$a_0\omega^4 - a_2\omega^2 + a_4 = 0 \tag{52}$$

$$\imath\omega(-a_1\omega^2 + a_3) = 0 \tag{53}$$

The case of $\alpha = 0$ is the critical case, which for flutter initiation ($\omega \neq 0$ because $a_4 \neq 0$) according to the condition (53) requires $\omega_F^2 = a_3/a_1$ which, substituted in the condition (52), gives the condition

$$H := a_1a_2a_3 - a_0a_3^2 - a_1^2a_4 = 0 \tag{54}$$

(As noted by Benjamin, 1961, the left-hand side of (54) constitutes the Hurwitz determinant, in this way constituting an alternative proof of this Hurwitz criterion.) The coefficients $a_0 - a_4$ are found directly from the matrices in (50). Then, with $a_i = a_i(\eta,\ \mu,\ \xi,\ \gamma,\ u)$, the Hurwitz determinant H can be written as an analytical function of our parameters $H = H(\eta,\ \mu,\ \xi,\ \gamma,\ u)$. The specific case of no concentrated mass and no damping with $\mu = \gamma = 0$ is

$$H = (\eta^2u^2/36)((13 - 24\eta^2)(u^2)^2 - (102 - 120\eta^2)u^2 + 169) \tag{55}$$

which, at the critical speed, gives $H = 0$ or, solved in terms of flutter fluid speed u_F and flutter frequency ω_F,

$$u_F = \sqrt{\frac{51 \quad 60\eta^2 - 2\sqrt{900\eta^4 - 51\eta^2 + 101}}{13 - 24\eta^2}} \quad and\ \omega_F = \sqrt{3(5 - u_F^2)/2} \tag{56}$$

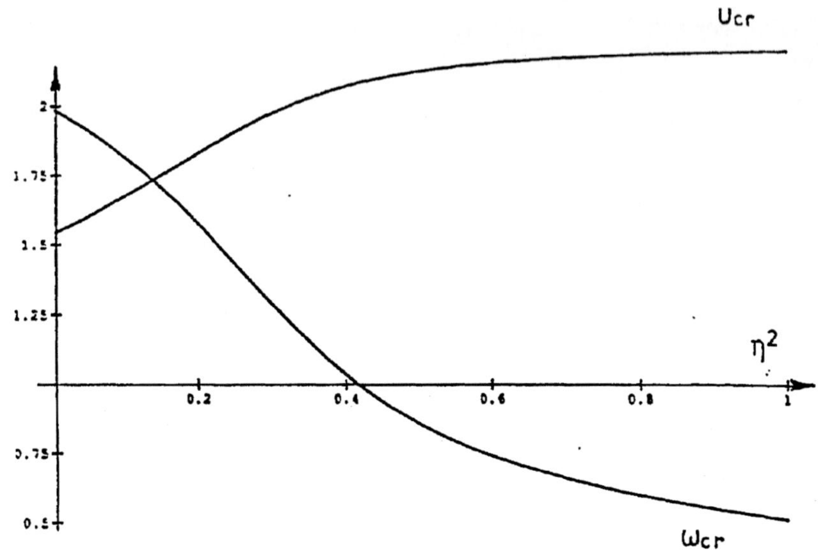

Figure 5. Critical non-dimensional fluid speed u_F and corresponding flutter frequency ω_F as a function of the non-dimensional fluid mass η^2.

The analytical solutions (56) are for a specific case tested with the results in Sugiyama and Noda, 1981. Figure 5 shows the results as a function of the non-dimensional fluid mass.

To illustrate the destabilizations as well as the stabilizations by the involved parameters, we show in figure 6 the parameter dependence of the critical speed more generally, i.e. $u_F = u_F(\eta,\ \mu,\ \xi,\ \gamma)$ in the domains of $0 < \eta < 1,\ 0 < \mu < 5,\ \xi = 0.25, 0.5, 0.75, 1$ *and* $\gamma = 0.00001, 0.001, 0.1$. The results in figure 6 are obtained by Newton-Raphson solutions to $H = 0$. A similar plot of the flutter frequency and plots of the flutter mode are also available, although not shown. We conclude that analytical stability analysis is possible for this two-degrees-of-freedom, four-parameter problem. All the analytical expressions were derived with the help of the Mathematica program, see Wolfram, 1991.

Transfer of instability between branches. Instability transference is one of those interesting phenomena that occur very often in parametric studies of stability of non-conservative systems, see figure 3. Transfer of instability between branches was discussed in the papers by Bunkov, 1969, and by Sugiyama and Noda, 1981. Bishop and Fawzy, 1976 suggested terming this phenomenon "dynamic interference". We shall here give a consistent analytical description and explanation. The major point in understanding the behaviour relates to the existence of an isolated double root to the characteristic equation (51). Only one corresponding eigenvector from (50) exists and thus we have the case of strong interaction between eigenmodes, see Seyranian, 1991, Seyranian, 1993.

Now, with a double root $\lambda = \alpha + \imath\omega$ we know all the four roots because also the complex conjugate is then a double root $\lambda = \alpha - \imath\omega$. Thus, the fourth order polynomial in terms of α, ω

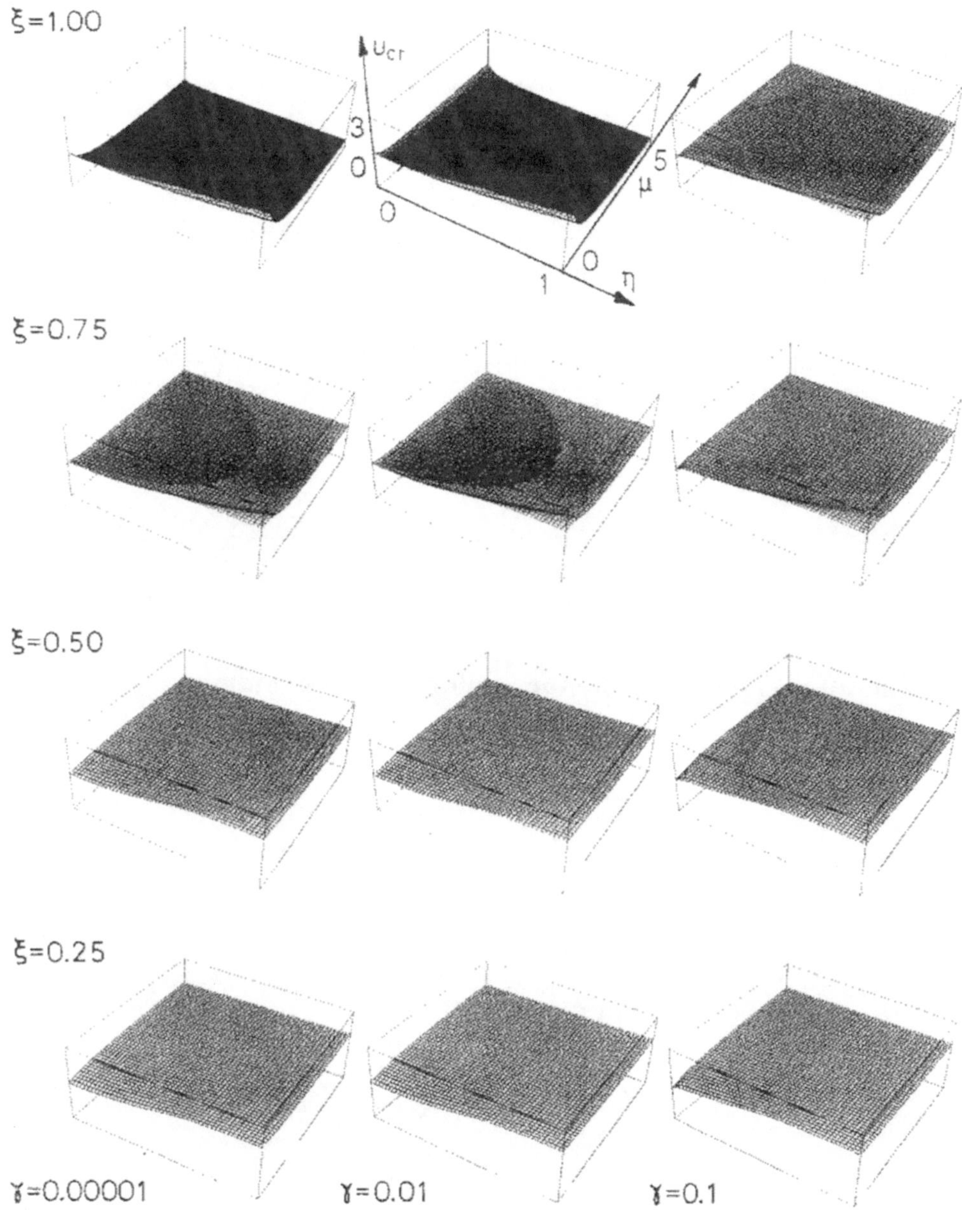

Figure 6. A four parameter plot for the non-dimensional critical fluid speed. Discrete parameters: relative damping $\gamma = 0.00001, 0.001, 0.1$ and position of concentrated mass $\xi = 0.25, 0.5, 0.75, 1$ Continuous parameters: non-dimensional fluid mass $0 < \eta < 1$ and relative concentrated mass $0 < \mu < 5$.

will be

$$\lambda^4 - 4\alpha\lambda^3 + 2(3\alpha^2 + \omega^2)\lambda^2 - 4\alpha(\alpha^2 + \omega^2)\lambda + (\alpha^2 + \omega^2) = 0 \tag{57}$$

The simple model without damping and without the concentrated mass, i.e. $\gamma = \mu = 0$, gives the polynomial

$$\lambda^4 - (24/7)\eta\mu\lambda^3 + (1/7)(108 - 30u^2 + 36\eta^2 u^2)\lambda^2 + (36/7)\eta u(5 - u^2)\lambda + 36/7 = 0 \tag{58}$$

and now, by directly equalizing the coefficients of (57) and (58), we can find the possible double root(s). We get only one possible solution which corresponds to the values

$$u_0 = \sqrt{\frac{5\sqrt{7} - 4}{\sqrt{7}}} = 1.867657, \ \eta_0 = \sqrt{\frac{7(7\sqrt{7} - 6)}{18(5\sqrt{7} - 4)}} = 0.726353,$$

$$\alpha_0 = -\sqrt{\frac{72(7\sqrt{7} - 6)}{7\sqrt{7}}} = -1.162782, \ \omega_0 = \pm\sqrt{\frac{2(27 - 7\sqrt{7})}{7\sqrt{7}}} = \pm 0.956936 \tag{59}$$

The values of (59) agree with the numerical results in Sugiyama and Noda, 1981. Note that $\alpha_0 < 0$ and thus we have a stable solution for the parameter values in (59). We have thus given an analytical explanation for the numerical results obtained by Sugiyama and Noda, 1981.

2.3 The simple model of a damped follower force column

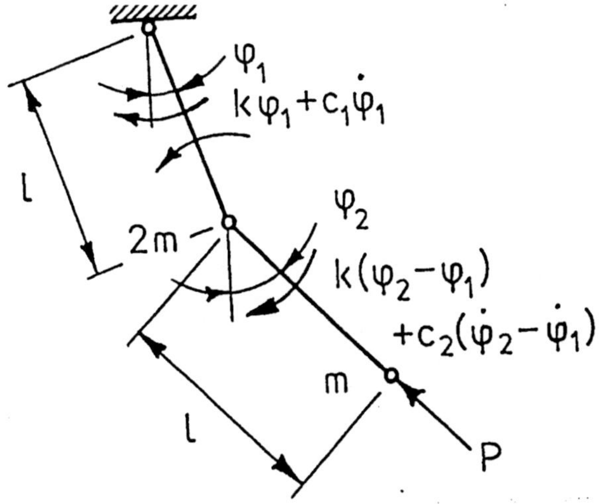

Figure 7. The Ziegler/Herrmann simple model of a damped follower force column.

A simple, two-degrees-of-freedom model is shown in figure 7. This column loaded with a follower force P has been studied by Ziegler, 1952 and in the present extended version by

Herrmann and Jong, 1965. We shall mainly concentrate on the effects known as destabilization by damping and show that a consistent explanation can be obtained with the sensitivity analysis. Also we shall point out the extended information that is available from the Hurwitz determinant.

The equations of free motion for the model are, with non-dimensional time, $\tau := t\sqrt{k/(ml^2)}$ given by

$$\begin{bmatrix} 3 & 1 \\ 1 & 1 \end{bmatrix} \begin{Bmatrix} \ddot{\phi}_1 \\ \ddot{\phi}_2 \end{Bmatrix} + \begin{bmatrix} \gamma_1+\gamma_2 & -\gamma_2 \\ -\gamma_2 & \gamma_2 \end{bmatrix} \begin{Bmatrix} \dot{\phi}_1 \\ \dot{\phi}_2 \end{Bmatrix} + \begin{bmatrix} 2-p & -1+p \\ -1 & 1 \end{bmatrix} \begin{Bmatrix} \phi_1 \\ \phi_2 \end{Bmatrix} = \begin{Bmatrix} 0 \\ 0 \end{Bmatrix} \tag{60}$$

with the non-dimensional parameters

$$\begin{aligned} damping\ at\ support: \gamma_1 &:= c_1\sqrt{kml^2}\ for\ 0 \le \gamma_1 \\ damping\ at\ midpoint: \gamma_2 &:= c_2\sqrt{kml^2}\ for\ 0 \le \gamma_2 \\ follower\ load:\ p &:= \ Pl/k\ for\ 0 \le \xi \le 1 \end{aligned} \tag{61}$$

The absolute quantities are: concentrated masses m and $2m$, length of half the column l, stiffness coefficient k, viscous damping coefficients c_1 and c_2, follower load P, all shown in figure 7.

Stability analysis with Hurwitz criteria. For the model given by (60) we directly get the Hurwitz determinant

$$H := (\gamma_1+6\gamma_2)(7-2p+\gamma_1\gamma_2)(\gamma_1+\gamma_2) - 2(\gamma_1+\gamma_2)^2 + (\gamma_1+6\gamma_2)^2 \tag{62}$$

and its derivative with respect to the load parameter

$$\partial H/\partial p = -2(\gamma_1+6\gamma_2)(\gamma_1+\gamma_2) \tag{63}$$

Without damping ($\gamma_1 = \gamma_2 = 0$) the Hurwitz determinant is identically equal to zero, so we have only little information. The parentheses in (62) show the individual terms of the characteristic equation. The Hurwitz criteria for stability are $H > 0$ and $7 - 2p + \gamma_1\gamma_2 > 0$ because the other criteria are always satisfied for $\gamma_1 > 0$ and $\gamma_2 > 0$. To be more specific we shall include the sensitivity of the Hurwitz determinant in our conditions for onset of flutter, meaning that stability is changed to instability when p becomes greater than p_F

$$H = 0\ \ and\ \ \partial H/\partial p < 0 \tag{64}$$

and we note from (63) that the last condition is always satisfied. From (64) and (62) we can get the critical flutter load for a damped system

$$p_F = \frac{4\gamma_1^2+33\gamma_1\gamma_2+4\gamma_2^2}{2(\gamma_1^2+7\gamma_1\gamma_2+6\gamma_2^2)} + \frac{1}{2}\gamma_1\gamma_2 \tag{65}$$

This expression was first obtained by Herrmann and Jong, 1965. Equation (65) describes a surface in a three dimensional space, see figure 8. It should be noted that the function of this surface is discontinuous at the point $\gamma_1, \gamma_2 = 0, 0$. However, Herrmann and Jong, 1965 showed that for any fixed ratio of damping coefficients $\zeta = \gamma_1/\gamma_2\ with\ 0 \le \zeta < \infty$ there exists a limit of the critical load p_F when damping tends to zero, which satisfies the inequality

$$p_F \le p_0 = \frac{7}{2} - \sqrt{2} \approx 2.08 \tag{66}$$

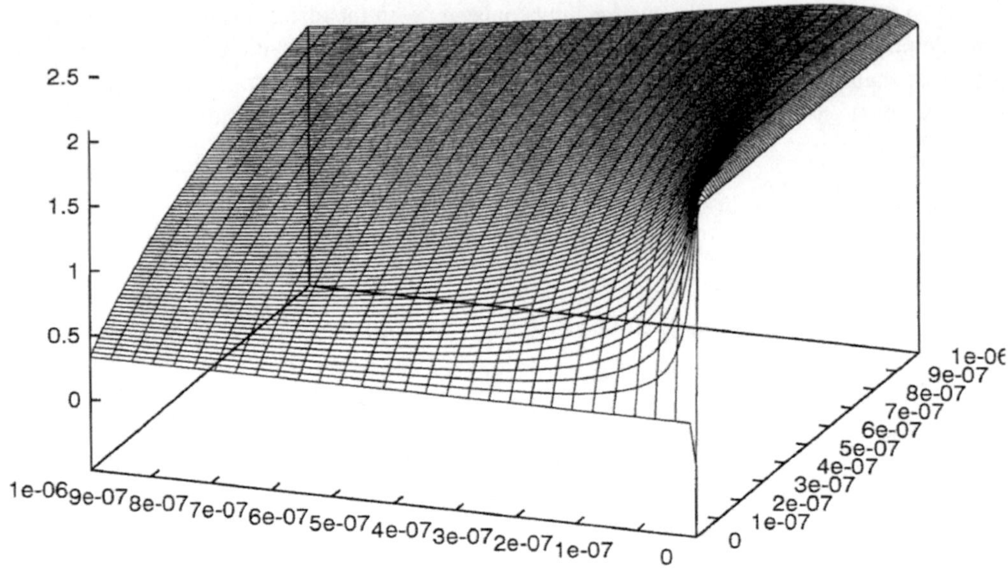

Figure 8. Stability surface in two-dimensional damping space.

Interaction of eigenvalues. Based on the knowledge that without damping the flutter instability is related to a point of strong interaction, we shall study the conditions for a double eigenvalue as we did for the pipes conveying fluid. The equation (57) agrees with the actual characteristic equation for

$$\alpha = -\tfrac{1}{8}(\gamma_1 + 6\gamma_2), \qquad \alpha^2 + \omega^2 = \sqrt{2}/2,$$
$$2(3\alpha^2 + \omega^2) = \tfrac{1}{2}(7 - 2p + \gamma_1\gamma_2), \qquad -4\alpha(\alpha^2 + \omega^2) = \frac{1}{2}(\gamma_1 + \gamma_2) \tag{67}$$

which is satisfied only for the constant ratio

$$\zeta_0 = \gamma_1/\gamma_2 = 5\sqrt{2} + 4 \ \ or \ \ for \ \ \gamma_1 = \gamma_2 = 0 \tag{68}$$

For damping ratio $\zeta \neq (5\sqrt{2} + 4)$ no points of double eigenvalues exist, i.e. $\lambda_0 = \imath 2^{-\frac{1}{4}}$ is the isolated double eigenvalue.

Destabilization. In order to relate the specific problem described by (60) for $\gamma_2 = \gamma, \gamma_1 = \zeta\gamma$, with the constant $\zeta \neq (5\sqrt{2}+4)$ we solve the eigenvalue problem for $\gamma = 0, p_0 = \frac{7}{2} - \sqrt{2}$ and obtain the double eigenvalue $\lambda_0 = \imath 2^{-\frac{1}{4}}$ and the corresponding vector. We find the asymptotic behaviour of eigenvalues near the double root and from this give an analytical explanation of destabilization due to small damping. For details see Seyranian and Pedersen, 1995.

Stabilization. For damping ratio $\zeta \neq (5\sqrt{2}+4)$ we obtain from the solution in (115) the load at which the double eigenvalue exists

$$p_1 = (7/2 - \sqrt{2}) - \gamma^2(59 + 30\sqrt{2})/8 \tag{69}$$

This load for the double root decreases with damping and then corresponds to a stable solution because from the solutions (115) and (68) we have

$$\alpha_1 = -\gamma(5\sqrt{2}+10)/8 < 0 \;\; and \;\; \omega_1 = \sqrt{\sqrt{2}/2 - \gamma^2 25(3+2\sqrt{2})/32} \tag{70}$$

This means that the double eigenvalue $\lambda_0 = \imath 2^{-1/4}$ is shifted to the left half plane when small damping with the specific ratio $\zeta \neq (5\sqrt{2}+4)$ is added.

Thus, at this point the strong interaction of eigenvalues takes place, i.e. with the increase of load p the eigenvalues approach, meet at $p = p_1$, and then split along the line which is in the first approximation parallel to the real axis. So, in this case inclusion of small damping leads to stabilization because the critical load of damped system found from Hurwitz criterion is equal to

$$p_F = (7/2 - \sqrt{2}) + \gamma^2(5\sqrt{2}+4)/2 \;\; i.e. \; p_1 < p_0 < p_F \;\; for \; \gamma > 0 \tag{71}$$

Thus, in this particular case the behaviour of the non-conservative system is similar with the behaviour of the conservative system, when damping is added.

We can find in the γ_1, γ_2 -plane the region of stabilization due to damping if we solve the inequality $p_F \geq p_0$ using (65). The boundaries of the stabilization region are described by the asymptotes, which gives a very narrow region

$$\gamma_1 = \gamma_2(4 + 5\sqrt{2}) \pm \gamma_2^2\sqrt{50(133 + 94\sqrt{2})} \tag{72}$$

2.4 Conclusions

We have considered two interesting phenomena that are often discussed in the literature. The first phenomenon is known as transfer of instability branch and the second is called destabilization due to small damping. Having derived the asymptotic behaviour of eigenvalues near a double root of multi-parameter non-conservative systems, we give a consistent analytical description and explanation of these effects.

Multiple eigenvalues play a very important role in non-conservative vibration problems because in many cases they describe the boundaries between stable and unstable domains. From a mathematical point of view, sensitivity analysis of multiple roots is a rather complicated problem. The sensitivity analysis of multiple roots is carried out by applying the perturbation method to multi-parameter vibrational systems.

It is shown that the interaction of the eigenvalue branches near a double root is described by hyperbolas. To find the coefficients of these hyperbolas it is necessary to solve the eigenvalue problem and then calculate eigenvectors of the system and of the adjoint system as well as the associated vector. This supplies not only qualitative but also quantitative information.

We carried out the analysis in relation to two specific physical problems. One of the important consequences of the paper Seyranian and Pedersen, 1995 is that destabilization or stabilization of a non-conservative system due to small damping depends on how many damping parameters are available: in the case of more than one it is possible to choose the parameters so that damping stabilizes the system. The other interesting feature when two independent damping parameters are considered , is that destabilization due to infinitely small damping, $p_d \leq p_0$, does not occur in the sense that the limit $p_d = lim\ p_F\ for\ \gamma_1, \gamma_2 \Rightarrow 0, 0$ does not exist. This shows that the nature of destabilization depends on the dimension of damping parameters.

3 Stability Analysis for Linear Periodic Systems

3.1 Introduction.

A large number of practical important problems can be modelled by linear differential equations with periodic coefficients. From direct formulation we may mention periodically varying stiffness, mass and loads (parametric excitation). Other problems are frequency modulation, warble tone room testing in acoustics and plasma physics. Also stability of non-linear vibrations and energy transfer from one vibration mode to another by non-linear coupling can be studied by periodic systems. Finally we should not forget the applications which originated the study of the periodic differential equations, including mean motion of the lunar pedigree and wave propagation in stratified media.

Consequently the stability analysis for the solutions to differential equations with periodic coefficients has been a challenge for more than one hundred years. From a historical point of view, the important early studies are by Mathieu, 1868, Floquet, 1883, Hill, 1886, Rayleigh, 1887, Liapunov, 1949 (1892). Several books have been written on the subject, and in these several hundred references may be found. Among the important ones are the books by Strutt, 1932, McLachlan, 1947, Meixner and Schaefke, 1954, Arscott, 1964, Magnus and Winkler, 1966, Smith, 1975, Yakubovitch and Starzhinskii, 1975, Yakubovitch and Starzhinskii, 1987. For numerical analysis different methods are available. We mention the method of infinite determinants (see Bolotin, 1964), the perturbation method (see Nayfeh and Mook, 1979), the Galerkin method (see Pedersen, 1985) in addition to the classical Floquet method (see Floquet, 1883, Cesari, 1964). Few of these methods can from a practical point of view be extended to multi-degree of freedom systems, i.e. to coupled, second order, linear systems with matrices containing periodic coefficients. For such extensions see Fu and Nemat-Nasser, 1972, Hansen, 1985, and Lindh and Likins, 1970. It is concluded that the Floquet method is a general and practical method for systems with multi-degrees of freedom, and for a single-degree of freedom problem general stability diagrams are available or can be made available.

In this lecture, we shall start with the single differential equation, known as the Hill equation, and relative to this classify the possible periodic functions as well as the solution functions. Then specific problems and specific stability diagrams are shown, and we present the Floquet method and define the Floquet matrix. Even with increasing computer power the large number of numerical integrations required in this method put limits to our possibilities, and research with the goal of carrying out these integrations in a most effective way have been performed, see Sinha and Wu, 1991. The multipliers of the system are the basis for the stability analysis, instability occurs when any of the multipliers has a numerical value greater than one.

In a following notes we add sensitivity analysis to the Floquet method and show that optimal design formulations can then be solved.

3.2 The Hill equation and stability of the solutions

As standard form for the general Hill equation with damping, we choose

$$\ddot{y} + 2c_0\dot{y} + (a_0 + 2q_0\phi_0(\tau))y = 0 \tag{73}$$

where all quantities are real. The parameters are $c_0 \geq 0$ for damping, a_0 we term the excitation level, q_0 the excitation amplitude, and $\phi_0(\tau)$ is a π-periodic excitation function. For some problems the term excitation may be rather abstract, but we choose to keep it anyhow. The unknown displacement function is $y = y(\tau)$ and dot indicates differentiation with respect to τ. The meaning of the sub-index $_0$ is explained later.

The specific case $\phi_0(\tau) = \cos 2\tau$ or $\phi_0(\tau) = \sin 2\tau$ constitutes the Mathieu equation with damping. In general, the results of the stability analysis are not influenced by the origo τ_0 of time, and thus we may change cos 2τ to sin 2τ without influence, In fact, this would be the form of the Mathieu equation which directly implies symmetry (relative to $q_0 = 0$) of the stability diagram.

Actual physical stability problems. As the first problem described by the Mathieu equation we take the simple pendulum mounted on a **vertically oscillating support**. This classical example, which is also studied experimentally (Kapitsa inverted pendulum) is found in many textbooks, e.g. Ziegler, 1968 or Panovko and Gubanova, 1964, and the equation is usually presented without damping

$$\ddot{y} + (\pm\frac{4g}{\Omega^2 l} + \frac{4a}{l}\cos 2\tau)y = 0 \tag{74}$$

Here, g is gravity, Ω is the frequency and a the amplitude of the oscillating support ($a\cos\Omega t$), while l is the pendulum length. Upper sign + refers to the hanging pendulum and lower sign − refers to the inverted pendulum, which, in accordance with experiments, we know can be stable for certain parameter combinations. This problem may be generalized to a double pendulum and furthermore to an n-degree of freedom pendulum, as explained by Hsu, 1961. The limiting case of $n \Rightarrow \infty$ then models the Indian rope trick as discussed by Otterbein, 1982.

The next class of problems are models with **periodically varying stiffness**. As a specific example, the stiffness of a shaft may vary with varying length. The undamped Mathieu equation for torsional vibrations according to Panovko and Gubanova, 1964, with the assumption of $a << l_0$ is

$$\ddot{y} + \frac{4GK}{Jl_0\Omega^2}(1 - \frac{a}{l_0}\cos 2\tau)y = 0 \tag{75}$$

where GK is the torsional stiffness of the shaft, which has length $l = l_0 + a\cos\Omega t$. The polar mass of inertia of the disc is J.

Problems of **periodically varying mass** are described by a specific differential equation which by Epstein and Barakat, 1977 is termed the Carson-Cambi equation

$$(1 + \epsilon\cos 2\tau)\ddot{y} + 4py = 0 \; for \; |\epsilon| < 1 \tag{76}$$

This equation also relates to the theory of frequency modulation, to warble tone room testing in acoustics, and to plasma physics, according to Epstein and Barakat, 1977. By Fourier expansion, it is transformed into the corresponding Hill-equation. The stability diagrams for this equation show the important property that every second domain of instability (relative to the general Hill diagram) vanishes.

Periodically varying loads may also give rise to the Hill equation. Transverse vibrations of a uniform, simple supported column with axial load $P = P_0 + P_1\phi_0(\tau)$ are described by

$$\ddot{y} + (4\eta(1 - \lambda_0) + 4\eta\lambda_1\phi_0(\tau))y = 0 \tag{77}$$

with the non-dimensional quantities: η for frequency, λ_0 for the column force P_0, and λ_1 for the pulsating force P_1. These problems belong to the subject termed **parametric excitation**. For more extended examples, see the books by Bolotin, 1964 and by Smith, 1975.

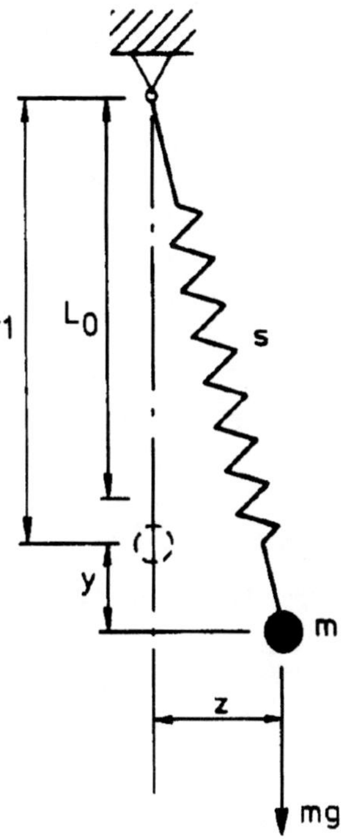

Figure 9. Elastic pendulum with possible energy transfer from longitudinal to angular vibrations.

The question of **stability for non-linear vibrations** may also to some extent be answered by stability diagrams for Hill-equations. In the classic book by Stoker, 1950, the stability of solutions to the Duffing equation is studied. Assuming the solutions to the Duffing equation

$$d^2x/dt^2 + \gamma(dx/dt) + \alpha x + \beta x^3 = A\cos\Omega t \tag{78}$$

to be $x = a\cos\Omega t + b\cos 3\Omega t$ (approximate for $\beta, \gamma \to 0, 0$). We want to find out whether this (approximate) solution is stable relative to small disturbances. Then we obtain a damped Hill equation for the variation δx relative to this solution. The specific parameters are: $c_0 = \gamma/(2\Omega), a_0 = (\alpha + \frac{1}{2}(a^2 + b^2)3\beta)/(\Omega^2), q_0 = (3\beta/(2\Omega^2)$ and $\phi_0(\tau) = (\frac{1}{2}a^2 + ab)\cos 2\tau + ab\cos 4\tau + \frac{1}{2}a\cos 6\tau$.

Problems related to **energy transfer** from one vibration mode to another undesired vibration mode due to a non-linear coupling are also for the most simple models described by Mathieu-Hill equations. For reference to these problems see Mettler, 1959, Nayfeh and Mook, 1979, Cartmell, 1990, Thomsen, 1997. Figure 9 shows the elastic pendulum with the describing differential equations containing quadratic terms. The initiation of pendulum vibrations from z =

0 is determined by the Mathieu equation with $c_0 = 0, a_0 = (4mg)/(L_0 s + mg)$ *and* $q_0 = (2L_0 A)/(L_0 + mg/s)^2$.

Finally, we should not forget the applications which originated the study of the periodic differential equations, see Mathieu, 1868, Hill, 1886 and Rayleigh, 1887, who studied problems including **vibration of a stretched membrane** with elliptical boundary; the **mean motion of the lunar pedigree**, and **wave propagation in stratified media**.

Normalizations and classifications. The purpose of a stability diagram is to provide results for a broad class of problems. This implies some degree of uniformity, and this uniformity is naturally more or less a matter of choice. First of all, it is necessary to deal with a reference equation in addition to the "physical" equation, which is already based on some normalizations. We choose to normalize time in such a way that ϕ_0 is π-periodic, which results in instability values of a_0 corresponding to $q_0 = 0$ being integers: $a_0 = n^2 = 0, 1, 4, 9, \ldots$ To get a clear definition of a_0 which is itself periodic (constant) , a basic assumption on the mean value of ϕ_0 is necessary. With a zero mean value the following Fourier expansion is in general possible

$$\phi_0 = \sum_{k=1,3,\ldots} a_k \sin((k+1)\tau) + \sum_{k=2,4,,\ldots} a_k \cos(k\tau) \; with \; a_1, a_2 \neq 0, 0 \tag{79}$$

whereby the idea of the notion a_0 is explained. To further normalize the factor q we choose two normalization conditions

$$\int_0^{2\pi} \phi d\tau = 0 \;\; and \;\; \int_0^{2\pi} \phi^2 d\tau = \pi/2 \tag{80}$$

which are both satisfied for $\phi(\tau) = \cos 2\tau$ as well as for $\phi(\tau) = \sin 2\tau$.

We now introduce the parameter $c = \alpha + \imath\omega$, which is not a physical quantity but c makes it possible to obtain a practical reference equation. The variable transformation is

$$y(\tau) = \exp\left((c - c_0)\tau\right) z(\tau) = \exp\left((\alpha - c_0)\tau\right) \exp\left(\imath\omega\tau\right) z(\tau) \;\; with \; c := \alpha + \imath\omega \tag{81}$$

and we get a new damped Hill equation, **the reference equation**,

$$\ddot{z} + 2c\dot{z} + (a + 2q\phi(\tau))z = 0 \;\; with \; a - c^2 = a_0 - c_0^2 \;\; and \;\; q\phi = q_0\phi_0 \tag{82}$$

So that ϕ satisfies the normalizations (160). The variable substitution should be seen in the light of the fact that $z = z(\tau)$ is a 2π-periodic solution function satisfying (82) and chosen so, because boundaries between stable and unstable parameter domains can be proven to correspond to 2π-periodic functions. A number of these functions are also π-periodic functions and in total we **classify the solution functions** as follows

$$\begin{aligned} &z(\tau) \;\; even \;\; \pi - periodic \\ &z(\tau) \;\; odd \;\; \pi - periodic \\ &z(\tau) \;\; even \;\; 2\pi - periodic \; but \; not \; \pi - periodic \\ &z(\tau) \;\; odd \;\; 2\pi - periodic \; but \; not \; \pi - periodic \end{aligned} \tag{83}$$

In Pedersen, 1985 a number of important simplifications are proven for different **classes of the excitation function Φ**, and we therefore also list the classification of these given functions

$$\begin{array}{lll} \phi & neither\ even & nor\ odd\ for\ any\ origo\ \tau_0 \\ \phi & even\ for\ some\ origo\ \tau_0 & but\ not\ odd\ for\ any\ origo \\ \phi & odd\ for\ some\ origo\ \tau_0 & but\ not\ even\ for\ any\ origo \\ \phi & even\ for\ some\ origo\ \tau_{0e} & and\ odd\ for\ some\ other\ origo\ \tau_{0o} \end{array} \quad (84)$$

Solution procedure. By the transformation from (73) to (82) we have in reality obtained an eigenvalue problem, where c is an eigenvalue, that directly give quantitative information about stability. Again referring to Pedersen, 1985 we find that the sign of c has no influence on the solution and the solutions c^2 will always be real. From this follows

$$\begin{array}{l} c^2\ real\ and\ > 0 \Rightarrow y(\tau) = \exp\left((\alpha - c_0)\tau\right) z(\tau)\ or \\ c^2\ real\ and\ < 0 \Rightarrow y(\tau) = \exp\left(-c_0\tau\right)\exp\left(\imath\omega\tau\right) z(\tau) \end{array} \quad (85)$$

In the first case we have stability for $\alpha = \sqrt{c^2} < c_0$, 2π-periodic solution for $\alpha = \sqrt{c^2} = c_0$ and instability for $\alpha = \sqrt{c^2} > c_0$. In the second case with $\omega = \sqrt{-c^2}$ we always have stability (assumed that the physical damping c_0 is non-negative). The factor $\exp\left(\imath\omega\tau\right) z(\tau)$ will in general be an almost periodic function.

To get the eigenvalue c^2 we need to solve (82) and this is done by a Galerkin approach, where the 2π-periodic function z is expanded in trigonometric functions with coefficients collected in a vector $\{B\}$. The discretized problem is then

$$[S]\{B\} = \{0\} \Rightarrow \|[S]\| = \Delta(c, q, a, \phi) = 0 \quad (86)$$

and we present the solutions as

$$c = c(q, a, \phi)\ or\ c = c(a_0 - c_0^2, q = q_0 A)\ for\ given\ \phi \quad (87)$$

A specific stability diagram from Pedersen, 1985 will illustrate the method for stability analysis. Such a diagram is shown in figure 10.

3.3 The Floquet method and the Floquet matrix

Consider a system of linear, homogeneous, differential equations with periodic coefficients

$$[M]\{\ddot{D}\} + [C]\{\dot{D}\} + [S]\{D\} = \{0\} \quad (88)$$

where $[M]$ is termed the mass matrix, being the coefficient matrix to the column matrix $\{\ddot{D}\}$ of acceleration. Thus partial derivative with respect to time t is denoted by the $\dot{}$ (dot) superscript. The matrix $[C]$ is termed the damping matrix, being the coefficient matrix to the column matrix $\{\dot{D}\}$ of velocity. The stiffness matrix $[S]$ is the coefficient matrix to the displacements $\{D\}$, and we are interested in analysing the stability of the solution $\{D\} = \{0\}$.

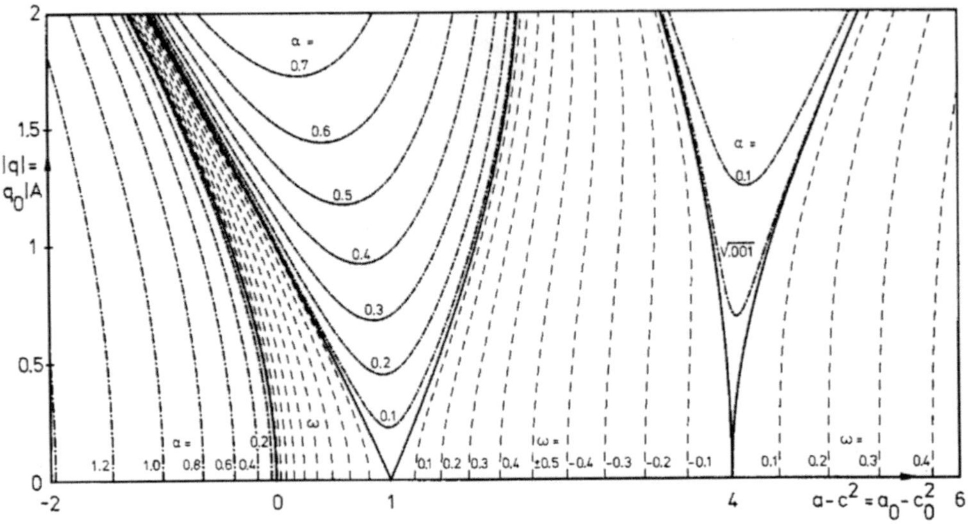

Figure 10. Detailed stability diagram for the specific Hill differential equation: $\ddot{y} + 2c_0\dot{y} + (a_0 + 2q_0(\cos 2\tau - \cos 6\tau/2)y = 0$.

The mass matrix, the damping matrix and the stiffness matrix are periodic with period T. We can write (88) in first order form

$$\{\dot{X}\} = [G(t)]\{X\} \; with\ the\ definitions$$

$$\{X\} := \begin{Bmatrix} \{\dot{D}\} \\ \{D\} \end{Bmatrix},\ [G] := \begin{bmatrix} [0] & [I] \\ -[M]^{-1}[S] & -[M]^{-1}[C] \end{bmatrix},\ [G(t+T)] = [G(t)] \tag{89}$$

where $[G(t)]$ is a real $m \times m$ matrix-function. The vector $\{X\}$ is a column vector of dimension m. Let $[G(t)]$ be periodic with minimum period T, that is, T is the smallest positive number for which $[G(t+T)] = [G(t)]$. Let $\{X_1(t)\},\ \{X_2(t)\},\ \ldots\{X_m(t)\}$ be any set of m solutions to the system (91), linearly independent for any t (and thus for all t). The matrix $[X]$ with columns $\{X_1(t)\},\ \{X_2(t)\},\ \ldots\{X_m(t)\}$ is called a fundamental matrix. If $[X(0)] = [I]$ where $[I]$ is the $m \times m$ identity matrix, then $[X(t)]$ is called a principal fundamental matrix. The fundamental matrix $[X(t)]$ is non-singular for all t.

If $[X(t)]$ is a principal fundamental matrix, the matrix given by

$$[F] := [X(T)] \tag{90}$$

is called the Floquet transition matrix or the monodromy matrix, see Yakubovitch and Starzhinskii, 1975. For short, we name it the Floquet matrix.

The eigenvalues of $[F]$ are called the characteristic multipliers for the system. To determine the stability of the solutions of the system (88) it is sufficient to consider these multipliers. As $[F]$ is a real $m \times m$ matrix, there are m (complex) multipliers, counting according to multiplicity. Every non-real multiplier has a complex-conjugate multiplier. None of the multipliers can take the value 0 . Let ρ_i for i = 1, 2, ..., m be the multipliers. The stability of the system (88) can be settled by using the following conditions, see Grimshaw, 1990 or Yakubovitch and Starzhinskii, 1975.

- If $|\rho_i| < 1$ for all i, all solutions are bounded and all the solutions turn to zero as $t \to \infty$. The trivial solution is (uniformly and asymptotically) stable in the Liapunov sense.
- If $|\rho_i| > 1$ for at least one i, solutions exist which are unbounded. The trivial solution is unstable in the Liapunov sense.
- If $|\rho_i| \leq 1$ for all i, and if for those multipliers, for which $|\rho_i| = 1$, the algebraic multiplicity equals the geometric multiplicity all solutions are bounded. The trivial solution is (uniformly) stable in the Liapunov sense.
- If $|\rho_i| \leq 1$ for all i, and if for any of the multipliers for which $|\rho_i| = 1$, the algebraic multiplicity is greater than the geometric multiplicity, solutions exist which are unbounded. The trivial solution is unstable in the Liapunov sense.

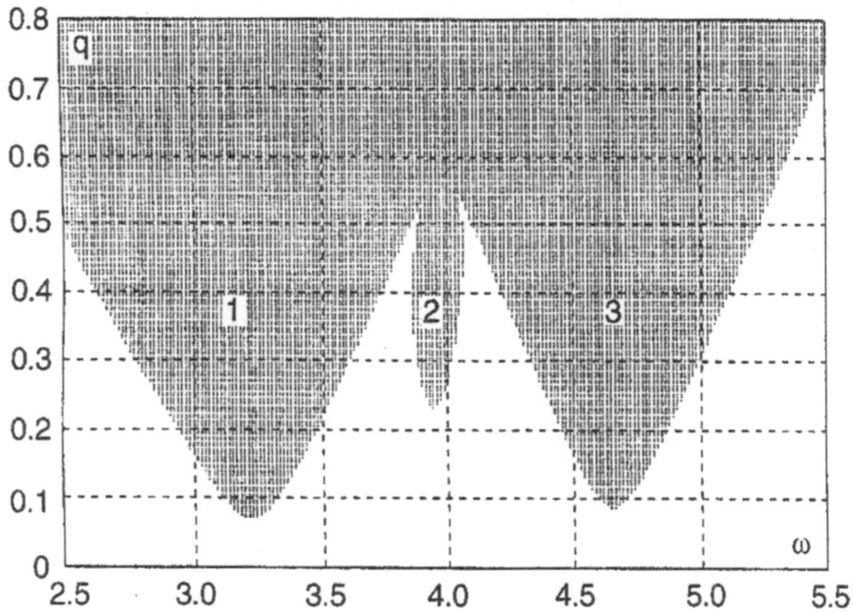

Figure 11. Stability diagram obtained by the Floquet method, corresponding to two coupled differential equations.

The conditions on the characteristic multipliers for the existence of periodic solutions can be summarized as follows (see again Grimshaw, 1990 or Yakubovitch and Starzhinskii, 1975): If at least one multiplier is equal to 1 , T-periodic solutions exist, if at least one multiplier is equal to -1 , 2T-periodic solutions exist, and if for at least one multiplier $\rho_i^k = 1$, where k is an integer, then kT-periodic solutions exist.

We conclude that the stability of the zero solution to a system of the form (88) is described by the eigenvalues of the Floquet matrix. If all the eigenvalues (characteristic multipliers) are situated inside the unit circle in the complex plane, all the solutions turn to zero as $t \rightarrow \infty$. If any of the characteristic multipliers are situated outside the unit circle, solutions exist which are unbounded. If all the multipliers are inside or on the unit circle, the stability conditions are determined by the difference between the algebraic and the geometric multiplicity of the multipliers situated on the unit circle.

For a problem described by Hansen, 1985 a stability diagram determined by the Floquet method is shown in figure 11.

3.4 Conclusion

Stability analysis for a single periodic differential equation (Hill equation) is effectively established by an approach based on Galerkin expansions. This, in fact, is a natural choice because each "point" in the stability diagram corresponds to an eigenvalue problem. We have in these notes focused on the classification of the periodic excitation functions as well as the solution functions, and then shown a specific stability diagram.

For systems of linear periodic differential equation an alternative approach must be used. It is concluded that the Floquet method is a general method for systems with multi-degrees of freedom. In this method we by numerical integration evaluate the Floquet transition matrix and then analyse the multipliers of this matrix.

This stability analysis for multi-parameter systems is most effectively carried out when sensitivity analysis with respect to problem parameters is an integrated part. This will be described in following notes.

4 Sensitivity Analysis for Linear Periodic Systems

4.1 Introduction

For stability analysis of linear periodic systems with more than one degree of freedom, the Floquet method is a general and valuable, practical method. In multi-parameter periodic systems, repeated numerical integration to obtain the Floquet matrix may be a limiting factor, and effective sensitivity analysis of stability characteristics is therefore needed. In this note we present analytical first and second order sensitivities of the Floquet matrix and its eigenvalues (multipliers). Some numerical applications are given. These include effective stabilization by proper change of parameters.

4.2 The Floquet matrix

Consider a system of linear, first order, homogeneous, differential equations

$$\{\dot{X}\} = [G(t)]\{X\} \quad with \quad [G(t+T)] = [G(t)] \tag{91}$$

where $[G(t)]$ is a real $m \times m$ matrix-function. The vector $\{X\}$ is a column vector of dimension m. Let $[G(t)]$ be periodic with minimum period T, that is, T is the smallest positive number for which $[G(t+T)] = [G(t)]$. Let $\{X_1(t)\}, \ \{X_2(t)\}, \ \ldots \{X_m(t)\}$ be any set of m solutions to the system (91), linearly independent for any t (and thus for all t). The matrix $[X]$ with columns $\{X_1(t)\}, \ \{X_2(t)\}, \ \ldots \{X_m(t)\}$ is called a fundamental matrix. If $[X(0)] = [I]$ where $[I]$ is the $m \times m$ identity matrix, then $[X(t)]$ is called a principal fundamental matrix. The fundamental matrix $[X(t)]$ is non-singular for all t.

If $[X(t)]$ is a principal fundamental matrix, the matrix given by

$$[F] := [X(T)] \tag{92}$$

is called the Floquet transition matrix or the monodromy matrix, see Yakubovitch and Starzhinskii, 1975. For short, we name it the Floquet matrix. The Floquet matrix is obtained by numerical integration.

4.3 Sensitivity of the Floquet matrix

We consider first order linear differential equations on fundamental $m \times m$ matrices X and Y with initial conditions

$$[\dot{X}] = [G][X], \ [\dot{Y}] = -[G]^T[Y] \quad with \quad [X(0)] = [Y(0)] = [I] \tag{93}$$

where $[G(t, \{p\})] = [G(t+T, \{p\})]$ is a real periodic $m \times m$ matrix with period T, continuously depending on time t and smoothly depending on the components of the vector of parameters $\{p\}$. The solutions $[X]$ and $[Y]$ to (93) satisfy the equality

$$[Y]^T[X] = [X]^T[Y] = [I] \quad i.e. \quad [X]^{-1} = [Y]^T \tag{94}$$

which can be verified by direct differentiation of (94) and use of (93). Note, for the examples considered it proves to be more efficient to invert the principal fundamental matrix than to solve the adjoint system $[\dot{Y}] = -[G]^T[Y]$.

Consider an increment of the vector of parameters $\{p\}$ in the form

$$\{p\} = \{p_0\} + \epsilon\{e\} \ \ with \ \ ||\{e\}|| = \sqrt{e_1^2 + e_2^2 + \cdots + e_n^2} = 1 \tag{95}$$

where ϵ is a small positive number and $\{e\}$ is an arbitrary vector in the parameter space. Then the $[G]$ matrix takes the increment

$$[G(t,\{p\})] = [G_0]+\epsilon[G_1]+\epsilon^2[G_2]+\cdots \ with \ [G_1] = \sum \frac{\partial[G]}{\partial p_k}e_k, \ [G_2] = \frac{1}{2}\sum \frac{\partial^2[G]}{\partial p_i \partial p_j}e_i e_j \tag{96}$$

where for convenience, the notation $[G_0] = [G(t,\{p_0\})]$ is used. The matrices $[G_1]$ and $[G_2]$ are defined by partial derivatives as shown.

Due to perturbation of the $[G]$ matrix the solution to (93) takes on an increment which is an analytical function of ϵ

$$[X(t,\{p\})] = [X_0] + \epsilon[X_1] + \epsilon^2[X_2] + \cdots \tag{97}$$

To find the coefficients $[X_1]$ and $[X_2]$ of this expansion a perturbation technique is used. Substituting the expansions (96), (97) into (93), we obtain a chain of equations

$$\begin{aligned}
&[\dot{X}_0] = [G_0][X_0] \ \ with \ \ [X_0(0)] = [I] \\
&[\dot{X}_1] = [G_0][X_1] + [G_1][X_0] \ \ with \ \ [X_1(0)] = [0] \\
&[\dot{X}_2] = [G_0][X_2] + [G_1][X_1] + [G_2][X_0] \ \ with \ \ [X_2(0)] = [0]
\end{aligned} \tag{98}$$

We then pre-multiply $[\dot{X}_1]$ from (98) by $[Y_0(t)]$ and integrate over the time from 0 to t

$$\int_0^t [Y_0]^T[\dot{X}_1]d\tau = \int_0^t [Y_0]^T[G_0][X_1]d\tau + \int_0^t [Y_0]^T[G_1][X_0]d\tau \tag{99}$$

Using integration by parts, we represent the left-hand side of (99) in the form

$$\int_0^t [Y_0]^T[\dot{X}_1]d\tau = [Y_0]^T[X_1]|_0^t - \int_0^t [\dot{Y}_0]^T[X_1]d\tau = [Y_0(t)]^T[X_1(t)] + \int_0^t [Y_0]^T[G_0][X_1]d\tau \tag{100}$$

Here we have used the initial condition $[X_1] = [0]$ and (93) for $[Y_0]$. Comparing (99) and (100), we finally find

$$[Y_0(t)]^T[X_1(t)] = \int_0^t [Y_0]^T[G_1][X_0]d\tau \Rightarrow [X_1(t)] = [X_0(t)] \int_0^t [Y_0]^T[G_1][X_0]d\tau \tag{101}$$

where also (94) is used. At the end of the period for $t = T$ this gives $[X_1(T)]$. As $[F_1] := [X_1(T)]$ we have found the first directional (Gateaux) derivative of the Floquet matrix.

To find the second order terms in the expansion (97) we pre-multiply $[\dot{X}_2]$ from (98) by $[Y_0(t)]$ and again integrate over time from 0 to t

$$[X_2(t)] = [X_0(t)](\int_0^t [Y_0]^T[G_1][X_1]d\tau + \int_0^t [Y_0]^T[G_2][X_0]d\tau) \tag{102}$$

We substitute here the expression (108) and obtain finally

$$[X_2(t)] = [X_0(t)](\int_0^t [Y_0(\tau)]^T[G_1][X_0(\tau)](\int_0^\tau [Y_0]^T[G_1][X_0]d\zeta)d\tau + \int_0^t [Y_0]^T[G_2][X_0]d\tau) \tag{103}$$

We now have the second order term in the expansion of the Floquet matrix, by the definition $[F_2] := [X_2(T)]$.

4.4 The multipliers (eigenvalues) of the Floquet matrix

The eigenvalues of $[F]$ are called the characteristic multipliers for the system. To determine the stability of the solutions of the system (91) it is sufficient to consider these multipliers. As $[F]$ is a real $m \times m$ matrix, there are m (complex) multipliers, counting according to multiplicity. Every non-real multiplier has a complex-conjugate multiplier. None of the multipliers can take the value 0 . Let ρ_i for i = 1, 2, …, m be the multipliers.

The stability of a trivial solution of system (91) is described by the eigenvalues of the Floquet matrix. If all the eigenvalues (characteristic multipliers) are situated inside the unit circle in the complex plane, all the solutions turn to zero as $t \to \infty$. If any of the characteristic multipliers are situated outside the unit circle, solutions exist which are unbounded. If all the multipliers are inside or on the unit circle, the stability conditions are determined by the difference between the algebraic and the geometric multiplicity of the multipliers situated on the unit circle.

4.5 Sensitivity of distinct multipliers of the Floquet matrix

Consider an eigenvalue problem for the Floquet matrix $[F] = [F(\{p\})]$ depending on a vector of real parameters $\{p\}$

$$[F]\{U\} = \rho\{U\} \;\; with \;\; [F] = [X(T)] \;\; where \;\; [F] = [F_0] + \epsilon[F_1] + \epsilon^2[F_2] + \cdots \tag{104}$$

As the coefficient matrix $[G]$ is a real matrix, the Floquet matrix is a real non-symmetric matrix. We assume that at fixed $\{p\} = \{p_0\}$ the eigenvalue ρ_0 is a simple eigenvalue of the Floquet matrix (a multiplier), complex or real. Our aim is to know what will happen to the multiplier ρ_0 with a change of parameters in the vicinity of the point $\{p_0\}$. For this purpose we take an increment of the form shown in (165) with $[F_1]$ and $[F_2]$ determined in the previous section.

Then the multipliers ρ and the corresponding eigenvectors $\{U\}$ are subjected to increments. According to the perturbation theory of non-selfadjoint operators in the case of simple eigenvalues, these increments can be expressed as series in integer powers of ϵ

$$\begin{aligned} \rho &= \rho_0 + \epsilon\rho_1 + \epsilon^2\rho_2 + \cdots \\ \{U\} &= \{U_0\} + \epsilon\{U_1\} + \epsilon^2\{U_2\} + \cdots \end{aligned} \tag{105}$$

For the following we need to introduce a left eigenvector $\{V_0\}$, corresponding to ρ_0, defined by

$$\{V_0\}^T[F_0] = \rho\{V_0\}^T \; with \; normalization \; \{V_0\}^T\{U\} = 1 \tag{106}$$

Substituting the expansions (105) for $\{U\}$ and ρ in (165) yields the following equations to be solved

$$\begin{aligned}
([F_0] - \rho_0[I])\{U_0\} &= \{0\} \\
([F_0] - \rho_0[I])\{U_1\} &= -([F_1] - \rho_1[I])\{U_0\} \\
([F_0] - \rho_0[I])\{U_2\} &= -([F_1] - \rho_1[I])\{U_1\} - ([F_2] - \rho_2[I])\{U_0\} \\
with \; normalization \; \{V_0\}^T\{U_1\} &= 1, \{V_0\}^T\{U_1\} = 0, \{V_0\}^T\{U_2\} = 0
\end{aligned} \tag{107}$$

Solving (107) we find first the unknown coefficients of the linear expansions in (105)

$$\begin{aligned}
\rho_1 &= \{V_0\}^T[F_1]\{U_0\} \\
\{U_1\} &= -[C_0]^{-1}([F_1] - \rho_1[I])\{U_0\} \\
&= -[C_0]^{-1}[F_1]\{U_0\} + \{V_0\}^T[F_1]\{U_0\}[C_0]^{-1}\{U_0\} \\
where \;\; by \;\; definition \; [C_0] &= [F_0] - \rho_0[I] - \{\bar{V_0}\}^T\{V_0\}
\end{aligned} \tag{108}$$

The vector $\{\bar{V_0}\}$ is the complex conjugate of the left eigenvector $\{V_0\}$. It is shown by Yakubovitch and Starzhinskii, 1987 that the matrix $[C_0]$ is non-singular, and the vector $\{U_1\}$ given by (108) satisfies the normality condition in (107) and is unique.

Multiplying the third equation in (107) by $\{V_0\}^T$ and using the normality conditions, we obtain

$$\begin{aligned}
\rho_2 &= \{V_0\}^T[F_1]\{U_1\} + \{V_0\}^T[F_2]\{U_0\} \\
&= -\{V_0\}^T[F_1][C_0]^{-1}[F_1]\{U_0\} + \{V_0\}^T[F_1]\{U_0\}\{V_0\}^T[F_1][C_0]^{-1}\{U_0\} + \{V_0\}^T[F_2]\{U_0\} \\
\{U_2\} &= -[C_0]^{-1}([F_1] - \rho_1[I])[C_0]^{-1}([F_1] - \rho_1[I])\{U_0\} - [C_0]^{-1}([F_2] - \rho_2[I])\{U_0\}
\end{aligned} \tag{109}$$

4.6 Sensitivity of bimodal and double multipliers

In Seyranian et al., 1999 a complete description of sensitivity analysis for bimodal multipliers (two different eigenvectors) and for double multipliers (only one eigenvector) is given. The most important aspects for the double multipliers is that the expansion of eigenvalue and of eigenvector must be in square roots of the small positive parameter ϵ, according to Vishik and Lyusternik, 1960 and furthermore that in addition to the eigenvector and the adjoint eigenvector we need to determine an associated vector.

4.7 Stabilization of an unstable solution: an example

We will show how to stabilize an unstable system by changing the values of the parameters. As an example the results are applied to the Carson-Cambi equation, treated in Pedersen, 1980. This equation is

$$(1 + p_1 \cos t)\frac{d^2y}{dt^2} + p_2 y = 0 \;\; with \;\; |p_1| < 1 \tag{110}$$

with p_1, p_2 as the problem parameters. Inside the unstable domains, at least one characteristic multiplier ρ is situated outside the unit circle. That is

$$\textit{at least one } |\rho| > 1 \tag{111}$$

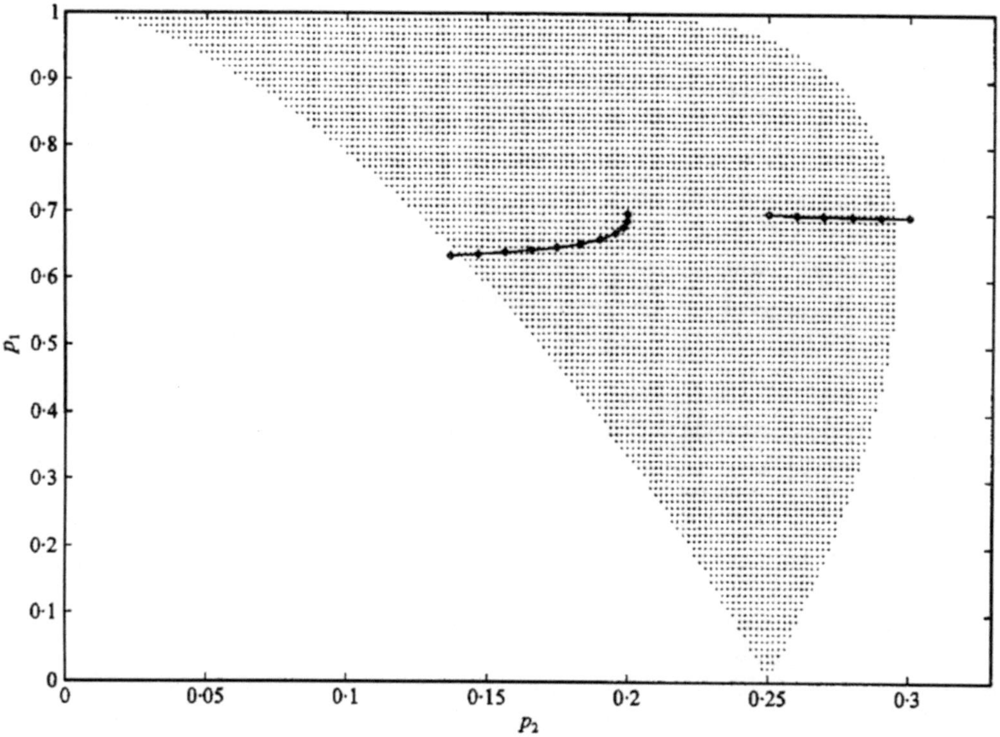

Figure 12. Part of the stability diagram for solutions to the Carson-Cambi equation, with instability indicated by dots. The paths from instability to stability for the two points $p_1, p_2 = 0.70, 0.25$ and $p_1, p_2 = 0.70, 0.20$ are shown.

To stabilize the system, all the multipliers outside the unit circle must be brought onto or inside the unit circle. Let ρ be a multiplier situated outside the unit circle. The modulus of ρ depends on the parameters p_1, p_2, and if δp_i is chosen according to

$$\delta p_i = -k \frac{\partial |\rho|}{\partial p_i} \tag{112}$$

where k is a real positive constant, then we get

$$\delta|\rho| = -k((\frac{\partial|\rho|}{\partial p_1})^2 + (\frac{\partial|\rho|}{\partial p_2})^2) < 0 \qquad (113)$$

Thus, by choosing the parameter change according to equation (112), the change of $|\rho|$ is smaller than or equal to zero and the system is made "more" stable.

The necessary formula for calculating the sensitivities is given by (108) for the simple multiplier case. The implementation in a computer programme gave the results shown in figure 12 with part of the stability diagram for the Carson-Cambi equation. Comparing these results with the detailed stability diagram in Pedersen, 1980 , we see that the paths follow the steepest descents of $|\rho|$. In a multi-degree of freedom system, the procedure will be the same as illustrated here.

4.8 Conclusion

For stability analysis of linear periodic systems with multi-degrees of freedom, the classic Floquet method seems to be a general and practical method. However, for a multi-parameter analysis, this procedure limits on our possibilities due to the extensive numerical integration necessary to obtain the Floquet transition matrix.

We derived explicit analytical expressions for the derivatives of the Floquet matrix and its eigenvalues (multipliers) with respect to problem parameters. This is what we call sensitivity analysis. It allows us to extract full information of dependence on parameters using only a single numerical integration. With these new results, optimization of multi-parameter, multi-degree of freedom systems is more feasible.

The example shows the application of sensitivity results in a procedure for effective stabilization of an unstable solution.

To keep the paper short, we have restricted the sensitivity analysis to simple multipliers. A more extensive analysis with multiple multipliers is presented in Seyranian et al., 1999. Adding sensitivity analysis, the Floquet method constitutes a practical method that can be applied also to multi-parameter, multi-degree of freedom systems.

5 Applications of Sensitivity Analysis

5.1 Introduction

The goal of this note is to show the versatility in application of sensitivity analysis. It has been argued that sensitivity is just a new word for partial derivative, and to same extent this is correct. When our response is measured by a scalar like an eigenvalue and we ask for the sensitivity with respect to another scalar then it is a partial derivative. However, we may also ask for the sensitivity with respect to a function, and then the notion of gradient-function is often used. Furthermore, the response may be a function like an eigenfunction or another displacement description, and then we need a notion for a change of a function with respect to another function. So sensitivity analysis is just a notion that covers the work of determining all this important information.

5.2 The use of sensitivity analysis in optimal design

As argued in Pedersen, 1983, sensitivity analysis is the "cornerstone" of almost any approach to optimal design. For problems with only a single constraint, the ratio between two sensitivities forms an explicit optimality criterion, as will be shown later. The conclusion to be drawn is that the quantities of the optimality criterion must all be obtained by sensitivity analysis.

More practical problems are formulated with a larger number of constraints, of which we do not in advance know the active ones. Let Φ be our objective and h_e the design variables, and let the constraints be described by the inequalities $g_j \leq 0$ for $j = 1, 2, \ldots, J$. The optimal design problem is highly nonlinear and we have to linearize around a given design. Thus the problem of optimal redesign is:

$$Min\ (\Phi(h_e) + \sum_e \frac{\partial \Phi}{\partial h_e} \Delta h_e) = Min\ (\sum_e \frac{\partial \Phi}{\partial h_e} \Delta h_e) \tag{114}$$

subjected to the constraints

$$g_j + \sum_e \frac{\partial g_j}{\partial h_e} \Delta h_e \leq 0\ for\ j = 1, 2, \ldots, J \tag{115}$$

This constitutes a linear programming problem, and we note the dominating importance of the sensitivity analysis to determine the partial derivatives.

5.3 Other domains of application of sensitivity analysis

A great deal of work on sensitivity analysis is presented in papers on optimal design, and this may give a wrong impression of the domains of application. In reality some of the most important applications are found outside the domain of optimal design.

First of all, knowledge of sensitivities adds to the **reliability of ordinary analysis**, whether this is stiffness, strength, stability or vibration analysis. The results of such analyses depend on the data used, some of which we are often forced to estimate. Especially for these data, it is of vital interest to determine the sensitivities, and from this estimate the reliability of our analysis.

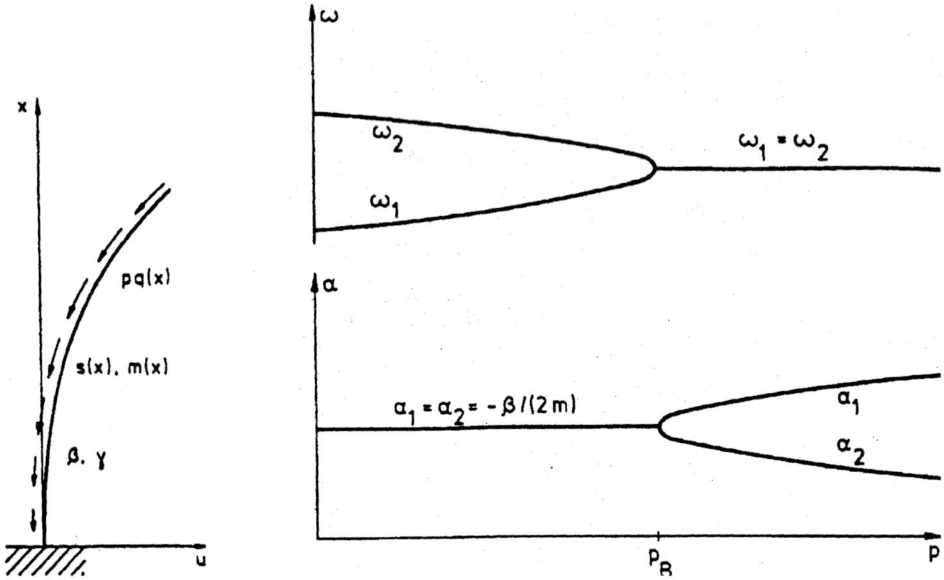

Figure 13. Extended Hauger column ($q[x] = 1 - x$) with characteristic curves corresponding to the first two frequencies.

Another class of problems is termed **inverse problems**. In contrast to ordinary analyses where the response of a given system is determined, then for these problems we prescribe the response of the system and then look for the system itself. These problems are immediately formulated as optimization problems with the objective of minimizing the difference between the calculated response and the prescribed response. With this formulation as an optimization problem, we thus see the need for sensitivity analyses. There are specific problems related to the inverse formulations, existence of solutions are not guaranteed and if solutions can be found, then they are often not unique. Thus engineering skills are important for the practical solutions.

The subject termed **automatic design** which is relevant in advanced CAD/CAM systems includes in reality problems of the inverse type, and the above arguments will therefore again be relevant.

Finally, the domain termed **identification and parameter estimation**, including the domain of **numerical/ experimental methods**, which have recently received large attention, belongs to the same category, with the specific property that the prescribed response is obtained by physical experiments.

5.4 Sensitivity to external damping

The present example relates to the use of sensitivity analysis in ordinary analysis, specifically we want to prove the characteristics of obtained numerical results. The full explanation is given in Pedersen, 1984.

Let us study the results of an ordinary analysis for the Beck/Leipholz/Hauger columns shown in figure 13 with the characteristic curves for frequency $\omega = \omega(p)$ and for stability parameter

$\alpha = \alpha(p)$ related to the first two modes. This case corresponds to a uniform Hauger column with stiffness to mass relation $s(x) = m^2(x)$ and uniform mass $m(x) \equiv 1$, without internal damping $\gamma = 0$ but with external damping by the real non-negative parameter β.

The principal look of this result is independent of β and the load distribution, i.e. it is also valid for Beck, Leipholz columns. Thus, the interesting question is whether we from sensitivity analysis can predict the constant solution $\alpha = -\beta/(2m)$ over the large load domain $0 \le p \le p_B$, where p_B is the load level at which we have a double eigenvalue. Note that only for $\beta = 0$ is p_B the flutter load, so it is normally a stable load level.

To obtain the results in figure 13 we have, as described in Pedersen and Seyranian, 1983, discretized the continuous problem and obtained the elements of the stiffness matrix $[S]$, the non-symmetric load matrix $[K]$, the damping matrix $[C]$, and the mass matrix $[M]$, that determine the eigenvalue $\lambda = \alpha + \imath\omega$ from the eigenvalue problem

$$[L]\{\Phi\} = \{0\},\ \{\Psi\}^T[L] = \{0\}^T \ with \ [L] := \lambda^2[M] + \lambda[C] + [S] + p[K] \tag{116}$$

For the specific case the damping matrix simplifies to $[C] = \frac{\beta}{m}[M]$ and the sensitivity to pure load level change will then be

$$\frac{\partial\lambda}{\partial p} = \frac{\partial\alpha}{\partial p} + \imath\frac{\partial\omega}{\partial p} = \frac{\mathcal{U}/\mathcal{T} + (\beta\lambda/m + \lambda^2)}{p(\beta/m + 2\lambda)} \tag{117}$$

where $\mathcal{U}/\mathcal{T}$ is the ratio between specific mutual potential and kinetic energies.

Let us **postulate** the solution $\alpha = -\beta/(2m)$, which makes $\beta\lambda/m + \lambda^2$ real, i.e.

$$\lambda = -\beta/(2m) + \imath\omega \Rightarrow \beta\lambda/m + \lambda^2 = -(\omega^2 + \beta^2/(4m^2)) \ and \ \beta/m + 2\lambda = \imath 2\omega \tag{118}$$

Eigenvector $\{\Phi\}$ and adjoint eigenvector $\{\Psi\}$ will then be real, and $\mathcal{U} := \{\Psi\}^T[S]\{\Phi\}$ and $\mathcal{T} := \{\Psi\}^T[M]\{\Phi\}$ will thus also be real. In relation to (117) this gives

$$\frac{\partial\alpha}{\partial p} = 0 \ and \ \frac{\partial\omega}{\partial p} = \frac{1}{2\omega p}(\omega^2 + \frac{\beta^2}{4m^2} - \frac{\mathcal{U}}{\mathcal{T}}) \tag{119}$$

and thus proves unchanged $\alpha = -\beta/(2m)$. The increase in ω_1 and decrease in ω_2 as seen in figure 13 is due to the different energy ratios $\mathcal{U}/\mathcal{T}$ for the two modes.

How is the simple behaviour described in (119) stopped at $p = p_B$? To understand this we return to an assumption behind (117), which is $\mathcal{T} \neq 0$ as it follows from the ratio $\mathcal{U}/\mathcal{T}$. Our goal now is to prove $\mathcal{T} = 0$ at $p = p_B$, where we have a double eigenvalue.

For $p < p_B$ let ω_i be the frequency corresponding to the eigenvector $\{\Phi\}_i$. Pre-multiplying equilibrium with the adjoint eigenfunction $\{\Phi\}_j$ corresponding to the frequency ω_j we get

$$\{\Psi\}_j^T[S]\{\Phi\}_i + p\{\Psi\}_j^T[K]\{\Phi\}_i - (\omega_i^2 + \beta^2/(4m^2))\{\Psi\}_j^T[M]\{\Phi\}_i = 0 \tag{120}$$

Interchanging indexes i and j we analogously get

$$\{\Psi\}_i^T[S]\{\Phi\}_j + p\{\Psi\}_i^T[K]\{\Phi\}_j - (\omega_j^2 + \beta^2/(4m^2))\{\Psi\}_i^T[M]\{\Phi\}_j = 0 \tag{121}$$

and the difference between these two last equations gives

$$(\omega_j^2 - \omega_i^2)\{\Psi\}_i^T[M]\{\Phi\}_j = 0 \tag{122}$$

which for $\omega_j^2 \neq \omega_i^2$ shows what for undamped systems is termed the **biorthogonality condition**. At a bimodal point the eigenfrequencies are equal and by arguments of continuity we get

$$\mathcal{T}_B = \{\Psi\}_B^T[M]\{\Phi\}_B = 0 \tag{123}$$

which explains the singularity points in figure 13. A deeper explanation is given in Seyranian and Pedersen, 1995 and in Seyranian, 1993, based on a description with not only equal eigenfrequencies but also equal eigenvectors.

For completely undamped systems we have

$$\beta = 0 \Rightarrow p_B = p_F, \ \{\Phi\}_B = \{\Phi\}_F, \ \{\Psi\}_B = \{\Psi\}_F \tag{124}$$

and (123) is termed the flutter condition.

5.5 Sensitivities in laminate inverse problems

In two dimensional formulations of laminates we deal with a number of 3×3 matrices: the constitutive (material stiffness) matrix $[C]$, the laminate membrane stiffness matrix $[A]$, the laminate bending stiffness matrix $[D]$, the laminate coupling stiffness matrix $[B]$, the laminate stress strength matrix $[F]$, and the laminate strain strength matrix $[G]$. In relation to the present inverse problem they are all treated analogously and as example we formulate the problem of finding a laminate that results in a given bending matrix $[\bar{D}]$. This example is described more detailed in Pedersen, 1999.

With the four index notation from the tensor formulations the matrix $[D]$ is defined

$$[D] = \begin{bmatrix} D_{1111} & D_{1122} & \sqrt{2}D_{1112} \\ D_{1122} & D_{2222} & \sqrt{2}D_{2212} \\ \sqrt{2}D_{1112} & \sqrt{2}D_{2212} & 2D_{1212} \end{bmatrix} \quad with\ [D]\ positive\ definite \tag{125}$$

or with matrix $[D]$ in contracted vector notation $\{D\}$

$$\{D\}^T = \{D_{1111}\ D_{2222}\ 2D_{1212}\ \sqrt{2}\{D_{1122}\ 2D_{1112}\ 2D_{2212}\} \tag{126}$$

The factors $\sqrt{2}$ and 2 are related to a moment vector definition $\{M\}^T = \{M_{11}\ M_{22}\ \sqrt{2}M_{12}\}$ and to a curvature vector definition $\{\kappa\}^T = \{\kappa_{11}\ \kappa_{22}\ \sqrt{2}\kappa_{12}\}$ that imply orthogonal rotation transformations and then, as an example, invariant trace of $[D]$ equal to $(D_{1111}+D_{2222}+2D_{1212})$.

The matrix $[D] = [D(p_i)]$ depends on the design parameters p_i and our goal is to find parameters that give $[D] = [\bar{D}]$ or at least obtain as close as possible agreement. This closeness is measured by the residual matrix $([D] - [\bar{D}])$. To get an objective that is invariant we define an error function Φ as the squared Frobenius norm of this matrix, i.e.

$$\Phi = (\{D\}^T - \{\bar{D}\}^T)(\{D\} - \{\bar{D}\}) \tag{127}$$

It follows that $\Phi \geq 0$, and if $\Phi = 0$, we have located an exact solution $[D] = [\bar{D}]$. A solution approach is then by iterations to minimize Φ and in such an approach we need the gradients

$$\frac{d\Phi}{dp_i} = 2(\{D\}^T - \{\bar{D}\}^T)\frac{d\{D\}}{dp_i} \tag{128}$$

where formulas for $d\{D\}/dp_i$ can be found in Pedersen, 1987.

A redesign described by

$$(p_i)_{next} = p_i + \Delta p_i \tag{129}$$

will then give

$$(\Phi)_{next} = \Phi + \sum_i (d\Phi/dp_i)\Delta p_i \tag{130}$$

and since $(d\Phi/dp_i)$ is only correct for small Δp_i a possible strategy would be in each iteration to

$$Minimize\ (\Phi)_{next}\ \ subjected\ to\ g = \sum_i (\Delta p_i)^2 - R = 0 \tag{131}$$

where R is a redesign limit. Problem (131) can be solved analytically by proportionality between gradients of constraint and of objective and we get

$$\Delta p_i = a\frac{d\Phi}{dp_i} = a2(\{D\}^T - \{\bar{D}\}^T)\frac{d\{D\}}{dp_i} \tag{132}$$

where the constant a is determined by the condition in (131)

$$a^2 \sum_i (d\Phi/dp_i)^2 = R \tag{133}$$

5.6 Sensitivities in material identification problems

A combined experimental-numerical method is used with the goal of obtaining material stiffnesses for composite materials. The identification is based on eigenfrequencies for a free rectangular plate, because excellent agreement between measured and calculated eigenfrequencies can be obtained. The numerical identification problem is formulated as an optimization problem and in one experiment we obtain estimates of all the involved moduli. The total approach consists of experiment, numerical eigenvalue analysis, eigenvalue sensitivity analysis, optimization, and error estimation. The general conclusion is that the method has several significant advantages compared to traditional determination of material moduli. The description here focus's on the importance of eigenvalue sensitivity analysis for this method.

Traditional laboratory tests are designed to give the desired quantities in a direct manner. As an example, determination of material parameters of constitutive models leads to special design of test samples and choices of applied load in an attempt to obtain homogeneous stress-strain fields. These idealized tests are not easy to perform, and especially with composite systems problems often arise. Furthermore, the tests have a local nature, e.g. the point of the strain gauge. So for more general material information a series of tests are required. The identification approach is an experimental strategy from an opposite point of view. The experiment is made as simple as possible to give reliable results. Although our aim is to find static material data, we decide to measure eigenfrequencies of a free rectangular plate because excellent agreement between measured and calculated eigenfrequencies can be obtained. A structural eigenfrequency is an integrated quantity and we thus obtain material quantities that are valid in the mean for the entire laminate.

Identification formulation. A structure can be described in many ways, directly or indirectly. The direct approach can be by parameters, like length, width, height, weight, topology and material parameters. Alternatively, we can describe it by its response like static displacements or dynamic eigenfrequencies. Combination approaches are also possible. A laminated plate is a structure where the ply stacking is an important aspect. The thickness of the plies and their positions in addition to ply orientations are the parameters that describe this stacking. Also boundary conditions are of vital importance for the response of the plate. The laminate membrane stiffnesses $[A]$, as well as the laminate bending stiffnesses $[D]$ are linearly depending on the material stiffness $[C]$ for each ply, i.e. upon the constitutive matrix. We can describe the identification problem as a problem where we do not have information enough about $[C]$, but we have information about the structural response, say in terms of measured eigenfrequencies ω_i. Let us be specific and formulate the identification problem as an optimization problem. A number of parameters p_k are unknown for the determination of the material stiffnesses

$$[C] = [C(p_k)] \tag{134}$$

and therefore a calculated eigenfrequency response will also depend on these parameters

$$\omega_i = \omega_i(p_k) \tag{135}$$

The functional relations (134) and (135) can be found in textbooks on laminates, say in Jones, 1975.

The optimization problem is to find best possible agreement between measured eigenfrequencies $\bar{\omega}_i$ and calculated eigenfrequencies ω_i. An objective could be to

$$Minimize\ \Phi = \sum_i \frac{(\omega_i - \bar{\omega}_i)^2}{\bar{\omega}_i^2} \tag{136}$$

i.e., a weighted least-squares estimator, but naturally other choices are possible. For example, more weight could be given to more accurate measurements, and a priori expectation on parameter values can be included. The solution must be obtained in an iterative process, where the parameters are simultaneously improved

$$(p_k)_{n+1} = (p_k)_n + (\Delta p_k)_n \ for\ n = 1, 2, \ldots \tag{137}$$

and constraints can be put on the absolute values of the parameters (side constraints) as well as on the change of the parameters (move limits)

$$(p_k)_{min} \leq (p_k) \leq (p_k)_{max} \ \ and \ \ (\Delta p_k)_{min} \leq (\Delta p_k) \leq (\Delta p_k)_{max} \tag{138}$$

With these kinds of constraints we do not in advance need to distinguish very clearly between known and unknown parameters. For orthotropic two dimensional models, two parameters are especially important. The **level of anisotropy** described by the parameter α_2

$$\alpha_2 := (1 - E_T/E_L)/2 \tag{139}$$

where E_T, E_L are the engineering moduli in transverse and fiber direction, respectively. The relative shear stiffness is described by the parameter α_3

$$\alpha_3 := (1 + (1 - 2\nu_{LT})E_T/E_L - 4\alpha_0 G_{LT}/E_L)/8 \tag{140}$$

where ν_{LT} is the major Poisson's ratio and G_{LT} is the shear modulus. The non-important parameter α_0 is given by $\alpha_0 = 1 - \nu_{LT}^2 E_T/E_L$, and a further not so important parameter is

$$\alpha_4 = \alpha_3 + \nu_{LT}(1 - 2\alpha_2) \tag{141}$$

i.e. only new information from the Poisson's ratio. Solving the problem (136) in the formulation

$$Minimize\ \Phi = \Phi(\alpha_2, \alpha_3, \alpha_4) \geq 0 \tag{142}$$

we can then invert the relations (139), (140) and (141) and identify E_L, E_T, G_{LT} and ν_{LT}.

Sensitivity analysis. For the identification we also need the partial derivatives of the eigenfrequencies with respect to the material moduli, here symbolized by a parameter p_k. The general result for distinct eigenfrequencies are

$$\frac{d\omega^2}{dp_k} = \{D\}^T(-\omega^2\frac{d[M]}{dp_k} + \frac{d[S]}{dp_k})\{D\} \tag{143}$$

when the normalization

$$\{D\}^T[M]\{D\} = 1 \tag{144}$$

is assumed. For the actual problem where $d[M]/dp_k = [0]$, because the material moduli have no influence on the mass matrix, we get

$$\frac{d\omega^2}{dp_k} = \{D\}^T\frac{d[S]}{dp_k}\{D\} \tag{145}$$

For multiple eigenfrequencies a special calculation is necessary to obtain the eigenvectors that uncouple the sensitivities. With $\{D\}_i, \{D\}_j$ being two eigenvectors corresponding to the same eigenfrequency ω_i, we determine these vectors uniquely by the conditions:

$$\{D\}_i^T[S]\{D\}_j = \delta_{ij}\omega_i^2 \ \ and \ \ \{D\}_i^T\frac{d[S]}{dp_k}\{D\}_j = \delta_{ij}\frac{d\omega^2}{dp_k} \tag{146}$$

where δ_{ij} is Kronecker's delta.

Results. Let us comment on results for weakly, moderately and strongly anisotropic materials. A rolled aluminum plate is identified in the early work by Markworth and Petersen, 1987 which based on ten frequencies found $E_L = 68.9GPa, E_T = 66.0GPa, G_{LT} = 24.8GPa, \nu_{LT} = 0.343$ and fairly good agreement with traditional uniaxial tests. Frederiksen, 1997 for another aluminum plate identified $E_L = 71.1GPa, E_T = 70.7GPa, G_{LT} = 25.9GPa, \nu_{LT} = 0.336, G_{LZ} = 27.9GPa, G_{TZ} = 28.0GPa$, thus showing the possibility for identifying the out-of-plane shear moduli G_{LZ}, G_{TZ}.

For glass-epoxy also the early studies report good agreement with static strain gauge test results in the identification of the in-plane moduli. Based on 14 frequencies, Frederiksen, 1997 identified the following parameters $E_L = 42.4GPa, E_T = 11.6GPa, G_{LT} = 4.68GPa, \nu_{LT} =$

$0.305, G_{LZ} = 4.55GPa, G_{TZ} = 4.07GPa$, with details on iteration history and the extremely small residuals from experimental and calculated frequencies.

Carbon-epoxy identification is properly of most interest and extended results are available, in early as well as recent studies. Let us here again comment the result of Frederiksen, 1997, who reports $E_L = 113.0GPa, E_T = 8.50GPa, G_{LT} = 4.45GPa, \nu_{LT} = 0.323, G_{LZ} = 4.43GPa, G_{TZ} = 2.97GPa$. We find almost transversely isotropic stiffness for this unidirectional laminate and the ratio $G_{TZ}/G_{LZ} = 0.666$ is in other studies reported between 0.548 and 0.624 . Note that the determination of out-of-plane shear moduli is very difficult to perform by other experimental techniques.

6 Optimal Design with Eigenvalue Constraints

6.1 Introduction

In the research area of optimal design we normally distinguish between two group of methods

- optimality criteria methods
- mathematical programming methods

The procedures resulting may look rather similar and the major difference relate to the point of view. For the optimality criteria methods we set up a necessary criterion to be fulfilled for the optimal solution, and then iterate to satisfy this criterion by using the information available for a given design. However, for the mathematical programming methods we just use the information for a given design with the goal of constantly improving the design in a sequence of redesigns.

The methods of mathematical programming, like linear programming, quadratic programming and the method of moving asymptotes (MMA) provide the most general and powerful tools, but for obtaining general knowledge of simplified problems the optimality criteria methods are most valuable. In this aspect it is important to recall the general result when only one constraint is active. Let Φ be the objective which we extremize and V the constraint to be kept constant ($dV = 0$). Then for a discretized problem with design parameters h_e for $e = 1, 2, \ldots$ (may be collected in a design vector or may be generalized to a design function) we have

$$d\Phi = \sum_e \frac{\partial \Phi}{\partial h_e} dh_e \ , \ dV = \sum_e \frac{\partial V}{\partial h_e} dh_e \tag{147}$$

from which follows directly that the necessary condition of proportionality between gradients (same constant A for all design variables h_e)

$$\frac{\partial \Phi}{\partial h_e} = A \frac{\partial V}{\partial h_e} \tag{148}$$

implies that $d\Phi = AdV = 0$ when $dV = 0$, and thus a necessary condition for an extremized Φ objective.

Four specific examples will show the versatility of sensitivity analysis when applied to optimal design with eigenvalue constraints. The four examples are chosen from projects where the author has been personally involved. They are

- Beam design with several eigenvalue constraints.
- Material orientation of composite plates for optimization of supersonic flutter.
- Column design with follower forces.
- Beam design with periodic loads.

6.2 Formulation of the optimal redesign problem

We assume the design to be characterized by the vector of design variables $\{h\}$ consisting of N real numbers h_e. A generalization to design functions is rather obvious. Let the **objective** of our design problem be

$$Minimize \ \Phi = \Phi(\{h\}) \tag{149}$$

To maximize we just change sign. The main point to be noted is that we can only deal with one objective.

We state the **constraints** of our design problem in general terms by the functionals g_j as

$$g_j = g_j(\{h\}) \leq 0 \; for \; j = 1, 2, \ldots, J \tag{150}$$

assuming constraints $g_j \geq 0$ to be transformed by a change of sign. The constraints (150) limit our **feasible design space**, and in fact make the problems easier to solve. Note that we shall make no attempts to convert this constrained problem into a non-constrained one by Lagrangian multipliers or penalty formulations.

As optimal design is a highly non-linear problem, we are forced to change the problem of optimal design to an iterative formulation of optimal redesigns

$$\{h\}_p \Rightarrow \{h\}_p + \{\Delta h\}_p = \{h\}_{p+1} \tag{151}$$

where $\{h\}_p$ is a given design after p redesigns, and the design changes $\{\Delta h\}_p$ are so small that linear expansions around $\{h\}_p$ are acceptable. According to (149) the objective of our redesign is to

$$Minimize \; \Phi(\{h\}_{p+1}) = Minimize \; \Delta\Phi = Minimize \; (\sum_e \frac{\partial \Phi}{\partial h_e} \Delta h_e = \{\nabla \Phi\}_p^T \{\Delta h\}_p) \tag{152}$$

The constraints (150) for the redesigned model will be

$$g_j(\{h\}_{p+1}) = g_j(\{h\}_p) + \sum_e \frac{\partial g_j}{\partial h_e} \Delta h_e \leq 0 \; for \; j = 1, 2, \ldots, J \tag{153}$$

which (in the last form omitting the index p of redesign) we can write in matrix form

$$g_j(\{h\}_{p+1}) = g_j(\{h\}_p) + [\nabla g]_p \{\Delta h\}_p \leq \{0\} \; or \; [\nabla g]\{\Delta h\} \leq -\{g\} \tag{154}$$

At this point it should be noted that at each redesign we attempt to cure possible non-feasibility of the current design (non-feasibility due to the linear expansion). This means that we do not assume $\{g\} \leq \{0\}$ before each redesign, which implies that rather large $\{\Delta h\}$ are acceptable except in the last redesigns.

Introducing the unknown non-negative **slack variables** $\{\tilde{g}\}$ we convert the inequality (154) into an equality constraint

$$[\nabla g]\{\Delta h\} + \{\tilde{g}\} = -\{g\} \tag{155}$$

The resulting value of the slack variable tells whether a constraint is active or not, because a constraint is only active if the optimization procedure returns $\tilde{g}_j = 0$. Comparing the resulting values of slack variables with the response corresponding to the changed design, we get a measure of the current non-linearity of the problem.

Except for the fact that Δh_e is not required to be non-negative, then (152) and (155) constitute a linear programming problem. Using **move limits** for a specific redesign

$$\{0\} \geq \{\Delta h\}^{min} \leq \{\Delta h\} \leq \{\Delta h\}^{max} \geq \{0\} \tag{156}$$

we transform to non-negative variables and at the same time take **absolute limits**

$$\{h\}^{min} \leq \{h\} \leq \{h\}^{max} \tag{157}$$

into account. Details of these transformations are given in Pedersen, 1982 83

6.3 Beam design with several eigenvalue constraints

A finite element discretization, combined with the method of subspace iterations has proven to be a unified approach to eigenvalue analysis of elastic solids. Treating the sensitivity analysis as an integrated part of this approach, one obtains gradients of the eigenvalues without any new eigenvalue analysis. This forms the necessary information for an optimal redesign which is formulated as a linear programming problem. By a sequence of optimal redesigns, one obtains a solution to the problem of optimal design or a solution to an inverse eigenvalue problem.

Taking as an example the vibration of straight Timoshenko beams, we in Pedersen, 1982 83 focus on the gradient functions, and on the dependence of slenderness. This case is described by a conservative, non-gyroscopic system with equilibrium

$$[S]\{D\} = \omega^2[M]\{D\} \tag{158}$$

With symmetric mass matrix $[M]$ and symmetric stiffness matrix $[S]$ we will always have a complete set of eigenvectors $\{D\}$, even when some of them have the same eigenvalue. All eigenvalues ω^2 and eigenvectors will be real with left and right eigenvectors being identical (self-adjoint). A straight forward sensitivity analysis gives for the specific solution $(\omega^2, \{D\})$

$$\frac{\partial \omega^2}{\partial h} = \{D\}^T(-\omega^2\frac{\partial[M]}{\partial h} + \frac{\partial[S]}{\partial h})\{D\} \tag{159}$$

when the normalization

$$\{D\}^T[M]\{D\} = \mathcal{T} = 1 \tag{160}$$

is assumed. This result can be found in Wittrick, 1962 and in many later references. The case of multiple eigenvalues is described in earlier notes and will not be repeated here. For a finite element model we have

$$[M] = \sum[M_e], \quad [S] = \sum[S_e] \tag{161}$$

This has to be read symbolically because $[M]$ and $[S]$ are matrices of order say $10^3 - 10^5$ while $[M_e]$ and $[S_e]$ are element matrices of order 10 . Often a parameter h_e only influences a specific element and then (159) is brought down to element level

$$\frac{\partial \omega^2}{\partial h_e} = \{D\}_e^T(-\omega^2\frac{\partial[M_e]}{\partial h_e} + \frac{\partial[S_e]}{\partial h_e})\{D\}_e \tag{162}$$

with element matrices only of the order 4 for the actual problem. Furthermore, for a Bernoulli-Euler beam model the h_e dependence is homogeneous, such that

$$\partial[M_e]/\partial h_e = m[M_e]/h_e, \quad \partial[S_e]/\partial h_e = k[S_e]/h_e \tag{163}$$

where m, k are given factors, that with beam area as design variable will be $m = 1, k = 2$. For Timoshenko beam theory the change in slenderness ratio can also be taken into account and analytical sensitivity formulas can be found in Pedersen, 1982 83. This reference also include a detailed study of the effect of incorporating this modification, which is only important for short beams (and/or shorter eigenmodes).

As an example we consider a simply supported beam whose volume we want to minimize subject to the constraints that the eigenfrequencies must be greater than or equal to those of

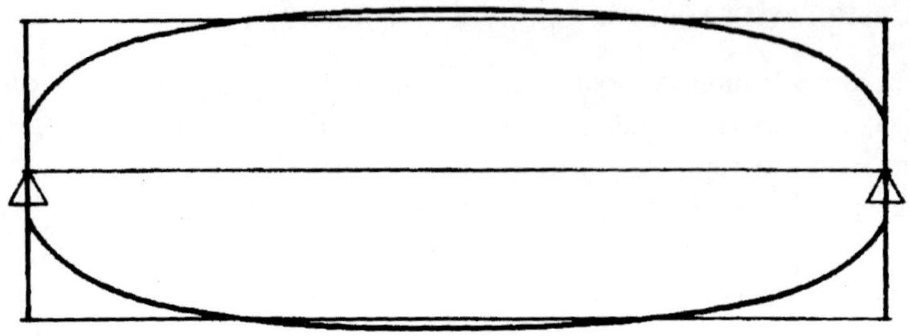

Figure 14. Optimal design of slender, simply supported beam for minimum volume with constraints on the first frequency.

the corresponding uniform beam. We choose a slender beam and at first only set constraint on the lowest frequency. The result in figure 14 is in good agreement with the pioneering result of Niordson, 1965, although we do not reach zero area at the supports when shear deformations are taken into account. The resulting volume is only decreased to 88.3 % of that of the uniform beam.

Figure 15 shows the results when two, three or four eigenfrequencies are constrained. Bearing in mind the shape of the eigenmodes, the trend in design changes is rather natural. Relative to the uniform beam the obtained optimal volumes are now 93.4 %, 95.5 %, 96.7 %, respectively. Results for other boundary conditions are shown in Pedersen, 1981.

6.4 Material orientation of composite plates for optimization of supersonic flutter

Panel flutter is known as one of the important problems in the design of air-crafts, space vehicles, and turbomachinery. Based on the paper by Jørgensen, 1991 we shall shortly discuss how optimal orientation of orthotropic materials can be used to maximize the stability.

Consider a flat panel exposed to a high supersonic flow field over its surface. In order to obtain the aeroelastic equilibrium equations, we assume

- The flexible response is modelled by a first order shear flexible laminated composite plate finite element.
- The free stream Mach number is sufficiently high so that linear piston theory aerodynamics is applicable. Aerodynamic damping is disregarded and it is assumed that the total lift is proportional to the local inclination
- Rotational inertia of transverse sections are included
- Structural damping is neglected.

These assumptions lead to the equilibrium equations

$$[M]\{\ddot{\tilde{D}}\} + p[K]\{\tilde{D}\} + [S]\{\tilde{D}\} = \{0\} \tag{164}$$

where the mass matrix $[M]$ is symmetric, the symmetric part of the stiffness matrix is $[S]$, and the non-symmetric aerodynamic coefficient matrix is $[K]$. The nodal degrees of freedom $\{\tilde{D}\}$

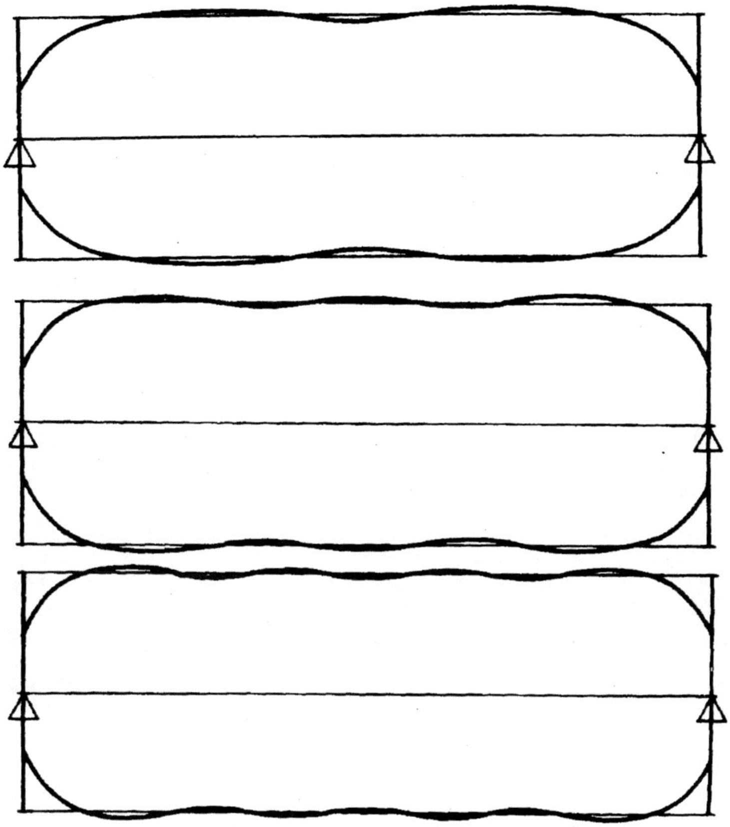

Figure 15. Optimal design of slender, simply supported beams for minimum volume with constraints on two, three and four frequencies, respectively.

is a function of time t. Assuming separation $\{\tilde{D}\} = \{D\}\exp(\alpha + \imath\omega)t$ we obtain the time independent eigenvalue problem

$$([M](\alpha + \imath\omega)^2 + p[K] + [S])\{D\} = \{0\} \tag{165}$$

with eigensolutions $((\alpha_i + \imath\omega_i)^2, \{D\}_i)$. Due to the circulatory forces described by the non-symmetric $[K]$ matrix we may obtain complex eigenvalues as well as complex eigenvectors. The general aspect of this is described in earlier notes, and the main point to remember is that an eigenvalue with positive α_i corresponds to instability.

Figure 16 shows two eigenvectors that converge toward the flutter vector by increasing load p, say increasing flow speed. Note that at flutter we have a double eigenvalue with only a single eigenvector.

For optimal design of material orientation described by θ_e in all elements e we need the sensitivities of the flutter load

$$\partial p_F / \partial \theta_e \; for \; e = 1, 2, \ldots, N \tag{166}$$

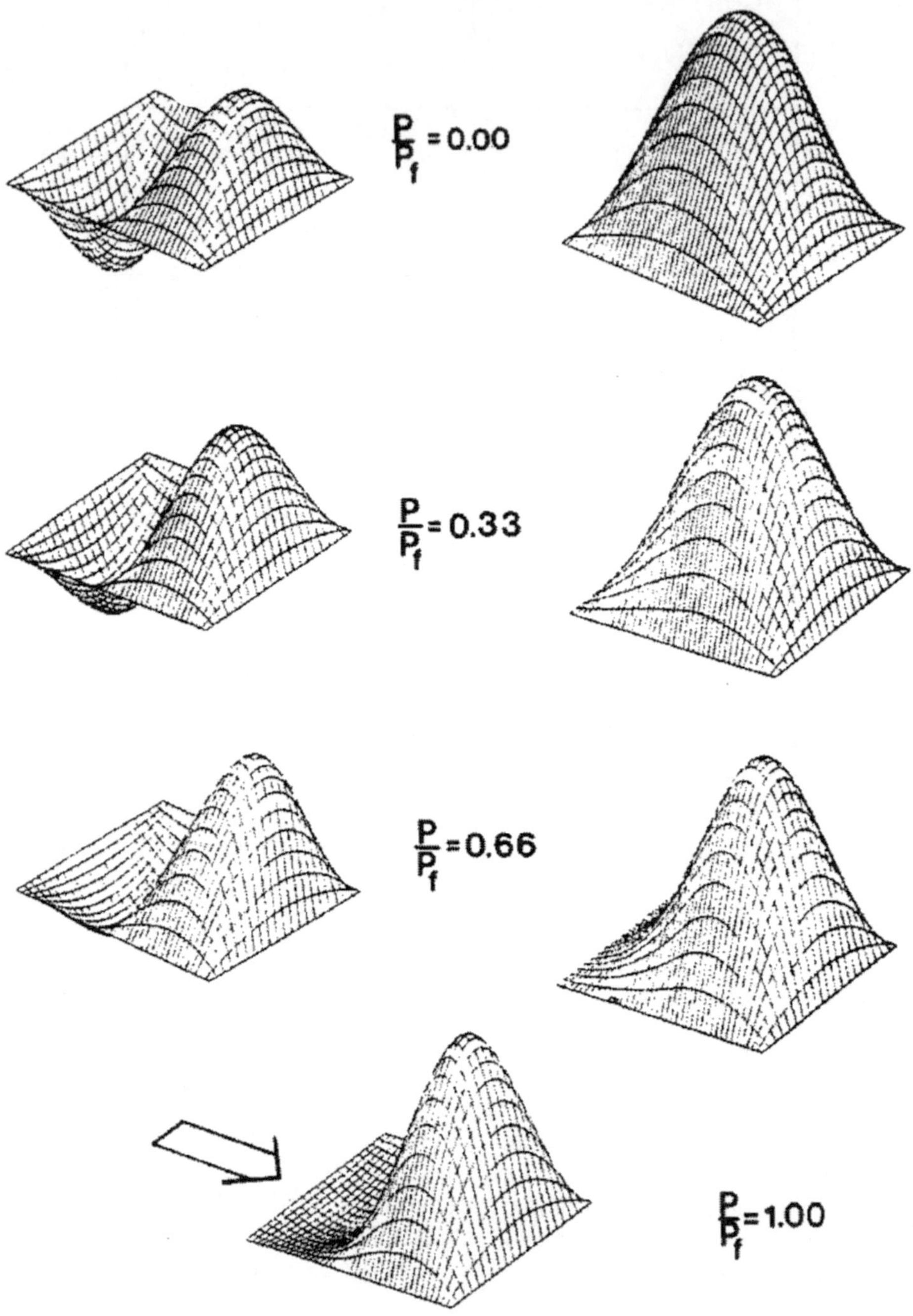

Figure 16. For an isotropic panel the two lowest eigenmodes are shown to converge toward the same mode for the flutter load. From Jørgensen, 1991

and because the mass matrix is independent of θ_e the general sensitivity analysis will for this case give

$$\frac{\partial p_F}{\partial \theta_e} = -\frac{\{\hat{D}\}_e^T(\partial[S]_e/\partial\theta_e)\{D\}_e}{\{\hat{D}\}^T[K]\{D\}} \tag{167}$$

where $\{\hat{D}\}_e^T$ is the adjoint eigenvector (left eigenvector). We have used that only the specific element stiffness matrix $[S]_e$ is depending on θ_e.

In some of the optimization problem the changed designs leads to divergent instability at lower loads than the optimized flutter load p_F. In order to control this non-wanted behaviour, a further constraint is included in the formulation, that now reads

$$Maximize\, p_F\; subjected\, to\, \omega_{min}^2 > 0 \tag{168}$$

where ω_{min}^2 is defined by

$$\omega_{min}^2 := Minimum\; of \omega_i^2\; for\; all\; i\; and\, p \in [o, p_F] \tag{169}$$

Solutions and interesting discussions can be found in Jørgensen, 1991

6.5 Column design with follower forces

In the thesis by Langthjem, 1996 and in the papers by Langthjem and Sugiyama, 1999a, 1999b, 2000b, and 2000c the often addressed problem of optimal column design with follower forces is presented. We shall here just comment on some general issues.

In the papers by Langthjem and Sugiyama the results of impressive parameter studies including experimental verification are presented. The influence of external and internal damping together with different levels of dominance of non-conservative loads are discussed, and all in all we learn that the response of these problems are very sensitive to the parameter values. Adding design optimization to the problem statement extreme sensitivity is found and thus the experimental comparisons are of vital importance.

A recent survey by Langthjem and Sugiyama, 2000a gives a valuable and balanced view of this field that has been so emotionally discussed in the journals.

6.6 Beam design with periodic loads

The actual optimization problem is described in Seyranian et al., 2000 to show application of sensitivity analysis for linear periodic systems. This sensitivity analysis is described earlier in these notes.

Consider the straight beam of length L in figure 17. The beam is in the axial direction loaded at one end with a force $p\cos(\omega t)$. The beam is externally damped with c as external viscous damping coefficient. The equation for transverse vibration $w = w(x,t)$ of the beam is

$$\rho A\frac{\partial^2 w}{\partial t^2} + c\frac{\partial w}{\partial t} + \frac{\partial^2}{\partial x^2}(EI\frac{\partial^2 w}{\partial x^2}) + p\cos(\omega t)\frac{\partial^2 w}{\partial x^2} = 0 \tag{170}$$

Figure 17. Beam with an axial, periodic load. The beam is externally damped.

where Youngs modulus E and the mass density ρ are constants, $I = I(x)$ is the cross-sectional moment of inertia, and $A = A(x)$ is the cross-sectional area. The beam has circular cross-section, and then we have

$$A(x) = \pi r^2(x) \ \ and \ \ I(x) = \pi r^4(x)/4 \tag{171}$$

where $r(x)$ is the radius of the beam at position x (the function to be designed). Boundary conditions correspond to simple supports.

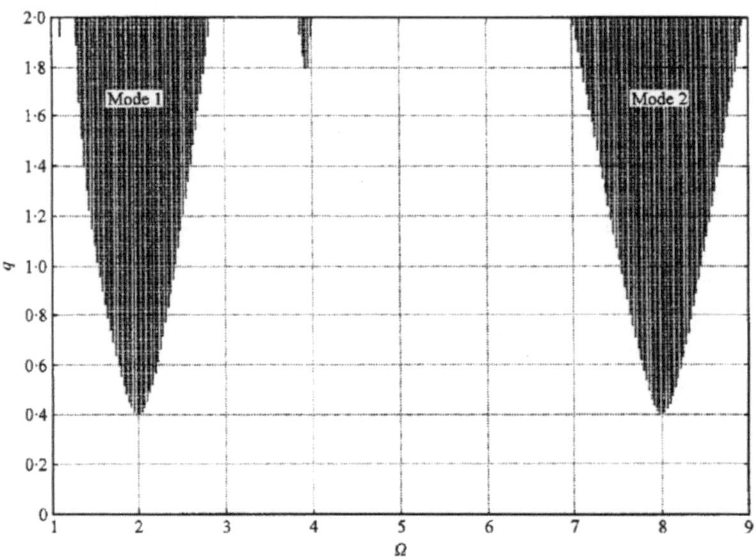

Figure 18. Part of the stability diagram for a uniform beam with external damping. The beam is subjected to an axial, periodic load.

The system is then written in non-dimensional form and discretized, so the design variable will be the relative radius of m elements of equal length. Floquet method with multipliers gives the stability diagram in figure 18 for a uniform beam. Note that the critical q (from load level p)

is active in mode 1 and very close to being active in mode 2. It is therefore to be expected that both modes are simultaneous active for the optimized design.

For a beam divided into 25 elements we minimize the volume with two active constraints or we maximize the minimum of critical q for a given volume. This last formulation involves a standard transformation of a max-min problem into a more simple max problem. The optimization is carried through using analytical sensitivity analysis (as derived in earlier notes). The final result is presented in figure 19, and we notice the similarity with the first solution in figure 15.

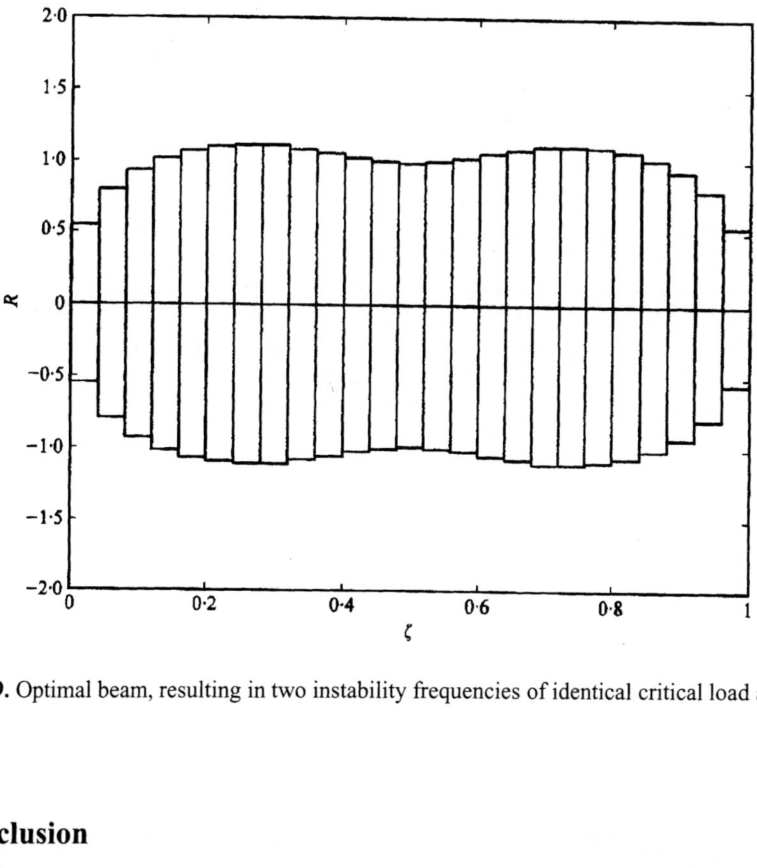

Figure 19. Optimal beam, resulting in two instability frequencies of identical critical load amplitude.

6.7 Conclusion

In this short note we put focus on the sensitivity analysis for both the methods mostly used to solve optimal design problems. In the optimality criteria methods the necessary condition of proportionality (148) should be kept in mind. For mathematical programming methods the reformulation to optimal redesign is also rather general and therefore is valuable to remember.

The specific examples are chosen based on personal experience and a literature study will give a much deeper insight into this extensive area of research. Review articles are also available, e.g. Olhoff, 1985.

References

Arscott, F. M. (1964). *Periodic differential equations*. Pergamon. 281 pages.

Benjamin, T. B. (1961). Dynamics of a system of articulated pipes conveying fluid. *Proc. Roy. Soc. Lond.* A 261:457–499.

Bishop, R. E. D., and Fawzy, I. (1976). Free and forced oscillation of a vertical tube containing a flowing fluid. *Phil. Trans. Roy. Soc. Lond.* A 284:1–47.

Bolotin, V. V., and Zhinzher, N. I. (1969). Effects of damping on stability of elastic systems subjected to non-conservative forces. *Int. J. Solids and Structures* 965–989.

Bolotin, V. V. (1963). *Nonconservative Problems of the Theory of Elastic Stability*. Pergamon. 324 pages.

Bolotin, V. V. (1964). *The Dynamic Stability of Elastic Systems*. Holden-Day. 451 pages.

Bunkov, V. G. (1969). Calculation of optimal flutter characteristics by gradient method (in russian). *TSAGI* 1166.

Cartmell, M. P. (1990). *Introduction to Linear, Parametric and Nonlinear Vibrations*. London: Chapman and Hall.

Cesari, L. (1964). *Asymptotic behaviour and stability problems in ordinary differential equations*. Academic Press.

Epstein, B., and Barakat, R. (1977). Perturbation solutions of the Carson-Cambi equation. *J. Franklin Institute* 303:177–188.

Floquet, G. (1883). Sur les equation differentielles lineaires coefficients periodiques. *Annales de Ecole Normal Superior* 2(12):47–89.

Frederiksen, P. S. (1997). Experimental procedure and results for the identification of elastic constants of thick orthotropic plates. *J. Composite Materials* 31:360–382.

Fu, F. C. L., and Nemat-Nasser, S. (1972). Stability of solution of systems of linear differential equations with harmonic coefficients. *AIAA J.* 10(1):30–36.

Gregory, R. W., and Paidoussis, M. P. (1966). Unstable oscillation of tubular cantilevers conveying fluid. ii. experiments. *Proc. of the Royal Society Lond.* A 293:528–542.

Grimshaw, R. (1990). *Nonlinear ordinary differential equations*. Blackwell.

Hansen, J. M. (1985). Stability diagrams for coupled mathieu-equations. *Ingenieur-Archiv* 55:463–473.

Herrmann, G., and Jong, I. C. (1965). On the destabilizing effect of damping in nonconservative elastic systems. *J. Appl. Mech.* 32:592–597.

Herrmann, G. (1967). Stability of equilibrium of elastic systems subjected to nonconservative forces. *Appl. Mech. Revs* 20:103–108.

Hill, G. W. (1886). On the part of the motion of the lunar perigee. *Acta Math.* 8:1–36.

Hsu, C. S. (1961). On a restricted class of coupled Hill's equations and some applications. *J. of Appl. Mech.* 28:551–556.

Huseyin, K. (1978). *Vibrations and stability of multiple parameter systems*. Sijthoff and Nordhoff.

Jones, R. M. (1975). *Mechanics of Composite Materials*. McGraw-Hill.

Jørgensen, O. (1991). Optimization of the flutter load by material orientation. *Mechanics of Structures and Machines* 19(3):411–436.

Langthjem, M. A., and Sugiyama, Y. (1999a). Optimal design of Beck's column with a constraint on the static buckling load. *Structural Optimization* 18:228–235.

Langthjem, M. A., and Sugiyama, Y. (1999b). Optimal shape design against flutter of a cantilevered column with an end-mass of finite size subjected to a non-conservative load. *J. of Sound and Vibration* 226(1):1–23.

Langthjem, M. A., and Sugiyama, Y. (2000a). Dynamic stability of columns subjected to follower loads: A survey. *J. of Sound and Vibration* 238(5):809–851.

Langthjem, M. A., and Sugiyama, Y. (2000b). Optimal design of cantilevered columns under the combined action of conservative and nonconservative loads. part i the undamped case. *Computers and Structures* 74:385–398.

Langthjem, M. A., and Sugiyama, Y. (2000c). Optimal design of cantilevered columns under the combined action of conservative and nonconservative loads. part ii the damped case. *Computers and Structures* 74:399–408.

Langthjem, M. A. (1996). *Dynamics, Stability and Optimal Design of Structures with Fluid Interaction.* DCAMM Sxx, Solid Mechanics, DTU - thesis for the Ph.D.

Liapunov, A. (1949). *Probleme general de la stabilite du mouvement*, volume 17 of *Ann. Math. Studies.* Princeton Univ. Press.

Lindh, K. G., and Likins, P. W. (1970). Infinite determinant methods for stability analysis of periodic-coefficient differential equations. *AIAA J.* 8(4):680–686.

Lottati, I., and Kornecki, A. (1986). The effects of an elastic foundation and of dissipative forces on the stability of fluid-conveying pipes. *J. of Sound and Vibration* 109:327–338.

Magnus, W., and Winkler, S. (1966). *Hill's equation.* 125 pages.

Markworth, N. J., and Petersen, C. (1987). *Identifikation af Materialeparametre for Fiberbaserede Laminatplader (McS thesis in Danish).* Solid Mechanics, DTU.

Mathieu, E. (1868). Memoire sur le mouvement vibratoire d'une membrane de forme elliptique. *J. de Math. Pures et Appliquees (J. de Liouville)* 13:137–203.

McLachlan, N. W. (1947). *Theory and application of Mathieu functions.* 401 pages.

Meixner, J., and Schaefke, F. W. (1954). *Mathieusche funktionen und sphaeroidfunktionen.* 414 pages.

Mettler, E. (1959). Stabilitaetsfragen bei freien schwingungen mechanicher systeme. *Ingenieur-Archiv* 28:213–228.

Nayfeh, A. H., and Mook, D. T. (1979). *Nonlinear oscillations.* Wiley.

Niordson, F. I. (1965). On the optimal design of a vibrating beam. *Quarterly of Applied Mathematics* 23:47–53.

Olhoff, N. (1985). Optimal design with respect to structural eigenvalues. In Rimrott, F. P. J., and Tabarrot, B., eds., *IUTAM Congress*, 133 –149. Canada: North-Holland.

Otterbein, S. (1982). Stabilisierung des n-pendels und der indische seiltrick. *Arch. for Rational Mechanics and Analysis* 78:381–393.

Paidoussis, M. P., and Li, G. X. (1993). Pipes conveying fluid: a model dynamical problem. *J. of Fluids and Structures* 7:137–204.

Panovko, Y. G., and Gubanova, I. I. (1964). *Stability and Oscillations of Elastic Systems.* Consultants Bureau. 290 pages.

Pedersen, P., and Seyranian, A. P. (1983). Sensitivity analysis for problems of dynamic stability. *Int. J. of Solids and Structures* 19(4):315–335.

Pedersen, P. (1980). A quantitative stability analysis of the solutions to the Carson-Cambi equation. *J. of the Franklin Institute* 309(5):359–367.

Pedersen, P. (1981). The integrated approach of fem-slp for solving problems of optimal design. In Haug, E., and Cea, J., eds., *Optimization Distributed Parameter Structures*, volume 49 of *NATO ASI series*, 757–780. Sijthoff and Noordhoff.

Pedersen, P. (1982–83). Design with several eigenvalue constraints by finite elements and linear programming. *J. Structural Mechanics* 10(3):243–271.

Pedersen, P. (1983). A unified approach to optimal design. In Eschenauer, H., and Olhoff, N., eds., *Optimization Methods in Structural Design*, 182–187. Siegen, Germany: B.I. Wissenschafsverlag.

Pedersen, P. (1984). Sensitivity analysis for non-selfadjoint problems. In Komkov, V., ed., *Sensitivity of Functionals with Applications to Engineering Sciences*, volume 1086 of *Lecture Notes in Mathematics*, 119–130. New York, USA: American Mathematical Society.

Pedersen, P. (1985). On stability diagrams for damped Hill equations. *Quarterly of Applied Mathematics* 42(4):477–495.

Pedersen, P. (1987). On sensitivity analysis of optimal design of specially orthotropic laminates. *Engineering Optimization* 11:305–316.

Pedersen, P. (1999). Sensitivity analysis and inverse problems for laminates and materials. In Mota Soares, C. A., Mota Soares, C. M., and Freitas, M. J. M., eds., *Mechanics of Composite Materials and Structures*, NATO ASI series, 453–463. Troia, Portugal: Kluwer.

Rayleigh, L. (1887). Maintenance of vibrations by forces of double frequency, and propagation of waves through a medium with a periodic structure. *P.M.* 24:145.

Seyranian, A. P., and Pedersen, P. (1995). On two effects in fluid/structure interaction theory. In Bearman, P. W., ed., *6th Int. Conference on Fluid-Induced Vibration*, 565–576. London, UK: Balkema.

Seyranian, A. P., Lund, E., and Olhoff, N. (1994). Multiple eigenvalues in structural optimization problems. *Structural Optimization* 8(4):207–227.

Seyranian, A. P., Solem, F., and Pedersen, P. (1999). Stability analysis for multi-parameter periodic systems. *Archive of Applied Mechanics* 69:160–180.

Seyranian, A. P., Solem, F., and Pedersen, P. (2000). Multi-parameter linear periodic systems: Sensitivity analysis and applications. *J. of Sound and Vibration* 229(1):89–111.

Seyranian, A. P. (1990). Destabilization paradox in stability problems of nonconservative systems. *Advances in Mechanics* 13(2):89–124.

Seyranian, A. P. (1991). *Stability and catastrophes of vibrating systems depending on parameters - Lecture notes*. DCAMM. 76 pages.

Seyranian, A. P. (1993). Sensitivity analysis of multiple eigenvalues. *Mechanics of Structures and Machines* 21(2):261–284.

Sinha, S. C., and Wu, D. H. (1991). An efficient computational scheme for the analysis of periodic systems. *J. of Sound and Vibration* 151(1):91–117.

Smith, G. (1975). *Parameterregte schwingungen*. VEB Berlin. 313 pages.

Stoker, J. J. (1950). *Nonlinear Vibrations in Mechanical and Electrical Systems*. Interscience. 273 pages.

Strutt, M. J. O. (1932). *Lamesche-Mathieusche-und verwandte funktionen in physik und technik*. Springer. 116 pages.

Sugiyama, Y., and Noda, T. (1981). Studies on stability of two-degrees-of-freedom articulated pipes conveying fluid. *Bulletin of JSME* 24(194-9):1354–1362.

Thomsen, J. J. (1997). *Vibrations and Stability, Order and Chaos*. McGraw-Hill. 323 pages.

Vishik, M. I., and Lyusternik, L. A. (1960). Solution of some perturbation problems in the case of matrices and self-adjoint or non-self-adjoint equations. *Russian Math. Surveys* 15(3):1–73.

Wittrick, W. H. (1962). Rates of change of eigenvalues, with reference to buckling and vibration problems. *J. Royal Aeronautical Soc.* 66:590–591.

Wolfram, S. (1991). *Mathematica: a system for doing mathematics by computer*. Addison-Wesley.

Yakubovitch, V. A., and Starzhinskii, V. M. (1975). *Linear differential equations with periodic coefficients*, volume 1. Wiley.

Yakubovitch, V. A., and Starzhinskii, V. M. (1987). *Parametric resonance in linear systems (in Russian)*. Nauka.

Ziegler, H. (1952). Die stabilitaetskriterien der elastomechanik. *Ing.-Arch.* 20:49–56.

Ziegler, H. (1968). *Principles of Structural Stability*. Blaisdell. 150 pages.

PART VII. **Experimental Approach to Nonconservative Stability Problems**

Yoshihiko Sugiyama

Department of Aerospace Engineering
Osaka Prefecture University, Sakai-shi, 599-8531 Japan
e-mail: sugiyama@aero.osakafu-u.ac.jp

ABSTRACT

The intended aim of this part is to give some experimental supports to nonconservative stability problems. Chapter 1 describes the state of art of dynamic stability of elastic systems subjected to nonconservative follower forces. Experimental works are specially emphasized. Chapter 2 is concerned with flutter of cantilevered pipes conveying fluid. The combined effect of a spring support and a lumped mass on stability of a tubular cantilever conveying fluid is discussed. It is shown theoretically and experimentally that for particular combinations of a spring support and a lumped mass, there exists a flutter peninsula, which juts out locally from the main flutter region. Chapter 3 aims at showing experimental verifications of the reality of follower force. Three experiments are presented on dynamic stability of cantilevered columns subjected to a follower force produced by a solid rocket motor. The first experiment shows that a horizontal cantilevered column subjected to a rocket thrust loses its stability by flutter. The second experiment gives an experimental support to the effect of an attached lumped mass on flutter of horizontal cantilevered column. The third experiment demonstrates the reality of sub-tangential follower force. Chapter 4 discusses the effect of damping configuration on flutter of two-degree-of-freedom mechanical models subjected to a Reut-type nonconservative force. The nonconservative force was produced by an impinging air jet. Damping due to an attached dash-pot is taken into account, as well as internal and external damping. Chapter 5 describes the general concluding remarks. In Appendix are shown some photographs of experiments on dynamic stability of cantilevered columns subjected to a rocket thrust.

1. General Introduction

A cantilevered column subjected to a tangential follower force has been referred to as Beck's column, since the critical force for the column was first found by Beck in 1952. However his result and its related topics have not been appreciated well by structural engineers. For example, Timoshenko (1961) commented on Beck's result that "no definite conclusion can be made (as yet) regarding the practical value of this result, since no method has been devised for applying a tangential force to a column during bending". Even at present there is a criticism to the reality of tangential follower forces (see, e.g., Koiter, 1996).

The origin of a follower force can be found in an end rocket thrust applied to flexible missiles and spacecraft (see Panovko and Gubanova, 1965). In this sense a follower force is a very realistic force in the field of aerospace structures engineering. Free-free beams under a follower force have been intensively studied by Beal (1965), Feodosiev (1965) and Mladenov and Sugiyama (1997). Beck's column is an imaginary version of the free-free beams under a follower force, as shown in Figure 1.1. It is noted that the column is not the exact model of the actual structures, but just an idealized model. The actual flexible structures have variable mass, stiffness and axial compressive stress distributions, while the column model has the corresponding uniform distributions. It is however emphasized that Beck's column keeps basically the same bending flutter characteristics as the corresponding free-free beams. In order to clarify the basic dynamics of free-free beams, it is a sharp way to work on Beck's column in place of free-free beams, since the experimental verification can be done on the cantilevered column, while it is very difficult to do so on the free-free beams.

Beck's column and modified Beck's columns have been studied by a vast body of researchers. The first review of this branch of applied mechanics has been made in the book form by Bolotin (1963). So far a lot of interesting results on the effect of follower forces have been drawn through analytical studies based on some mathematical models. However, it has been the case that some of the results are confusing, because the mathematical models on which the results were brought are greatly and sometimes unduly simplified (see, e.g., Sugiyama, 1976). Hence it has been suggested by several authors (see, e.g., Sugiyama, 1987) that the stability problems of nonconservative systems should be studied in connection with the corresponding engineering practice and that theoretical results on nonconservative systems must be checked by experiments.

As to the experimental demonstrations of nonconservative systems, Herrmann and his colleagues (1966) made the first systematic experimental demonstrations of the systems. Another first experimental study was made by Yagn and Parshin (1966). First review of the experimental studies on instability of elastic systems subjected to nonconservative forces was reported by Sugiyama and Sekiya (1971). However a thorough experimental study of the tangential follower force problem has not been conducted.

As to reality of a follower force by itself, it has been known well in the field of fluid-structure interaction systems that a momentum flux discharged from the free end of the cantilevered pipe conveying fluid makes a follower thrust (see, e.g., Paidoussis and Li, 1993, Sugiyama et al., 1985). Even the experiment of flutter caused by a follower force can be conducted easily at swimming pool when a long hose from a tap is immersed in water and the tap is opened full to supply water to the pool. The long hose can dance in the pool to show flutter motion due to a follower force.

One of the interesting topics in nonconservative stability problems has been the destabilizing effect of damping (Bolotin, 1963). It is noted that this effect has not been verified experimentally to the best of author's knowledge. The destabilizing effect is obtained when the asymptotic stability condition is applied to nonconservative systems. The asymptotic stability condition implies that the dynamical system is unstable if the amplitude of the disturbed motion of the system becomes infinite as the time goes to infinity. However it is of vital importance to watch the behavior of complex eigenvalues in case of the destabilizing effect of damping. It is found that the most dangerous eigenvalue mostly runs in parallel with the imaginary axis. The growth-rate of the amplitude of the motion is thus very small indeed. Mathematically the asymptotic stability criterion gives the theoretical flutter limit, however in practice a follower force caused by a rocket motor can act upon elastic structures only for a finite time interval. In the latter case, the theoretical flutter limit obtained by neglecting damping can predict the experimental flutter limit (see Sugiyama et al, 1995a, 1995b).

Finally it is also worthwhile to note that the follower force is not caused only by jet or rocket thrust. Considering automobile disk brakes, for example, the "squeezing" force acting on the rotating disk is again a nonconservative follower force. The force is caused by dry friction and the squeal is a flutter type instability. The onset of brake squeal is completely equivalent to the passage through the stability boundary in Beck's column (Nishiwaki, 1993, Mottershead and Chan, 1996).

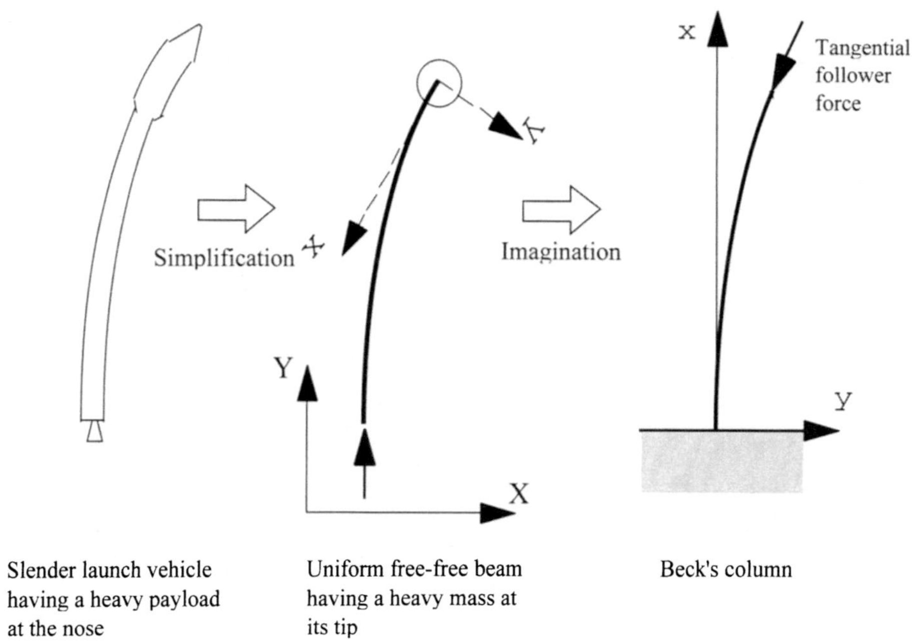

Figure 1.1 Beck's column

2. Cantilevered Pipes Conveying Fluid

2.1 Introduction

Recently, slender pipes have become widely used to convey fluid in various plants (Paidoussis and Li, 1993, Paidoussis, 1998). They have, however, the possibility of divergence or flutter. There exists the demand for knowledge necessary for supporting such pipes properly. In particular, the cantilever pipes conveying fluid are nonconservative systems (see, e.g., Paidoussis, 1998), and their dynamical behavior can not be predicted from a knowledge of conservative systems. As shown by Sugiyama et al. (1985), addition of a lateral spring support to a fluid-conveying cantilever pipe may destabilize the system.

In this chapter we consider the combined effect of a spring support and a lumped mass on the stability of cantilevered pipes conveying fluid. In addition to critical flutter velocity, critical flutter configurations are also determined theoretically. In determining the configuration, the number of terms in the Galerkin's approximate solution will be examined. The local flutter region in the stability map, which is being referred to as a flutter peninsula, is discussed theoretically and experimentally.

2.2 Stability Analysis

A fluid-conveying pipe of length l is cantilevered and is assumed to move only in the horizontal x - y plane, as shown in Figure 2.1. The pipe has a mass per unit length m_l and conveys incompressible fluid of mass per unit length m_f, flowing through the pipe at a constant velocity U. The material of the pipe is viscoelastic and may be characterized by the Young's modulus E and the viscous resistance coefficient E^*. Moreover, a spring support with a spring constant K and a lumped mass M_1 are attached to the pipe at a distance l_1 from the clamped end. For simplicity, it is convenient to introduce the following dimensionless quantities

$$\alpha = \frac{M_1}{\left(m_f + m_l\right) l}, \quad \beta = \frac{m_f}{m_f + m_l}, \quad \gamma = \frac{1}{l^2}\frac{E^*}{E}\left(\frac{EI}{m_f + m_l}\right)^{1/2}, \quad \kappa = \frac{Kl^3}{EI}, \quad \eta = \frac{y}{r},$$

$$\nu = \frac{Cl^2}{\left[EI\left(m + m_f\right)\right]^{1/2}}, \quad \xi = \frac{x}{l}, \quad \xi_1 = \frac{l_1}{l}, \quad \tau = \frac{t}{l^2}\left[\frac{EI}{m_f + m_l}\right]^{1/2}, \quad u = Ul\left(\frac{m_f}{EI}\right)^{1/2}, \tag{2.1}$$

where I is the area moment of inertia of the pipe, C the coefficient of external viscous damping, r the radius of gyration of the pipe section and t is the time. With quantities defined in Equations (2.1), the governing equation for small motion of the pipe is given by

$$\left[1+\alpha\delta(\xi-\xi_1)\right]\frac{\partial^2\eta}{\partial\tau^2}+\gamma\frac{\partial^5\eta}{\partial\tau\partial^4\xi}+2u\beta^{1/2}\frac{\partial^2\eta}{\partial\tau\partial\xi}+\nu\frac{\partial\eta}{\partial\tau}$$
$$+\frac{\partial^4\eta}{\partial\xi^4}+u^2\frac{\partial^2\eta}{\partial\xi^2}+\kappa\delta(\xi-\xi_1)\eta=0, \qquad (2.2)$$

where $\delta(\xi)$ is the Dirac delta function. The dimensionless boundary conditions are

$$\left.\begin{aligned} &\eta(0,\tau)=\frac{\partial\eta(0,\tau)}{\partial\xi}=0,\\ &\frac{\partial^2\eta(1,\tau)}{\partial\xi^2}+\gamma\frac{\partial^3\eta(1,\tau)}{\partial\tau\partial\xi^2}=0,\\ &\frac{\partial^3\eta(1,\tau)}{\partial\xi^3}+\gamma\frac{\partial^4\eta(1,\tau)}{\partial\tau\partial\xi^3}=0.\end{aligned}\right\} \qquad (2.3)$$

Now the Galerkin method can be applied to solve Equation (2.2) approximately. Put the solution to Equation (2.2) in the form

$$\eta(\xi,\tau)=\sum_{n=1}^{M}a_n(\tau)\varphi_n(\xi), \qquad (2.4)$$

where $\varphi_n(\xi)$ is the n-th normalized eigenfunction of the cantilevered beam under the boundary conditions

$$\varphi_n(0)=\frac{d\varphi_n(0)}{d\xi}=\frac{d^2\varphi_n(0)}{d\xi^2}=\frac{d^3\varphi_n(0)}{d\xi^3}=0. \qquad (2.5)$$

Because of Equation (2.5), the trial solution (2.4) satisfies the boundary conditions (2.3). Substituting Equation (2.4) into Equation (2.2), and performing the Galerkin integral, we obtain a system of equations of motion with finite degrees of freedom

$$\sum_{n=1}^{M}\left[A_{mn}\ddot{a}_n+B_{mn}\dot{a}_n+C_{mn}a_n\right]=0 \quad (m=1,2,\mathrm{L}\ ,M), \qquad (2.6)$$

where dot ( )means differentiation with respect to dimensionless time τ. Equation (2.6) may be cast into the matrix form

$$\boldsymbol{A}\ddot{\boldsymbol{a}}+\boldsymbol{B}\dot{\boldsymbol{a}}+\boldsymbol{C}\boldsymbol{a}=\boldsymbol{0}, \qquad (2.7)$$

where

$$\left.\begin{aligned} &\boldsymbol{A} = [A_{mn}], \quad \boldsymbol{B} = [B_{mn}], \quad \boldsymbol{C} = [C_{mn}], \\ &\boldsymbol{a} = [a_1, a_2, \ldots, a_m]^T. \end{aligned}\right\} \tag{2.8}$$

Equation (2.7) may be written in the form

$$\dot{\boldsymbol{Z}} = \boldsymbol{P}\boldsymbol{Z}, \tag{2.9}$$

where

$$\boldsymbol{Z} = \begin{bmatrix} \dot{\boldsymbol{a}} \\ \boldsymbol{a} \end{bmatrix}, \qquad \boldsymbol{P} = \begin{bmatrix} -\boldsymbol{A}^{-1}\boldsymbol{B} & -\boldsymbol{A}^{-1}\boldsymbol{C} \\ \boldsymbol{I} & \boldsymbol{0} \end{bmatrix}. \tag{2.10}$$

I and ***O*** are the unit matrix and zero matrix, respectively. Now let us put $a_n(\tau)$ in the form

$$a_n(\tau) = h_n e^{\lambda\tau}. \tag{2.11}$$

With Equations (2.9), (2.10) and (2.11), the condition for existence of a non-trivial solution yields the characteristic equation

$$|\mathbf{P} - \lambda\mathbf{I}| = 0. \tag{2.12}$$

Generally, characteristic roots λ may be written as

$$\lambda = \sigma \pm j\omega, \tag{2.13}$$

where σ and ω are real numbers, respectively, and j= √-1. If all real parts of the roots λ are negative, $a_n(\tau)$decays exponentially with time, and thus the system is stable. Whereas, if at least one of the roots has a positive real part, the system is unstable because of a divergent $a_n(\tau)$. From the above discussion, when any real part of the roots becomes zero at first, the corresponding flow velocity u is the critical velocity u_{cr}; if $\omega \neq 0$, u_{cr} is the flutter limit, if $\omega = 0$, the divergence limit, respectively.

2.3 Critical Flutter Configurations

In this section, the configurations of stationary flutter of the pipe conveying fluid at the critical flow velocity u_{cr} are calculated as follows:

Let us assume that when $u = u_{cr}$, the i-th order branch cuts across the imaginary axis of the σ - ω plane, and the corresponding characteristic root is

$$\lambda_i = \pm\omega_{i*}, \tag{2.14}$$

In this case, any other branches but the i-th are in the left half $\sigma - \omega$ plane, so that every oscillatory components associated with such branches decay with time. Thus, only an oscillatory component of the i-th order branch under the critical condition is interesting. Equation (2.11) is reduced to

$$a_n(\tau) = h_n e^{\pm j\omega_{i*}\tau}. \qquad (2.15)$$

Substituting Equation (2.15) into Equation (2.6), we have

$$\sum_{n=1}^{M} [(C_{mn} - \omega_{i*}^2 A_{mn}) \pm j\omega_{i*} B_{mn}] h_n = 0 \quad (m = 1, 2, \cdots, M). \qquad (2.16)$$

Dividing Equation (2.16) by h_1 and transposing the terms with n=1 to the right hand side, we have

$$\sum_{n=2}^{M} [(C_{mn} - \omega_{i*}^2 A_{mn}) \pm j\omega_{i*} B_{mn}] g_n = -(C_{m1} - \omega_{i*}^2 A_{m1}) \mp j\omega_{i*} B_{m1}$$

$$(m = 1, 2, \cdots, M-1), \qquad (2.17)$$

where

$$g_n = \frac{h_n}{h_1} \qquad (n = 2, 3, \cdots, M). \qquad (2.18)$$

Though Equation (2.16) are M homogeneous equations, only (M-1) equations are required to calculate the set of the (M-1) ratios g_p. Since Equation (2.17) comprises (M-1) simultaneous linear equations with (M-1) complex unknowns, (M-1) g_p's may be determined. Since g_p is generally complex, it can be written as

$$g_p = f_p e^{\pm j\theta_p}, \qquad (2.19)$$

where f_p is the amplitude ratio of the p-th mode (p>2) to the 1st mode, and θ_p is the phase difference between the p-th mode (p>2) and 1st mode. When subscripts of f and θ are 1, f_1=1 and θ_1=0, respectively. In this way, we can add g_p to the set of (m-1) g_p (p=2,3, $\ldots$, M). Equations (2.18) and (2.19) yield

$$h_n = h_1 f_n e^{\pm j\theta_n} \qquad (n = 1, 2, \cdots, M). \qquad (2.20)$$

Moreover h_1 is denoted by using exponents as

$$h_1 = h_0 f_n e^{\pm j\theta_0}, \qquad (2.21)$$

where h_0 and θ_0 correspond to the initial amplitude and the initial phase, respectively. They can be determined by the initial condition. Equations (2.20) and (2.21) give the solution (2.4) at the critical flow velocity as

$$\eta(\xi, \tau) = h_0 \left[\sum_{n=1}^{M} f_n \varphi_n(\xi) \cos(\omega_* \tau + \theta_n + \theta_0) \right], \qquad (2.22)$$

in which only the real part of $\eta(\xi, \tau)$ is given. In the actual calculation, it is assumed that $\theta_0 = -\pi/2$ and $h_0 = 1$.

2.4 Discussion on the Number of Terms in the Approximate Solution

Let us consider how many terms should be taken in the approximate solution from the view point of the eigenmode spectrum in critical flutter configurations. For this purpose, the critical flutter configurations of the cantilevered pipe with neither a spring support nor a lumped mass are investigated.

From the procedure described in section 2.3, the amplitude ratio f_n in Equation (2.20) of the above pipe may be obtained. Figure 2.2 shows the variation of the amplitude ratio for n-th eigenmode, f_n/f_{max}, with n terms for an approximate solution, where f_{max} is the maximum value of f_n. From this figure, it is easy to see how each eigenmode contributes the critical flutter configurations.

Figure 2.2(a) shows the effect of the mass ratio β on the amplitude ratio f_n/f_{max}. When $\beta = 0.25$, the 2nd eigenmode is the most predominant, and the 1st and 3rd come after in turn. When $\beta = 0.50$, the 3rd eigenmode is the most predominant and is followed by the 2nd. When $\beta = 0.75$, the 3rd eigenmode is the most predominant and is followed by the 4th and the 2nd, having almost similar values. Therefore, it is seen that the higher eigenmodes become predominant as β increases. It should be noted that the effect of an eigenmode of more than the 5th order is considerably small and decreases with increasing n.

Figure 2.2(b) shows the variation of f_n/f_{max} with internal damping γ. The effect of γ on f_n/f_{max} is discussed, for example, for $\beta = 0.75$, at which the higher order eigenmodes are rather significant. When $\gamma = 0$ or $\gamma = 10^{-3}$, the 3rd eigenmode is the most predominant, and the 2nd and the 4th are comparable to the 1st. When $\gamma = 10^{-2}$, the 2nd and 3rd order eigenmodes are similarly predominant, and the influence of the 4th order is very small compared with the 2nd and 3rd. It is seen that the higher order eigenmodes of the amplitude ratio f_n/f_{max} decrease with increasing γ.

It is seen from Figures. 2.2(a) and (b) that the amplitude ratio of 8th order eigenmode is only 2 to 3% of that of the most predominant mode. From the above discussion, therefore, we can conclude that the 8-term approximation is accurate enough for the present stability analysis.

Figure 2.3 shows the critical flutter configurations of some cases in Figure 2.2(b). The flutter configurations are drawn every 1/12 period, where the maximum amplitude is taken as 1/10 of the pipe length. In Figure 2.3(a) for $\beta = 0.75$ and $\gamma = 10^{-3}$, there are 3 oscillatory nodes and the flutter configuration is like that of the 4th mode. On the other hand, Figure 2.3(b) shows that the flutter configuration for $\gamma = 10^{-3}$ is similar to that of the 3rd mode because of having 2 nodes. Thus it is seen that the flutter configuration may vary drastically depending on the internal damping γ.

2.5 Discussion on the Critical Velocity

For several values of the parameter α, the effect of the parameter κ on the critical flow velocity u_{cr} is theoretically evaluated and is shown in Figures. 2.4 to 2.7, where $\beta = 0.5$ and the abscissa denoted the dimensionless spring constant κ. It is seen that there exists the combined effect of a spring support and a lumped mass if Figures 2.4 to 2.7 are compared with the stability maps for the case of only a spring support (see Sugiyama et al., 1985).

Because the divergence velocity is in effect independent of κ, therefore, we shall discuss only the flutter velocity, depending on the combined effect of a lumped mass and a spring support.

Figure 2.4 shows the stability map of the cantilever pipe having a spring support and a lumped mass at the discharge end ($\xi_1 = 1.0$). As the parameter α increases from 0.1 to 1.0, the flutter velocity drastically falls. When $\alpha > 1.0$, however, the critical velocity becomes almost independent of α and the pipe loses its instability by flutter instead of by divergence over a considerably wide range of κ with increasing α.

Figure 2.5 shows the stability map of the cantilever pipe having a spring support and a lumped mass at the quarter span from the free end ($\xi_1 = 0.75$). In this figure, the flutter velocity decreases with increasing α.

Figure 2.6 shows the stability map of the cantilever pipe having a spring support and a lumped mass at its midspan. Similarly with Figures. 2.4 and 2.5, the flutter velocity decreases with increasing α. A drop of u_{cr} from 0.1 to 1.0 is nearly equal to that of u_{cr} from 1.0 to 10.0.

Figure 2.7 shows the stability map of the pipe having a spring support and a lumped mass at the quarter span from the fixed end ($\xi_1 = 0.25$), which loses its stability by flutter only. When $\alpha = 0.1$ and $\alpha = 1.0$, respectively, the critical velocity gradually increases in the moderate region of κ. When $\alpha = 0.5$, however, it should be noted that the critical velocity decreases locally for only a region from $\kappa = 10^2$ to $\kappa = 10^3$. As α increases, the decreasing zone develops more and more: finally, when $\alpha = 10$, the overhang AB, which is here referred to as the flutter peninsula, appears. The peninsula is a particular region of flutter instability produced by the combined effect of a lumped mass and a spring support for particular combinations of them.

Now as for the reason why the flutter peninsula occurs, the root loci for $\alpha = 10$ and $\kappa = 600$ as shown in Figure 2.8 shall be examined. Here the branch which starts from λ_i (i =1,2,3, K)K K when $u = 0$ is referred to as the i-th branch. Three points A, B, and C on the imaginary axis of Figure 2.8, therefore, correspond to three points of the flutter limit A, B, and C in Figure 2.7, respectively. From the above, it is seen that the points B and C are taken place on the 2nd branch, while the point A is on the 1st. It should be noted that the 2nd branch BC runs along the imaginary axis, although the 1st branch cuts that axis obliquely. Figures 2.9(a) and (b) draw the flutter configurations at the points A and C, respectively. Not the magnitude of the amplitude but the relative distribution is important, since the flutter configurations in Figure 2.9 are obtained on the assumption of small oscillation. In the flutter configuration of point A, the amplitude of the pipe at the discharge end is considerably larger than that at the point ($\xi_1 = 0.25$) where a spring support and a lumped mass are attached. On the other hand, in the configuration of point C, the amplitude is similar throughout its length to that at the point with a spring support and a lumped mass. However, the lumped mass of $\alpha = 10.0$ is much larger than the mass of the pipe itself; and the spring support of $\kappa = 600$ is much stiffer than that of the pipe. Therefore the amplitude at the point with both the spring support of $\kappa = 600$ and the lumped mass of $\alpha = 10$ can not become larger. From the above consideration, it is concluded that the amplitude of the flutter configuration of the point C can not become larger along the pipe axis, and thus the flutter at the point C must be mild. We conducted an experiment to verify the above discussion on the flutter peninsula. The experimental procedure is omitted because it has been described in detail in the paper by Sugiyama et al. (1985). The flutter velocities obtained experimentally for $\alpha = 0.96$ and $\alpha = 10.1$, respectively, are plotted in Figure 2.7. It is noted that the mass ration $\beta = 0.505$.

For $\alpha = 0.96$ it is seen that the experimental values are much lower than the theoretical predictions, and the dependence of the former on κ is quite similar to that of the latter. Violent flutter motions were observed in the experiment with $\alpha = 0.96$. It is found for $\alpha = 10.1$ that the experimental values are much lower than the theoretical ones, and that the experimental flutter velocity drops suddenly around the dimensionless spring constant $\kappa = 10^3$. Furthermore, the flutter configurations observed at rather small velocities had small amplitudes as shown in Figure 2.10. The value of κ as determined by experiment, in which the flutter velocities drop, differs considerably from the value as determined by theory.

2.6 Concluding Remarks

In the present chapter, the combined effect of a spring support and a lumped mass on the stability of fluid-conveying cantilever pipes has been discussed. Some interesting conclusions are summarized as follows:

1. The validity of the number of terms in the approximate solution by Galerkin method has been investigated from the view point of the influence of each eigenmode on the critical flutter configuration. Thus it is concluded that the eight terms approximation in this study may be very accurate.
2. It has been found from the stability maps in Figures 2.4, 2.5, and 2.6 that the increase of a lumped mass reduces the flutter velocity, as in the stability map of a pipe having only a spring support.
3. It has been shown, for the special combinations of a spring support and a lumped mass, for example $\kappa = 600$ and $\alpha = 10$, that there exists an instability region which juts out locally from the main flutter domain when both the lumped mass and the spring support are attached to the location $\xi_1 = 0.25$. This local region is referred to as the flutter peninsula. It is shown theoretically and experimentally that the instability in the region is of the mild flutter type, not the violent type.

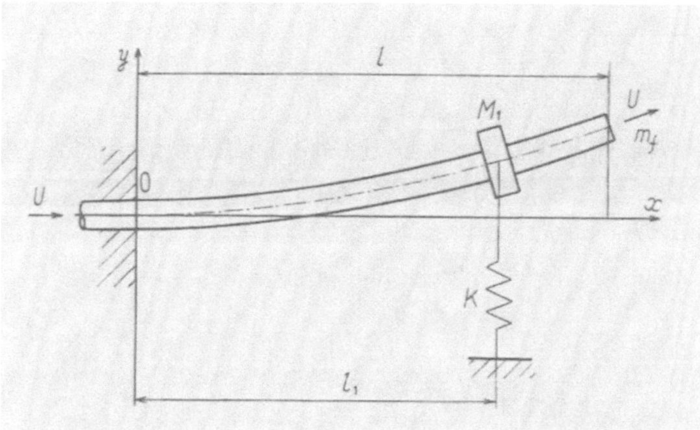

Figure 2.1 Fluid conveying pipe having a spring support and a lumped mass.

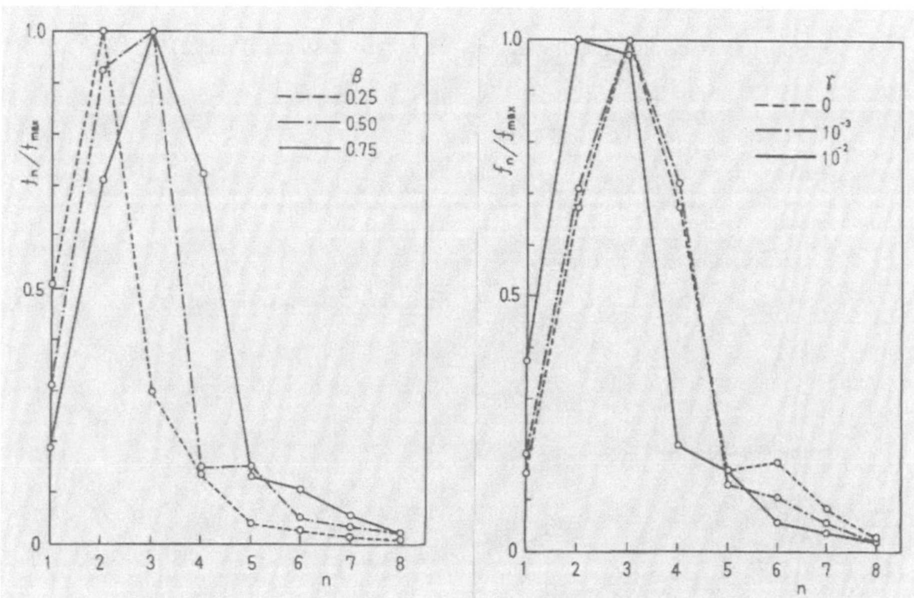

(a) Effect of the parameter β (γ = 10^{-3}) (b) Effect of the parameter γ (β = 0.75)

Figure 2.2 Variation of the amplitude ratio for the n-th eigenmode, $fn/fmax$.

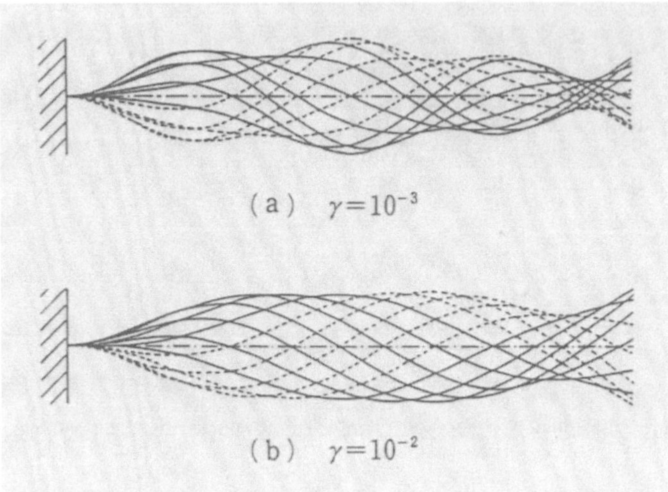

Figure 2.3 Critical flutter configurations ($M = 8$, $\alpha = \kappa = \nu = 0$, $\beta = 0.75$).

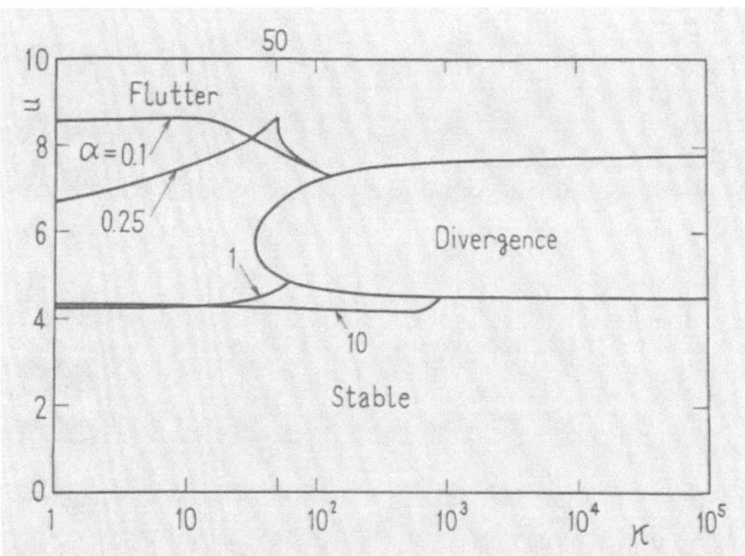

Figure 2.4 Stability map of cantilevered pipe having a spring support and a lumped mass at the discharge end ($\xi_1 = 1.0$, $\beta = 0.5$, $\gamma = 0.02$, $\nu = 0$).

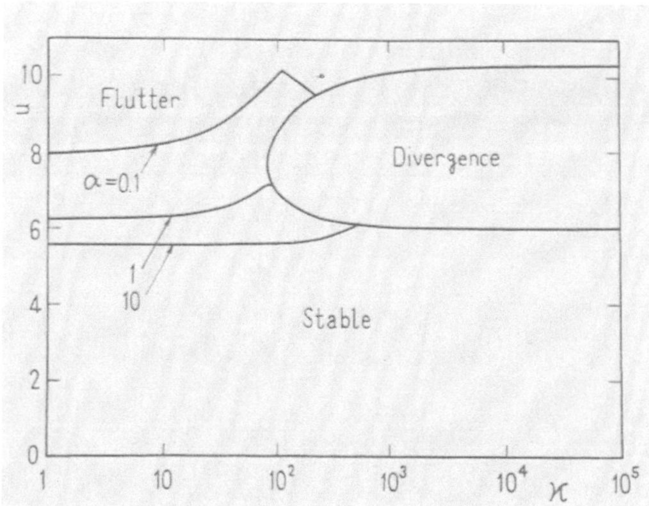

Figure 2.5 Stability map of cantilevered pipe having a spring support and a lumped mass at the quarter span from the free end ($\xi_1 = 0.75$, $\beta = 0.5$, $\gamma = 0.02$, $\nu = 0$).

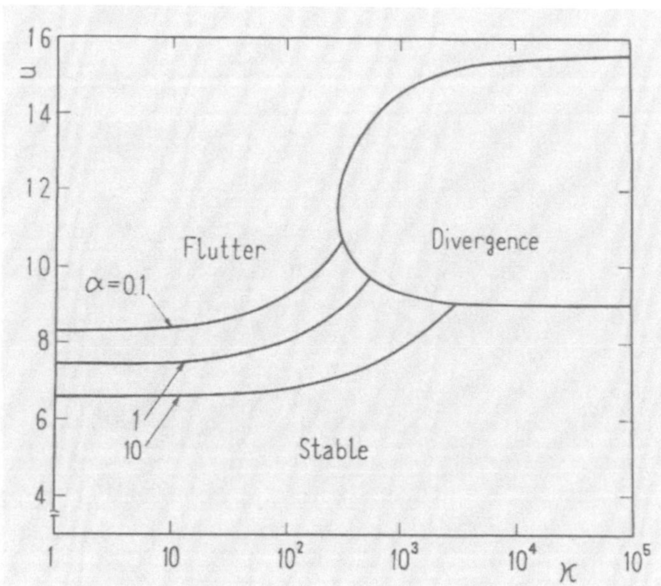

Figure 2.6 Stability map of cantilevered pipe having a spring support and a lumped mass at the midspan ($\xi_1 = 0.5$, $\beta = 0.02$, $\nu = 0$). 5, $\gamma = 0$.

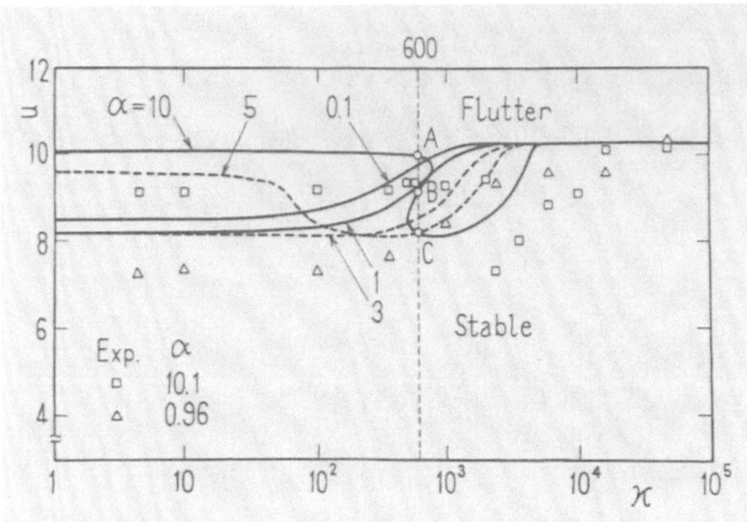

Figure 2.7 Stability map of cantilevered pipe having a spring support and a lumped mass at the quarter span from the fixed end ($\xi_1 = 0.25$, $\beta = 0.5$, $\gamma = 0.02$, $\nu = 0$).

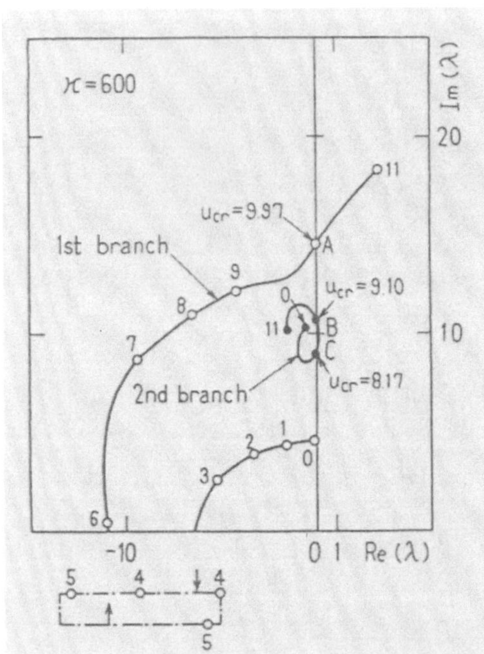

Figure 2.8 Root locus diagram for the case $\alpha = 10$, $\kappa = 600$ in Figure 2.7.

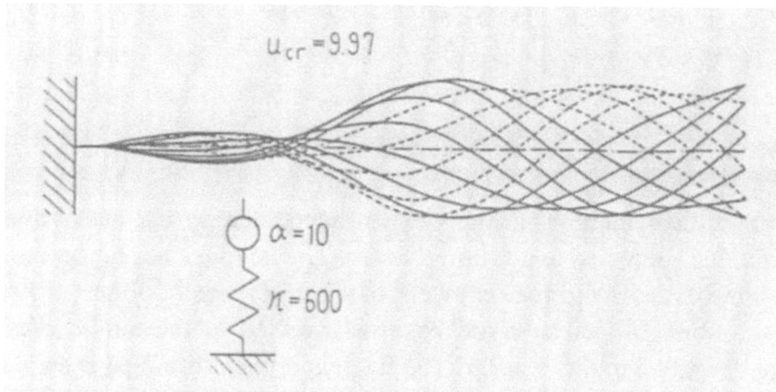

(a) Flutter configurations at the point A ($\alpha=10, \kappa=600, u_{cr}=9.97$).

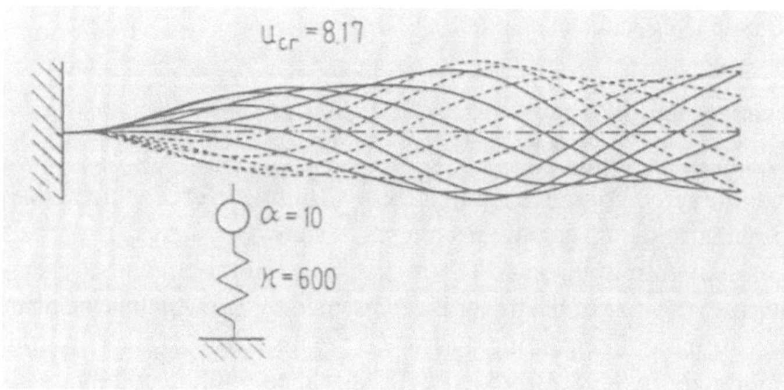

(b) Flutter configurations at the point C ($\alpha=10, \kappa=600, u_{cr}=8.17$).

Figure 2.9 Critical flutter configurations.

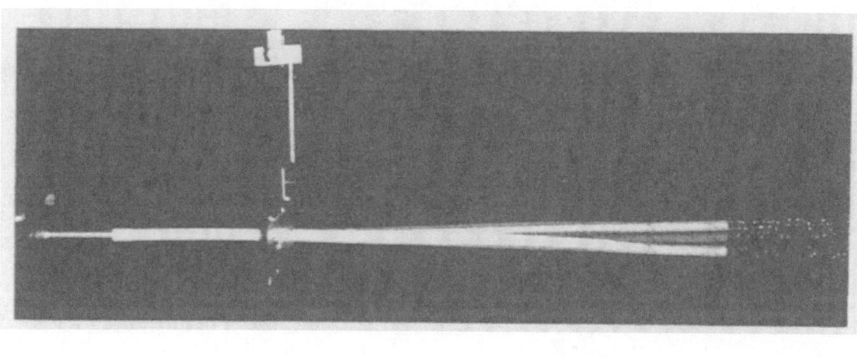

Figure 2.10 Observed mild flutter ($\alpha=10.1, \kappa=3700, \beta=0.505$)

3. Dynamic Stability of Cantilevered Column Subjected to a Rocket Thrust

3. 1 Introduction

The intended aim of this chapter is to show experimental demonstrations of dynamic stability of cantilevered columns subjected to a follower force. The key idea to realize a follower force is to mount a real small-sized solid rocket motor to a cantilevered column at its tip end. Thus a follower force can be realized as a rocket thrust applied to the cantilevered column. Three experiments are described in this chapter. The first experiment is aimed at showing experimental demonstration of the flutter caused by a rocket thrust (Sugiyama et al., 1995). The second is to verity the quantitative theoretical prediction of flutter force (Sugiyama et al., 1995, 1996). The third is to demonstrate the stabilizing effect of the follower thrust on the dynamics of a vertical cantilevered column (Sugiyama et al,. 2000).

3. 2 Experiment on Flutter of Cantilevered Column

3. 2. 1 Statement of problem

A horizontal cantilevered column is subjected to a follower force which is devised by a solid rocket motor installed at the tip end of the column, as shown in Figure 3.1. A mathematical model of the column is shown in Figure 3.2. The rocket motor at the tip end is assumed to be a rigid body of a finite size. The size of the motor is represented by L_R, which is the distance from the tip of the column to the motor's center of gravity. *M* is the mass of the motor, and *J* is the moment of inertia of the motor about the centroid axis. It is assumed that the column is viscoelastic in the sense of the Voigt-Kelvin model. The mass per unit length of the column is denoted by *m*. *E* and E^* are Young's and the viscosity modulus, respectively. Equation of small motion of the column is given by

$$E\,I\frac{\partial^4 y}{\partial x^4} + E^* I\,\frac{\partial^5 y}{\partial t \partial x^4} + P\frac{\partial^2 y}{\partial x^2} + K\frac{\partial y}{\partial t} + m\frac{\partial^2 y}{\partial t^2} = 0, \tag{3.1}$$

where *I* is the moment of inertia of the column, and *K* is the coefficient of external damping. The boundary conditions are given by

$$y = \frac{\partial y}{\partial x} = 0 \qquad \text{at } x = 0, \qquad (3.2a$$

$$EI\frac{\partial^2 y}{\partial x^2} + E^* I\frac{\partial^3 y}{\partial t \partial x^2} = -J\frac{\partial^3 y}{\partial t^2 \partial x} - M\,L_R\frac{\partial^2}{\partial t^2}\left(y + L_R\,\frac{\partial y}{\partial x}\right) \qquad \text{at } x = L, \qquad (3.2b)$$

$$EI\frac{\partial^3 y}{\partial x^3} + E^* I\frac{\partial^4 y}{\partial t \partial x^3} = M\frac{\partial^2}{\partial t^2}\left(y + L_R\frac{\partial y}{\partial x}\right) \qquad \text{at } x = L, \tag{3.2c}$$

For the sake of simplicity, it is convenient to introduce the dimensionless quantities

$$\xi = \frac{x}{L},\ \eta = \frac{y}{L},\ \alpha = \frac{L_R}{L},\ \mu = \frac{M}{mL},\ \nu = \frac{J}{mL^3},\ \rho = \frac{PL^2}{EI}, \tag{3.3 a-f}$$

$$\gamma = \frac{E^*}{EI^2}\left(\frac{EI}{m}\right)^{1/2}, \qquad \kappa = \frac{KL^2}{(mEI)^{1/2}}, \qquad \tau = \frac{t}{L^2}\left(\frac{EI}{m}\right)^{1/2}, \tag{3.3 g-i}$$

For the first theoretical prediction of the flutter thrust, it is convenient to consider the simplified model having only a mass point, assuming that $\alpha = \nu = \kappa = 0$. The model is known as Pflüger's column with internal damping (Bolotin, 1963). The stability map in the μ, ρ plane predicts that the flutter limit obtained by taking small internal damping ($\gamma = 0.001$) into account is about half that by neglecting the damping ($\gamma = 0$). Because of this it is necessary to verify experimentally the theoretical prediction of the effect of internal damping on flutter of columns under a tangential follower force produced by a rocket thrust.

3.1.2 Experimental system

Rocket motor and test columns. Small-sized solid rocket motors were used in the present experiments. The rocket motor consists of solid propellant and hardware including an ignitor, a motor case, a headcap, and a nozzle. For reusability and safety, the motor case was designed strong enough and thus very heavy. The mass of the hardware was 13.73 kg, while the mass of the propellant was 0.90 kg. The average thrust of the rocket motor was about 40 kgf (392 N). The burning time was about 4 seconds. The typical thrust curve of the rocket motor is shown in Figure 3.3. Though the thrust curve has a tendency to increase during burning, the thrust is assumed to be constant and equal to the average thrust of 40 kgf (392 N) during the whole burning period. The measured data of the rocket motor are as follows; average mass of the rocket motor M = 14.18 kg, distance L_R = 0.20 m and moment of rotary inertia J = 0.120 kg.m. In consideration of the constant thrust of 40 kgf, test columns were made of aluminum bars with the cross-section of 6 mm x 3 0 mm, and the length of L = 800 - 1,400 mm. The measured data of the test columns are as follows; mass per unit length m = 0.481 kg/m, bending stiffness $EI = 3.43 \times 10^6$ kgf.m^2, dimensionless coefficient of internal damping $\gamma = 4.94 \times 10^{-4}$, and dimensionless coefficient of external clamping $\kappa = 1.19 \times 10^{3}$.

The dimensionless quantities defined in Equations (3.3) vary with the length of test columns. The dimensionless mass μ of the rocket motor is 29.2. that is $1/\mu = 0.0343$. for the column of L = 1,000 mm. The dimensionless flutter thrust ρ was taken as 20.19 and 7.54 for the elastic and viscoelastic columns, respectively. The mass of the rocket motor is assumed to be constant and equal to the average mass during burning, as the mass of the propellant is small in comparison with the total mass of the motor.

Experimental setup. The schematics of the experimental setup is depicted in Figure 3.4. The test column was cantilevered horizontally. The rocket motor, mounted to the tip end of the column, was suspended in space, from the ceiling, by a thin 3 meter-long wire. The length of the column was adjustable at the clamped end. The column's flutter motion was recorded by a motor-driven camera and a video camera fixed on the ceiling, as shown in Figure 3.4. An axial strain was recorded by strain gauge as a measure of the thrust. The target disk was placed on one side of the column to observe a deflection of the test column. Since the thrust is nominally constant, the load parameter ran the length of the test column.

3. 1. 3 Preliminary experiment

The stability of the nonconservative system is determined by the dynamic criteria of stability. So it is desirable to know the dynamic bending stiffness. The vibration tests were conducted to obtain the dynamic bending stiffness,. The eigenfrequencies were measured to find the dynamic bending stiffness. The measurement of internal damping coefficient was made. The damped motion of a test column was picked up by a non-contact type deflection sensor and recorded on a pen-recorder. The damping ratio ζ and the first natural frequency (ω_1) were measured by the wave shape of decaying motion. Then the internal damping coefficient E' was obtained to find the dimensionless internal damping coefficient γ.

3. 1. 4 Experiment on flutter

The experiments with rocket thrust were conducted in a rocket motor functional test facility in Harima Plant, Daicel Chemical Industries, Ltd.. Though handling a rocket motor is generally prohibited by the law, the test facility is officially licensed by the government to carry on the rocket motor combustion test. The first test was conducted by using the longest test column. The length was determined by the first theoretical prediction based on Pflüger column. Then the critical flutter length of a column was searched for by considering the sequential test results and the stability map. The actual rocket thrust acted upon the columns was calibrated by checking the measured axial strain.

The test results are summarized in Table 3. 1. Figure 3.5 shows the corresponding axial thrust and displacement curves observed in the run No. 7.

Table 3.1 Summary of test results.

Run No.	Length L (mm)	Average thrust P (kgf)	Stability
1	1,400	40	Instantaneous buckling
2	800	37	Stable
3	1,100	34	Flutter
4	1,000	40	Stable
5	1,050	39	Flutter
6	1,025	40	Flutter
7	1,000	43	Flutter

3. 1. 5 Discussions on test results

The first theoretical flutter predictions were made by assuming the rocket motor as a mass point. Now, the effect of size of the rocket motor should be discussed. The theoretical flutter predictions obtained by taking into account the size of the rocket motor are compared with the test results. It is found that both the distance L_R and the inertia moment J reduce the flutter boundary.

Another important discussion to be made, in both theory and experiment, is the stability criteria. Theoretical flutter limit comes from the asymptotic stability, and the stability is considered in an infinite time interval. If the real part (σ) of the eigenvalue (λ) is even slightly positive, regardless of its magnitude, the system is determined to be unstable mathematically during an infinite time period. On the other hand, the stability in experiment is determined by the stability in a finite interval (Leipholz, 1980). Stability in a finite time interval states that a system is stable if its response, during a specified finite time interval, does not exceed a prescribed amount of response; otherwise it is considered unstable.

In the present flutter test, the column is determined to be stable if its dynamic deflection (η) does not exceed the prescribed value in a prescribed finite time interval τ_1. The criteria are written as follows:

$$\text{Stable} \quad \text{if} \quad \eta < n\,\eta_0 \qquad \text{for} \quad 0 \le \tau \le \tau_1. \tag{3.4a}$$

$$\text{Unstable} \quad \text{if} \quad \eta \ge n\,\eta_0 \qquad \text{for} \quad 0 \le \tau \le \tau_1. \tag{3.4b}$$

where η_0 is initial deflection at $\tau = 0$, and n is a specified amplification factor.

It is found that the flutter analysis made by neglecting internal damping can apply to flutter tests conducted by using a rocket motor with a finite (really short) burn-out time period.

3. 1. 6 Conclusion

To verify the flutter of cantilevered columns subjected to a rocket thrust, the experiment was first conducted by the direct installation of a solid rocket motor to the tip end of the columns. It was confirmed that the columns lost their stability by flutter. The concept of stability in a finite time interval was successfully introduced to predict the theoretical flutter.

3. 3 Experiment on Effect of Lumped Mass

3. 3. 1 Statement of problem

The aim of this section 3 is to give an experimental support to theoretical stability analysis on the effect of concentrated mass on the dynamic stability of a cantilevered column subjected to a rocket thrust. The experimental results are to be compared with theoretical predictions made by taking account of the mass and size of the rocket motor as well as an intermediate lumped mass. The stability problem of uniform cantilevered columns under a follower force has been studied by a lot of researchers. However actual slender rockets and missiles have some concentrated masses, such as payload and fuel tank. So their mass distribution are no more uniform. Therefore it is very significant from a engineering standpoint to study the uniform columns having a concentrated mass at its intermediate span.

Figure 3.6 depicts a mathematical model of the column subjected to a follower force produced by solid rocket motor, which is installed at the tip of a cantilevered column. The column has a total length L, bending stiffness EI, mass per unit length m. The rocket motor is now considered as a rigid body having mass M_2 and rotary inertia J_2. The distance from the tip of the column to the gravity centre of the rocket motor is denoted by L_R, and called the motor distance. An intermediate lumped mass (hereafter may be called the intermediate mass) having magnitude M_1, and rotary inertia J_1 is attached to the column at x_{ml} apart from the fixed end.

Extended Hamilton's principle for the nonconservative system under consideration can be written in the form

$$\int_{t_1}^{t_2} \{\delta (T - V + W_C)\}\, dt + \int_{t_1}^{t_2} \delta W_N dt = 0, \tag{3.5}$$

where T denotes the kinetic energy of the system, V the potential energy, W_c is the work done by conservative component of the applied force. δW_N is a virtual work done by nonconservative component of the applied force.

To apply the finite element method (FEM) to the system, the column structure is divided into N segments having an equal length of l.
Now the following dimensionless parameters are introduced;

$$\Omega^2 = mL^4 s^2 / EI, \quad P^* = PL^2 / EI, \quad \alpha_1 = M_1 / mL, \quad \alpha_2 = M_2 / mL,$$
$$\beta_1 = J_1 / mL^3, \quad \beta_2 = J_2 / mL^3, \quad a = L_R / L. \tag{3.6}$$

In Equation (3.6), dimensionless parameters Ω and P^* mean the eigenfrequency and follower thrust, respectively. α_1 and β_1 are the dimensionless magnitude and rotary inertia of the intermediate mass, respectively. α_2 and β_2 depict the dimensionless mass and rotary inertia of the rocket motor, respectively. a means the dimensionless motor distance. Finally the characteristic equation can be obtained in the form

$$\left\{\Omega^2 [M] + [k]\right\} \{U\} = \{0\}. \tag{3.7}$$

The stability is determined by investigating the eigenfrequency Ω of the Equation (3.7).
Only flutter-type instability can take place in the present problem.

3. 3. 2 Rocket motor

For the present experiments a new light-weighted motor case was designed to reduce total mass of the rocket motor. The thick motor case was machined out into a thin case to reduce the mass. The total mass has been reduced by about 10 kg in comparison with the rocket motor used in the section 2. Masses of the new rocket motor before the burning was 4.50 kg. The mass of the propellant was 0.90 kg as before. The average thrust of the rocket motor was about 62.1 kgf (609 N) and the burning time was about 3.2 seconds. Though the thrust curve has a tendency to increase with time and then decrease afterward at half of burn time, the thrust was assumed to be constant and equal to the average thrust of 62.1 kgf during entire burning period. The measured data of the refined rocket motor are as follows; average mass $M_2 = 4.05$ kg, average thrust $P =$ 62.1 kgf, bum out time $t_1 = 3.2$ sec, rotary inertia $J_2 = 0.0284$ kg m^2, motor distance $L_R = 0.153$ m.

3. 3. 3 Intermediate mass and test columns

Two different lumped masses (tentatively named here A and B) were designed for the present experiment. The lumped masses were made of brass block with hollow rectangular cross section. The details of the two masses are as follows; lumped mass A: M_I = 1. 0 kg, J_I = 5.74x10^{-4} kg m^2, lumped mass B: M_I = 2.3 kg, J_I = 2.03x10^{-2} kg.m^2. The mass was attached carefully to the test column by bolts, not to cause an additional stiffness to the column. The lumped mass was located at position μ_I= 0.6. The reason why the intermediate lumped mass was attached to the column at μ_I= 0. 6 can be seen in Figure 3.7. Figure 3.7 shows the effect of the intermediate lumped masses on the ratio of critical follower force having an intermediate lumped mass and that having no lumped mass. It is noted that the importance of the mass position is emphasized in the figure. The lumped masses mounted to the column at μ_I = 0. 6 have the highest destabilizing effect on the flutter of the column. This was the reason why the mass was fixed to the test column at μ_I = 0. 6.

The dimensions of the test columns can be determined on the basis of the theoretical predictions made by FEM formulation. Theoretical predictions of the critical follower forces for different candidate test columns were made extensively. The final dimensions of the cross section of the test columns were determined to have the cross section of 7.0 mm x 30.0 mm. The length was taken to be 1,100 to 1,200 mm. The details of the test columns are as follows; thickness h = 7.0 mm, width b = 30.0 mm, length L = 1,100 - 1,200 mm, mass per unit length m = 0.567 kg/m, bending stiffness EI = 6.00 kgf/m^2.

3. 3. 4 Experiment

The schematics of the experimental setup is shown in Figure 3.8. The test column and the rocket motor mounted to the tip end of the column were suspended in space by a thin steel wire hung from the ceiling. Thus the small motion of the column was restricted in the horizontal plane.

Four test runs were conducted in the present experiment. Figures 3.9 (a) - (d) show the corresponding dynamic displacements in the four test runs. The stable, flutter and critical responses of the column were observed in the present experiment.

3. 3. 5 Results and discussions

Theoretical flutter boundaries for the test columns without and with an lumped mass at μ_I = 0.6 are shown in Figure 3.10. The theoretical boundaries were obtained by taking account of the measured data of the test column. It is noted that internal damping was neglected in the theoretical flutter predictions. The test results are summarized in Table 3. 2.

The experimental results agreed well with the theoretical predictions on the condition that all the three parameters M_2, J_2 and L_R of the rocket motor are taken into account. Therefore, it is confirmed that all the three parameters of the rocket motor are considerably important in the analysis of the follower force induced by rocket thrust devised by a real solid rocket motor.

Table 3.2 Test results

Test run number	Length of column L (mm)	Intermediate mass M_l (kg)	Position of the intermediate mass X_{ml} (mm)	Average thrust P (kgf)	Stability
1	1,200	0.0	No mass	62.1	Violent flutter
2	1,100	0.0	No mass	62.1	Stable
3	1,100	2.3	660	62.1	Flutter
4	1,100	1.0	660	62.1	Critical

3. 3. 6 Conclusion

The present section has described the experimental verification of flutter phenomena of cantilevered columns subjected to an end rocket thrust and having an intermediate lumped mass. The position of a lumped mass associated with the lowest destabilizing effect ranges from $\mu_l = 0.7$ to $\mu_l = 0.3$, as the magnitude of the lumped mass is increased. The experimental results agreed well with the theoretical ones if the magnitude of the rocket motor, the rotary inertia of the rocket motor and the motor distance are all taken into account. Therefore, consideration of the rocket motor as a rigid body is of vital importance in predicting the critical follower thrust of columns subjected to a realistic follower force devised by an end rocket motor.

3. 4 Experiment on Stabilizing Effect of Rocket Thrust

3. 4. 1 Statement of problem

This section describes the experimental observation of the effect of a subtangential follower force on the stability of vertical cantilevered columns. Self-weight of a solid rocket motor fixed at the free end of the column makes a conservative loading which acts on the column initially, while rocket thrust of the motor yields a nonconservative loading which acts on the column in addition to the conservative one, thus the total compressive load is much greater than the buckling load of the considered column. The present section is going to give an experimental observation that application of rocket thrust to a column under the action of critical conservative force can make the column stable as long as the thrust acts on the column.

Figure 3.11 shows a mathematical model of a vertical cantilevered column subjected to a subtangential force. The column is assumed to be a slender uniform column. The force can be produced by the combined action of rocket thrust and self-weight of the rocket motor. In Figure 3.11, ρ is the density of the uniform column, A is the cross-sectional area of the column, M is the mass of the rocket motor, J is the rotary inertia of the rocket motor, and L_R means the distance between the free end of the column and the mass center of the rocket motor. The rocket motor is considered as a rigid body, not a mass point. The subtangential force can be represented as the resultant vector P of a vertical force Mg and a pure tangential follower thrust P_t. The direction of the resultant subtangential force is specified by $\alpha\varphi$ as shown in Figure 3.11, where φ is the angle of inclination of the tangent at the tip end. The parameter α can specify the angle between the direction of the resultant force and the z-axis and it is called "tangency coefficient". When $\alpha = 0$, the direction of the force is vertical, i.e. the force is conservative. When $\alpha = 1.0$, it is tangential to the tip end, i.e. the force is purely nonconservative. Thus the coefficient α is sometimes referred to as the nonconservativeness parameter.

The energy expressions for the above mathematical model can be written in the following extended Hamilton's principle;

$$\delta\int_{t_1}^{t_2}(T - V + W_c + W_{sw})\,dt + \int_{t_1}^{t_2}\delta\, W_{nc}\,dt = 0. \tag{3.8}$$

In order to obtain a characteristic equation related to the flexural vibration of the column, let us rely on finite element method (FEM). Now, the column is divided into N segments having an equal length L. Now the dimensionless parameters are defined as follows;

$$\lambda^2 = \frac{\rho A L^4 s^2}{EI_y}, \quad T = \frac{PL^2}{EI_y}, \quad \overline{M} = \frac{M}{\rho A L}, \quad \overline{J} = \frac{J}{\rho A L^3},$$

$$\overline{L_R} = \frac{L_R}{L}, \quad \alpha = \frac{P_t}{Mg + P_t} = \frac{P_t}{P}. \tag{3.9}$$

3. 1. 2 Rocket motor and test columns

The rocket motors used in the present experiment are small-sized solid rocket motors designed for the combustion test of propellant, which are same as used in the first experiment. The nominal initial mass of the motor is assumed to be 14.65 kg. The rotary inertia of the rocket motor J=0.120 kg m^2, and the distance between the centre of gravity of the rocket motor and free end of the column L_R = 0.20 m.

Test beam in the present experiment had the width b = 30 mm, and the thickness h = 8 mm, 9 mm. The beam was made of aluminum. The density ρ = 2672 kg/m^3, and the Young's modulus E = 6.903 x 10^3 kg/mm^2, which was found by the bending test.

3. 1. 3 Numerical results

Eigenfrequency of the test columns. As to the basic dynamics of the columns in the present experiment, the first and second eigenfrequencies of the columns with and without the rocket thrust were calculated by the characteristic equation. The first eigenfrequencies of the column with and without the thrust for the test run No. 2 are plotted in Figure 3.12.

Stability maps of the test columns. Type of instability of the columns depends on the tangency coefficient α of the total compressive force $(Mg + P_t)$. When the tangency coefficient of the rocket thrust is changed from zero to unity, the divergence-type instability in the first mode can take place for the coefficient $\alpha \leq 0.5$, while for $\alpha > 0.5$ flutter-type instability can occur. Typical stability map of the test run No. 4 is shown in Figure 3.13. In the actual experiment, the rocket thrust P_t is assumed constant and 40 kgf. Thus the tangency coefficient is taken as α = 40/(14.2+40)= 0.74. where the average weight of the rocket motor during burning is taken as 14.2 kgf, since it was assumed that the initial mass of the motor is 14.65 kg and the mass of the propellant is 0. 9 kg (thus the final mass of the motor is 14.65 - 0.9 = 13.75 kg).

Design of the test columns. Since one run of a rocket motor is expensive, only four test runs were planned in the present experimental project. Thus detailed considerations on the design of the test columns were needed. The buckling length of four test columns were calculated as the reference data for the experiment. It is noted that the applied conservative force was fixed as 14.65 kgf.

3. 1. 4 Test runs with rocket thrust

Outline of experiment. As seen in Figure 3.12, the first eigenfrequency decreases as the conservative load ($\alpha = 0$) increases, while it decreases slowly as the subtangential load of $\alpha = 0.74$ increases. It is now confirmed that the term "stabilizing" effect here implies "higher" frequency. Stability map in Figure 3.13 depicts straightforward the state of stability of the column when it was subjected to a conservative load of $\alpha = 0$ and to a subtangential load of $\alpha = 0.74$. In Figures 3.12 and 3.13, the point A means the column under a conservative load ($\alpha = 0$), while the point B the column under a subtangential load of $\alpha = 0.74$. When the tangency coefficient $\alpha = 0$, the column loses its stability by divergence i.e., buckling. When the coefficient $\alpha = 0.74$, the column loses its stability by flutter. It is noted that the flutter load is much more higher than the divergence one, as seen in Figure 3.13. Application of rocket thrust to the column in addition to a conservative load can change the type of instability. This is the physical mechanism of the stabilizing effect of rocket thrust. The present experiment is focused on demonstrating the following two cases;

Case 1: Initially the column under a conservative load sways with a low frequency, since the load is close to but lower than the buckling load. Then the rocket thrust of 40 kgf will be applied to the column. The column may oscillate with a higher frequency during the burning of the rocket motor. After the burn out of the motor, the column shall again sway with the low frequency.

Case 2: Initially the column under the buckling load is at rest in a bent configuration. Then the rocket thrust of 40 kgf is applied to the column in addition to the conservative buckling load. The application of the rocket thrust may make the column dynamically stable as long as the thrust is alive. After the burn out, the column shall again have the bent configuration.

Figure 3.14 shows a sketch of the experimental setup. A vertical column was cantilevered upward and equipped with a solid rocket motor at its free end. The free end was loosely harnessed by two thin wires which prohibited the column to sway out extremely, but allow it to oscillate freely with small and moderate amplitude. Measuring devices were installed for axial compressive strain and lateral displacement of the test columns. Dynamic behavior of the columns was recorded by a video camera and a motor-driven camera.

Results of experiments. Four test runs Nos. 1-4 were conducted in the present experiments. Test run Nos. 1 and 2 were planned to apply to the Case 1, while Nos. 3 and 4 to the Case 2.

Test run No. 2: The dimensions of the column were as follows; L =1130 mm, b = 30 mm and h = 9 mm. Since the test column of the run No. 2 had a longer length than that of No. 1, the former oscillated with a lower frequency than the latter. Recorded displacement and axial strain were shown in Figure 3.15(a). It was observed that the column under a conservative load oscillated with a low frequency, and that, under a rocket thrust in addition to the conservative load, it oscillated with a higher frequency.

Test run No. 4: The stability map shown in Figure 3.13 depicts the stability behavior in this run. The dimensions of the column were 1125 x 30 x 8 mm. The dimensions were chosen to realize the supper-critical state for buckling. Theory has shown that the effect of offset moment due to the motor can reduce the buckling load of the column. Under the combined action of the motor's weight and rocket thrust, the column oscillated with a moderate frequency around the undeformed configuration. After the bum out of the motor, the column retained its buckled configuration. Figure 3.15(b) depicts the records of the axial strain and dynamic displacement in the test run No. 4. The records demonstrate the stabilizing effect of rocket thrust on the dynamics of the column subjected initially to the conservative buckling load.

3. 1. 5 Conclusion

The present experiment has demonstrated the first experimental verification of the stabilizing effect of rocket thrust on the dynamics of vertical cantilevered columns having a tip rigid body. The test run Nos. 1 and 2 have demonstrated that rocket thrust can cause the column's oscillation with a higher frequency than when it was subjected only to a conservative load. More dramatic demonstrations were made by the test runs Nos. 3 and 4, where the buckled column under a conservative load regained their straight configurations during the application of rocket thrust. That is, it is observed experimentally that the cantilevered columns were stabilized by the action of rocket thrust when the columns were subjected to rocket thrust in addition to a dead load, in spite of the fact that the resultant total compressive load was increased greatly, more than three times as much as the buckling load.

3. 2 General Conclusion

Follower forces can be devised by mounting a solid rocket motor to a cantilevered column at its free end. A thorough experimental verification of the effect of follower force has been shown by the laboratory experiments.

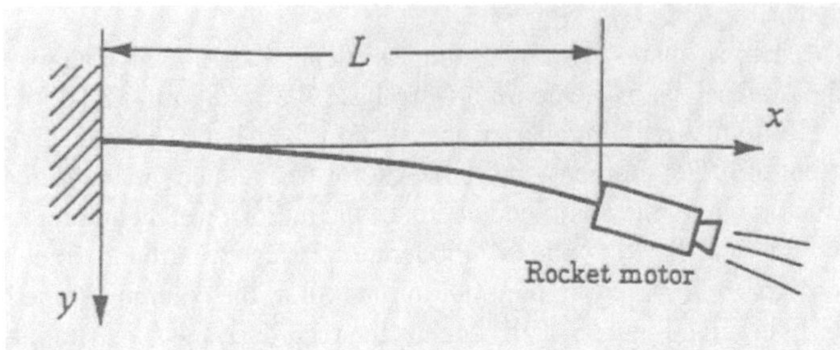

Figure 3.1 Cantilevered column with a rocket motor at tip end (Sugiyama et al., 1995 a).

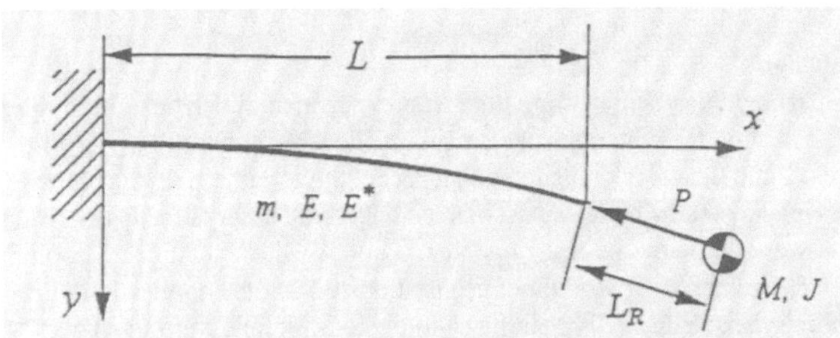

Figure 3.2 Mathematical model of column (Sugiyama et al., 1995 a).

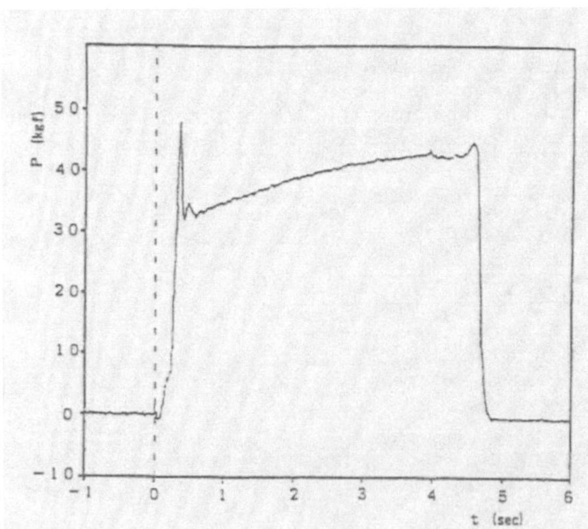

Figure 3.3 Thrust curve of the rocket motor (Sugiyama et al., 1995 a).

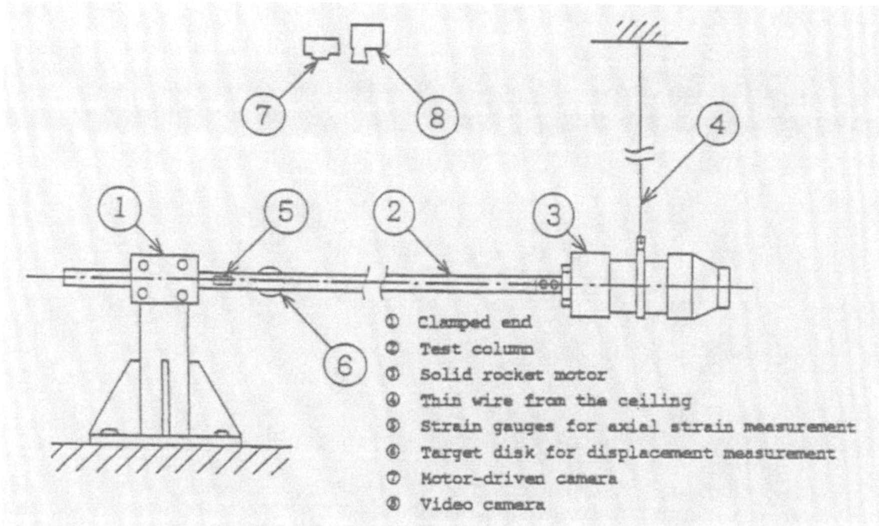

Figure 3.4 Schematics of experimental setup (side view) (Sugiyama et al., 1995 a).

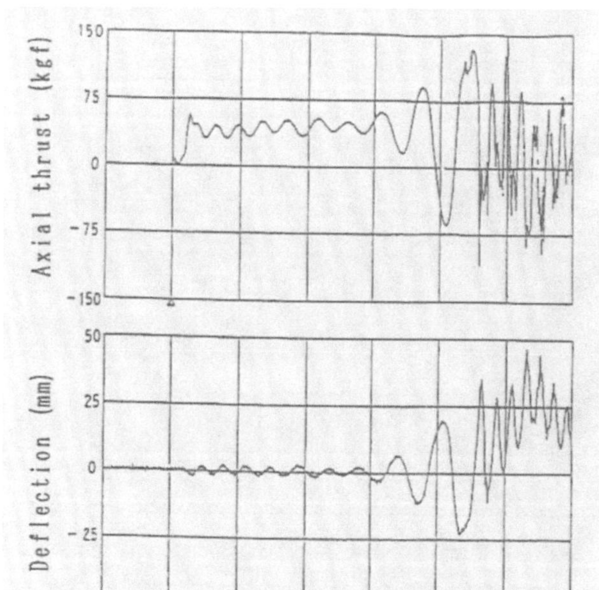

Figure 3.5 Records of axial thrust and deflection in test run No.7 (Sugiyama et al., 1995 a).

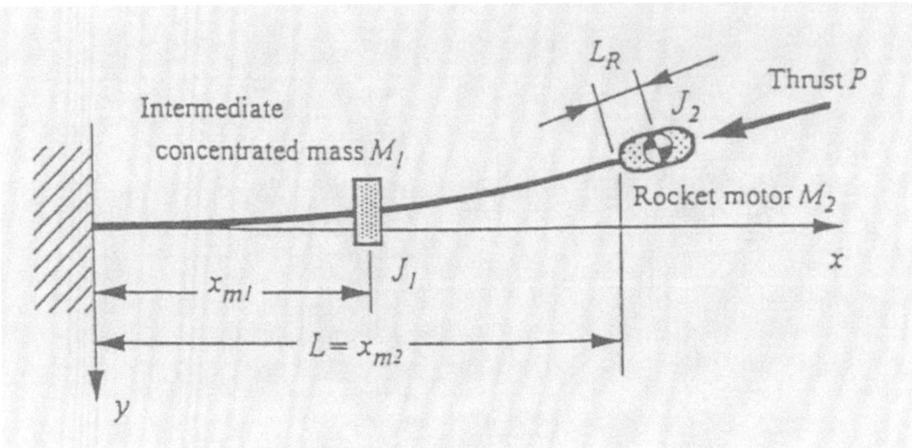

Figure 3.6 A mathematical model of a cantilevered column subjected to a rocket thrust at its free end and having an intermediate lumped mass (Sugiyama et al., 1995 b).

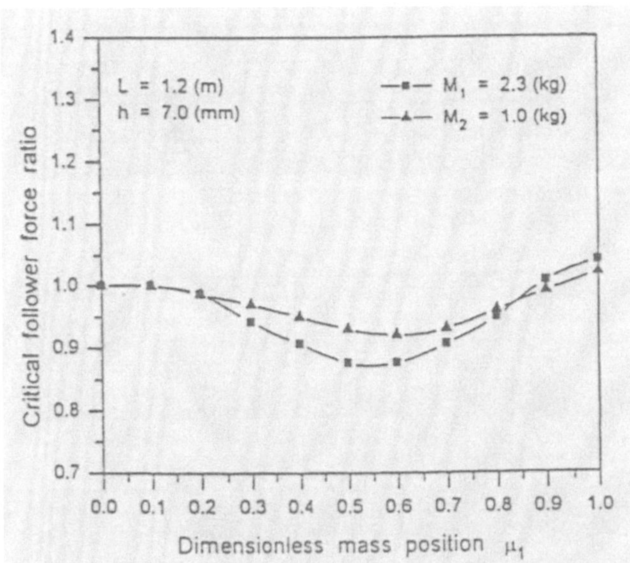

Figure 3.7 Critical follower force depending on the position of the intermediate lumped mass: L =1,100mm, h =7.0 mm, b =30.0 mm (Sugiyama et al., 1995 b).

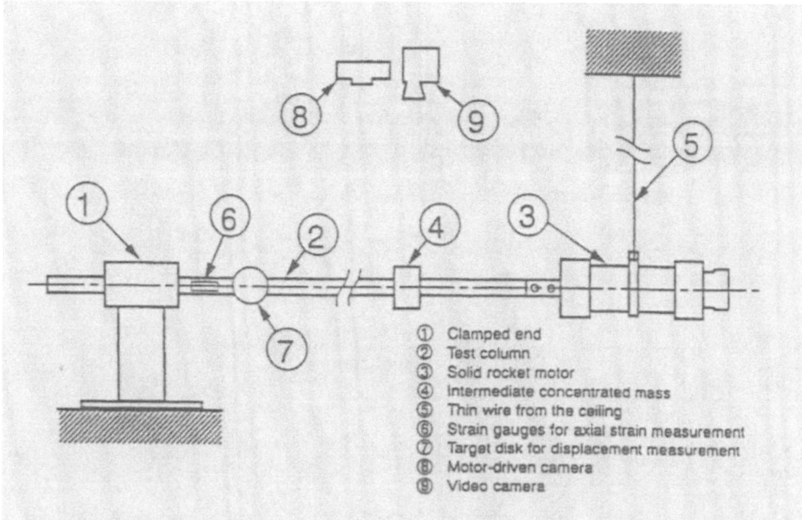

Figure 3.8 Schematics of experimental setup (side view) (Sugiyama et al., 1995 b).

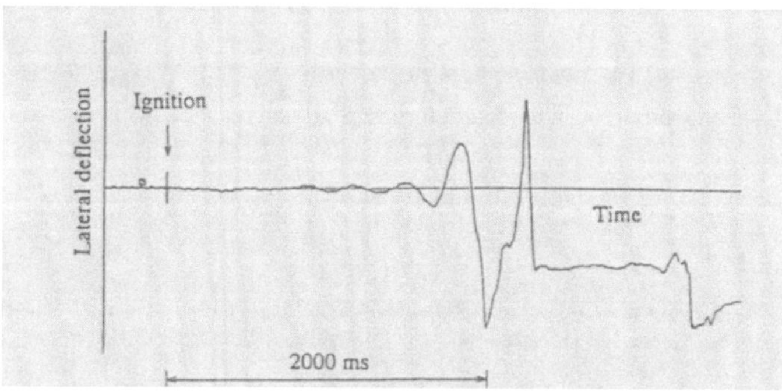

(a) Test run No.1 (Violent flutter)

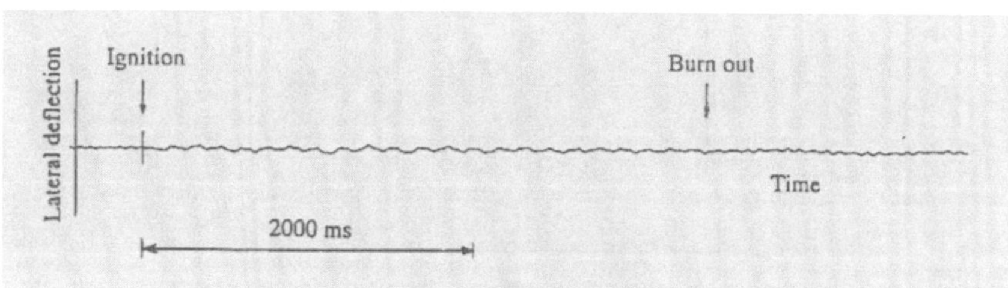

(b) Test run No.2 (Stable motion)

Figure 3.9 Dynamic responses in the four test runs (Sugiyama et al., 1995 b) (continued).

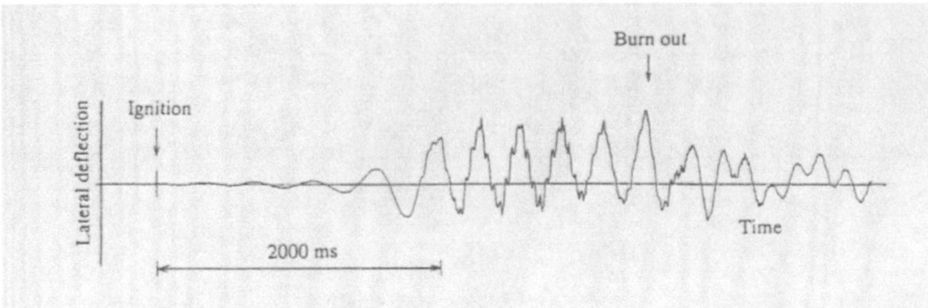

(c) Test run No.3 (Flutter)

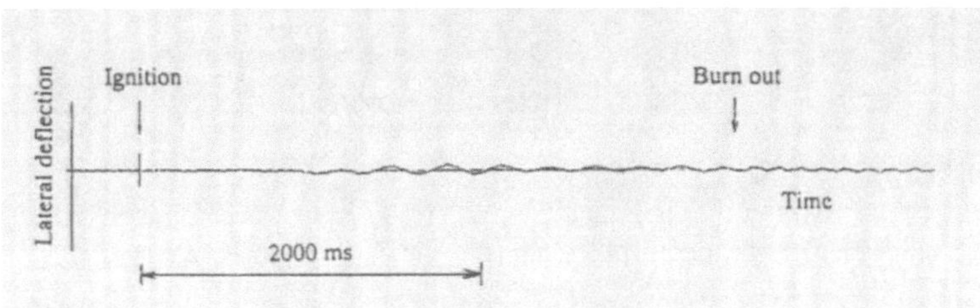

(d) Test run No.4 (Critical motion)

Figure 3.9 Dynamic responses in the four test runs (Sugiyama et al., 1995 b).

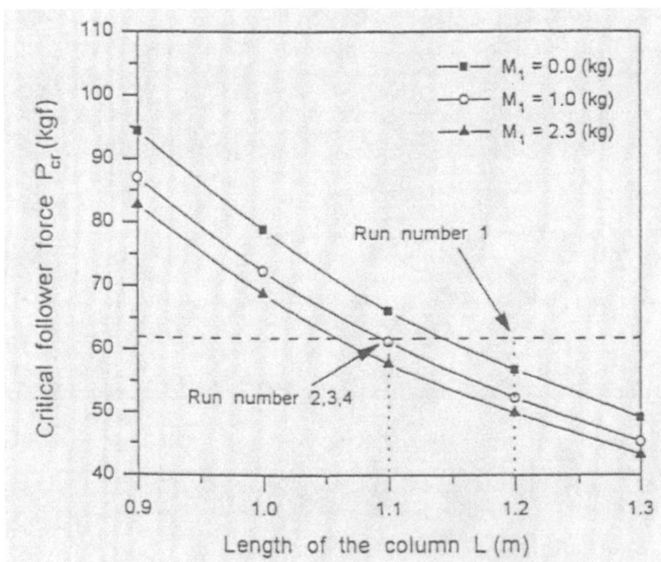

Figure 3.10 Theoretical predictions of the critical follower thrust and the test run numbers (μ_1 = 0.6) (Sugiyama et al., 1995 b).

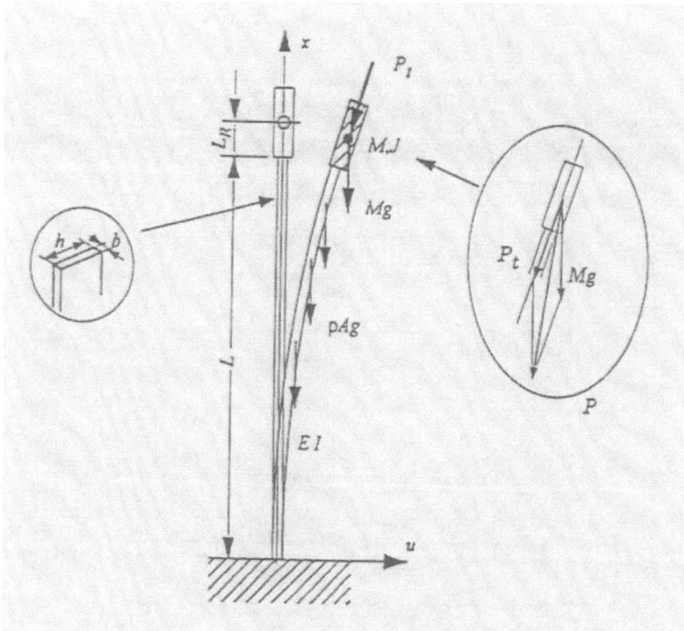

Figure 3.11 Mathematical model of column subjected a subtangential force (Sugiyama et al., 2000).

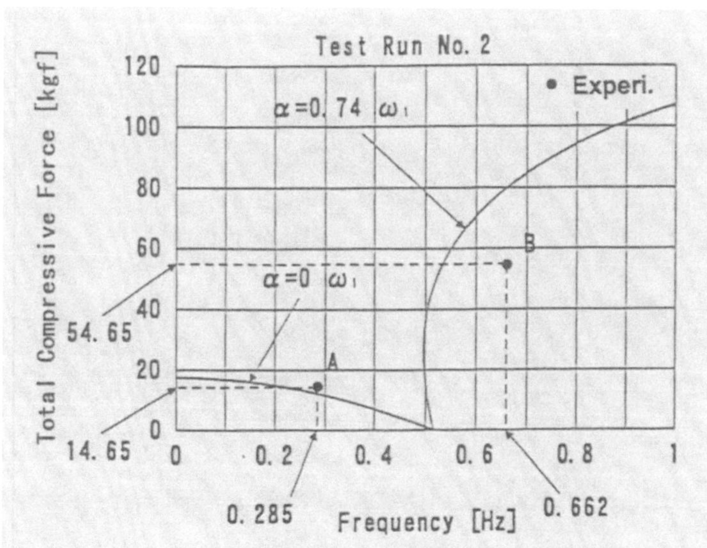

Figure 3.12 First eigenfrequency curves for test run No.2 (Sugiyama et al., 2000).

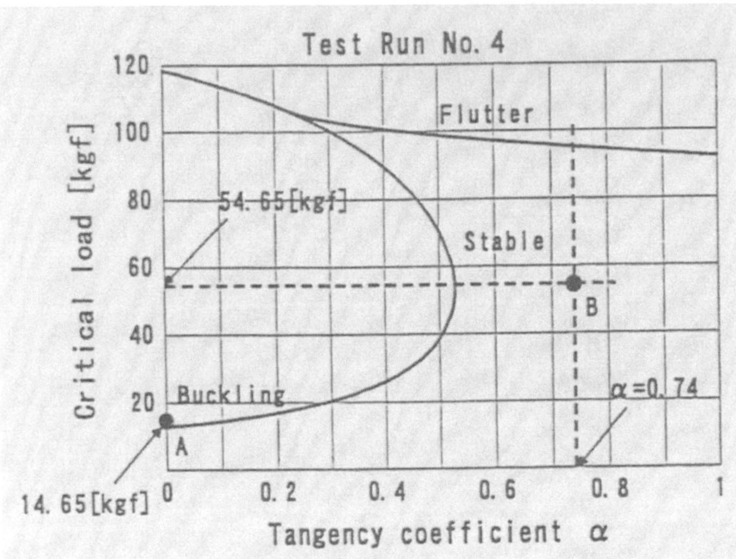

Figure 3.13 Stability map in load-tangency coefficient plane for test run No.4 (Sugiyama et al., 2000).

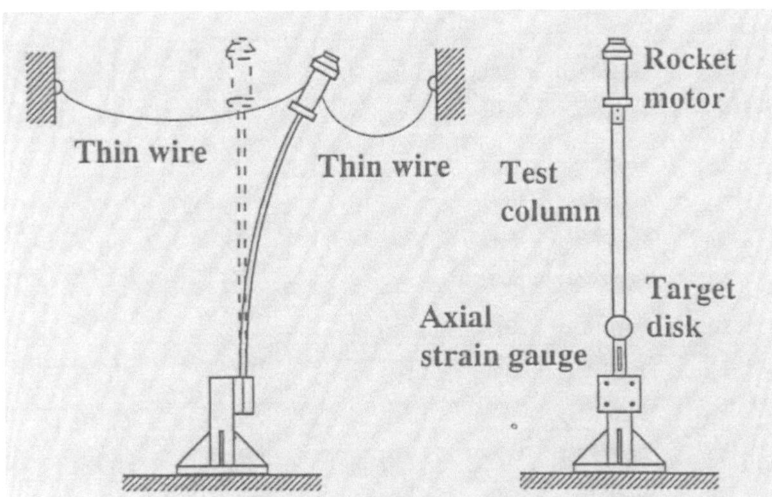

Figure 3.14 Sketch of experimental setup (Sugiyama et al., 2000).

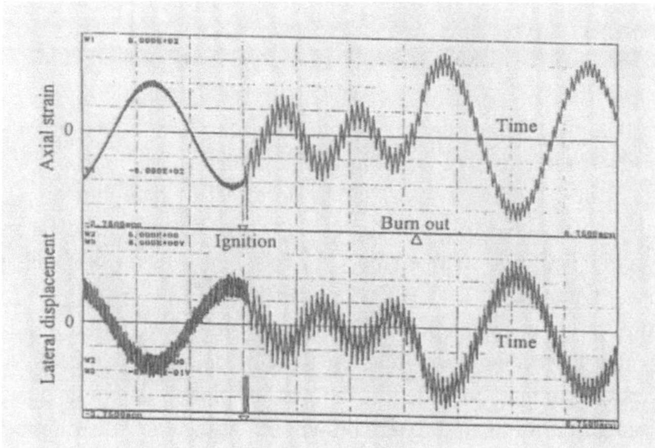

(a) Test run No.2

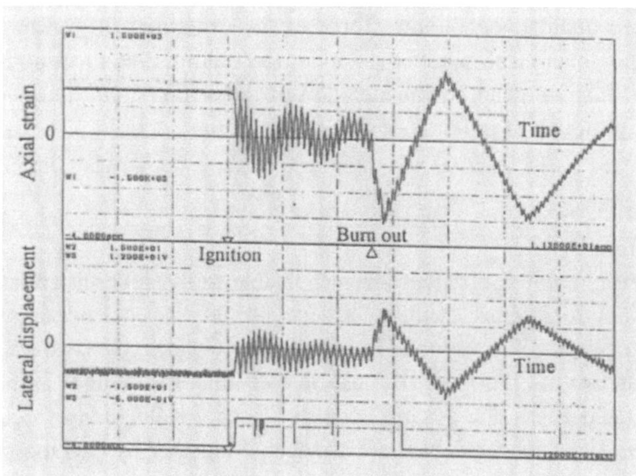

(b) Test run No.4

Figure 3.15 Records of axial strain and displacement (Sugiyama et al., 2000).

4. Effect of Damping Configuration on Flutter of Nonconservative Systems

4.1 Introduction

It has been well known that the effect of damping on the stability limit is important in the analysis of nonconservative stability problems (see Bolotin, 1963). So far a vast body of papers have been published on the effect of destabilizing and/or stabilizing effect of damping on flutter of elastic systems subjected to nonconservative follower forces. However the papers so far published have been concerned with some fictitious mathematical models being in weak touch with physical reality (Langthjem and Sugiyama, 2000).

In the present lecture, let us consider nonconservative systems, which are realistic at least in laboratory. Let us consider a two-degree-of-freedom Reut model having a dash-pot. This means that there are three types of damping included in the present model, damping due to a dash-pot, internal damping at viscoelastic joints and external damping along the rods due to air resistance. Relation between the damping configuration (structure of damping matrices) and the effect of damping will be discussed. Effect of the location of the attached dash-pot on the flutter limit will be discussed theoretically and experimentally. Experimental counterpart to the mathematical model is setup in laboratory. The coefficients of damping of the considered three types of damping were measured qualitatively. Flutter force of the experimental system was determined and compared with the theoretical predictions made by taking the measured coefficients of damping into account. Finally a general consideration of the damping configurations is discussed for standard equations of motion of multiple-degree-of-freedom systems.

4.2 Mathematical Model

4.2.1 Basic equations.

In case where only the first two eigenvalues are responsible to dynamic instability of mechanical systems, it is convenient to discuss qualitatively the stability by considering the mechanical systems having two degrees of freedom. Here let us consider the two-degree-of-freedom mechanical system as shown in Figure 4.1. It must be noted that the system can be realized in the laboratory as demonstrated later. The system consists of three rigid rods of equal length l and mass per unit length m. The three rigid bars are connected together by two massless viscoelastic joints 1 and 2. The two joints are assumed to have the same elastic restoring moment coefficient k and viscous damping coefficient b. The first bar is fixed on a base and the other two are free to perform a small motion in the horizontal plane x, y. Thus the system has only two degrees of freedom. As shown in Figure 4.1, a rigid massless attachment is mounted to the cantilevered model at its tip end. It is assumed that an applied force acts upon the model through the attachment. It is also assumed that the point of application of the force lies always on the undeformed axis of the mechanical system. It is here assumed that the direction of the force is inclined to the undeformed axis by $\alpha\,\phi_2$, where ϕ_2 is the angle of rotation of the end section of the system. The force is referred to as a generalized Reut force as the force has a similar character to Reut force (see Bolotin, 1963). If the parameter α is null, the force is purely nonconservative to make Reut force, while if it takes unity, then it is conservative. It is assumed that the parameter α is kept constant during the motion of the system. It is noted that Reut's column can be derived through a variational calculus as an adjoint system to Beck's column (see Bolotin, 1963).

Since the joints are viscoelastic, then there exists internal damping in the joints. Air resistance of the rods during the motion may yield external damping. It is assumed that a dash-pot with the damping coefficient C_1 is attached to the first rod l_1 apart from the joint 1, otherwise a dash-pot with C_2 is attached to the second bar l_2 apart from the joint 2.
The governing equations of small motion of the system are given by

$$ml^3\begin{bmatrix}4/3 & 1/2\\1/2 & 1/3\end{bmatrix}\begin{bmatrix}\ddot{\varphi}_1\\\ddot{\varphi}_2\end{bmatrix}+\left[el^3\begin{bmatrix}4/3 & 1/2\\1/2 & 1/3\end{bmatrix}+b\begin{bmatrix}2 & -1\\-1 & 1\end{bmatrix}+C_1l^2\begin{bmatrix}\rho_1^2 & 0\\0 & 0\end{bmatrix}\right.$$

$$\left.+C_2l^2\begin{bmatrix}1 & \rho_2\\\rho_2 & \rho_2^2\end{bmatrix}\right]\begin{bmatrix}\dot{\varphi}_1\\\dot{\varphi}_2\end{bmatrix}+\left[k\begin{bmatrix}2 & -1\\-1 & 1\end{bmatrix}+Pl\begin{bmatrix}-1 & \alpha\\1 & \alpha\end{bmatrix}\right]\begin{bmatrix}\varphi_1\\\varphi_2\end{bmatrix}=0, \qquad (4.1)$$

where

$$\rho_i = l_i/l \quad (i=1,2). \qquad (4.2)$$

The dot (•) means the differentiation with respect to time t. For simplicity, the following dimensionless quantities are defined;

$$F=\frac{Pl}{k},\quad \gamma=\frac{b}{l\sqrt{kml}},\quad \kappa=\frac{el^2}{\sqrt{kml}},\quad \eta_i=\frac{C_i\,l}{\sqrt{kml}}\ (i=1,2),\quad \tau=\frac{t}{l}\sqrt{\frac{k}{ml}}. \qquad (4.3)$$

With the dimensionless quantities defined in Equations (4.3), and normalizing the mass matrix, Equation (4.1) can be written in the form

$$\begin{bmatrix}1 & 0\\0 & 1\end{bmatrix}\begin{bmatrix}\ddot{\varphi}_1\\\ddot{\varphi}_2\end{bmatrix}+\left[\kappa\begin{bmatrix}1 & 0\\0 & 1\end{bmatrix}+\gamma\begin{bmatrix}6 & -30/7\\-12 & 66/7\end{bmatrix}+\frac{6}{7}\rho_1^2\eta_1\begin{bmatrix}2 & 0\\-3 & 0\end{bmatrix}\right.$$

$$\left.+\frac{6}{7}\eta_2\begin{bmatrix}2-3\rho_2 & (2-3\rho_2)\rho_2\\8\rho_2-3 & (8\rho_2-3)\rho_2\end{bmatrix}\right]\begin{bmatrix}\dot{\varphi}_1\\\dot{\varphi}_2\end{bmatrix}+\left[\begin{bmatrix}6 & -30/7\\-12 & 66/7\end{bmatrix}+\frac{6}{7}F\begin{bmatrix}-5 & -\alpha\\11 & 5\alpha\end{bmatrix}\right]\begin{bmatrix}\varphi_1\\\varphi_2\end{bmatrix}=0, \qquad (4.4)$$

where the dot(•) means the differentiation with respect to dimensionless time τ. If the stiffness matrix is normalized, then Equation (4.4) is reduced to the following normalized form;

$$\begin{bmatrix}1 & 0\\0 & 1\end{bmatrix}\begin{bmatrix}\ddot{\theta}_1\\\ddot{\theta}_2\end{bmatrix}+\left[\kappa\begin{bmatrix}1 & 0\\0 & 1\end{bmatrix}+\gamma\begin{bmatrix}\omega_1^2 & 0\\0 & \omega_2^2\end{bmatrix}+\eta_1\rho_1^2\begin{bmatrix}\mu_{11} & \mu_{12}\\\mu_{21} & \mu_{22}\end{bmatrix}+\eta_2\begin{bmatrix}\nu_{11} & \nu_{12}\\\nu_{21} & \nu_{22}\end{bmatrix}\right]\begin{bmatrix}\dot{\theta}_1\\\dot{\theta}_2\end{bmatrix}$$

$$+\left[\begin{bmatrix}\omega_1^2 & 0\\0 & \omega_2^2\end{bmatrix}+F\begin{bmatrix}f_{11} & f_{12}\\f_{21} & f_{22}\end{bmatrix}\right]\begin{bmatrix}\theta_1\\\theta_2\end{bmatrix}=0, \qquad (4.5)$$

where ω_1 and ω_2 are the first two eigen-frequencies of the system without loading; $\omega_1^2=0.341$

and $\omega_2^2 = 15.09$.

It is noted that the elements μ_{ij} (i, j = 1,2) in the damping matrix are constant numbers, while ν_{ij} are given by the quadratic form of ρ_2. It is also noted that the elements of the applied force matrix is constant number for a fixed value of α. In the followings, we will consider only the case when the parameter $\alpha = 0$, that is the force is a pure Reut force.

4. 1. 2 Stability analysis

Equation (4.5) can be put in the form

$$\begin{bmatrix} 1 & 0 \\ 0 & 1 \end{bmatrix}\begin{bmatrix} \ddot{x}_1 \\ \ddot{x}_2 \end{bmatrix} + \begin{bmatrix} c_{11} & c_{12} \\ c_{21} & c_{22} \end{bmatrix}\begin{bmatrix} \dot{x}_1 \\ \dot{x}_2 \end{bmatrix} + \begin{bmatrix} k_{11} & k_{12} \\ k_{21} & k_{22} \end{bmatrix}\begin{bmatrix} x_1 \\ x_2 \end{bmatrix} = 0. \tag{4.6}$$

The solution can be written in the form

$$x_i = B_i \exp(\lambda\tau) \qquad (i = 1,2). \tag{4.7}$$

Substitution of Equation (4.7) into Equation (4.6) and application of the condition for non-trivial solution result in the following form of characteristic equation;

$$\lambda^4 + a_1\lambda^3 + a_2\lambda^2 + a_3\lambda + a_4 = 0, \tag{4.8}$$

where

$$\left.\begin{aligned} a_1 &= c_{11} + c_{22}, \\ a_2 &= k_{11} + k_{22} + c_{11}c_{22} - c_{12}c_{21}, \\ a_3 &= c_{11}k_{22} + c_{22}k_{11} - c_{12}k_{21} - c_{21}k_{12}, \\ a_4 &= k_{11}k_{22} - k_{12}k_{21}. \end{aligned}\right\} \tag{4.9}$$

Stability of the dynamical systems described by Equation (4.8) can be investigated directly without determining the roots of the characteristic equation by applying the Routh-Hurwitz criteria. The criteria requires the following condition;

$$a_1a_2a_3 - a_3^2 - a_1^2a_4 = 0. \tag{4.10}$$

4. 1. 3 Special case(I)

Now let us consider the case where there is no dash-pot attached to the system, that is $\eta_i = 0$. This means that we discuss the combined effect of internal and external damping on flutter of the system.

The elements of the damping matrix are

$$c_{11} = \kappa + \gamma\omega_1^2, \qquad c_{22} = \kappa + \gamma\omega_2^2, \qquad c_{12} = c_{21} = 0. \tag{4.11}$$

Thus the damping matrix can be written in the form

$$\begin{bmatrix} c_{11} & c_{12} \\ c_{21} & c_{22} \end{bmatrix} = c_{11} \begin{bmatrix} 1 & 0 \\ 0 & \delta \end{bmatrix}, \tag{4.12}$$

where

$$\delta = \frac{\kappa + \gamma\omega_2^2}{\kappa + \gamma\omega_1^2}. \tag{4.13}$$

It is obvious that $1 < \delta < \omega_2^2/\omega_1^2$.

Figure 4.2 shows the combined effect of internal damping and external damping. Figure 4.2(a) depicts the effect of internal damping γ on flutter limit F_{cr} for some constant values of external damping κ. It is confirmed theoretically in Figure 4.2(a) that the increasing internal damping leads to the decreasing flutter limit. Figure 4.2(b) demonstrates the effect of external damping on the flutter limit for some fixed values of internal damping. It is confirmed that the increasing external damping results in the increasing flutter limit. Therefore it is confirmed that the internal damping may have a destabilizing effect on the flutter, while the external one a stabilizing effect.

Let us now discuss the effect of damping from the viewpoint configuration of the damping matrix given by Equation (4.12). It is noted that the configuration of damping matrix is given by Equation (4.12). First if the order of internal damping γ is greater than that of external damping κ, then the order of δ becomes larger than unity, say at the maximum up to 44 ($\delta = \omega_2^2/\omega_1^2 = 15.09/0.341$). The configuration of damping matrix is diagonal and the element of (2,2) is larger than that of (1,1). This type of damping configuration may result in the destabilizing effect.

On the other hand, if the order of the external damping is much larger than that of internal damping, then the value δ may take at its minimum value, unity. Then the configuration of the damping matrix will be the unit matrix. Thus if the damping matrix may take this type of configuration, then the damping may have stabilizing effect.

4.1.4 Special case (II)

Let us consider a special case where there might be no internal and external damping, that is $\gamma = \kappa = 0$. We now consider only damping due to an artificially attached dash-pot. It is now assumed that damping coefficient of the dash-pot is small. It is also assumed that there exists only one dash-pot attached to the system. Figure 4.3 shows the variation of the flutter limit depending the location of a dash-pot along the system (when the dash-pot is on the first rod, then $\eta_2 = 0$, while on the second rod, $\eta_1 = 0$). It is noted in Figure 4.3 that the flutter limit of damped system is normalized by that of the undamped system.

If the dash-pot is attached to the system on the first rod, that is $\eta_1 = 10^{-3}$, $\eta_2 = 0$. Then the damping matrix can be written in the form

$$\begin{bmatrix} c_{11} & c_{12} \\ c_{21} & c_{22} \end{bmatrix} = c_0 \begin{bmatrix} \mu_{11} & \mu_{12} \\ \mu_{21} & \mu_{22} \end{bmatrix}, \tag{4.14}$$

where $c_0 = \eta_1 \rho_1^2$. This means that the configuration of damping matrix is constant when there is attached a dash-pot to the first rod. The magnitude of damping can be changed more effectively by changing the location of the dash-pot than the damping coefficient of the dash-pot. It is noted in this special case that the following relation holds

$$\mu_{11}\mu_{22} - \mu_{12}\mu_{21} = 0. \tag{4.15}$$

Equation (4.15) implies that the determinant of the damping matrix is equal to zero($c_{11}c_{22}$ - $c_{12}c_{21}$ = 0) and that the coefficient a_2 in the characteristic Equation (4.8) is the function of only the stiffness of the mechanical system. Finally Equation (4.10) can be written in the form

$$c_0 f(\mu_{11}, \mu_{12}, \mu_{21}, \mu_{22}) = 0. \tag{4.16}$$

It is now concluded that the magnitude of damping of the dash-pot does not affect the flutter limit as seen in Figure 4.3, as long as a dash-pot is attached to the first rod.

Now let us consider the case where the dash-pot is attached to the second rod, that is $\eta_1 = 0$, $\eta_2 = 10^{-3}$. The variation of four elements in the damping matrix depending on the changing location of the dash-pot is shown in Figure 4.4. It is interesting to find that the configuration of damping matrix can be controlled by the location of the dash-pot for $0 \le \rho_2 \le 1.0$.
Peculiar futures of flutter limit for $0 \le \rho_2 \le 1.0$ can be explained by investigating the structure of the damping matrix. For example, when the dash-pot is located at $\rho_2 = 0.472$, then $\mu_{12} = \mu_{21} = \mu_{22} = 0$. This results in the damping matrix in the form

$$\begin{bmatrix} c_{11} & c_{12} \\ c_{21} & c_{22} \end{bmatrix} = \eta_2 \begin{bmatrix} \nu_{11} & 0 \\ 0 & 0 \end{bmatrix}. \tag{4.17}$$

This form of damping configuration results in the vanishing flutter limit at $\rho_2 = 0.472$.
When the dash-pot is attached at $\rho_2 = 0.598$, then it does not affect the flutter limit, since the critical flutter mode of undamped model (that is $\gamma = \kappa = 0, \eta_1 = \eta_2 = 0$) has its node at $\rho_2 = 0.598$. However it is noticed that the considered special case of damping configuration is purely artificial and thus unrealistic.

4.1.5 Special case (III)

Let us consider a case where a dash-pot is attached to the system having a small internal damping, say $\gamma = 10^{-3}$. Flutter limit is depicted in Figure 4.5. Comparison of Figure 4.3 and 4.5 reveals that there is found no unrealistic aspect in Figure 4.5, not as in Figure 4.3. This can conclude that some of the peculiar futures found in Figure 4.3 was caused by unrealistic assumption that there exists an attached dash-pot, whereas no internal damping. It is found that the most stabilizing position for the dash-pot is $\rho_2 = 0.31$ and that the addition of a dash-pot can stabilize the system considerably if the dash-pot is attached to the system at an appropriate position.
Figure 4.6 shows the flutter configuration of the system when a dash-pot is attached to the first rod at its tip (Figure 4.6(a)) and to the second one at its tip (Figure 4.6(b)), respectively. The flutter configurations can be calculated by solving Equation (4.4).

4.3 Experiment

4.3.1 Aims of the experiment

Demonstration of the experimental evidence of the effect of damping due to a dash-pot on the flutter limit of a nonconservative mechanical system is the intended aim of the experiment. For the aims to be achieved, there will be planned the following three actions.

1. To show an experimental counterpart to the mechanical system shown in Figure 4.1.
2. To measure damping coefficients of the three types of damping; internal and external damping, and damping due to an attached dash-pot.
3. To compare the experimental results with the corresponding theoretical predictions.

4.3.2 Outline of the experimental setup

A schematic diagram of the experimental setup is shown in Figure 4.7. The test mechanical system was cantilevered in a horizontal plane, and thus the small motion of the system was restricted in the same plane. An attachment of basically a rigid flat plate was mounted to the system at its tip end. An impinging air jet from the nozzle was applied to the attachment.

The attachment was a rectangular balsa plate of 90 mm x 25 mm, the surface of which was covered by unwoven material. The mass of the attachment $M = 2.1 \times 10^{-3}$ kg. Mass ratio between the mass of the attachment and that of a rod was small enough, that is $M/ml = 1.8 \times 10^{-2}$, it was assumed therefore that the effect of the mass of the attachment was negligible in flutter analysis.

The test piece was composed of two equal brass pipes of length l, outer diameter of the pipes $d_0 = 12.0$ mm, thickness of the pipe wall $h = 1.00$ mm, mass per unit length $m = 0.289$ kg/m. Short rubber tubes of outer diameter of 14.5 mm and of the wall thickness 2.30 mm were used as the two viscoelastic joints. It is noted that there was an offset gap in the joint. The gap was 8.0 mm, while there is no gap in the mathematical model shown in Figure 4.1. The elastic restoring moment coefficient was measured to find $k = 0.233$ Nm/rad.

4.3.3 Measurement of damping

Coefficient of external damping *e*: The external damping due to an air resistance is strictly dependent on the external shape of the test piece. One of the two rods was hung down vertically having the upper end pinned to conduct a pendulum motion in the air. The decayed pendulum motion was recorded to find the logarithmic decrement which gave the coefficient of external damping $e = 8.2 \times 10^{-4}$. This value yielded the dimensionless coefficient of external damping $\kappa = 8.0 \times 10^{-4}$.

Coefficient of internal damping *b*: Internal damping was due to viscoelastic restoring moment at the joints. Since the two joints were assumed to have the equal mechanical properties, so only the viscoelastic property of the second joint was tested for internal damping. The first rod was fixed to allow the second rigid pipe (without a dash-pot) to conduct a horizontal pendulum motion around the second joint. The damped motion was recorded to find the logarithmic decrement. The decrement was attributed to both the air resistance and the viscocity of the joint. Since the coefficient of external damping e was known in the earlier test, then the contribution of the external damping could be deducted from the logarithmic decrement to find the coefficient of internal damping $b = 1.2 \times 10^{-3}$ Nms/rad. The value of b could be reduced into the dimensionless quantity $\gamma = 1.8 \times 10^{-2}$.

Coefficient of damping of dash-pot: The dash-pot was realized as a thin plate-viscous fluid system. A thin acrylic plate of 55.0 mm x 44.0 mm was immersed in a tap water in a beaker. The similar test as that for internal damping was conducted to find the coefficient of damping due to the dash-pot. The first rod was fixed to allow the second rod accommodated with a dash-pot to perform a horizontal damped motion. The logarithmic decrement in the test was attributable to external damping due to the air resistance, internal damping due to the viscocity of the joint and the dash-pot. As the coefficients of internal and external damping were known through the earlier two tests, then the coefficient of damping of the dash-pot could be obtained as $C_i = 1.0 \times 10^{-2}$ Nsm^{-1}, that is $\eta_i = 2.5 \times 10^{-2}$ (i = 1, 2) in the dimensionless quantity.

4. 3. 4 Flutter test

Figure 4.8 shows the photographs of flutter motion observed in the flutter test. The photographs were taken with the aid of electronically controlled flash illumination. The observed flutter configuration agrees at least qualitatively to the predicted flutter mode shown in Figure 4.6.
Figure 4.9 shows the experimental results compared with the theoretical prediction made by taking measured damping into account. Quantitatively the experimental results are higher than the theory. However knowing that the physical model can not be the same with the mathematical one, it is found that the dependence of experimental flutter limits on the location of the dash-pot agrees qualitatively to that of the mathematical one.

4. 4 General discussions

We have considered the particular two-degree-of-freedom mechanical system and discussed the effect of damping configuration on flutter limit. Now let us discuss the effect of damping configuration in the generalized forms.
Damped mechanical systems having multiple degrees of freedom can be written in the form

$$\mathbf{M}\ddot{\mathbf{x}} + \mathbf{C}\dot{\mathbf{x}} + \mathbf{K}\mathbf{x} = \mathbf{0}, \tag{4.18}$$

where **M** is the mass matrix, **C** the damping matrix and **D** the stiffness matrix. **x** is the vector of generalized coordinates.
Now it is assumed that the stiffness matrix can be written in the form

$$\mathbf{K} = \mathbf{K}_s + \mathbf{K}_f, \tag{4.19}$$

where $\mathbf{K}_s$ is the structural stiffness, while $\mathbf{K}_f$ is attributable to the stiffness due to applied external forces. If we assume only proportional types of damping, then the damping matrix **C** may be written in the form

$$\mathbf{C} = \varepsilon_m \mathbf{M} + \varepsilon_s \mathbf{K}_s + \varepsilon_d \mathbf{D}, \tag{4.20}$$

where **D** is the damping, which can not be related to **M** and **K**.
Substitution of Equations (4.19) and (4.20) into Equation (4.18) leads to the following equation of motion

$$\mathbf{M}\ddot{\mathbf{x}} + (\varepsilon_m \mathbf{M} + \varepsilon_s \mathbf{K}_s + \varepsilon_d \mathbf{D})\dot{\mathbf{x}} + (\mathbf{K}_s + \mathbf{K}_f)\mathbf{x} = \mathbf{0}. \tag{4.21}$$

After the normalization and the transformation of the above matrix form, we obtain the following equation

$$\mathbf{I}\ddot{\mathbf{X}} + (\varepsilon_m \mathbf{I} + \varepsilon_s \Lambda + \varepsilon_d \mathbf{B})\dot{\mathbf{X}} + (\Lambda + \mathbf{F})\mathbf{X} = \mathbf{0}, \tag{4.22}$$

where $\mathbf{\Lambda}$ is a diagonal matrix having the diagonal elements ω_i^2(i = 1, 2,...,n). ω_i (i = 1, 2,...,n) are

the eigen-frequencies of elastic systems under no loading.
It is seen that Equations (4.1) and (4.5) correspond to Equations (4.18) and (4.22), respectively. If we derive Equation (4.22) by starting with Lagrange's equations of motion, we will be able justify the damping configuration in the form of Equation (4.20).

4.5 Concluding remarks

We have discussed a realistic mechanical system subjected to a nonconservative force produced by an impinging air jet. Considering a dash-pot attached to the system, we have discussed the effect of damping configuration on flutter limit. The experimental counterpart to the mathematical model was set up in laboratory to demonstrate flutter type instability due to the nonconservative nature of the applied force. Three coefficients of damping involved in the experimental system were measured experimentally. Effect of the location of an attached dash-pot on flutter limit was investigated experimentally. The experimental results were compared with the corresponding theory to find a qualitative agreement between them. Finally the generalized form of damping configuration was discussed for mechanical systems having multiple degrees of freedom.

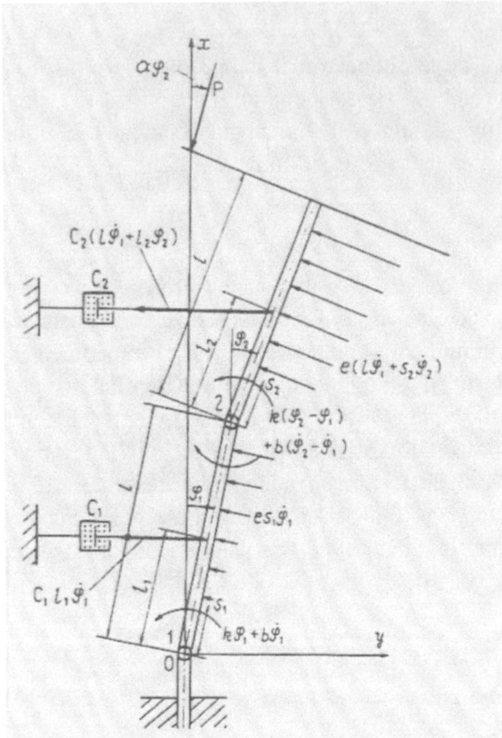

Figure 4.1 Two-degree-of-freedom mechanical system subjected to a generalized Reut force.

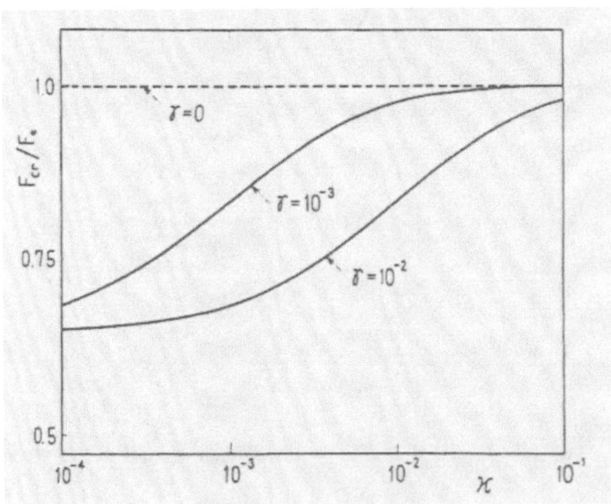

(a) Effect of internal damping

Figure 4.2 Combined effect of internal and external damping.(continued)

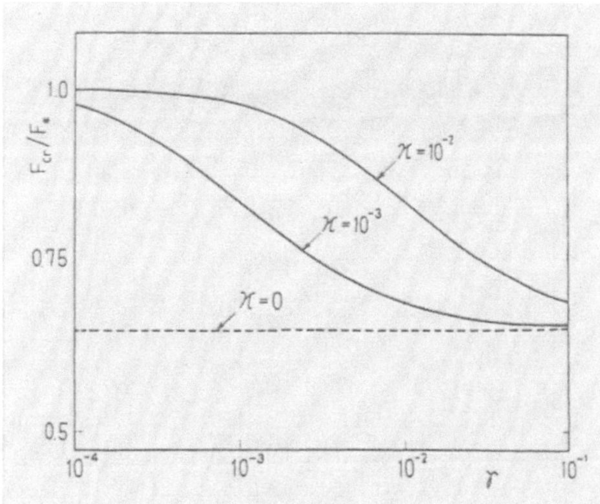

(b)Effect of External damping

Figure 4.2 Combined effect of internal and external damping..

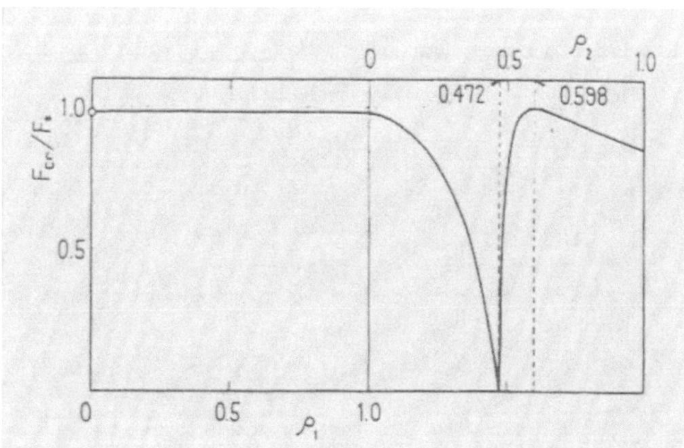

Figure 4.3 Effect of location of a dash-pot when internal and external damping are neglected ($\alpha=0, \gamma=\kappa=0, \eta_l=10^{-3}$) $F_*=2.542$:Flutter limit for the undamped system.

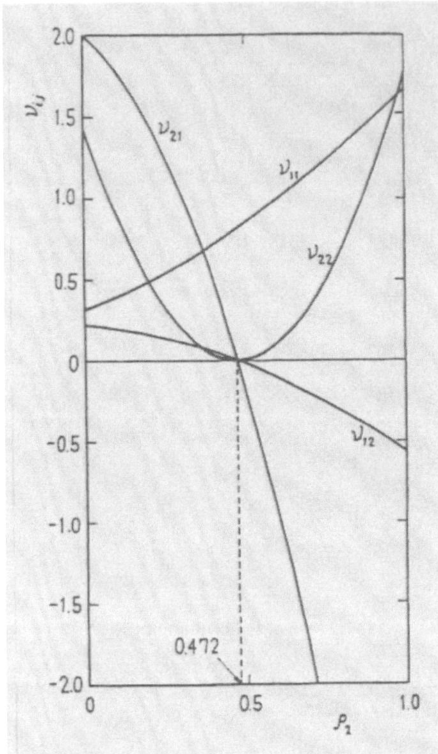

Figure 4.4 Elements of damping matrix, when a dash-pot is attached to the second rod.

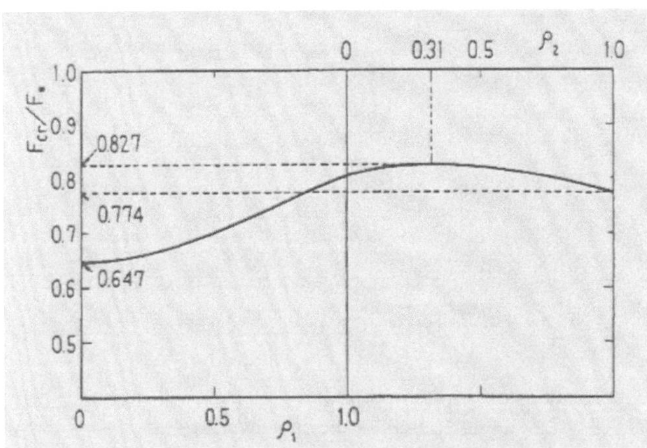

Figure 4.5 Effect of location of a dash-pot when internal damping is taken into account (α=0, γ=10^{-3}, κ=0, η_i = 10^{-3}) F_* = 2.542:Flutter limit for the undamped system.

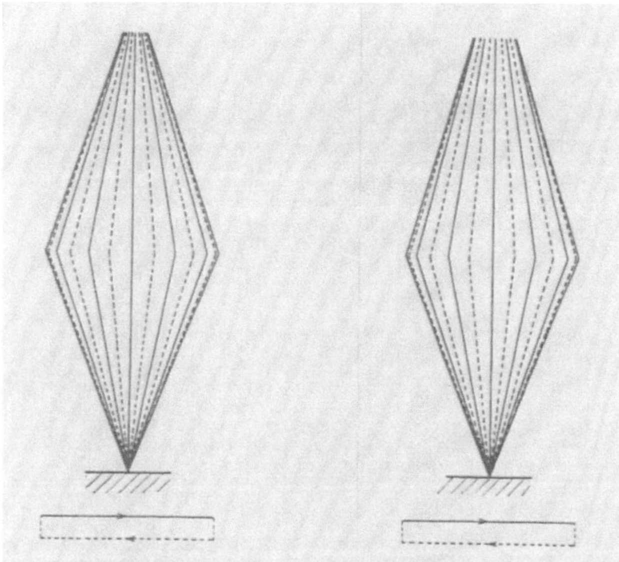

(a) ρ_1=1.0 (F_{cr}=2.057, ω_{cr}=0.949) (b) ρ_2=1.0 (F_{cr}=1.969, ω_{cr}=0.914)

Figure 4.6 Flutter configuration at the critical force.

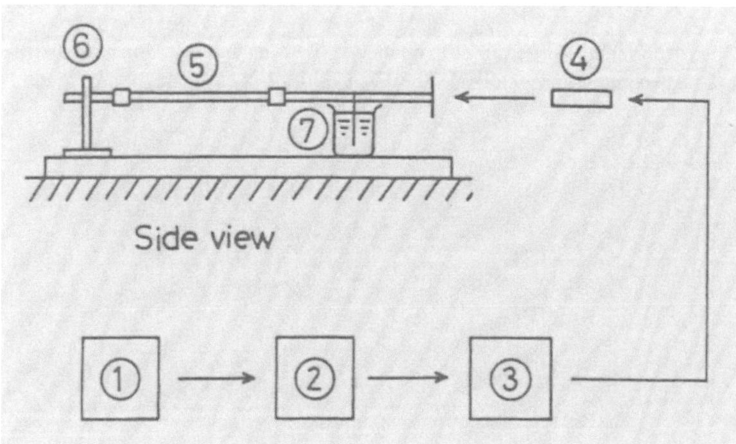

Figure 4.7 Schematic diagram of the experimental system; ① : Air compressor, ② : Air reservoir, ③ : Rotameter, ④ : Nozzle, ⑤ : Test piece, ⑥ : Fixed support block, ⑦ : Dash-pot.

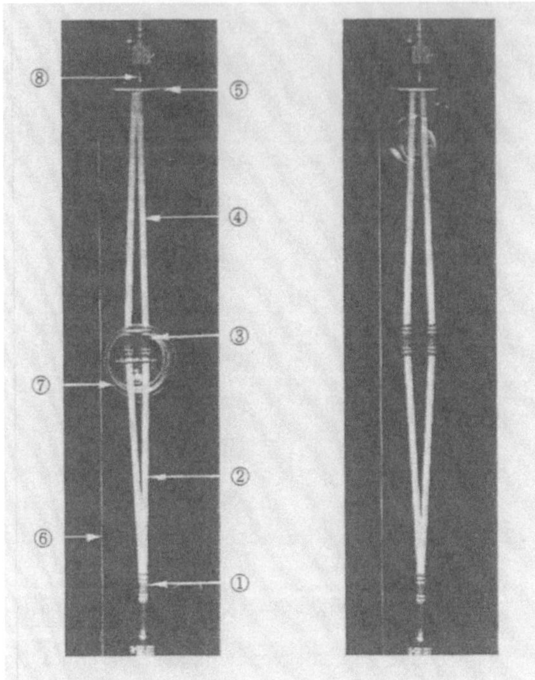

(a) $\rho_1 = 0.94$ (b) $\rho_2 = 0.90$

Figure 4.8 Observed flutter configurations (Top view) : ① : First joint, ② : First rod, ③ : Second joint, ④ : Second rod, ⑤ : Attachment, ⑥ : Reference line, ⑦ : Dash-pot, ⑧ : Nozzle for an air jet.

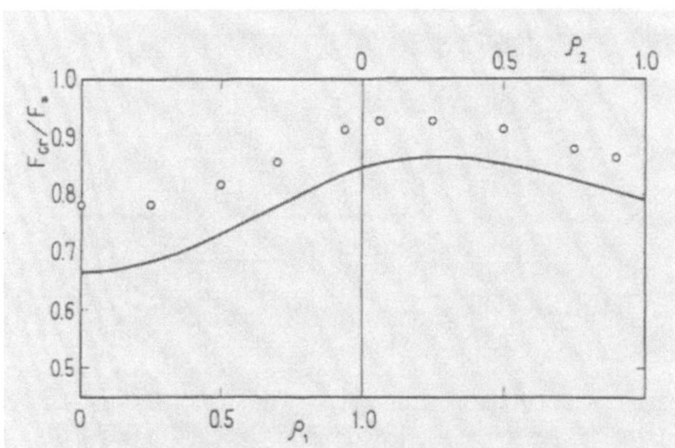

Figure 4.9 Comparison between experiment and theory on the effect of location of a dash-pot (α = 0.001, F_* = critical flutter limit of undamped system).

5. General Concluding Remarks

Flutter, as instability phenomenon of structures, has been well known in the field of aircraft structures engineering since the first artificial flight was done in 1903. Flutter and divergence phenomena of aircraft structures have been termed aeroelasticity. Flutter of aircraft wings can take place when structural deformations of wings induce additional aerodynamic forces (Bisplinghoff, Ashley and Halfman, 1955). Thus it is noted that the aerodynamic forces in aeroelasticity are another group of follower forces. When flutter arises in a flexible structure, there any type of nonconservative force is applied upon the structure. When an applied force is conservative, the structure subjected to the force would never lose its stability by flutter. Therefore occurrence of flutter is the most obvious evidence that there is a nonconservative force applied upon the structure.

The present lecture has discussed theoretically and experimentally three types of nonconservative stability problems; cantilevered pipes conveying fluid, cantilevered columns subjected to a rocket thrust, and two-degree-of-freedom mechanical systems subjected to a Reut force. It has been shown by experimental demonstrations that flutter-type instability may arise, otherwise statically unstable structure can be stabilized dynamically. These interesting dynamical phenomena are attributable to nonconservativeness of the applied forces.

For the future development of the theory of nonconservative stability problems, efforts for coordination between theory and experiment ought to be intensified.

References

Beck, M. (1952). Die Knicklast des einseitig eingespannten, tangential gedrückten Stabes. *Zeitschrift für Angewandte Mathematik und Mechanik* 3: 225-228.

Bisplinghoff, R. L., Ashley, H., and Halfman, R.L. (1955). *Aeroelasticity*. Cambridge, Mass.: Addison-Wesley Publishing Co..

Bolotin, V. V. (1963). *Nonconservative problems of the theory of elastic stability*. New York: Pergamon Press.

Beal, T. R. (1965). Dynamics stability of a flexible missile under the constant and pulsating thrust. *AIAA Journal* 3: 486-494.

Feodosiev V. I. (1965). On a problem of stability (in Russian). *Prikladnaya Matematika I Mekhanika*, 29: 391-392.

Herrmann, G., Nemat-Nasser, S. and Prasad, S. N. (1966). Models demonstrating instability of nonconservative mechanical systems. *Technical Report* No. 66-4, Department of Civil Engineering, Structural Mechanics Laboratory, Northwestern University.

Koiter, W. T. (1996). Unrealistic follower forces. *Journal of Sound and Vibration* 194: 636-638.

Leipholz, H. (1980). *Stability of elastic systems*. Alphen aan den Rijn, The Netherlands, Sijthoff & Noordohoff.

Langthjem, M. A. and Sugiyama, Y. (2000). Dynamic stability of columns subjected to follower loads: A survey. *Journal of Sound and Vibration* 238: 809-851.

Mottershead, J. E., and Chan, S. N. (1995). Flutter instability of circular discs with frictional follower loads. *Transactions of the ASME, Journal of Vibration and Acoustics* 117: 161-163.

Mladenov, K. A. and Sugiyama, Y. (1997). Stability of a jointed free-free beam under end rocket thrust. *Journal of Sound and Vibration* 199: 1-15.

Nishiwaki, M. (1993). Generalized theory of brake noise. In *Proceedings of the Institution of Mechanical*

Engineers 207: 195-202.

Paidoussis, M. P., and Li, G. X. (1993). Pipes conveying fluid: A model dynamical problems. *Journal of Fluids and Structures* 7: 137-204.

Paidoussis, M. P. (1998). *Fluid-Structure Interactions; Slender Structures and Axial Flow* Volume 1. New York, Academic Press.

Panovko, Ya. G., and Gubanova, I. I. (1965). *Stability and Oscillations of Elastic Systems, Paradoxes, Fallacies, and New Concept.* New York, Consultants Bureau Enterprises, Inc. (see page 63).

Ryu, B.-J., Katayama, K., and Sugiyama, Y. (1998). Dynamic stability of Timoshenko columns subjected to subtangential forces. *Computers and Structures* 68: 499-512.

Sugiyama, Y., and Sekiya, T. (1971). Surveys of the experimental studies of instability of the elastic systems subjected to nonconservative forces (in Japanese). *Journal of the Japan Society for Aeronautical and Space Sciences* 19: 61-68.

Sugiyama, Y., Kashima, K., and Kawagoe, H. (1976). On an unduly simplified model in the non-conservative problem of elastic stability. *Journal of Sound and Vibration* 45: 237-247.

Sugiyama, Y., Tanaka, Y., Kishi, T., and Kawagoe, H. (1985). Effect of a spring support on the stability of pipes conveying fluid. *Journal of Sound and Vibration* 100: 257-270.

Sugiyama, Y. (1987). Experiment on nonconservative problem of elastic stability. *ESP**24**. 87035*, In *24th Annual Technical Meeting*, Society of Engineering Sciences, University of Utah, Salt Lake City, USA.

Sugiyama, Y., Katayama, K., and Kinoi, S. (1995). Flutter of cantilevered column under rocket thrust. *Journal of Aerospace Engineering*, The American Society of Civil Engineers 8: 9-15.

Sugiyama, Y., Matsuike, J., Ryu, B.-J., Katayama, K., Kinoi, S., and Enomoto, N. (1995). Effect of concentrated mass on stability of cantilevers under rocket thrust. *AIAA Journal* 33: 499-503, see also 1996 *AIAA Journal* 34: 212.

Sugiyama, Y., Katayama, T., Kanki, E., Chiba, M., Shiraki, K., and Fujita, K. (1996). Stability of vertical fluid-conveying pipes having the lower end immersed in fluid. *JSME International Journal* 39: 57-65.

Sugiyama, Y., Langhtjem, M. A., and Ryu, B.-J. (1999). Realistic follower forces. *Journal of Sound and Vibration* 225:779-782.

Sugiyama, Y., Katayama, K., Kiriyama, K., and Ryu, B.-J. (2000). Experimental verification of dynamic stability of vertical cantilevered columns subjected to a sub-tangential force. *Journal of Sound and Vibration* 236:193-207.

Timoshenko, S. P., and Gere, J. M. (1961). *Theory of elastic stability*. New York: McGraw-Hill Book Co..

Yagn, Yu. I., and Parshin, L. K. (1966). Experimental study of stability of a column loaded by follower force. *Doklady Academy Nauk* 167: 49-50.

Appendix:

Photographs of Experiments on Dynamic Stability of Cantilevered Columns Subjected to a Rocket Thrust

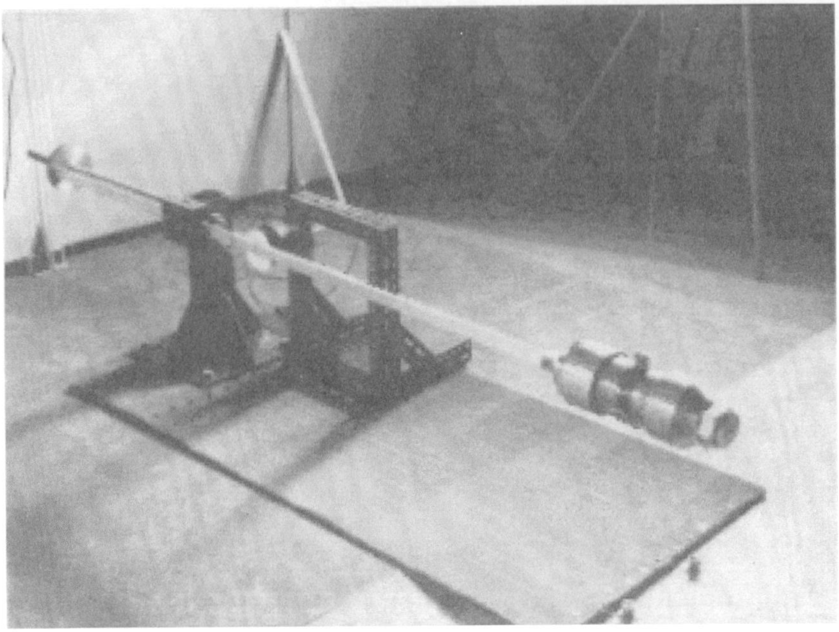

(a) Bird's-eye view

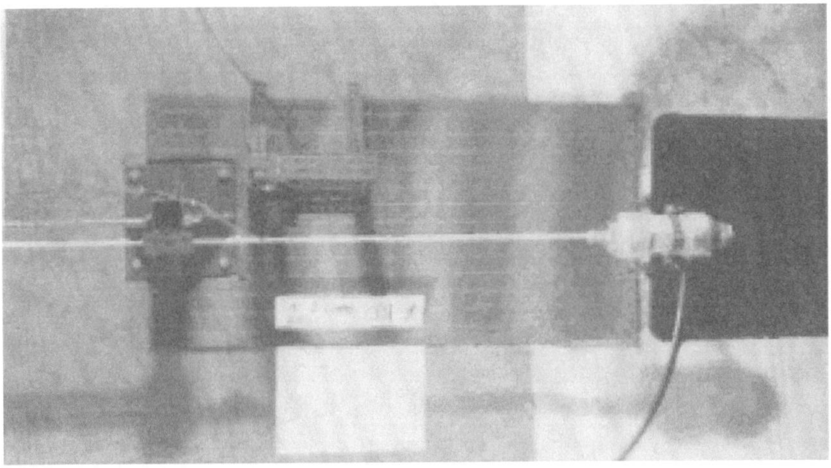

(b) Top view

Photograph 1 Experimental setup of a horizontal cantilevered column subjected to a rocket thrust (Section 3.1).

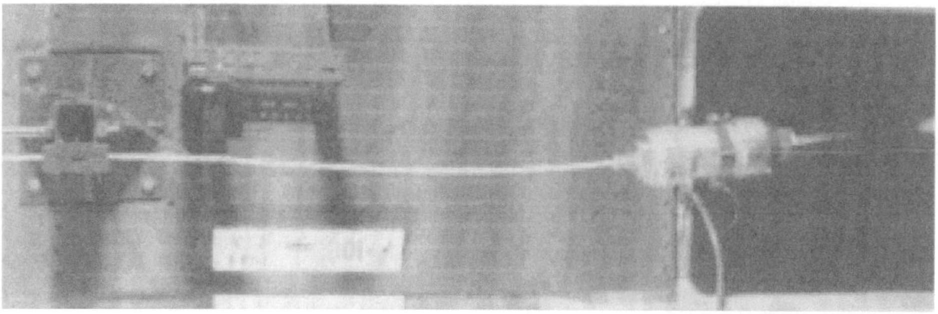

(a) t = 3.12 sec

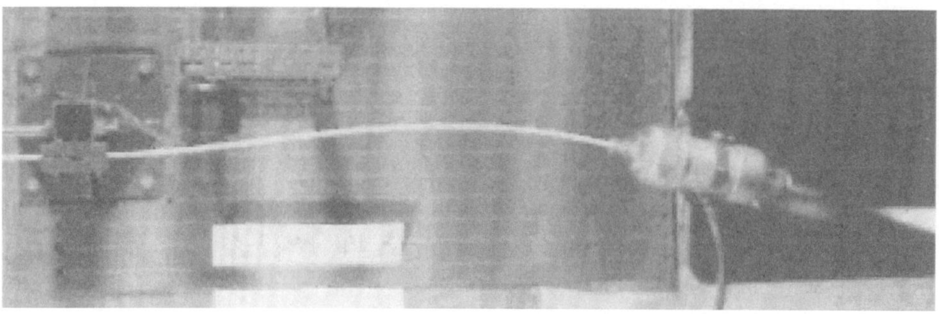

(b) t = 3.36 sec

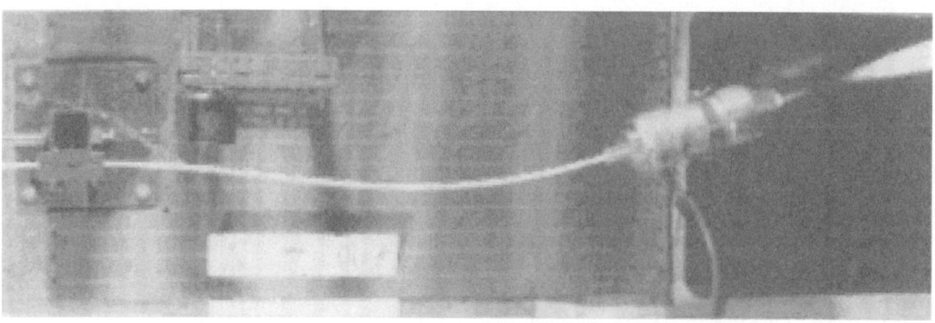

(c) t = 3.60 sec

Photograph 2 Flutter motion observed in Test run No.7 (Section 3.1).

Photograph 3 Experimental setup of a horizontal cantilevered column having a lumped mass (Section 3.2).

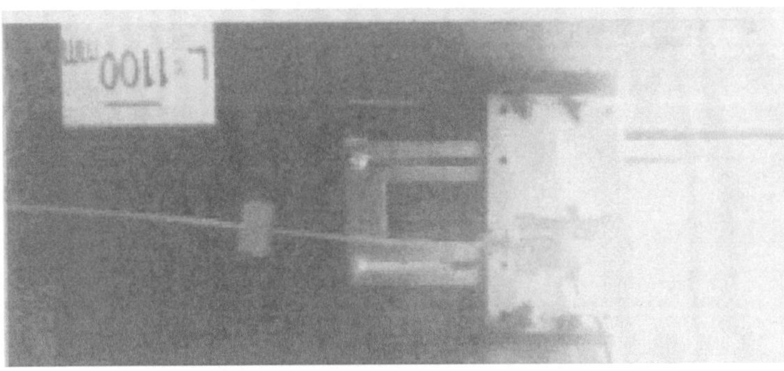

(a) t = 2.0 sec

(b) t = 2.17 sec

Photograph 4 flutter motion observed in Test run No. 3 (Section 3.2) (continued).

(c) t = 2.33 sec

Photograph 4 Flutter motion observed in Test run No. 3 (Section 3.2).

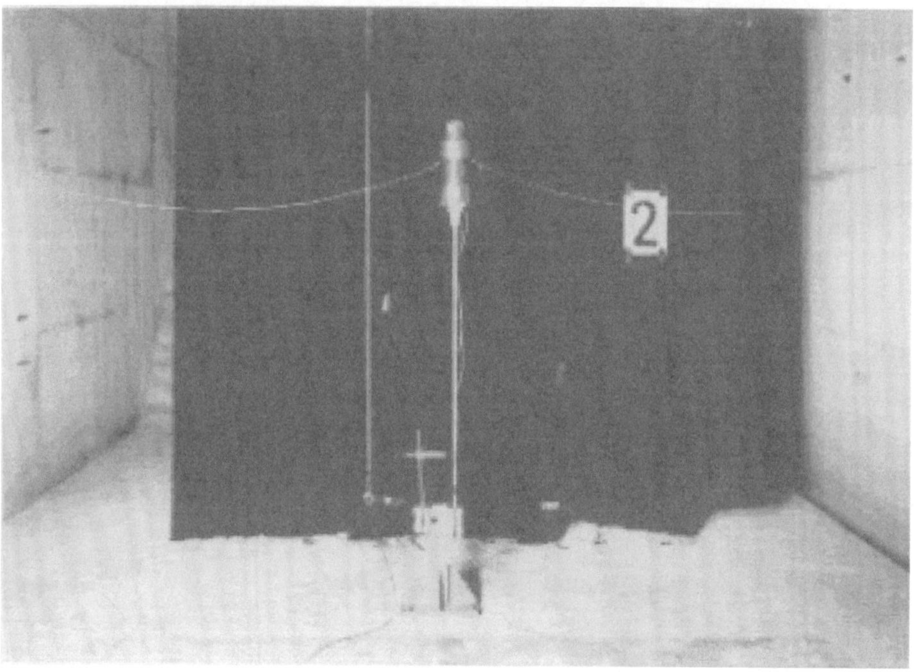

Photograph 5 Experimental set up of a vertical cantilevered column accommodated with a solid rocket motor at its tip end (Section 3.3).

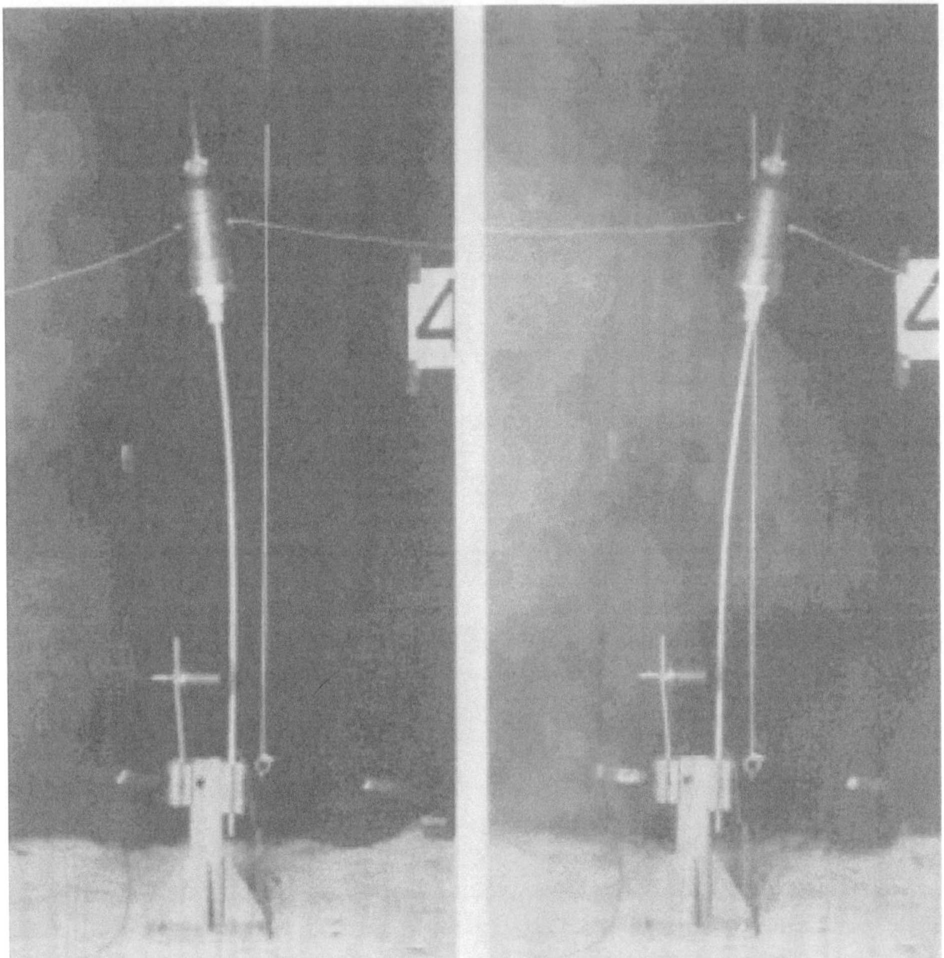

(a) t = 2.0 sec (b) t = 2.5 sec

Photograph 6 Dynamically stable motion of the vertical column under a rocket thrust (Section 3.3).

Kolibri 360 GmbH
Ferdinand-Jühlke-Straße 7
99095 Erfurt, Deutschland
produktsicherheit@kolibri360.de